Proceedings of the Second International Conference on

Calorimetry in High Energy Physics

Proceedings of the Second International Conference on

CALORIMETRY IN HIGH ENERGY PHYSICS

Capri, Italy 14 – 18 October 1991

Editor

Antonio Ereditato

INFN Napoli

World Scientific

Singapore • New Jersey • London • Hong Kong

Published by

World Scientific Publishing Co Pte Ltd
5 Toh Tuck Link, Singapore 596224
USA office: 27 Warren Street, Suite 401-402, Hackensack, NJ 07601
UK office: 57 Shelton Street, Covent Garden, London WC2H 9HE

British Library Cataloguing-in-Publication Data
A catalogue record for this book is available from the British Library.

Proceedings of the Second International Conference on
CALORIMETRY IN HIGH ENERGY PHYSICS

ISBN-13 978-981-02-0916-2
ISBN-10 981-02-0916-9

INTERNATIONAL ADVISORY COMMITTEE

G. Bellettini, Pisa
C. Fabjan, CERN
F. Eisele, DESY
W. Frisken, York
Yu. Galaktionov, ITEP Moscow
D. Hitlin, Caltech
S. Iwata, KEK
G. Trilling, LBL

ORGANIZING COMMITTEE

A. Ereditato, INFN Napoli (Chairman)
P. Jenni, CERN
V. Palladino, Univ. of Napoli
A. Para, Fermilab

SESSION ORGANIZERS

J. Colas, Annecy
M. Della Negra, CERN
M. Derrick, ANL
A. Ferrari, Milano
P. Lecoq, CERN
A. Menzione, Pisa
S. Rescia, BNL

SCIENTIFIC SECRETARIES

S. Buontempo
G. Fiorillo
F. Riccardi

CONFERENCE SECRETARIAT

M. Arienzo
C. Cutillo
M. Russo
C. Tornatore

ORGANIZED BY

Istituto Nazionale di Fisica Nucleare, Sezione di Napoli
Dipartimento di Scienze Fisiche dell'Università di Napoli

SPONSORED BY

Università di Napoli
Istituto Nazionale di Fisica Nucleare
Regione Campania
Azienda Soggiorno e Turismo di Capri
CAEN, *Viareggio*
C.S.M. Centro Servizi Metalli, *Milano*
Golden Education, *Napoli*
Hamamatsu Italia, *Milano*
Industrie Politecniche Meridionali, *Napoli*
Pol. Hi. Tech., *Carsoli*

Special thanks to Mimmo Iodice, who took the photograph of Capri appearing on the poster and on the proceedings.

CONTENTS

Electronics

Realization of Large Systems

Papers Not Received in Time

PREFACE

Following last year's conference at Fermilab, the Second International Conference on Calorimetry in High Energy Physics was held on October 14–18 in Capri, Italy, in the charming atmosphere of the island of Augustus and Tiberius.

The Conference began with a series of review talks on general issues of calorimetry, given by J. Virdee (CERN) and R. Wigmans (CERN) on the performance limitations of electromagnetic and hadron calorimeters, D. Green (Fermilab) on the physics requirements on calorimetry, H. Oberlack (MPI) on the experience with existing large calorimeter systems, J. Pohlen (Martin Marietta) on large scale engineering, A. Ferrari (Milano) on the radiation dose calculations, S. Rescia (BNL) on readout electronics for calorimeters, R. Brun (CERN) on the simulation of showers and calorimetric detectors, and M. Spiro (Saclay) on calorimeters for nonaccelerator physics.

The Conference then continued as a workshop devoted to the topical problem of calorimetry for experiments at the future hadron colliders LHC and SSC. The workshop was structured in parallel sessions, covering the physics subjects of requirements on calorimetry, scintillation calorimeters, ionization calorimeters, homogeneous calorimeters, electronics, radiation hardness and performance of large systems. About 80 contributed papers animated the discussions, which were summarized on the last day of the Conference by the parallel session conveners J. Colas (Annecy), M. Della Negra (CERN), M. Derrick (ANL), A. Ferrari (Milano), P. Lecoq (CERN), A. Menzione (Pisa) and S. Rescia (BNL).

The picture which emerged from the workshop was that of a large variety of ongoing R&D activities, covering almost all the calorimeter technologies and often involving experimental setups with the scale of small experiments. Interesting results were presented on the different options for LHC and SSC detectors: scintillating fibers and tiles, ionization calorimeters and crystals, for very accurate energy measurements. Moreover, important issues were discussed on radiation hardness, fast readout electronics and actual realization of complex systems. The session on the physics requirements on calorimeters provided for a direct link between the physics motivation of future experiments and the choice of specific calorimetric detectors. Concluding the Conference, Klaus Pretzl summarized the highlight results and stressed the enormous progress achieved over the last few years in the understanding and development of calorimetric techniques.

On behalf of the Organizing Committee, I would like to thank very warmly all the participants for their contributions and for the vivid debate which took place at the Conference. I would also like to acknowledge the excellent and pleasant collaboration of the Conference Secretaries, of the Scientific Secretaries and of all the others who contributed to the organization of the Conference. In addition, I greatly appreciate Chiara Tornatore's and Maria Arienzo's help with the editing of the proceedings.

Antonio Ereditato, 1991

Invited Papers

PERFORMANCE AND LIMITATIONS OF ELECTROMAGNETIC CALORIMETERS

Tejinder S. VIRDEE
Blackett Laboratory, Imperial College, London SW7 2BZ, UK.

ABSTRACT
The factors which determine and limit the performance of electromagnetic calorimeters are considered. Their performance can be characterized by the resolution attained in the measurement of the energy, the direction, the position and the timing of electrons or photons. Further factors such as two shower separation, electron-hadron separation and sensitivity to radiation damage are also considered.

1. Introduction

A high energy electron or photon incident on a thick absorber initiates a cascade of electrons, positrons and photons via pair production and bremsstrahlung. With increasing depth in the absorber material the energy of these particles is degraded and the number of shower particles increases. When the energy of the electrons and positrons falls below the critical energy, ϵ, the energy at which the radiative energy losses become equal to ionization energy losses, further dissipation of energy is dominated by ionization and atomic or molecular excitation rather than by the generation of more shower particles.

The longitudinal development of the showers is governed by the radiation length, X_o, defined to be the mean distance over which a high energy electron loses all but 1/e of its energy by bremsstrahlung. The lateral shower development is characterized by the Moliere radius which gives a measure of the lateral deflection of electrons of an energy of ϵ after traversal of $1X_o$. For rough estimation $X_o \approx 180A / Z^2$ in g cm^{-2} and R_M(cm) $\approx 10/\rho$ (ρ in g cm^{-3})†. The shower development is approximately material independent when defined in terms of X_o and R_M.

The longitudinal development of 10 GeV electron showers in Al, Fe and Pb[2c], generated using the EGS4 computer code, is shown in Fig.1. The shower maximum is deeper for higher Z materials due to the lower critical energy leading to particle multiplication down to lower energies. The slower decay beyond the maximum is due to the lower energies at which electrons can still radiate in higher Z materials.

Two effects are responsible for the lateral spread of an e.m. shower namely

- multiple scattering of electrons or positrons away from the axis,
- minimally attenuated photons depositing their energy a long way from their point of origin especially when emitted from electrons that travel at large angles w.r.t. shower axis.

---

† For more accurate estimation the following expressions[1] can be used
$X_o = 716.4\ A / [Z(Z+1) \ln(287/\sqrt{Z})] \approx 180\ A/Z^2$ g cm^{-2}
$R_M = 21.2$ (MeV) X_o/ϵ where $\epsilon = 800/(Z+1.2)$ MeV.

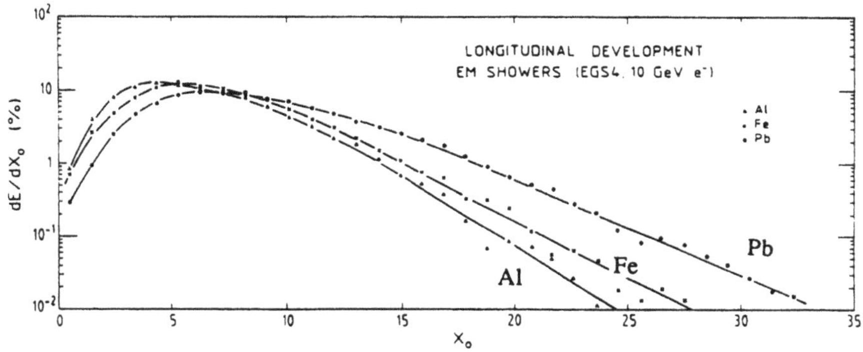

Fig.1 The longitudinal development of 10 GeV electron showers in different absorbers.

The former process dominates in the early part of the shower whilst the latter dominates deeper in. For example, the FWHM of the lateral profile of energy deposition is ~ 500 μm at a depth of 5 X_o in CeF_3 for 30 GeV photon showers (Fig. 2a). The e.m. showers are narrow as can be seen from Fig. 2b. In liquid xenon 90% of the shower energy is contained in a cylinder of radius ≈ 7 cm (~ 1.75 R_M). The corresponding numbers for CeF_3 are ≈ 4.25 cm or ≈ 1.65 R_M.

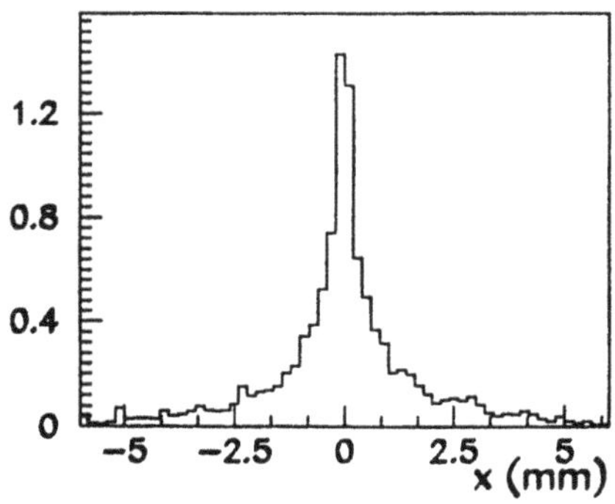

Fig. 2a. Lateral profile of the energy deposited in 300 μm thick Si at a depth of $5X_o$ in CeF_3 for a 30 GeV e^- shower.

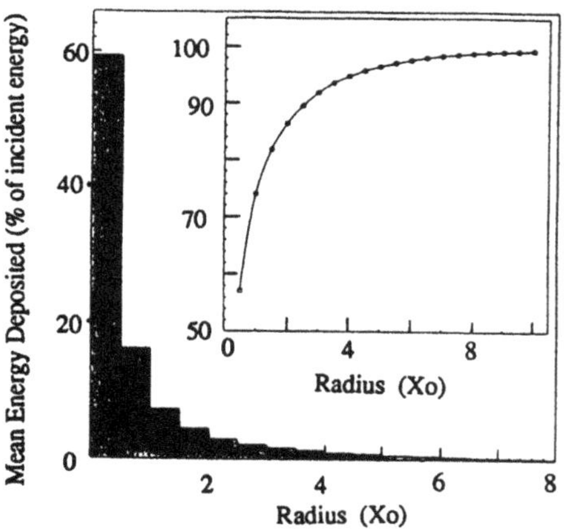

Fig. 2b. % of incident energy deposited in a long cylinder of radius r.

The performance of a stand–alone e.m. calorimeter can be quantified by considering :

- the energy resolution,
- impact position resolution,
- angular resolution,
- two shower separation,
- electron-hadron separation,
- the speed of response and,

- sensitivity to radiation damage

When integrated into an overall detector further parameters need to be considered, eg.

- jet energy resolution when combined with a hadronic calorimeter
- electron identification (with charged particle tracking and hadron calorimetery)
- effect of a magnetic field.

The relative emphasis placed on these parameters determines the calorimeter choice.

This report relies heavily on previous reviews[2] of electromagnetic calorimeters.

2. Energy Resolution

Conventionally the energy resolution of calorimeters is parameterized as

$$\frac{\sigma}{E} = \frac{a}{E} \oplus \frac{b}{\sqrt{E}} \oplus c$$

where the symbol $\oplus$ implies the square root of the quadratic sum of the three terms on r.h.s. The "noise" term [coefficient a] includes

- the energy equivalent of the electronics noise and,
- the fluctuation in the energy carried by particles, other than the one of interest, entering the measurement area during the time over which the energy measurement is made.

The "sampling" term [coefficient b] accounts for

- the statistical fluctuation in the number of primary and independent signal generating processes, or any further process that limits this number eg. conversion of scintillation light into photo-electrons by a photosensor.

The "constant" term [coefficient c] takes into account

- the fluctuation in the amount of energy leakage from the front, the rear and the sides of the volume used for the energy measurement,
- the fluctuation in the amount of energy deposited in dead materials in front of or inside the calorimeter,
- the cell to cell inter-calibration error and,
- the non-uniformity in the generation or collection of the signal.

Such a parmetrization allows the identification of the various causes of resolution degradation. The assumption that the terms be added quadratically implies that the contributions are independent which may not always be the case. When comparing different calorimeters it is, of course, the actual value of the resolution, σ/E, at the relevant energies, that is important.

2.1 Intrinsic Energy Resolution

2.1.1 Homogeneous Calorimeters

When considering energy resolution it is instructive to start by looking at homogeneous calorimeters. All the energy of an incoming electron or photon is deposited in the active medium. If this is assumed to be laterally and longitudinally infinite then the resolution in the measured energy would only arise from the fluctuation in the produced number of ion pairs or photons. This is given by $n = E/W$ where W is the mean energy required to produce an ion pair (or a photon). As the production process is random in nature the fluctuation on n will be given by

$$\sigma/E = \sqrt{n} / n = \sqrt{(W/E)}$$

However the fluctuation is reduced because the total energy deposited in the medium does not fluctuate. This reduction is characterized by the Fano factor, F, after Fano who first evaluated the size of such fluctuations [3]. The resolution will then be given by :

$$\sigma / E = \sqrt{(FW / E)} \qquad (1)$$

Photon energy resolution in Ge, operated at 77 °K, has been measured[4] to be $\sigma \approx 178$ eV at 100 keV, $\sigma \approx 390$ eV at 661 keV and $\sigma \approx 550$ eV at 1000 keV. It takes 2.98 eV to create an electron-hole pair. The energy difference between the band gap (Table 1) and the W value goes into phonons. For an energy deposit of 661 keV $n \approx 2.2.10^5$ and Poisson statistics lead to a resolution of ~1.4 keV. This is considerably worse than the measured resolution. Application of Fano's factor for Ge gives the observed value of ~ 0.4 keV. A comparison[4] of the pulse height spectra in Ge and NaI crystals is shown in Fig. 3.

Material	W_e (eV)	W_γ (eV)	F	Band Gap (eV)
Si	3.6	-	0.14	1.1
Ge	2.9	-	0.13	0.7
LAr	23.7	-	0.11	-
LKr	20.5	53	0.06	-
LXe	16.4	38	0.05	-

Table 1. Some properties of materials used for low energy total absorption calorimeters.

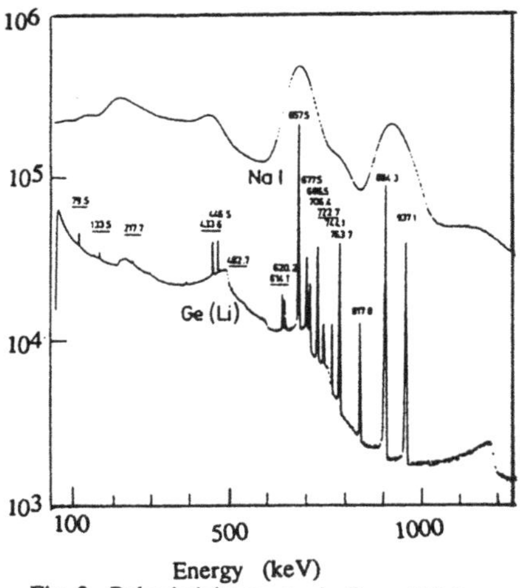

Fig. 3. Pulse height spectra in Ge and NaI.

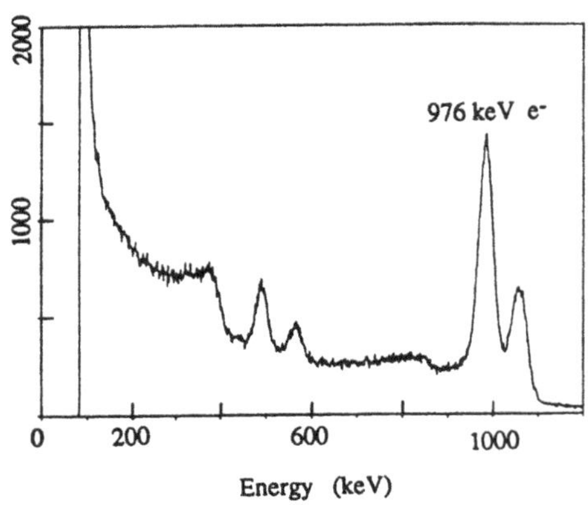

Fig. 4. Pulse height spectrum in liquid argon.

Noble liquid calorimeters should, in principle, allow measurement of ~ MeV photons to a similar precision. The energy resolution for 976 keV electron conversion line of ^{207}Bi is observed[5] to be only $\approx$ 11 keV (Fig. 4) whereas the use of Eq. 1 leads to 1.6 keV. It has been suggested[6] that the poorer energy resolution is due to the production of δ-rays. Approximately 30% of the energy deposited by a 976 keV electron is deposited by densely ionizing δ-rays[5b] carrying a kinetic energy of less than 10 keV. The electron-ion pair recombination is higher for densely ionizing tracks. This decreases the observed charge and the fluctuation in the amount of energy deposited by δ–rays then dominates the overall resolution.

Another source of fluctuation is in the amount of energy going into mechanisms other than the one being measured eg. scintillation when only ionzation charge is collected. The noble liquids listed in Table 1 also generate scintillation light. An anti-correlation between the amount of the charge collected and the luminescence intensity is seen[7] (Fig. 5). The scintillation light can arise from recombination and if it is detected together with the ionization then the energy resolution improves. Although this may not play a significant role for high energy measurements it could allow such liquids to be used for precise measurements at low energies.

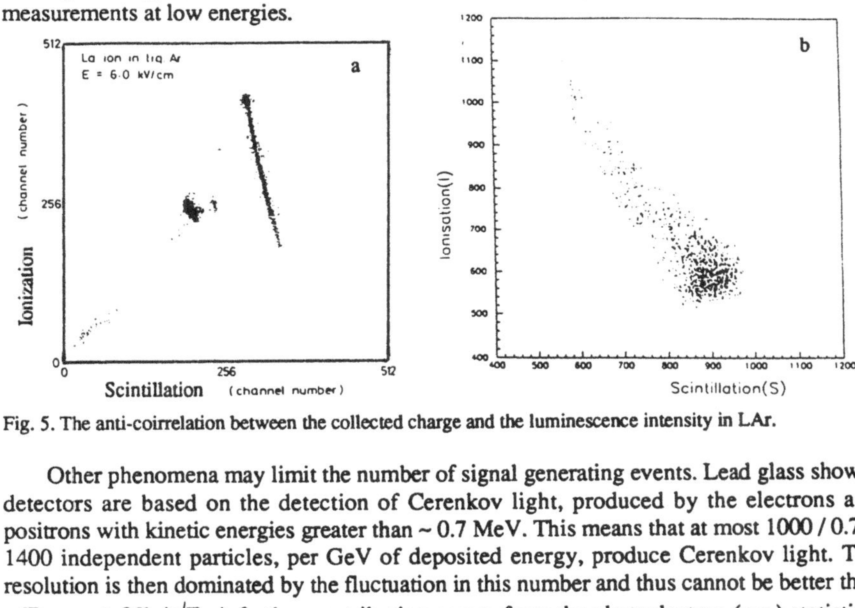

Fig. 5. The anti-coirrelation between the collected charge and the luminescence intensity in LAr.

Other phenomena may limit the number of signal generating events. Lead glass shower detectors are based on the detection of Cerenkov light, produced by the electrons and positrons with kinetic energies greater than ~ 0.7 MeV. This means that at most 1000 / 0.7 ~ 1400 independent particles, per GeV of deposited energy, produce Cerenkov light. The resolution is then dominated by the fluctuation in this number and thus cannot be better than $\sigma/E = \sigma_n \geq 3\% / \sqrt{E}$. A further contribution comes from the photoelectron (p.e.) statistics. Approximately 1000 p.e. are generated per GeV of deposited energy. The observed resolution then becomes $\sigma/E \sim \sqrt{(\sigma_n^2+\sigma_p^2)} \sim 4.4\%/\sqrt{E}$. An energy resolution of $\sigma/E \sim 5\% / \sqrt{E}$ for e.m. showers in the energy range 1-20 GeV has been attained in a large lead glass array [8].

Scintillation light from crystals has been used to attain good resolution at high energies. Their energy resolution is usually limited not by the statistics of primary light producing processes but by leakage of energy and by the non-uniformity in the generation and the collection of light. NaI(Tl) was the first type of crystal to be used in a large scale high energy physics experiment [9]. The energy resolution for 8 keV γ–rays is $\sigma/E \sim 15\%$, ~ 5% for 1 MeV and ~ 1% for 1 GeV electrons. Hence the resolution is worse by a factor ≥ 3 at 1 MeV and ~ 30 at 1 GeV if it is scaled by $\sigma/E \propto 1/\sqrt{E}$ from the lowest energy. This indicates that the resolution is not dominated by statistical fluctuations but by the systematic effects mentioned above. Fig. 6 shows the observed energy resolution for high energy electrons[10]. A scaling law more like $\sigma/E \propto 1/E^{1/4}$ describes the data and is ascribed to energy leakage. Variation in light yield or collection efficiency can also yield a similar dependence as shown by a simulation[11] in which the light output was varied with σ=5% per cc.

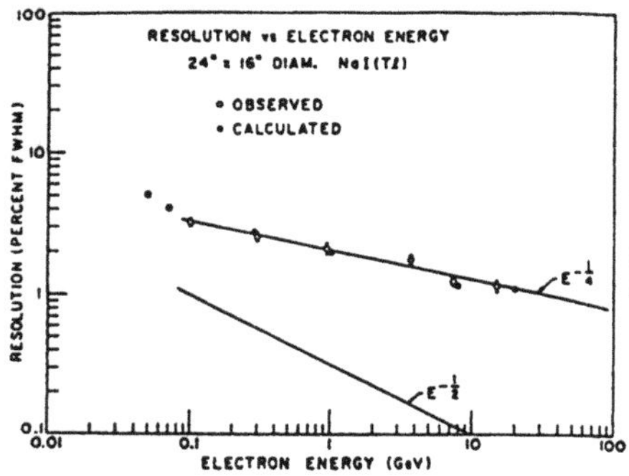

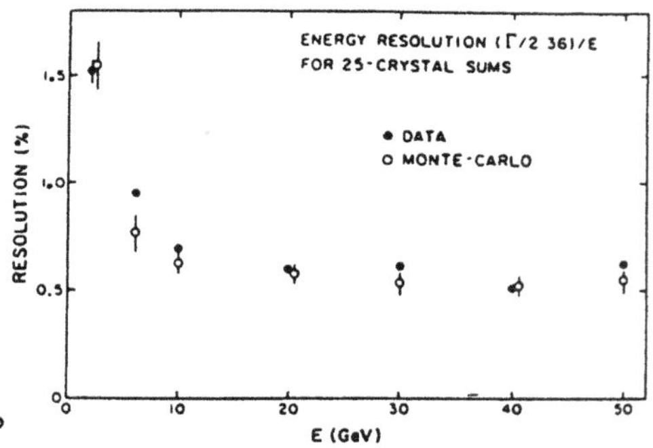

Fig. 6 The energy resolution for an array of NaI crystals. Fig. 7. As in Fig. 6 but for BGO crystals.

A very good energy resolution of $\sigma/E \sim 2\%/\sqrt{E} \oplus 0.5\%$ has been achieved[12] for a large number of BGO crystals in a test beam (Fig. 7). This has yet to be achieved in the L3 experiment where the resolution for electrons (~ 45 GeV) from Bhabba events is presently ~1.2%. The uncertainty in the energy of the electrons due to intial state radiation has not been unfolded. The poorer than expected resolution is believed to be due to cell to cell intercalibration error, poorer energy resolution for electrons entering near the crystal boundaries (section 2.5) and the variation of temperature inside the crystals. The light yield in BGO varies by −1.55% per °C of temperature change. This experience underlines the difficulty of attaining very high energy resolution in large systems.

2.1.2 *Sampling Calorimeters*

When the very best energy resolution is not required, sampling calorimeters are employed. The shower energy is measured in active layers, often of low Z, sandwiched by passive absorber layers of high Z materials. The energy resolution is determined by the fluctuation in the fraction of shower energy deposited in the active layers. If the energy loss in an active layer is much smaller than that in the absorber the number of independent charged particles crossing an active layer can be approximated by $n=E/\Delta E$ where ΔE is the energy lost by a minimum ionizing particle (m.i.p.) in the absorber layer. If t_{abs} is the thickness of the absorber layer and dE/dx is the energy lost per unit path length traversed by a m.i.p then $\Delta E = t_{abs}\, dE/dx$. The sampling fluctuation is then given by the familiar form

$$\sigma/E = \sqrt{n}/n \;\propto\; \sqrt{(\Delta E/E)} \;\propto\; k\sqrt{t_{abs}}/\sqrt{E} \quad (E \text{ in GeV}). \qquad (2)$$

The dependence of the resolution on ΔE can be seen in Fig.8 for various absorber and active materials[2d]. The constant k is ≈ 15% (with some material dependence) for $t_{abs} = 1\ X_0$.

A large fraction of the energy in an e.m. shower is deposited by low energy (few MeV)

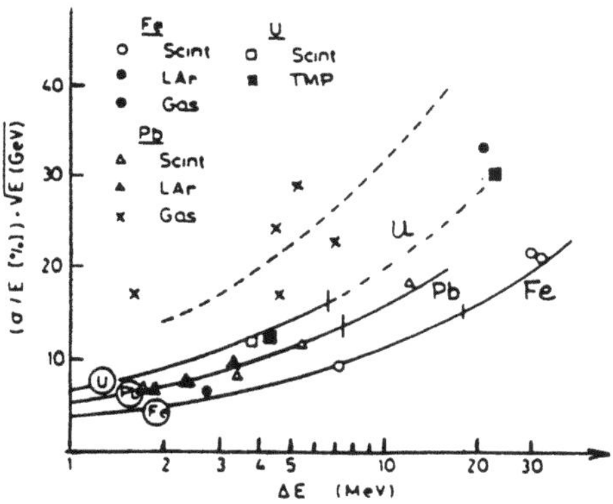

Fig. 8 Electron energy resolution as a function of the energy lost (ΔE) in one absorber layer.

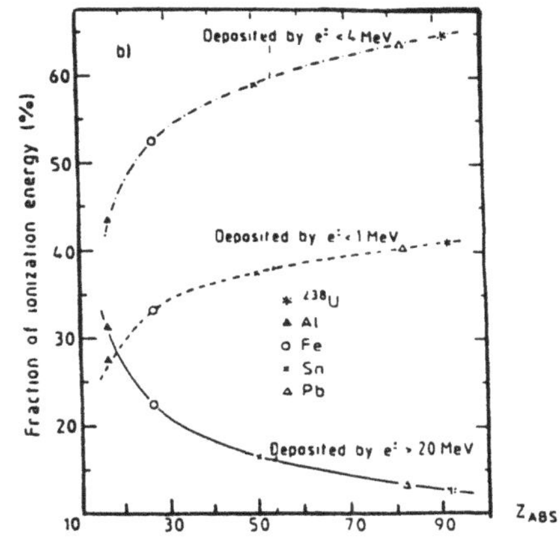

Fig. 9 Fraction of the ionization energy deposited by particles of different momenta in a 10 GeV e⁻ shower.

electrons especially in absorbers with high Z (Fig.9). These electrons are generated by either the photoelectric effect or Compton scattering. These have a high power dependence on Z and hence usually take place in the absorber layers. When the electron, from the process γ (of low energy) -> e, is created near the surface of an absorber layer it can enter the active layers. Usually it cannot traverse more than one active layer because, for the absorber thicknesses usually employed, the energy loss in one absorber layer is too large. This leads to an energy resolution dependence $\alpha \sqrt{t_{abs}}$.

Deviations from the scaling formula (2) can be understood by considering two cases. The first is where the absorber thickness is kept fixed but the thickness of the active layer is increased. The fraction of the incident energy deposited in the active medium increases. The improvement in the resolution[13] with increasing t_{act} can be seen in Fig. 10. Lead absorber plates of a thickness of 1/3 X_0 sandwich a liquid argon active layer of varying thickness. The magnitude of this effect depends on the difference between the Z of the active and the passive layers

In the second case scaling according to Eq. 2 will not continue when t_{abs} becomes very small as the number of crossings in successive active layers now become correlated. This is clearly evident in a liquid argon calorimeter[14] with very fine sampling : 400 cells of 1mm stainless steel followed by a 1.05 mm gap of LAr (total 25.7 X_0). A resolution of $\sigma \sim 1.4\%$ is observed for 25 GeV electrons ($\sigma/E \approx 7\%/\sqrt{E}$). Scaling using Eq. 2 yields $\sigma \sim 0.8\ \%$ ($\sigma/E \approx 4\%/\sqrt{E}$).

These effects can be taken into account by parametrizing the energy resolution in terms of the sampling structure and the amount of energy deposited in the active medium. If ΔE_c is the energy, in MeV, that is deposited (eg. approximately by a m.i.p.) in one active and one passive layer and A is the fraction of the incident energy deposited in the active medium then the energy resolution can be written as

$$\frac{\sigma_s}{E} = \frac{5\%}{\sqrt{E}}(1-A)\ \Delta E_c^{0.5\,(1-A)} \tag{3}$$

This is an extension of "Waseda's Formula"[15] which gives a worse resolution than M.C. simulation for the sampling effects for large ΔE_c and A. Some experimental data and results of M.C. simulations are compared in Fig. 11 with the resolution computed using Eqn.3. The contribution of the limited photoelectron statistics of the scintillator calorimeters has been unfolded. When A is small the familiar form for sampling calorimeters is recovered. We believe that this expression can predict the e. m. energy resolution to within 20%.

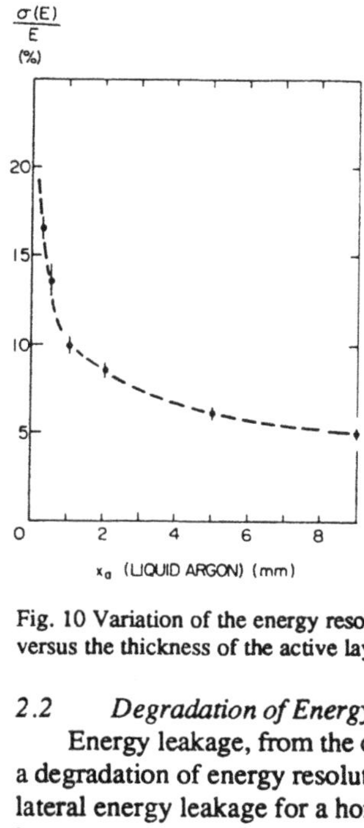

Fig. 10 Variation of the energy resolution versus the thickness of the active layer.

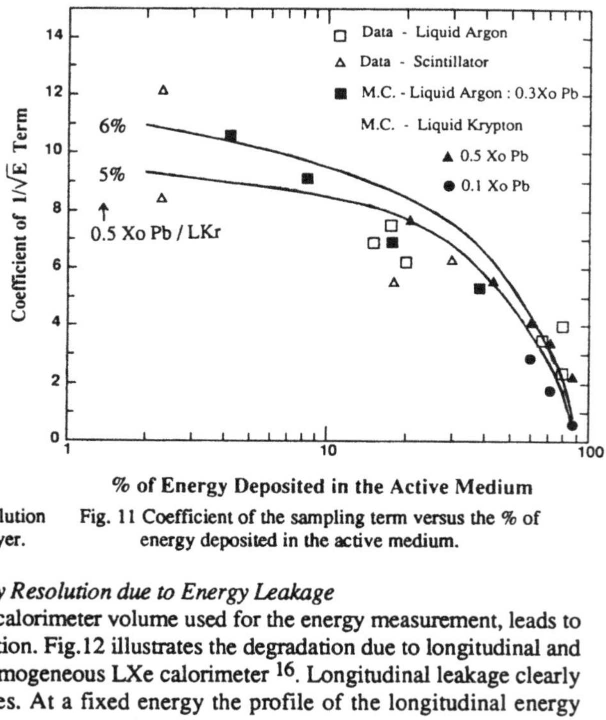

Fig. 11 Coefficient of the sampling term versus the % of energy deposited in the active medium.

2.2 *Degradation of Energy Resolution due to Energy Leakage*

Energy leakage, from the calorimeter volume used for the energy measurement, leads to a degradation of energy resolution. Fig.12 illustrates the degradation due to longitudinal and lateral energy leakage for a homogeneous LXe calorimeter [16]. Longitudinal leakage clearly has more serious consequences. At a fixed energy the profile of the longitudinal energy deposition differs from one shower to another. Showers developing late lead to a larger energy leakage from the rear of a finite depth calorimeter. The fraction of the incident energy leaking out, and the fluctuation on it, increases with energy since the depth at which the shower maximum occurs increases with energy, albeit logarithmically. Fig.13 shows the longitudinal development of photon showers, of different energies, in liquid xenon[16]. The resolution will no longer scale according to $1/\sqrt{E}$ as already seen in Fig.6 for NaI. The r.m.s. deviation, as a fraction of the mean deposited energy, versus the fractional longitudinal energy leakage (f) is plotted in Fig. 14. For a quick estimation[2b] $\sigma_{rms}/\langle E_{dep}\rangle \approx f/2$ for $f \leq 20\%$.

The loss of energy resolution due to the lateral energy leakage is smaller since the lateral profile of energy deposition differs much less from one shower to another. The energy dependence of the fluctuation is also weak as the lateral shower shape is almost independent

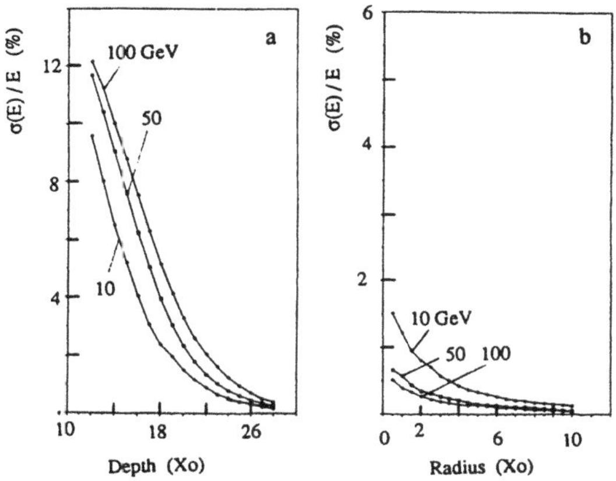

Fig. 12 The effect of a) longitudinal and b) lateral energy leakage on energy resolution of a LXe calorimeter.

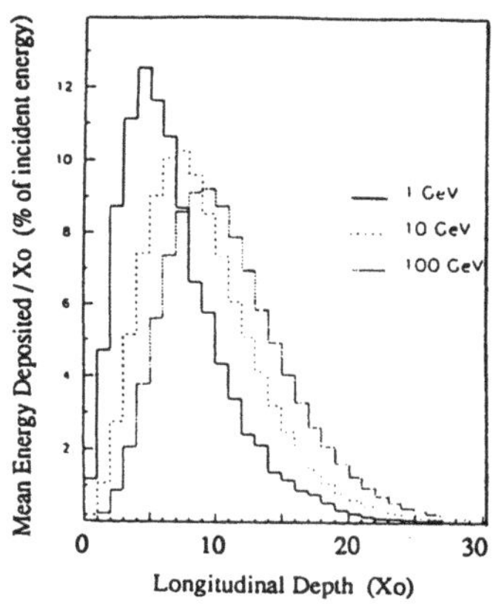

Fig. 13 The longitudinal profile of energy deposition in LXe for 1, 10 and 100 GeV γ's.

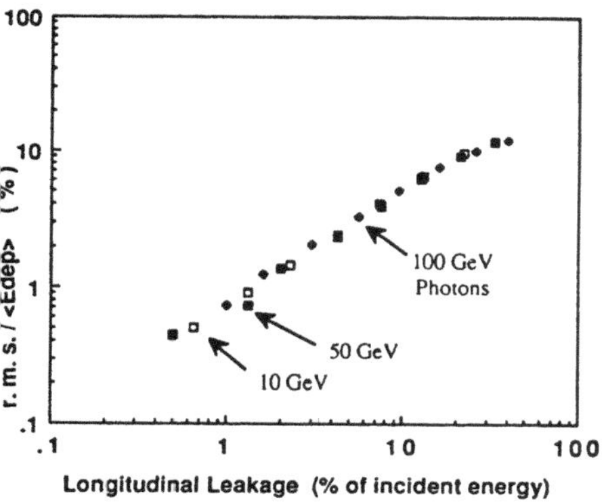

Fig. 14 Fractional r.m.s. of mean deposited energy versus longitudinal leakage in LXe.

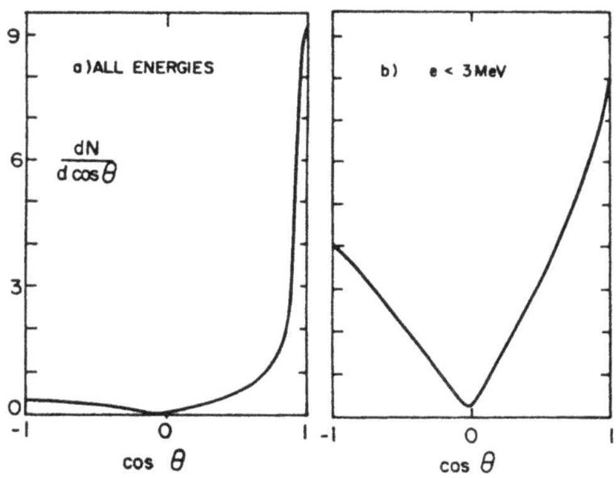

Fig. 15 The angular distribution of secondary electrons a) of all energies b) with < 3 MeV in a 1 GeV e^- shower.

of energy especially at high energies.

Energy leakage from the front of calorimeters can affect energy resolution, especially at low energies. The backscattered photons and electrons typically have energies lower than the critical energy. The fluctuation in the energy carried away by these particles results in a resolution loss[17] of ~ few percent for 100 MeV photons and ~ 0.1% for ones with energy greater than 10 GeV.

2.3 Resolution Worsening due to Path Length and Landau Fluctuations

In sampling calorimeters with relatively thin (≤ 0.01 g cm^{-2}) active layers the energy resolution is considerably worse than equivalent ones with thicker active layers. Gas sampling calorimeters fall into this category. The poor resolution is due to two effects: path length and Landau fluctuations[13].

The angular distribution of shower electrons, plotted in Fig.15, shows that about 12% of all the electrons are back scattered into the sampling gaps. For low energy electrons (< 3 MeV) this percentage is considerably higher. The wide spread of angles leads to large fluctuations in the length of tracks in the active layers and hence in the deposited energy. In thick active layers the energy deposited by low energy electrons going forward at small angles is not much different from those at larger angles since their range, at low energies, is roughly similar to the thickness of the active layer. Fig. 10 shows that thick in this context would mean a few mm of LAr. In a gas sampling calorimeter the active layer usually has a thickness of ~ 10^{-3} g cm^{-2}. Low energy electrons that travel in the plane of the active layer deposit considerably more energy than the ones moving perpendicularly to the plane.

The energy deposited by a m.i.p. traversing an active layer is described by the Landau distribution which is an asymmetric distribution with a long tail towards high deposited energy. Therefore the resolution scaling is worse than $1/\sqrt{n}$, expected for n crossings. An order of magnitude estimate for this effect is given by [2a]

$\sigma/E = (1/\sqrt{n})\,[2/\ln(10^4\, t\ (\mathrm{g.cm^{-2}}))]$.

In a calorimeter with active layers of t ~1 g.cm^{-2}, the Landau fluctuation worsens the resolution by ≤ 3% whereas in a gas sampling calorimeter (t ~ 10^{-3} g.cm^{-2}) the resolution worsens by ~ 40%.

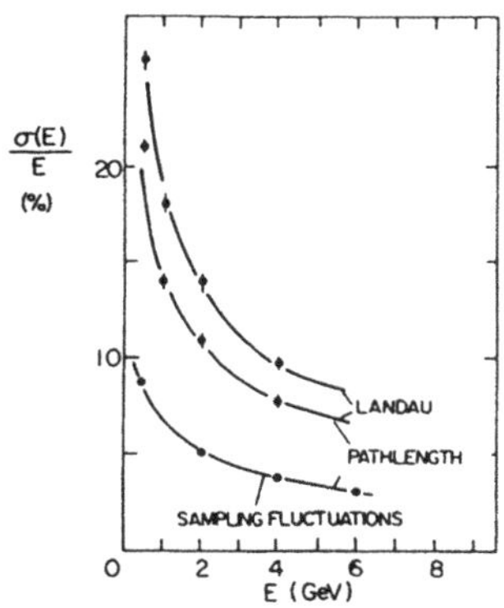

Fig. 16 Contribution of the Landau and path length fluctuations to the energy resolution

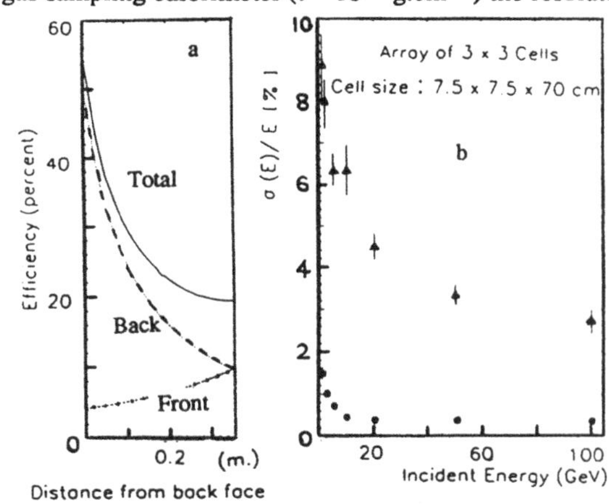

Fig. 17 a) Light collection efficiency b) energy resolution in a rectangular cell filled with liquid xenon.

For gas calorimeters the above two effects give similar contributions[13] (~12% at E=1 GeV). Combining with the sampling fluctuation (for a calorimeter with 1/3 X_0 absorber layers) leads to an overall resolution of ~ 18% $/\sqrt{E}$ (Fig.16). This can be compared with experimentally measured resolutions for gas calorimeters shown in Fig. 8.

2.4 *Resolution Worsening due to Non-uniformity of Signal Generation and Collection*

Longitudinal non-uniformity of signal generation and collection, when folded with the fluctuation in the longitudinal profile from one shower to another (at a fixed energy), leads to a loss of energy resolution. This will also lead to energy non-linearity due to the energy dependence of the average profile for longitudinal energy deposition.

The effect on the energy resolution of non-uniformity of light collection in a homogeneous liquid xenon calorimeter is shown in Fig.17[16]. Electron showers were generated in a setup of dimensions 22.5 x 22.5 x 70 cm^3 (8x 8x 25 X_o), split into 3 x 3 rectangular cells. The scintillation light is generated isotropically and the non-uniformity of light collection is due only to the imperfect reflectivity of the cell walls (with a reflectivity of 80% for 175nm light). The result is clearly unacceptable. Graded reflectivity of the side walls should enable more uniform light collection and consequently better energy resolution.

Light inside a crystal is usually internally reflected and hence the non-uniformity of light collection is not as severe as in the case mentioned above. The scintillation light in BGO crystals for the L3 experiment is detected by Si photodiodes mounted at the rear of the crystals. The efficiency of light collection from the front of the crystals is increased by the tapered geometry of the crystals. Fig. 18 shows the light collection uniformity as measured with a collimated ^{137}Cs source running along the main dimension of the crystal. The non-uniformity observed for crystals wrapped in aluminized mylar is greatly reduced by coating the polished crystals with a 40-50 μm thick layer of high reflectivity NE560 white paint. The effect of the non-uniformity can be parameterized by

$$S(x) = S(L)\,\{1 - R(1 - x/L)\} \qquad (4)$$

where S(x) is the observed light signal, x is the distance from the diode plane and S(L) is amount of light detected when the source is at a distance L from the diode plane (L is the length of the crystal). Fitting the data in Fig.18 to Eq. 4 yields R≈0.1 and 0.53 for the white paint coating and mylar wrapping respectively. Fig. 19 shows the Monte Carlo estimation of

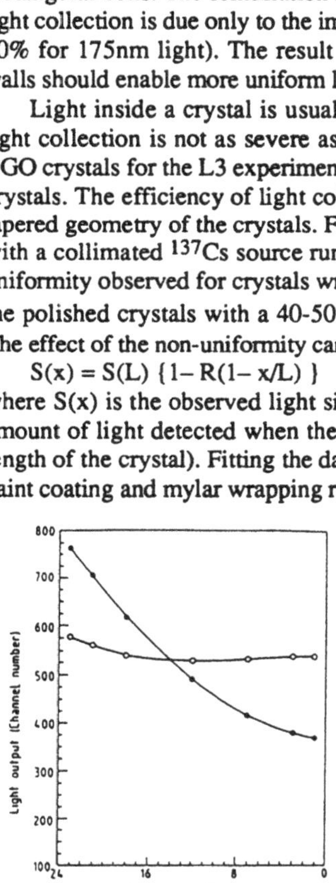

Fig. 18 Light collection versus distance from diode plane o - white paint coating, • aluminized mylar wrapping.

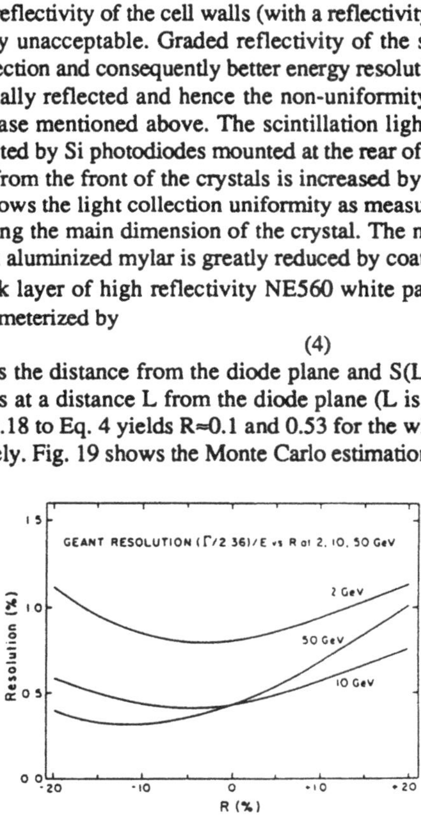

Fig. 19 Energy resolution as a function of the parameter R (see text).

the resolution degradation for different values of R. Without the coating the energy resolution of the BGO crystals would have been unacceptable. At the future colliders radiation damage will not be uniform and its effect on the uniformity of light collection for the crystal calorimeters must be estimated.

Calorimeters based on ionization chambers with a large drift gap also suffer from an effective non-uniformity of signal collection. An electron created near the cathode plane contributes 1e to the induced charge whereas one created next to the anode contributes nothing. The signal induction efficiency thus ranges from ≈ 100% close to the cathode plane and ≈ zero near the anode palne. The calorimeter therefore appears to have a coarser sampling. If the shaping time is made smaller than the drift time a more uniform efficiency is obtained but the size of the signal is reduced. The magnitude of this effect has been estimated[18] for a LAr calorimeter with a structure of 500 μm G10 and 1 cm LAr gap using a peaking time > drift time. The estimated contribution, from this effect, to σ/E is $\approx 1\%/\sqrt{E}$ whilst the sampling contribution is estimated to be $\approx 0.5\%/\sqrt{E}$.

Scintillating plastic fibre calorimeters, eg. SPACAL which has a lead to scintillator volume ratio of 4 and 1mm diameter fibres, show a large constant term in the electromagnetic energy resolution function[19]. This decreases when the angle of incidence w.r.t. the fibre axes gets larger. The sizeable constant term is due to at least two effects. Both the effects lead to lateral non-uniformity at a scale smaller than the overall lateral shower size and hence to a variation of response that is independent of incident energy. These are :

- dependence of the size of the observed signal on the impact point. The lateral size of the shower in the early part is smaller than the inter fibre separation. Consequently electrons entering the calorimeter in a fibre will, on average, produce larger signals than those entering lead.
- the number of fibres that contribute significantly to the shower measurement in this calorimeter is small because of the narrowness of e.m. showers. Hence fibre to fibre variations will affect the resolution. If the distribution of the fibre to fibre response has an r.m.s. of σ_f and the number of fibres contributing significantly to the signal is n_f then the size of the constant term will be $c = \sigma_f/\sqrt{n_f}$. For the test described in ref 19, $\sigma_f \approx 6\%$ and $n_f \approx 25$ leading to $c \approx 1.2\%$. A recent test[20] using fibres with a diameter of $\phi = 0.5$ mm, instead of $\phi = 1$mm, yielded a constant term of $\approx 0.7\%$. The Pb scintillator volume ratio was kept the same thus increasing n_f by a factor of 4. Since the inter-fibre separation also decreased the impact on the energy resolution of both the effects described above is reduced.

The sampling term in the resolution also improves by a factor of $\sqrt{2}$ from 12.9 % for the calorimeter with 1 mm diameter fibres to 9.3% for the one with ϕ=0.5mm.

2.5 *Effect of Dead Material in Front or Inside e.m .Calorimeters*

Dead material in front of e.m. calorimeters can be present because of eg. support structures or vacuum and cryostat walls for noble liquid calorimeters. Side wall supports may also be necessary to hold in place the calorimeter elements eg. crystals. The energy lost in the dead materials decreases the mean and increases the width of the observed energy distributions. This is shown in Fig 20 where a varying thickness of Al absorber was placed in front of an e.m. calorimeter[21] with a performance parametrized by $\sigma/E = 18\%/\sqrt{E}$. Sizeable deterioration takes place for absorber thicknesses of $\geq 2X_0$. This degradation can be reduced by measuring the energy flow just after the dead material.

The presence of dead material, consisting of 200 μm Carbon fibre walls, the crystal wrapping and the distributed air gap, in between BGO crystals contributes to the degradation of the energy resolution. The energy resolution[12] with the beam centred on a

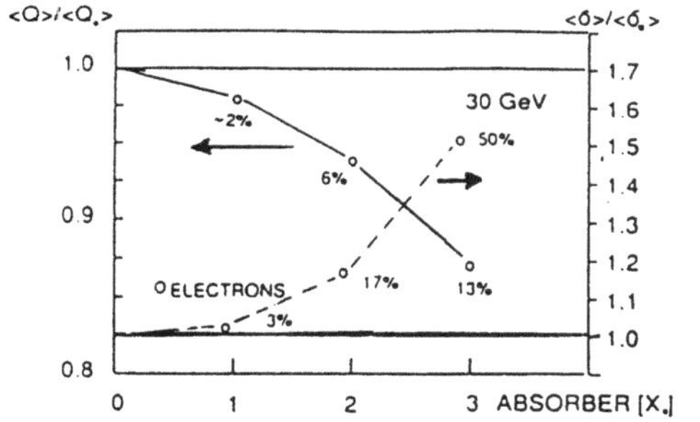

Fig. 20 Ratio of std. dev. with, $<\sigma>$, and without absorber, $<\sigma_o>$, versus the absorber thickness.

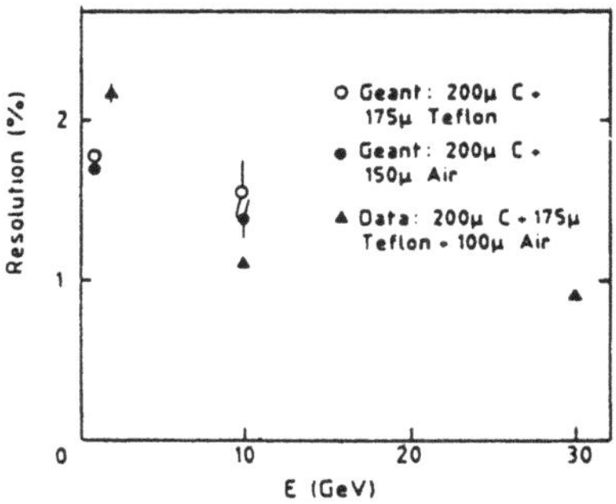

Fig. 21 Influence of walls on energy resolution. Beam is centred on the wall between two crystals.

wall is shown in Fig. 21. Comparison with Fig 7 shows that the degradation in the resolution, close to the crystal boundaries, is significant. Since the lateral shower shape does not vary much for high energy e.m. showers, such degradation of resolution is independent of incident energy and thus contributes to the constant term.

2.6 *Deterioration of Energy Resolution due to Inter-calibration Errors*

Electromagnetic showers are narrow (Fig.2b) and usually the central cell, or at most the central four cells, contains ≈ 80% of the shower energy. The reconstructed energy, Erec, can be expressed as

$$E_{rec} = \sum_{i=1}^{n} g_i E_i$$

where E_i is the energy in the i th cell, g_i is the calibration constant of the i th cell and n is the number of cells used to measure the energy. The lateral shape of e.m. showers does not change much from one shower to another (or with energy). Hence any deterioration of the energy resolution, due to inter-calibration errors, will be independent of energy. If the r.m.s. error on g_i is δ then the uncertainty on E_{rec} will be $c = \delta/\sqrt{n}$ leading to $c \approx \delta/2$ for n=4.

The cell to cell intercalibration error should therefore be considerably smaller than the desired constant term i.e. ≤ 0.3 % if the overall constant term is 0.5%. Although it is possible to achieve such accuracy in the calibration of electronics, it is more difficult in calorimeter system as a whole. The task is easier for calorimeters using unit gain and temporally stable primary signal generation and collection. An example of the excellent performance that is possible in such devices is the CERN-NA31 liquid argon calorimeter [22].

CDF collaboration has used their central magnetic field to measure the momenta of electrons originating from b, W and Z decays. These electrons are used to calibrate the electromagnetic calorimeter. An r.m.s. error of approximately 1.7% has been achieved[23] from a data sample yielding ~ 35 electrons per cell. At the future colliders electrons from Z decays can probably be used for calibrating the e.m. calorimeters. The rate of $Z \to e^+e^-$ pairs, in a rapidity interval of $|\eta| \leq 2$ and at a luminosity of $L = 10^{34}$ cm^{-2}s^{-1} is ≥ 10 Hz. Appoximately 100 $e^{\pm}$ tracks per cell (size $\Delta\eta \times \Delta\phi \approx 0.025 \times 0.025$) per week could be recorded. This should be sufficient to achieve a few per mille calibration error, provided that both the momentum and energy measurements are made with a precision of ≈ 1%. Electrons from b's and W's could be used at lower luminosities.

2.7 *Noise due to the Electronics and Energy Pileup*

Electronics noise, noise from radioactivity or from energy pileup make a constant contribution to the measured energy. This contribution can limit the precision of energy measurements up to ≤ 50 GeV at the future hadron colliders.

Electronics noise has to be considered whenever gain is obtained by the use of amplifiers. This can become a problem when very fast signals are desired. Consider the case of a liquid ionization chamber. If the electron lifetime in the liquid is very large, compared with the drift time (t_d), the current pulse is triangular in shape for uniform ionization across the gap. The observed charge for a peaking time of t_p is given by

$$Q_p = Q_T \{ 2t_p / t_d - (t_p / t_d)^2 \} \qquad 0 < t_p < t_d$$
$$Q_p \sim 2\, Q_T\, t_p / t_d \qquad \text{and } t_p << t_d.$$

Since the equivalent noise charge varies as $1/\sqrt{t_p}$, the S/N ratio deteriorates as $(t_p)^{3/2}$. The CERN RD3 Collaboration[24] have measured an electronics noise equivalent (ENE) of 3.6 MeV/cell for $t_p \sim 100$ ns and $t_d \sim 400$ns. The energy of high energy showers is measured in 5 x 5 cells, with a two-fold depth segmentation. This leads to a total ENE of ~ 25 MeV. A peaking time of $t_p \sim 20$ ns was tried[25] and the resulting ENE became ~ 350 MeV for a three-fold depth segmentation. For a high energy resolution calorimeter such a noise level will dominate the energy resolution up to ≈100 GeV (if the constant term turns out to be ≤0.5%)

At a high luminosity hadron collider the spray of energy from the many overlapping minimum bias events can spoil the energy resolution. This can be a serious problem for high energy resolution e.m. calorimeters placed too close to the interaction vertex. For a calorimeter with a perfect energy resolution the smearing of the reconstructed $\gamma\gamma$ mass (=100 GeV) due to energy pileup from <10> interactions is shown[26] in Fig 22. The pileup is shown as a positive contribution though negative pileup, due to pulse shaping, can also occur. The magnitude of the effect of energy pileup can be obtained by considering a calorimeter with a resolution parameterized by $\sigma/E = 2\% /\sqrt{E} \oplus 0.5\%$. The size of the optimized mass window is 1.5 GeV for the search of a 100 GeV mass object decaying into 2 γ's. At a luminosity of $L = 10^{34}$ cm^{-2}s^{-1}, and for a measurement area of $\Delta\eta\Delta\phi$=0.1x0.1, about 10% of the signal events will fall outside the mass window due to energy pileup. In CeF_3 and for 95% lateral energy containment this situation corresponds to a distance from the interaction vertex of 1.3m.

2.8 *Effects of Radiation Damage*

High radiation levels are expected at the future colliders. The radiation dose in e.m. calorimeters will be predominantly due to photons from the decays of low momenta π^0's. For integrated luminosities of 10^{41} cm^{-2}, the expected[27] radiation dose and neutron levels are shown in Fig 22. Calorimeters based on the detection of light are particularly prone to radiation damage due to:

- a loss of light emission efficiency and

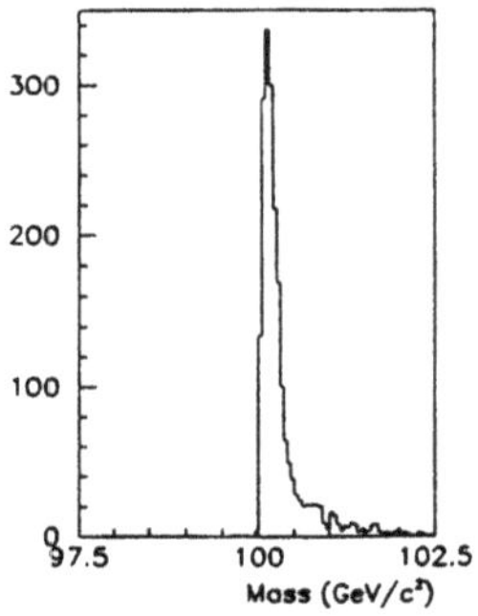

Fig.22 Reconstructed $\gamma\gamma$ mass for M_H=100 GeV with smearing due to energy pileup from <10> interactions.

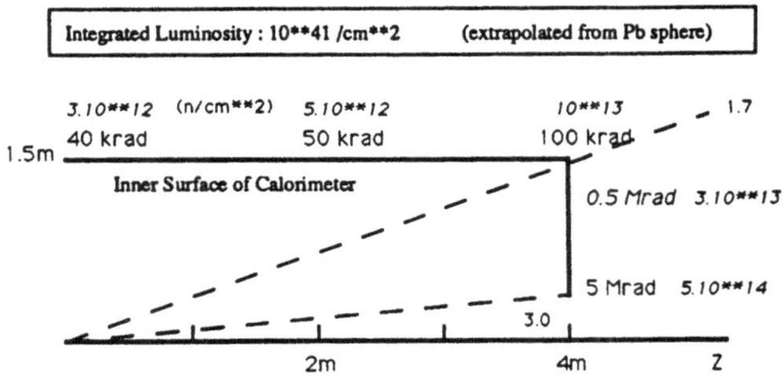

Fig. 23 Radiation doses and neutron fluences in e.m. calorimeters ($\int Ldt=10^{41}$ cm^{-2})

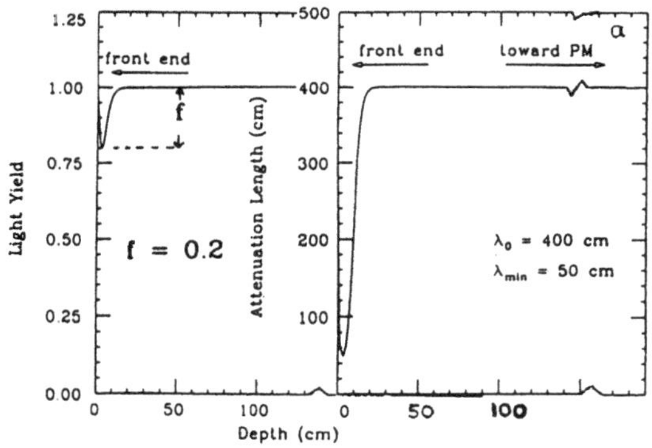

Fig. 24a Change in attenuation length and light yield as a function of depth due to radiation damage.

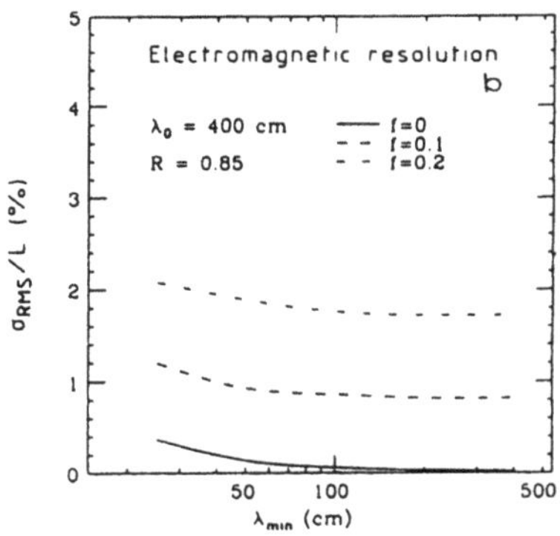

Fig. 24b Contribution to the energy resolution due to decreased light yield and attenuation length.

- an increase in the light attenuation in the damaged regions.

As an example we consider the studies carried out for a scintillating fibre calorimeter. Damage to plastic scintillating fibres results in the loss of both the emission and the transmission efficiency and leads to energy non-linearity and resolution degradation. Changes in attenuation length (λ) and light yield (f, which is the maximum fractional loss in light yield), as shown[28] in Fig. 24a, lead to a contribution to the energy resolution of ~ 1.69 % for 150 GeV electrons Fig.24b. The loss in efficiency of light emission is the more serious. If a contribution of 0.8% to the constant term is acceptable then the recieved dose should be ≤7 Mrad. Crystal scintillators will probably be less affected by radiation damage since the scintillation process is not affected up to doses of ~ 1 Grad.

3. Angular and Spatial Resolution

3.1 Angular Resolution

The two photon effective mass resolution is given by

$$\frac{\delta m}{m} = \frac{1}{2}\sqrt{\left(\frac{\delta E_1}{E_1}\right)^2 + \left(\frac{\delta E_2}{E_2}\right)^2 + \left(\cot\left(\frac{\theta}{2}\right)\delta\theta\right)^2}$$

where E_1, E_2 are the energies of the two photons and θ is opening angle between them. Good angular resolution can be obtained by measuring the impact position of the photons and linking them to the primary vertex. Running at high luminosities leads to many primary vertices per crossing. In order to reconstruct precisely the mass of a purely neutral final state, eg. H–> 2γ, measurement of the direction of the photons, with a good angular resolution, by the calorimeter itself is required. Fig. 25 illustrates[26] the degradation of mass resolution due to the lack of the knowledge of the vertex position, for a Higgs of a mass of 100 GeV/c^2, in a calorimeter with a perfect resolution starting at a radial distance of 2m from the interaction vertex. Reduction of this radius, without knowling the vertex position, would clearly worsen the resolution. The loss of resolution is equivalent to that due to energy resolution in a calorimeter with $\sigma/E = 7\%/\sqrt{E} \oplus 1.0\%$. An angular resolution of ~ 5 mrad is required to match an energy resolution of $\sigma/E = 2\%/\sqrt{E} \oplus 0.5\%$.

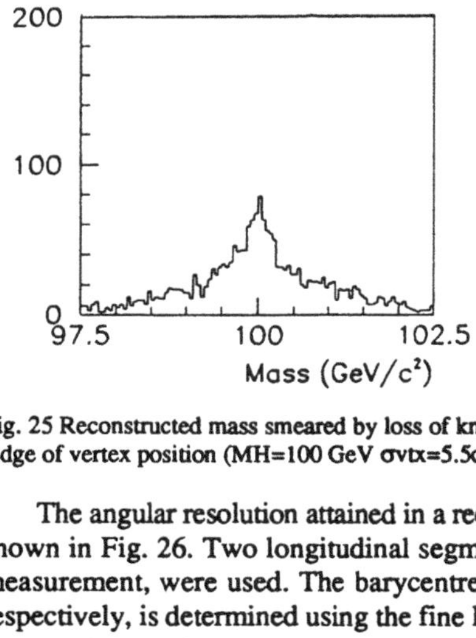

Fig. 25 Reconstructed mass smeared by loss of knowledge of vertex position (MH=100 GeV σvtx=5.5cm)

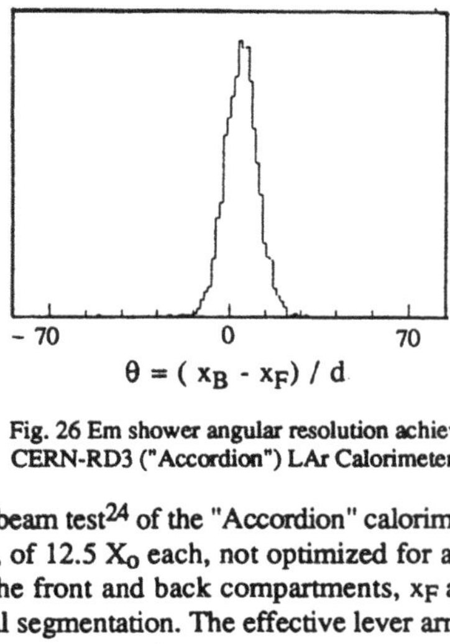

Fig. 26 Em shower angular resolution achieved by CERN-RD3 ("Accordion") LAr Calorimeter.

The angular resolution attained in a recent beam test[24] of the "Accordion" calorimeter is shown in Fig. 26. Two longitudinal segments, of 12.5 X_o each, not optimized for angular measurement, were used. The barycentre in the front and back compartments, x_F and x_B respectively, is determined using the fine lateral segmentation. The effective lever arm, d, is extracted from the data and the angle is then given by $\theta = (x_B - x_F) / d$. An angular resolution of $\Delta\theta \approx 6.1$ mrad for 120 GeV electrons is obtained. It should scale $\approx 1/\sqrt{E}$ as the accuracy of the barycentre determinations improve similarly.

3.2 Spatial Resolution

The position of impact of a photon or an electron can be deduced from

i) the centre of gravity of the transverse energy deposition in the calorimeter cells,
ii) a preshower detector, placed after some amount of dead material ($\approx 1.5\ X_o$),
iii) a position detector, placed after a first sampling of $\approx$ 3-5 X_o.

The last approach relies on the narrowness of the shower in the early part of its longitudinal development. The simulated transverse shower profile in 300 μm thick Si counter placed at a depth of $5X_0$ is shown in Fig. 2a. A spatial resolution for photon showers of ≈ 2 mm /√E (E in GeV) should be possible using Si-strips with a pitch of 2mm. The precision improves as 1/√n where n is the effective number of m.i.p.'s passing through Si. As n is roughly proportional to the incident energy the resolution improves as 1/√E. Similar precision, for a similar granularity, should be possible in the approach (ii) once the signal is clearly above the noise level.

In the first approach the size of the cells, and less importantly the noise contributions from either the electronics or the energy pileup, limits the precision of the spatial measurement. The spatial precision attained[29] versus cell size, 2Δ, using square lead glass blocks (X_0=2.4 cm) is shown in Fig. 27. A 25 GeV electron beam was used in this study. Recent results[24] from the "Accordion" LAr calorimeter (Fig. 28) show a position resolution of 4.4mm /√E (E in GeV) for single showers. The cells had a size of ≈ 2.7 cm (≈ 2.1 X_0).

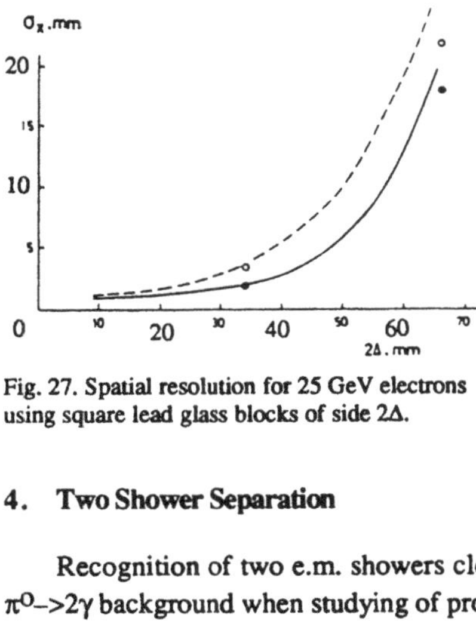

Fig. 27. Spatial resolution for 25 GeV electrons using square lead glass blocks of side 2Δ.

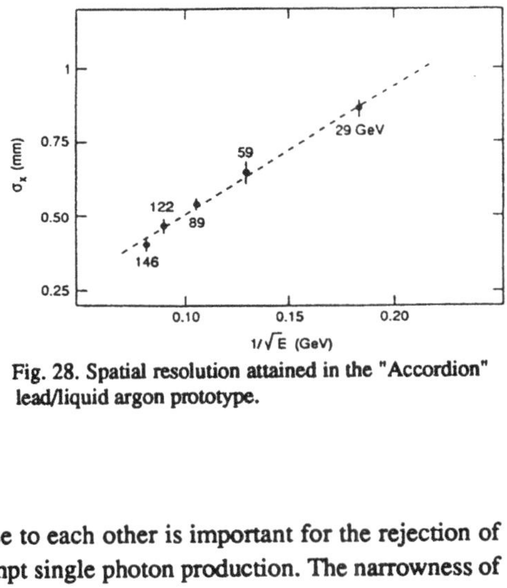

Fig. 28. Spatial resolution attained in the "Accordion" lead/liquid argon prototype.

4. Two Shower Separation

Recognition of two e.m. showers close to each other is important for the rejection of $\pi^0 \to 2\gamma$ background when studying of prompt single photon production. The narrowness of

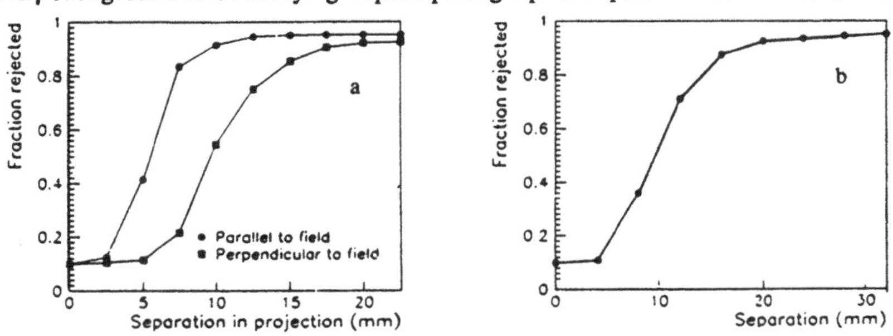

Fig. 29 Rejection of photon pairs versus their separation in a) one projection b) both projections.

the e.m. shower in the early part can be used to isolate events consisting of genuine single photons. The search for the intermediate mass Higgs via its two photon decay will require both a very high efficiency for the reconstruction of single photons and a high rejection power for two photon impacts (from jet -> π^o -> 2γ). A simulation[30] using two orthogonal planes of Si strips (2mm pitch) leads to typical rejection power of ~ 6–10 for two shower separation of ~ 1.5 cm (Fig.29). An efficiency of 90% for a single shower is retained.

Two shower rejection is also possible using the mosaic structure of towers and pre-shower detectors. However a high two shower rejection power and a high single shower efficiency may be compromised by the presence pileup of low energy photons.

5. Electron – Hadron Separation

A high energy pion faking an electron leads to the contamination of signals using prompt electrons. At LHC in order to bring the rate of fake electrons from this source to a factor ~ 10 below that from the genuine sources (b –> e X, W –> eν etc.) an e-π separation of $\geq$ 1000 is required for p_T's $\geq$ 10 GeV/c.

The electron–hadron separation is usually based on the difference in the longitudinal and lateral development of showers intiated by electrons and hadrons. One or more of the following can be used to achieve the desired pion rejection power when detecting electrons :

i) a preshower detector at $\approx$ 1.5 X_o
ii) a position detector at $\approx$ 3-5 X_o
iii) lateral segmentation
iv) longitudinal segmentation including a hadron calorimeter
v) energy - momentum matching

The ultimate rejection power is limited by the charge exchange process (or first interaction), which results in one or several π^o's taking most of the energy of the incoming hadron. The shower from such hadrons then looks like an e.m. shower. Therefore sampling of showers early in their longitudinal development is imprtant.

The separation power for single particle[31], using (ii - iv) is shown in Fig.30. The structure of the calorimeter consisited of :

- towers of a lateral size of ~11 x 11cm (effective $X_o \approx$ 8mm),
- 8-fold longitudinal segmentation, the first four samplings (2mm U / 2.5 mm TMP) with thicknesses of 3, 6, 10, 7 X_o leading to a total of 1λ, the next two (5mm U/ 2.5mm TMP) with thicknesses of 0.7 λ and the last two (5cm Fe/ 1cm scint) with thicknesses of 2.5 λ.
- a position detector placed at a depth of 3 Xo.

The rejection power, as a function of energy, using (ii), (iii) and (iv) individually and then all combined is shown.

6. Timing Resolution

Machines running at high luminosities, and with short inter-crossing times, will require very fast processing of signals. Consider the example of LHC (or SSC) where the crossing

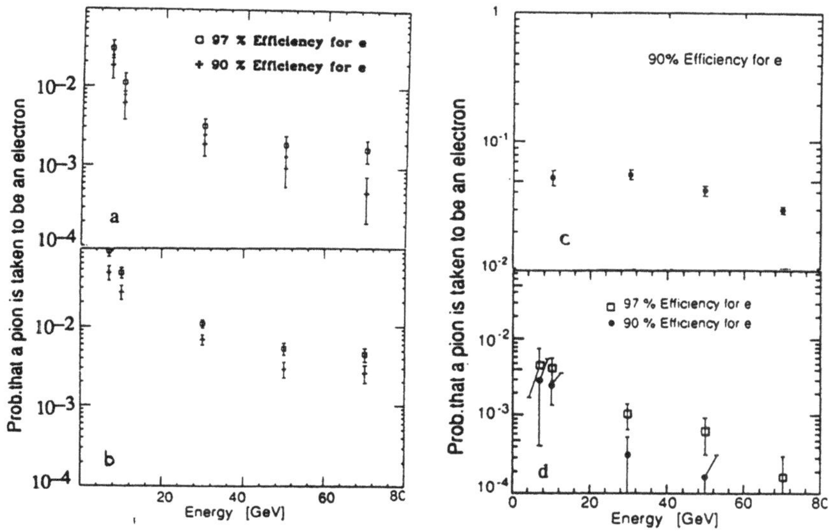

Fig. 30 The probability that a single pion is taken to be an electron using a) longitudinal profile only, b) lateral profile only, c) position detector only and d) all combined.

times are ~15ns. The consequences for a calorimeter that integrates over more than one crossing are due to the amount of extra pile-up energy that is introduced. Amongst these are

- a degradation of energy (and hence mass) resolution due to increased energy pileup from soft photons or hadrons from the many minimum bias events that occur within the resolution time of the calorimeter (section 2.7).
- a decreased effectiveness of isolation criteria due to extra energy entering the area used for isolation. Isolation is usually assured by making a cut on the sum of energy that falls in a cone around the track of interest in $\eta-\phi$ space. Integration over more than one crossing, even at a luminosity of $L = 10^{33}$ cm^{-2}s^{-1} may mean that cuts on energy sum, to ensure isolation, become inefficient. It may turn out that track isolation at high luminosities has to be carried out by requiring no extra tracks (photons or charged particles) with high p_T or E_T (> 2–5 GeV) inside the isolation cone.
- poorer shower track matching for electrons resulting in a decreased electron efficiency[2e] if only the transverse energy deposition, eg. in cells of a size $\Delta\eta x \Delta\phi \approx 0.03 \times 0.03$, is used to localize high energy electrons.

7. Jet Energy Resolution

The design of the electromagnetic and hadronic calorimeters has to be different if the physics emphasizes the need for a very good e.m. energy resolution. This usually excludes the possibility of obtaining an overall compensating hadronic calorimeter, assuming that the use of a weighting technique is not possible. The important issue is not only the jet or single particle energy resolution but also the shape of the measured energy distributions. Tails on these can fake large missing transverse energies which may compromise physics that requires cuts on such quantities. The performance of a calorimeter with different e.m. and hadronic sub-calorimeters has only been studied in simulations which usually cannot simulate accurately the systematic effects that lead to tails. Confirmation by beam tests of such systems should clearly be contemplated by groups designing such calorimeter systems.

8. Conclusion

The effect on the energy resolution of many of the sources described above can be seen[32] in Fig. 31. The results from beam tests of a liquid krypton filled calorimeter are shown. The calorimeter had a cylindrical shape with a depth of 76 cm (16.5Xo) and a diameter of 42 cm (9.5Xo) calorimeter. The ionization charge was collected on 500μm G10 boards separated by drift gap of 2 cm. Also shown are the calculated contributions to the energy resolution of different effects.

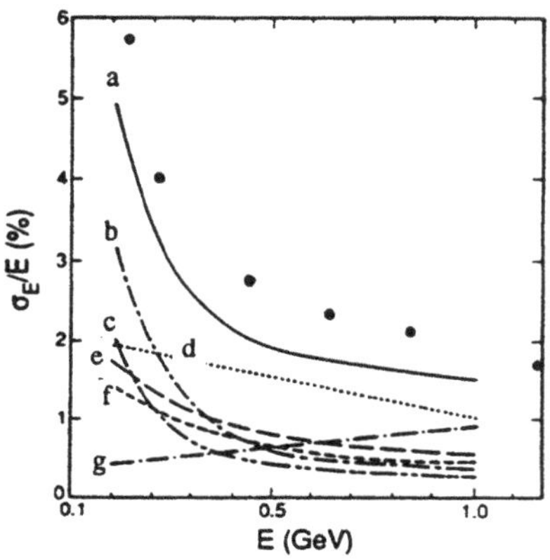

Fig. 31 Data from beam tests and calculated resolution of the KEDR liquid krypton e.m. calorimeter: a) total, b) elctronic noise, c) radioactivity, d) lateral leakage, e) charge induction efficiency, f) sampling fluctuations and g) longitudinal leakage

Acknowledgements

I would like to thank C. Fabjan and C. Seez for reading the manuscript and making several useful suggestions.

References

1. Particle Properties Data, M. Aguilar-Benitez et al., *Phys. Lett. B239.*

2a. U. Amaldi, *Phys. Scripta* **23** (1981) 409.

2b. C. W. Fabjan, *CERN-EP/85-54* (1985).
[2a] and [2b] also appear in *Experimental Techniques in High Energy Physics* ed. T.Ferbel (Addison-Wesley 1987) p 257 and p 325 respectively.

2c. C.W. Fabjan and R. Wigmans, CERN-EP/89-64, Reports on Progress in Physics

2d. R. Wigmans, *Proc. of ICFA School on Instrumentation in Elementary Particle Physics*, Trieste (1987) eds. C.W. Fabjan and J. E. Pilcher (World Scientific, Singapore 1988).

2e. J. Colas et al., p370, CERN 90-10 , ECFA LHC Workshop, Aachen (1990).

3. U. Fano, *Phys. Rev*. **72** (1947) 26.
4. G. F. Knoll, *Radiation Detection and Measurement*, John Wiley and Sons.
5a. E. Aprile et al., *Nucl. Instr. and Meths.* **A261** (1987) 519.
b. K. Giboni, *Nucl. Instr. and Meths.* **A269** (1988) 554.
6. A.E. Bolotnikov et al., *Prib. Tekh. Eksp.* **4** (1986) 42.
7a. H. Crawford et al., *Nucl. Instr. and Meths.* **A256** (1987) 47.
b. J. Seguinot et al., CERN-LAA-PC/90-24 (1990)
8. R. M. Brown et al., *IEEE Trans. Nucl. Sci.* **NS-32** (1985) 136.
9. E. Bloom and C. Peck, *Ann. Rev. Nucl. and Part. Sci.* **33**, (1983) 143.
10. E. B. Hughes et al., *IEEE Trans. Nucl. Sci.* **NS-19** (1972) 126.
11. E. Lorenz, p 621 CERN 89-10, ECFA Study Week, Barcelona (1989).
12. J. A. Bakken et al., *Nucl. Instr. and Meth.*, **A254** (1987) 535.
B. Adeva et al., *Nucl. Instr. and Meth.*, **A289** (1990) 35.
13. H. G. Fischer, *Nucl. Instr. and Meth.*, **156** (1978) 81.
14. C. Cerri and F. Sergiampietri, *Nucl. Instr. and Meth.*, **141** (1975) 207.
15. Y. Hoshi et al., *Japan. J. Appl. Phys.* **21** (1982) 1086.
16. E. Clayton et al., IC/HEP/90-9, Blackett Lab., Imperial College, London SW7 2BZ.
17. M. Kobayashi et al., *Nucl. Instr. and Meth.*, **A308** (1980) 47.
18. T. Doke et al., *Nucl. Instr. and Meth.*, **A237** (1985) 475.
19. D. Acosta et al., CERN-PPE/91-85 (1991).
20. J.-M. Gaillard, RD1 Status Report, CERN/DRDC 91-50 (1991).
21. J. Kruger, ZEUS Collaboration, CERN 90-10, ECFA LHC Workshop, Aachen (1990)
22. H. Burkhardt et al., *Nucl. Instr. and Meth.*, **A268** (1988) 116.
23. F. Abe et al., *Phys. Rev.*, **D43** (1991) 2070.
24. B. Aubert et al., CERN PPE/91-73 (1991), Submitted to Nucl. Instr. and Meths.
25. D. Fournier, RD3 Status Report to the CERN-DRDC Committee (1991)
G. Costa, these proceedings.
26. C. Seez et al., CERN 90-10 , ECFA LHC Workshop, Aachen (1990).
27. G. Stevenson, CERN 90-10 , ECFA LHC Workshop, Aachen (1990).
28. D. Acosta et al., CERN-PPE/91-45 (1991), Submitted to Nucl. Instr. and Meths.
29. G. A. Akopdjanov et al., *Nucl. Instr. and Meths.*, **140** (1977) 441.
30. C. Seez, these proceedings.
31. R. Apsimon et al., *Nucl. Instr. and Meth.*, **A305** (1991) 331.
32. V. M. Aulchenko et al., *Nucl. Instr. and Meth.*, **A289** (1990) 468.

Performance and Limitations of Hadron Calorimeters

Richard Wigmans
CERN, Geneva, Switzerland

ABSTRACT

The average response of a calorimeter to hadrons and jets depends in general strongly on the type of event and in particular on its electromagnetic content. This problem is avoided in compensating calorimeters. Strongly noncompensating calorimeters provide very poor hadronic performance, in particular for jet detection and energy flow measurements. Compensation and ultimate electromagnetic energy resolution cannot be achieved with the same calorimeter system.
The factors determining the energy resolution of sampling calorimeters are discussed. A model yielding some simple formulae describing the effects of sampling fraction and sampling frequency on the e.m. energy resolution works remarkably well. The hadronic energy resolution of compensating calorimeters is much less sensitive to the sampling frequency than the e.m. one.

1. Introduction

Traditionally, the term hadron calorimeter is reserved for a detector backing up an electromagnetic (e.m.) calorimeter in an experiment and intended for measuring the energy of those particles that are not completely absorbed by this first major obstacle they meet on their way. If interpreted in this way, the title of this talk may be misleading. In this contribution, I will discuss more generally the calorimetric detection of hadrons and fragmenting quarks (jets). And since these particles may deposit a major fraction of their energy in what is usually called the e.m. calorimeter, I will have to concentrate on the properties of these devices as well, thus risking some overlap with issues discussed by the previous speaker.

In the scientific literature, there is unfortunately much less experimental data available on the calorimetric detection of hadrons than for e.m. showers. Also, the spread in the quality of the data that does exist is much larger. It is very difficult to appreciate and compare the published results. The effects of containment cuts and weighting schemes that are used in practice to minimize the widths of the signal distributions are frequently large and many questions may arise about the legitimacy of such techniques. The absence of good-quality experimental data is even more striking for jets. There exist practically no published results that give a realistic indication about the accuracy with which jet energies may be measured by the various detectors employed in particle physics experiments.

Nevertheless, it is sometimes possible to extract such relevant information in an indirect way from the data that is available. I will exploit these possibilities. The picture that emerges from this is not very positive. There is a strong tendency to overestimate the quality of calorimetric hadron detection.

Calorimeters are instruments intended to measure the energy of particles. Their quality is therefore determined by the accuracy with which this is done. A good accuracy does not only imply a good energy resolution, but it also requires that the energy be, on average, measured correctly. Although the latter requirement may seem trivial, it is the basic cause of the problems with hadronic energy detection, as will be illustrated in the following sections.

Because of the importance of accurate energy measurements, I will mainly concentrate on this aspect of calorimetry. This means that I will not have the time to elaborate on other features that may play an important role in the design of calorimeter systems, such as position and angular resolution, signal speed, particle identification, calibration and monitoring, etc. In section 2, I will review the properties of existing or planned calorimeters and discuss the problems mentioned before. In section 3, I will indicate how calorimeter systems, which have in practice several tasks other than detecting hadrons, can be optimized. This is very relevant in view of the crucial role that these systems will play in future experiments at the multi-TeV hadron colliders LHC and SSC. In particular, I will discuss the factors that determine and limit the accuracy with which the energy can be measured. Conclusions are given in sect. 4.

2. Energy resolution and response of hadron calorimeters

2.1 An indication of the problem

All calorimeter systems intended for the detection of e.m. and hadronic showers and used in current or planned experiments are longitudinally segmented in an e.m. and hadronic section. Sometimes, e.g. in the H1 experiment[1], a much finer longitudinal segmentation is applied. Typically, the e.m. calorimeter section in 20-30 radiation lengths (X_0) deep, which, depending on the (average) Z of the materials of which the calorimeter consists, is equivalent to about one nuclear interaction length (λ_I). Hadrons entering such a calorimeter deposit on average a considerable fraction of their energy in the e.m. section. Typical results, obtained by the ALEPH[2] and ZEUS[3] Collaborations are shown in fig. 1. At 10 GeV, the average fraction amounts to $\sim$ 30% for ZEUS and $\sim$ 60% for ALEPH. The difference between these two sets of results is very large. Moreover, it has the wrong sign. The ALEPH e.m. calorimeter is thinner than the ZEUS one ($0.7\lambda_I$ vs. $0.96\lambda_I$).

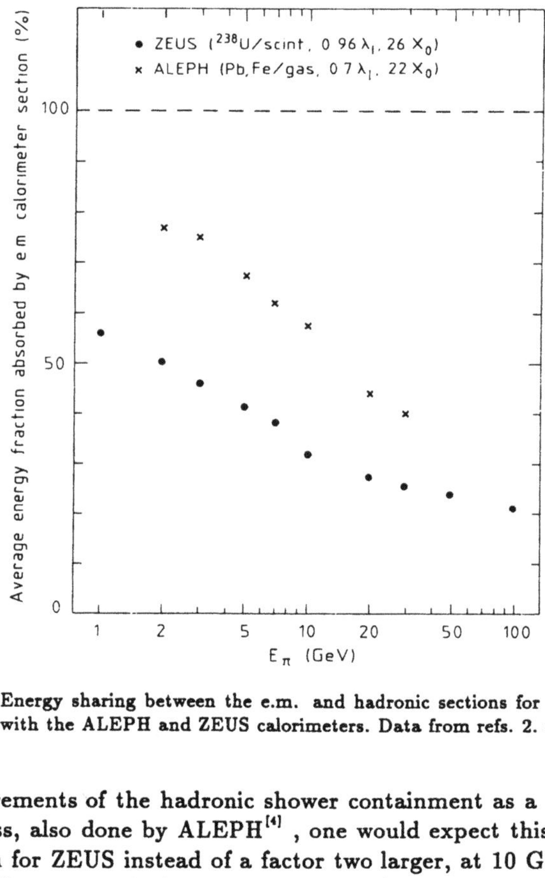

Figure 1 **Energy sharing between the e.m. and hadronic sections for pions detected with the ALEPH and ZEUS calorimeters. Data from refs. 2. and 3.**

Based on measurements of the hadronic shower containment as a function of the absorber thickness, also done by ALEPH[4] , one would expect this fraction to be 40% *smaller* than for ZEUS instead of a factor two larger, at 10 GeV (see fig. 2). Given this large discrepancy (a factor 2.5 at 10 GeV), we can draw the conclusion that one of the two experiments (or maybe both) measures the energy deposited by pions in the e.m. calorimeter section wrong by a considerable factor.

There are reasons to assume that this problem comes from ALEPH. Unlike the ZEUS detector, the ALEPH calorimeter is strongly noncompensating, which means that the response to e.m. and hadronic showers is very different. The e/π signal ratio was measured to range from 1.7 at 2 GeV to 1.3 at 30 GeV (fig. 3). Usually, showering pions deposit some fraction of their energy in the form of e.m. showers, generated by π^0's. Such π^0's are mainly produced in the early stage of the shower development and, therefore, the energy deposited in the e.m. calorimeter section contains on average a much larger π^0 component, compared to the hadronic section. If the response (the average calorimeter signal per unit energy) is much

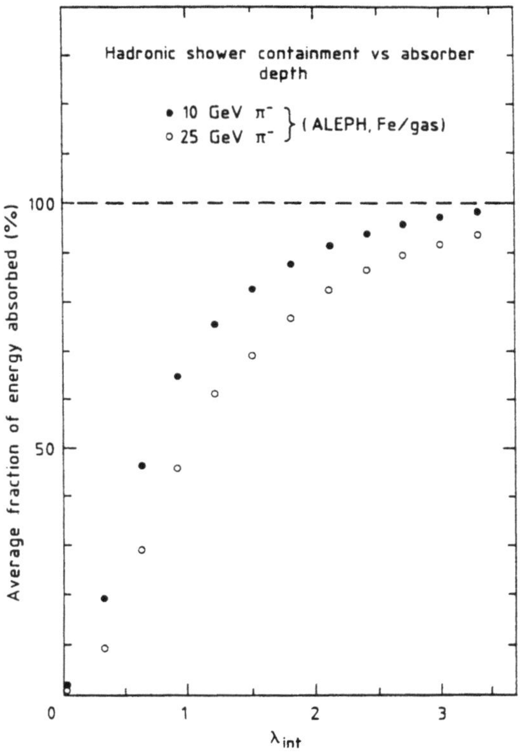

Figure 2 The energy fraction deposited by 10 and 25 GeV pions in an iron absorber as a function of the thickness of this absorber. Data from ref. 4.

larger for the e.m. shower component than for the non-e.m. one, as in the ALEPH detector, the energy deposited in the e.m. calorimeter section will be overestimated accordingly. Morever, ALEPH applies weighting factors to the signals from the different calorimeter sections. In order to minimize the width of the total hadronic signal distributions, the signals from the e.m. calorimeter setion are multiplied by a factor 1.5 with respect to the signals from the hadronic section. This operation wil further boost the calculated fraction of the energy deposited in the e.m. calorimeter section. These combined factors may well explain the large discrepancy with the ZEUS results.

The weighting factors mentioned before are a direct consequence of the fact that the calorimeter is not compensating and can be understood as follows. The fraction of π^0 production in the hadronic shower development gradually increases with the pion energy. This is illustrated by the fact that the e/π signal ratio decreases (fig.

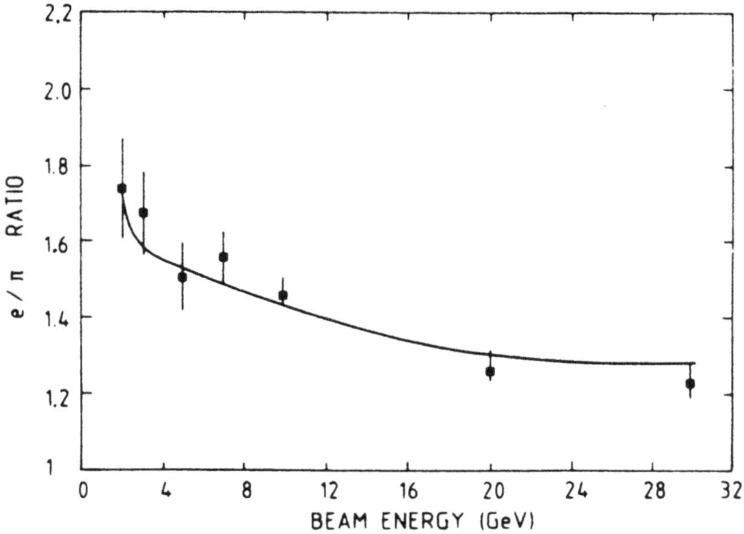

Figure 3 The e/π signal ratio as a function of energy for the ALEPH calorimeter. Data from ref. 2.

3). Since the calorimeter signal is proportional to the energy of e.m. showers, this phenomenon will result in a signal nonlinearity for pion detection. The response for high-energy pions will be larger than for low-energy ones. Signal linearity can be restored by giving the signals from the e.m. section a larger weight than the signals from the hadronic section, since high-energy pions deposit on average a larger fraction of their energy in the latter. The effect of the weighting is, however, that the energy deposited in the e.m. section is even more overestimated than one might expect from the different response to the e.m. and non-e.m. shower components.

The example discussed in this subsection illustrates that the signals from non-compensating calorimeters may lead to severe misinterpretations and that the energy of individual particles may be easily off by a factor of 2 or more in such detectors. I will come back to this problem when discussing jets.

2.2 The energy resolution

It has become customary to quote the energy resolution of calorimeters in terms of a percentage divided by the square root of the particle energy ($\sigma/E = x\%/\sqrt{E}$, with E expressed in GeV). While this relation gives a rather accurate description of the performance of many e.m. calorimeters (except at high energies where instrumental effects may cause deviations), it does in general not hold at all for hadronic shower detection. This may be illustrated with some representative examples (fig. 4). The energy resolution for single-pion detection is given as a function of the particle energy, which is plotted on a scale linear in $1/\sqrt{E}$. In this way, $1/\sqrt{E}$ scaling implies that the experimental points should follow a straight line through the bottom right corner of the graph. Only the resolutions for the (almost) compensating ZEUS and SPACAL calorimeters (fig. 4d) fulfill this condition approximately. The resolution scales at $30 - 35\%/\sqrt{E}$, with a small (1%) constant term for the slightly undercompensating SPACAL device.

The data in figs. 4a-c exhibit clear deviations from such scaling behavior. The resolution for the undercompensating ($e/h \approx 1.5$) iron/plastic-scintillator calorimeter operated by the CDHS Collaboration (fig. 4a) amounts to $\sim 55\%/\sqrt{E}$ at low energies, but for energies beyond $\sim$ 50 GeV the resolution barely improves any further and levels off at $\sim$ 8%. Similar results were obtained with the overcompensating ($e/h \approx 0.7$) uranium/plastic-scintillator calorimeter of the WA78 experiment (fig. 4b).

The constant term dominating the high-energy behavior is clearly smaller for the uranium/liquid-argon calorimeter being commissioned by the D0 Collaboration (fig. 4c), which is to be expected since the detector is closer to compensation ($e/h \approx 1.2$). However, this calorimeter exhibits another feature, typical for calorimeters using liquid argon (LAr) readout, namely the rapid deterioration of the energy resolution at low energies. Because of the very small signals generated by the LAr medium, which operates in the ionization-chamber mode, electronic noise is a major problem and tends to dominate the energy resolution at low energies through the contribution of a term inversely proportional to the energy. This noise problem is of course much worse for hadronic showers than for e.m. ones, since the latter require a much smaller detector volume to develop. In first approximation, the signal/noise ratio is inversely proportional to the sqare root of the volume over which the signals have to be integrated in order to measure the energy.

As a side remark, it should be noted that the signal/noise ratio deteriorates when fast signal shaping is applied, a technique advocated for the application of LAr detectors in the LHC/SSC environment. The collection of the ionization charge in these detectors is an intrinsically slow process, with a time constant of $\sim 0.5\mu$s. Bipolar pulse shaping techniques allow one to extract fast signals from these devices, but in that case only a fraction of the produced charge is collected, which results in smaller signals and a correspondingly worse signal/noise ratio. This may be illustrated by the results from the RD3 Collaboration, who found that at 20 ns shaping time, even the energy resolution for electrons (the most favorable case) was already dominated by noise up to energies of $\sim$ 40 GeV[10] .

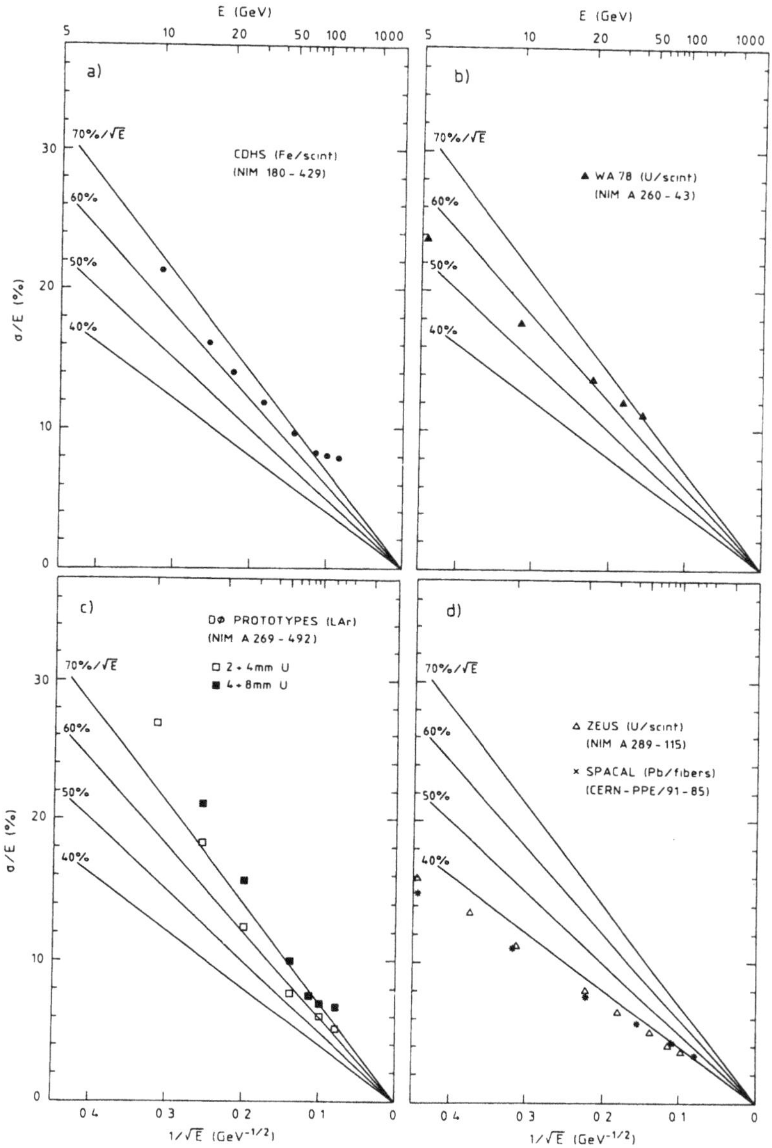

Figure 4 The energy resolution for the detection of single pions, for some representative calorimeters. The energy resolution is given as a function of the pion energy, plotted on a scale which is linear in $1/\sqrt{E}$. The straight lines represent $1/\sqrt{E}$ scaling. Results from WA1[3] (*a*), WA78[6] (*b*), D0[7] (*c*), ZEUS[3] and SPACAL[8] (*d*).

Such problems do not exist when the much faster scintillation process is used to produce the signals. The SPACAL Collaboration studied the performance of their detector as a function of the charge collection time[11] and found that the energy resolution, both for e.m. and hadronic showers, was practically unaffected down to $\sim$ 40 ns, which is equivalent to the 20 ns shaping time mentioned before for LAr calorimeters.

2.3 Jet detection

The real challenge in modern experiments is not so much the measurement of the properties of single hadrons, but rather of jets, the signatures of fragmenting quarks and gluons. Especially at high energies, these jets are characterized by a rather collimated bunch of particles, hadrons, photons (from π^0 decay) and sometimes also leptons (from fragmenting c or b quarks). A typical LHC or SSC jet may consist of several tens of particles. The leading particle (containing the fragmenting quark) usually carries a considerable fraction of the jet energy.

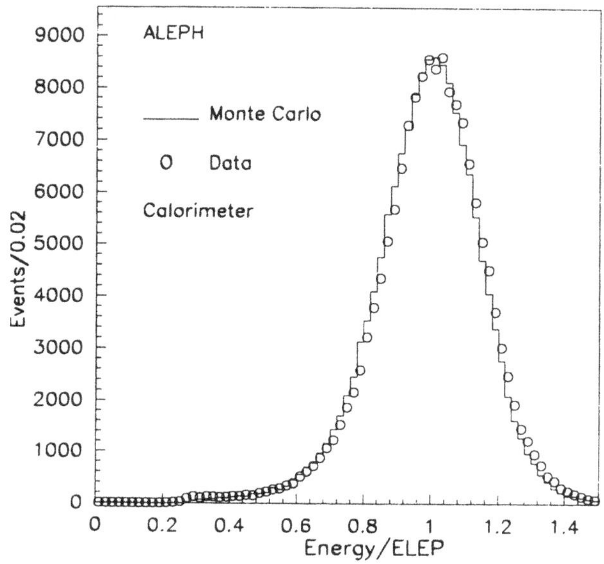

Figure 5 The total energy distributionof hadronically decaying Z^0 particles, measured with the calorimeters of the ALEPH experiment at LEP[9]. The distribution is asymmetric. A Gaussian fit in the region $E/E_{LEP} = 0.85 - 1.5$ gave a σ of 14.3%. This result improved to 9.3% if the calorimeter information on the charged decay products was replaced by the measured momenta and Z^0's were selected producing jets in the central detector region.

Because of their e.m. content, these jets will deposit on average an even larger fraction of their energy in the e.m. calorimeter section than single pions. Because of the specific characteristics of jets, the problems that arise when the calorimeter (or its e.m. section) is not compensating, may be very large.

Let us consider a high-energy jet initiated by a fragmenting u quark. Let us assume that the leading particle (carrying the original u quark) carries half of the jet energy. This particle has then equal probabilities of being a π^+ or a π^0. If the leading particle is a π^0, all its energy is deposited in the e.m. calorimeter section, if it is charged most of its energy will go to the hadronic section. The ALEPH detector would measure a very different value for the energy of these two jets. Because of the noncompensation, the signal from the leading π^+ would be smaller, e.g. by 20%. In addition, the signal from the leading π^0 would be multiplied by a factor 1.5. Assuming that the other particles from the jet would produce the same signal in both cases, the energy of the jet with the leading π^0 would then be measured to be $\sim$ 35% larger than the energy of the jet with the leading charged pion.

Evidence for the problems that this phenomenon causes in practice for jet detection may be derived from fig. 5. This figure shows the total energy distribution for hadronically decaying Z^0 particles measured with the ALEPH calorimeter system[9] and may therefore be considered as a jet measurement at 90 GeV. Of course, some of the jets initiated by b and c quarks may contain neutrinos and muons which are not (completely) absorbed by the calorimeters. Also, some energy may leak out due to the fact that the detector does not cover the complete 4π solid angle. Such phenomena cause a low-energy tail in the signal distribution, which is actually observed. When ignoring this tail and fitting the distribution in the area $E/E_{LEP} = 0.85 - 1.5$ ($[-1\sigma, +3\sigma]$) to a Gaussian function, a σ of 14.3% is obtained. This has to be compared to the 8.8% expected on the basis of the energy resolution quoted for single hadrons ($\sigma/E = 84\%/\sqrt{E}$[2]). Thus the fact that not one, but a multitude of hadrons (an unknown and fluctuating mixture of π^0's and other particles) are measured simultaneously by the detector introduces an extra, presumably energy independent, term of $\sim$ 11% ($\sqrt{14.3^2 - 8.8^2}$) to the energy resolution.

Similarly, problems arise in the measurement of the missing transverse energy. If a Z^0 decays in a $q\bar{q}$ pair and the two quarks fragment in the way described before, i.e. one jet has a leading π^0, the other a leading charged pion, the difference in the signals from these, otherwise identical, jets would mimic a missing transverse energy corresponding to 15% of the total transverse energy. Indeed, I have not yet seen claims from ALEPH, nor from any other LEP experiment for that matter, that missing transverse energy can be measured more accurately than this on the basis of calorimeter infomation alone.

Since the argument that was given as an explanation of these phenomena is energy independent, there is no reason to believe that the accuracy for jet and energy flow measurements will be much better at much higher energies, if strongly noncompensating calorimeters were to be used. The example illustrates that the signals from such calorimeters strongly depend on the topology of the event (i.c. its π^0 content). This is typical for hadronic shower detection, where the sampling fraction, and thus the calorimeter response, may be very different for the different shower components.

The mentioned problems may be avoided with a compensating calorimeter system, which has equal response to charged and neutral pions. No such calorimeter is yet operating at a collider experiment, but some indication about the correctness of the previous statement comes from data obtained by the SPACAL collaboration, who tried to simulate jet detection by measuring the reaction products from pions interacting in an upstream target[8] . As illustrated in fig. 6, the energy resolution for multiparticle detection at 150 GeV was found to be excellent, about 3%, for a "jet" defining cone with an opening angle of about 45 degrees (0.8 r).

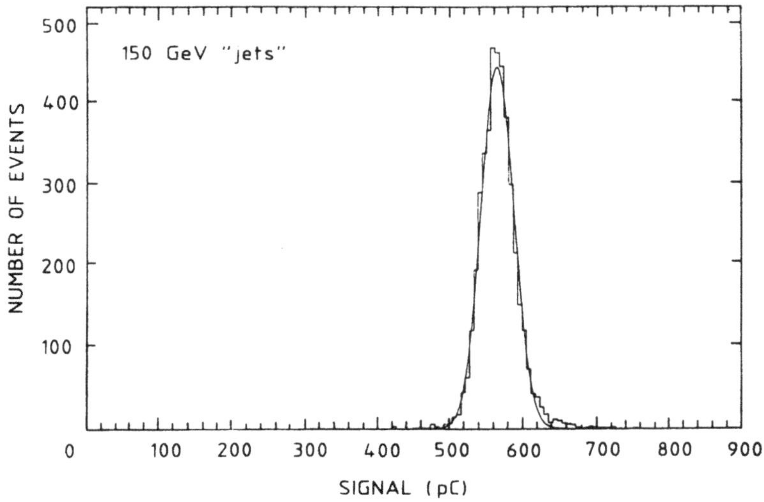

Figure 6 **Signal distribution for the secondary particles produced by 150 GeV π^- in an upstream target, measured with the SPACAL detector. Data from ref. 8.**

It should be stressed that the example given before was *not* chosen to criticize ALEPH. After all, their detector offers powerful alternatives to calorimetry for studying the Z^0 physics, for example an excellent tracking system. It has been demonstrated that a distribution as in fig. 5 can be considerably improved when the calorimetric information on hadrons is replaced by the measured track momenta. The same holds for missing E_T distributions. However, this possibility exists because of the very clean nature of the $e^+e^- \rightarrow Z^0$ events and it is extremely unlikely that it will be exploitable in LHC/SSC experiments. Such experiments will therefore have to rely completely on the quality of the calorimetry for the measurement of jets and energy flow and the example provided by the ALEPH data may show that serious problems are to be expected if the calorimeter system is strongly noncompensating.

3. Optimization of calorimeter systems

3.1 How to achieve compensation?

The secrets of the compensation mechanisms are now well understood[12] . In general, calorimeters produce different signals for e.m. and hadronic particles of the same energy, due to the fact that a large fraction of the non-e.m. energy ($\sim$ 40% on average) is used to break up atomic nuclei and therefore does not contribute to the measured signal (invisible energy). The degree of (non)compensation is given by the energy independent e/h value, which is the ratio of the calorimeter responses to the e.m. and non-e.m. components of hadronic showers. The e/h ratio cannot be measured directly, but it can be inferred from the energy dependent e/π signal ratios. By taking into account the energy dependence of the average fraction of e.m. energy generated by π^0's in hadronic shower development, in the way proposed by Groom[13] or Wigmans[14] , the relation between e/h and $e/\pi(E)$ becomes

$$e/\pi(E) = \frac{1}{1 + (h/e - 1)\, E^{-0.15}} \qquad (1a)$$

or

$$e/\pi(E) = \frac{e/h}{1 + (e/h - 1)\, 0.11 \ln E} \qquad (1b)$$

with E in GeV. In the energy range from 5 - 200 GeV, where most beam tests are made, both formulae give similar results.

Every material or combination of materials is characterized by its own specific e/h value. Because of the invisible energy phenomenon, calorimeters tend to be undercompensating ($e/h > 1$). Compensation can either be achieved by reducing the response to the e.m. shower components, or by boosting the response to the non-e.m. shower components (in particular the neutrons), or by a combination of both. A very specific way of boosting the non-e.m. calorimeter response consists of using depleted uranium as absorber material, which yields extra energy from nuclear fission, but this is neither a necessary nor a sufficient condition for achieving compensation.

The e/h value of non-uranium calorimeters with high-Z absorber material is satisfactorily described by the following simplified formula:

$$e/h = \frac{e/mip}{0.41 + 0.12n/mip} \tag{2}$$

in which e/mip and n/mip represent the calorimeter response to e.m. showers and to MeV-type neutrons, respectively. These responses are normalized to the one for minimum ionizing particles, which is directly given by the dE/dx values of the active and passive calorimeter components and by their relative abundance. Table 1 lists the calculated and (when available) the measured e/h values for some calorimeters that are being considered for future experiments. The measured values agree very well with the calculated ones.

The formula and the table show that both e/mip and n/mip are very important for achieving compensation in lead calorimeters. It is also clear that compensation becomes increasingly hard to achieve when the sampling frequency is increased, since the e.m. response tends to increase as well. This has the following reason.[18] A large fraction of the e.m. energy is deposited through very soft electrons ($E <$ 1 MeV) produced by Compton scattering or the photoelectric effect. The cross sections for these processes depend very strongly on Z and, therefore, practically all these photon conversions occur in the absorber material. The range of the electrons produced in these processes is very short, ~ 0.7 (0.03) mm for a 1 (0.1) MeV electron in lead. Therefore, such electrons only contribute to the calorimeter signal if they are produced near the boundary between the the lead and the active

Table 1 The calculated values of the e/h value, using the formula from eq. (2), and the e/mip (EGS4) and n/mip[14] values for some calorimeters considered for future experiments. Where possible, the results of experimental measurements are given as well.

Table 1 The (expected) e/h value of some calorimeters

Configuration	Reference	$(e/mip)_{calc}$	$(e/mip)_{meas}$	$(n/mip)_{calc}$	$(e/h)_{calc}$	$(e/h)_{meas}$
Pb/scint. tiles 10/2.5 mm	ZEUS[15]	0.62	0.63 ± 0.05	1.7	1.01	1.05 ± 0.04
Pb/scint. fibers 4/1 mm	SPACAL[8]	0.69	0.71 ± 0.05*)	1.7	1.12	1.15 ± 0.02
2 mm Pb/Lar 20 ns	RD3[10]	0.78	-	< 0.5	> 1.6	-
1 mm Pb/LAr 20 ns	GEM[16]	0.86	-	< 0.5	> 1.8	-
3+12 mm Pb/LAr 2 μs	SLD[17]	0.69	-	0.7	1.40	1.37 ± 0.12**)
BaF_2 crystals	GEM[16]	1.0	-	< 0.5	> 2.1	-

*) Preliminary result
**) Based on a measurement of $e/\pi = 1.24 \pm 0.10$ at 11 GeV, using eqs. (1)

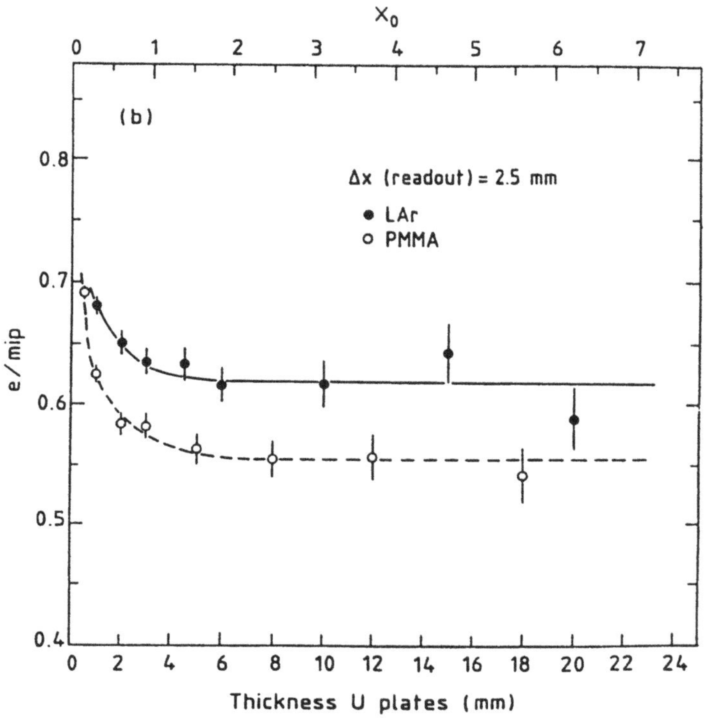

Figure 7 The *e/mip* ratio as a function of the thickness of the passive layers, for uranium/LAr and uranium/plastic-scintillator calorimeters. The thickness of the active layers is 2.5 mm. Results from EGS4 Monte Carlo calculations[18].

material. However, if the absorber material is made thinner and thinner, this effective boundary layer becomes a larger and larger fraction of the total absorber mass and the calorimeter response goes up. This effect was already predicted five years ago by EGS4 Monte Carlo calculations[18] (see fig. 7). It explains the fact that the e/h values of the ZEUS and SPACAL calorimeters, which contain the same relative amounts of lead and plastic, differ by about 10% (table 1).

The contribution of these soft electrons to the calorimeter signals, and thus the e/mip and e/h values, can be reduced by placing low-Z passive material between the absorber and the active medium. This was experimentally demonstrated by the SICAPO Collaboration[19], confirming Monte Carlo predictions of this effect[14,20].

The Monte Carlo calculations also predict that the e.m. response for LAr calorimeters is consistently larger than for calorimeters with plastic-scintillator readout. This effect, which is presumably due to the larger Z value of argon, makes it more difficult to approach $e/h = 1$ in LAr calorimeters. Another, more fundamental difficulty in that respect is the impossibility of boosting the calorimeter response to the low-energy neutrons.

In calorimeters with hydrogenous readout media, these neutrons tend to lose most of their kinetic energy through elastic $n-p$ collisions and, therefore, they are sampled more efficiently than the charged shower particles by a large factor. Although saturation phenomena in the active medium (e.g. Birk's law in scintillators) strongly reduce the signals from the densily ionizing recoil protons, the neutrons, which carry on average $\sim$ 12% of the non-e.m. energy, may account for 30-40% of the signals delivered by such calorimeters. This phenomenon may be used to tune the e/h ratio: When the sampling fraction for charged particles is reduced, the relative contribution of neutrons to the signal increases and e/h decreases. The small undercompensation observed in the SPACAL detector might be eliminated in this way. By reducing the sampling fraction (the number of fibers per unit volume), n/mip would increase and e/mip would decrease (because of the increased fiber-to-fiber distance). Both effects reduce e/h. The latter effect would also be obtained if the sampling frequency were reduced (thicker fibers, for a constant lead/fibre volume ratio).

The elastic scattering products do not contribute to the signals of LAr calorimeters. These detectors only observe the γ's produced by inelastic neutron scattering and from thermal neutron capture. But since the thermalization is a process with a time scale of $\sim 1\mu$s, the latter component may only be observed if the charge is integrated over a sufficiently long period of time (see also table 1).

3.2 Electromagnetic energy resolution and compensation

Calorimeter systems for future experiments at ultrahigh energies will have many tasks, among which accurate detection of the electrons, photons and jets produced in the collisions and measurement of the energy flow parameters (e.g. missing E_T). It is generally believed that good energy resolution is more important for e and γ detection than for jets and missing E_T. One reason for this is the process $H^0 \to \gamma\gamma$, which is believed to be one of the most promising channels for detecting the Higgs particle if $m_W < m_H < 2m_W$, and for which e.m. calorimeter resolution is important. There is a tendency to give absolute priority to the e.m. energy resolution in the design of the calorimeter systems for these experiments, ignoring the consequences for the hadronic performance. Without entering into a discussion about the justification of this approach, I want to point out that this will lead to extremely noncompensating calorimeter systems. In section 2 I have tried to indicate what the practical consequences for jet and missing E_T measurements are likely to be in that case.

Compensation and ultimate e.m. energy resolution are mutually exclusive. Compensation requires a small sampling fraction, e.g. $\sim$ 3.5% (for mips) in the ZEUS and SPACAL protoypes (table 1). Ultimate e.m. energy resolution requires a very large sampling fraction, e.g. 100% in the case of crystals and more than 25% in the accordion-type LAr detectors cited in table 1.

The only way to improve the e.m. energy resolution for compensating calorimeters, given the small sampling fraction required, is to increase the sampling frequency. This philosophy gave birth to the fiber calorimeters. Recently, the RD1 Collaboration measured that the energy resolution for electrons improved from $12.9\%/\sqrt{E} + 1.2\%$ to $9.3\%/\sqrt{E} + 0.67\%$ when 0.5 mm fibers were used instead of 1 mm ones, keeping the lead/fiber volume ratio the same at 4:1[21] . However, also this approach has its limits, since the e/mip and therefore the e/h value goes up when the fiber-to-fiber distance is reduced (sect. 3.1). I conclude that it will be very hard to achieve e.m. energy resolutions better than $10\%/\sqrt{E}$ for a reasonably compensating ($e/h < 1.2$) calorimeter system.

As a side remark, it should be mentioned that calorimeters with a small sampling fraction may have additional advantages, which may be (at least) as crucial as excellent energy resolution. As an example, I mention the extremely compact construction. The very small shower parameters (radiation length, Molière radius, interaction length) will both limit the size, and thus the cost, of the experiment, and the lateral shower dimensions, which makes e, γ and μ identification a lot easier and reduces pileup effects. To set the scale, the radiation length of the scintillator calorimeters listed in table 1 is about a factor 3 smaller than for the accordion and crystal calorimeters.

3.3 Factors determining the energy resolution

Calorimetry is based on statistical processes and, therefore, the energy resolution usually contains a term that scales like $1/\sqrt{E}$. Among the fluctuations that contribute to this term can be mentioned:

- Sampling fluctuations, to be discussed in the following.
- Socalled "intrinsic" fluctuations, typical for hadronic shower development. These are fluctuations in the fraction of the initial energy converted into a measurable quantity. It has been shown that such fluctuations ultimately limit the achievable hadronic energy resolution of compensating calorimeters to $\sim 13\%/\sqrt{E}$[22] .
- Photon statistics, which ultimately limit the achievable e.m. energy resolution of scintillation and Čerenkov counters.

Apart from that, there is usually a variety of effects that contribute in another way to the energy resolution, for example:

- Noise, typical for LAr calorimeters. This noise dominates the low-energy performance of such calorimeters through the contribution of a resolution term that is inversely proportional to the signal/noise ratio, i.e. scaling like $1/E$ (see also sect. 2.2).
- Nonuniformities, for example coming from gap width variations in LAr calorimeters or fiber-to-fiber response fluctuations in spaghetti calorimeters. Such effects contribute an energy independent term to the energy resolution.
- Other instrumental effects giving energy independent contributions, such as calibration uncertainties, stability of electronics, temperature effects on the light yield of crystals, etc.
- Upstream or downstream energy losses, due to material in front of the calorimeter, or to incomplete shower containment. The exact energy dependence of such effects has not been studied in great detail, but depends presumably on many parameters, e.g. the type of particle being detected, the type of calorimeter, its size, etc. Upstream material mainly affects the resolution at low energies, shower leakage deteriorates the high-energy performance. The SPACAL Collaboration measured the effect of side leakage on the hadronic energy resolution to scale as $f/\sqrt[4]{E}$, f being the average leakage fraction[8] .

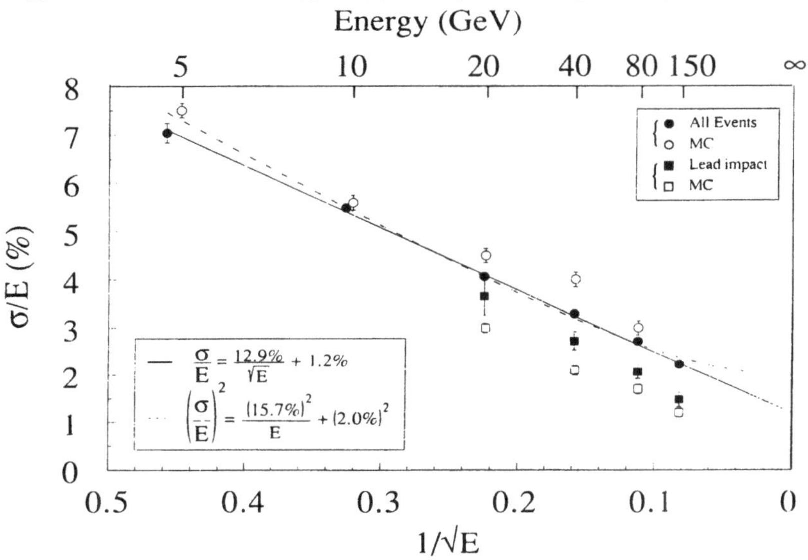

Figure 8 The energy resolution for electrons as a function of energy, for the SPACAL calorimeter. The closed circles encompass all events, the closed squares are for electrons that enter the detector in the lead. Monte Carlo predictions (GEANT) are shown as open circles and squares. Data from ref. 8.

In general, the terms describing the contributions from such effects have to be added in quadrature to the $1/\sqrt{E}$ term. This is, however, not a correct procedure if a subset of particles produces signals whose distribution has a different average response. A typical example concerns electron detection at small angles in the SPACAL calorimeter[8]. Electrons entering the detector in a fiber generate quite some light in this particular fiber through δ-ray emission. As a consequence, the signal distribution for these events is clearly shifted to larger values than for electrons entering in the lead. The total signal distribution is the sum of Gaussian distributions with different mean values and, therefore, the energy distribution does not improve as $1/\sqrt{E}$. As illustrated in fig. 8, the energy resolution is in this case much better described by an expression of the type $\sigma/E = c_1/\sqrt{E} + c_2$ than by $\sigma/E = c_1/\sqrt{E} \oplus c_2$. The figure also shows that the constant term practically vanishes when events are selected in which the electron enters the absorber material. This type of constant term is therefore typical for electron detection and will not play a role for γ's, to whom the plastic is transparant and which all will undergo conversion in the lead, irrespective of their entrance point.

It is important to realize that high-energy electrons and γ's behave quite differently in small amounts of material ($\sim 1X_0$). Electrons will always lose some energy (thin preshower detectors are based on this principle), while γ's may or may not convert. If they do convert, they may lose twice as much energy as single electrons. The spread in the energy loss distribution, and thus the effect of a cryostat or a magnet coil on the energy resolution of the calorimeter, is therefore larger for γ's than for electrons.

In a well-designed sampling calorimeter, the energy resolution is dominated by sampling fluctuations, at least in the energy range that is most relevant for the physics goals of the experiment. If this is not the case, money is wasted. How do these sampling fluctuations depend on the structure of the calorimeter? Figure 9a depicts the "classical" concept of an e.m. shower. An electron enters the detector, radiates γ's, which convert into e^+e^- pairs, which radiate γ's, etc. It should be emphasized that this is not the relevant picture when it comes to understanding the signals from sampling calorimeters, and in particular the effects of sampling fluctuations on the energy resolution. Evidence for this comes both from experimental data and from Monte Carlo simulations:

- Experimentally, one has observed that even in very fine sampling calorimeters, there is almost no correlation between the signals in subsequent sampling layers, which means that each layer is hit by *different* shower particles[22,23].
- Monte Carlo simulations (EGS4) tell us that in lead calorimeters only $\sim 5\%$ of the signal is caused by positrons, 95% comes from electrons.

The reason for this is simple. Most of the bremsstrahlung photons radiated

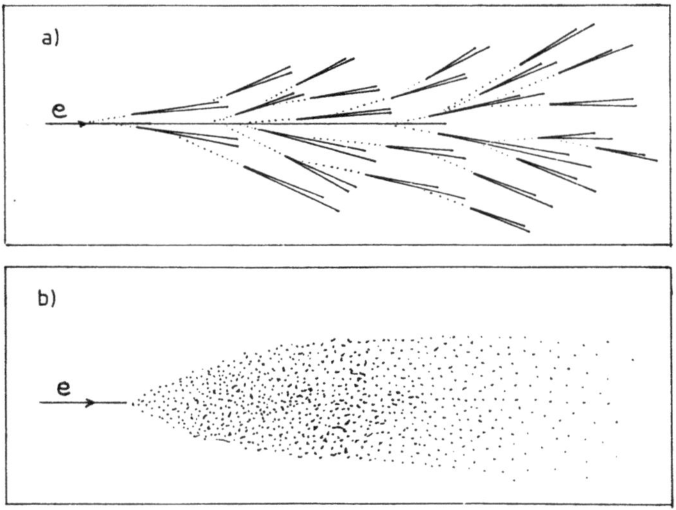

Figure 9 The classical concept of a developing e.m. shower (*a*) and a picture that is more relevant for understanding the signals from sampling calorimeters (*b*). See text for details.

by electrons are soft, with energies below the pair production limit of 1022 keV. These photons will produce electrons through Compton scattering or photoelectric effect. As was pointed out in sect. 3.2, the range of such electrons is usually only a small fraction of an absorber layer. A much more relevant picture of e.m. shower development is therefore the one shown in fig. 9b. The energy of the initial particle is deposited through a very large number of very soft electrons, with a range much smaller than the distance between subsequent sampling layers. Sampling fluctuations can then be interpreted as the fluctuations in the number of these short stubs contributing to the signal.

From this model, one may derive the relations between the energy resolution and the sampling parameters of the calorimeter. Since the number of different shower particles contributing to the signal is proportional to the sampling fraction $f_{\rm samp}$ (the fraction of the energy deposited in the active material), the energy resolution may be expected to be inversely proportional to the square root of that number:

$$\sigma/\sqrt{E} = c_1/\sqrt{f_{\rm samp}} \qquad (3)$$

Figure 10a summarizes the experimental results for a number of representative plastic-scintillator calorimeters. The energy resolution is shown versus the sampling fraction in such a way as to test eq. (3). The sampling fraction is plotted on a

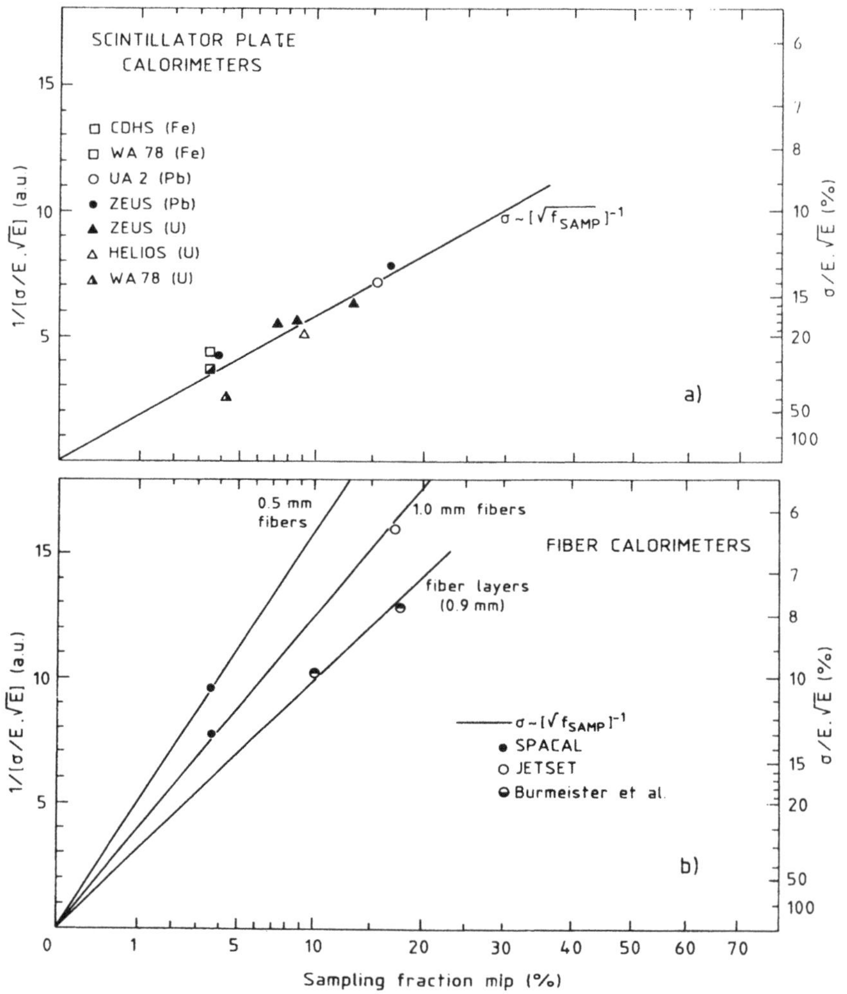

Figure 10 **The e.m. energy resolution as a function of the sampling fraction, for various representative plastic-scintillator plate calorimeters (*a*). The same, for various fiber calorimeters (*b*). Data from refs. 3,5,6,15, and 24-28. See text for details.**

scale linear in its square root and the vertical axis is linear in $\sqrt{E}/\sigma$. In this way, eq. (3) corresponds to a straight line through the origin, which turns out to give a reasonable description of the experimental data.

For a more refined description, one should realize, that the $\gamma \rightarrow e^-$ conversions occur overwhelmingly in the high-Z absorber material and that only a contribution to the signal may be expected if such a conversion takes place in the vicinity of a boundary layer (see sect. 3.2). The number of different shower particles contributing to the signal will, therefore, be proportional to the total surface of the interface between the active and passive calorimeter material in the region in which the shower develops. This leads to the following dependence of the resolution on the sampling frequency, for a fixed sampling fraction:

$$\sigma/\sqrt{E} = c_2\sqrt{\phi_{\text{fib}}} \tag{4a}$$

$$\sigma/\sqrt{E} = c_3\sqrt{d} \tag{4b}$$

for fiber (diameter ϕ_{fib}) and plate (thickness d) calorimeter structures, respectively. It is interesting to note that active elements surrounded by absorber material should lead to better energy resolutions than for plate structures, which offer less boundary surface for the same volume of active material. In particular, one expects $c_3 = c_2\sqrt{2}$.

Evidence for the validity of these relations can be obtained from fig. 10b, in which the energy resolution is given as a function of the sampling fraction, for various calorimeters employing scintillating plastic fibers as active material. The scales are identical to the ones in fig. 10a. For a given sampling fraction, the energy resolution improves considerably if fibers are used instead of plates. And indeed, the energy resolution improves by a factor $\sqrt{2}$ when 0.5 mm fibers are used instead of 1.0 mm ones[21] . Although the number of different fiber calorimeters that has been built is rather small, the ones that can be compared confirm the behaviour expected on the basis of our simple model. The points for the SPACAL[8] and JETSET[27] calorimeters are linked by a line that goes through the origin (3). The slope of this line is much steeper than for the line describing the plate calorimeters, reflecting the effect of the sampling frequency. The slope of the line for the 2 calorimeters built by Burmeister et al.[28] is less steep than for the 1 mm Spaghetti calorimeters, in spite of the fact that fibers with 0.9 mm were used by them. This is because of the fact that these detectors were constructed with fiber *ribbons*, comparable with plates for this purpose, and this result therefore illustrates the validity of the conclusion that there is a considerable resolution advantage in using single fibers surrounded by absorber material.

The resolution for LAr calorimeters follows almost the same pattern as for scintillator plate calorimeters. In particular, it turns out that, because of these effects, a LAr calorimeter matching the resolution obtained with the 0.5 mm fibers needs to have a sampling fraction that is about 7 times as large[10,21] . As a consequence, such a LAr calorimeter will need to be about 2.5 times longer to contain the showers, neglecting the space taken by the cryogenics.

Sofar, I have only discussed the effects of sampling fluctuations on the e.m.

energy resolution. However, since a major fraction of the hadronic energy may be dissipated by γ's from π^0 decays, these effects are very relevant for the hadronic energy resolution as well. The energy deposited through π^0's generated in the hadronic shower development is measured with the same accuracy as for e.m. showers. However, the non-e.m. shower component is measured much less precisely than the e.m. one. As pointed out by Groom[29], one should therefore expect the *total* hadronic energy resolution of a compensating calorimeter to improve *faster* than $1/\sqrt{E}$ with energy, since the e.m. shower fraction increases. However, since the average e.m. fraction changes only very slowly with the energy, this effect will be very hard to disentangle experimentally. Moreover, there is at least one effect that works in the opposite direction, as we will see later on.

The non-e.m. shower component contributes to the calorimeter signal through neutrons and charged particles. The energy carried by soft neutrons is presumably measured very accurately by compensating calorimeters, since these particles deposit a very large fraction of their energy directly in the active material. Also, the precision of the measurement should be very insensitive to changes in the sampling frequency (the sampling fraction should not be changed in this case because of the compensation condition). The charged particles must be distinguished in two classes: Mesons and nuclear fragments (p, α, etc.) The mesons travel in general over an interaction length through the detector before causing a nuclear reaction and they traverse a large number of active detector planes in doing so. The precision with which their energy loss is measured is probably comparable to that for low-energy muons and again insensitive to changes in the sampling frequency. Because of the Landau effect, the fluctuations in the energy loss measurement are non-Gaussian and therefore the resolution for this shower component may be expected to improve more *slowly* than $1/\sqrt{E}$ with energy. The nuclear fragments have a range comparable to the one for the photoelectrons dominating the e.m. response. Therefore, the sampling fluctuations which determine the precision with which the energy of this shower component is measured depend presumably in the same way on the sampling frequency as for the e.m. component (see eq. 4).

The situation is thus definitely more complicated for hadrons than for e.m. showers. Nevertheless, one may conclude that the hadronic energy resolution of a compensating calorimeter is much less sensitive to changes in the sampling frequency than the e.m. resolution. The signals from neutrons and mesons, which contribute sizeably to the total signal, are unaffected by such changes. Therefore, the hadronic energy resolution of the SPACAL calorimeter would deteriorate much more slowly than by a factor $\sqrt{\phi_{\text{fib}}}$ if the fiber diameter were to be increased. This is confirmed by a comparison of the hadronic energy resolutions measured for the SPACAL and ZEUS lead calorimeters (table 1). Based on the measurement of $30\%/\sqrt{E}$ for the 1 mm fiber configuration, the $\sqrt{\phi_{\text{fib}}}$ rule would predict $47\%/\sqrt{E}$

for a device with 2.5 mm fibers, whereas the 2.5 mm plate calorimeter, which should in any case be worse because of the boundary surface effect discussed before, already gave $44\%/\sqrt{E}$. I conclude that the hadronic energy resolution of a compensating fiber calorimeter is very insensitive to a change in the fiber diameter, which means that a lot of money can be saved if one is willing to relax a bit on the hadronic energy resolution requirements.

4. Conclusions

In calorimetric detection of hadrons and jets, the relation between the energy deposit and the (average) calorimeter signal depends in general strongly on the type of event, unless the compensation condition is met. If the e/h value is large, e.g. larger than 1.6, very poor hadronic performance has to be expected. Jet resolutions will be at least 10% at all energies and the missing E_T cannot be measured more accurately than to $\sim 15\%$. Another direct consequence of such noncompensation concerns the fact that severe trigger biases may be expected. Missing E_T triggers are likely to be saturated by events in which a concentration of jets with leading π^0's occurs in one hemisphere of the detector. Unless the new physics signalled by a missing E_T signature manifests itself as clearcut as in the case of W production at the $p\bar{p}$ colliders, it will be very hard to discover. Multiparticle measurements, whose results are sensitive to the effects of noncompensation, are a strongly recommended component of any test beam program.

The requirements for compensation and for ultimate e.m. energy resolution are mutually exclusive. Given the small sampling fraction needed for compensation, the e.m. resolution can be considerably improved by increasing the sampling frequency, but in practice it will be very hard to achieve better than $10\%/\sqrt{E}$ for a reasonably compensating detector.

Sampling fluctuations, which dominate the energy resolution of well-designed calorimeters, depend in a relatively simple way on the sampling fraction and the sampling frequency, at least for e.m. showers and hence for the e.m. component of hadron showers. The total surface between the passive and active material is a crucial parameter for the energy resolution of sampling calorimeters, leading to considerably better performance for fiber than for plate structures. The hadronic energy resolution for compensating calorimeters is relatively insensitive to changes in the sampling frequency, which is good news from a financial point of view.

References

1. W. Braunschweig *et al.*, Nucl. Instr. and Meth. **A265** (1988) 419 and **A275** (1989) 246.
2. G. Bagliesi *et al.*, Nucl. Instr. and Meth. **A286** (1990) 61.
3. U. Behrens *et al.*, Nucl. Instr. and Meth. **A289** (1990) 115.
4. M.G. Catanesi *et al.*, Nucl. Instr. and Meth. **A292** (1990) 97.
5. H. Abramowicz *et al.*, Nucl. Instr. and Meth. **180** (1981) 429.
6. M.G. Catanesi *et al.*, Nucl. Instr. and Meth. **A260** (1987) 43.
7. S. Aronson *et al.*, Nucl. Instr. and Meth. **A269** (1988) 492.
8. D. Acosta *et al.*, Nucl. Instr. and Meth. **A308** (1991) 481.
9. ALEPH Collaboration, private communication (1991).
10. C. Costa, contribution to these Proceedings.
11. C.V. Scheel, contribution to these Poceedings; D. Acosta *et al.*, The Performance of a Lead/Scintillating-Fiber Calorimeter at LHC/SSC Compatible Gate Widths, preprint CERN-PPE/91-195 (1991), submitted to Nucl. Instr. and Meth.
12. R. Wigmans, Ann. Rev. Nucl. Part. Sci. **41** (1991) 133 and references therein.
13. D.E. Groom, Proc. Workshop on Calorimetry for the Supercollider, Tuscaloosa 1989, eds. R. Donaldson and M.G.D. Gilchriese, p. 59.
14. R. Wigmans, Nucl. Instr. and Meth. **A265** (1988) 273.
15. E. Bernardi *et al.*, Nucl. Instr. and Meth.**A262** (1987) 229.
16. The GEM Collaboration, Letter of Intent for an Experiment at the Superconducting Supercollider (1991).
17. D. Hitlin, SLD Liquid Argon Prototype Tests, Report CALT-68-1305, Pasadena: Caltech (1985).
18. R. Wigmans, Nucl. Instr. and Meth. **A259** (1987) 389.
19. E. Borchi *et al.*, Phys. Lett. **B222** (1989) 525 and preprint CERN-PPE/91-18 (1991); A.L.S. Angelis *et al.*, Phys. Lett. **B242** (1990) 293.
20. R. Wigmans, Proc. Workshop on Experiments, Detectors and Experimental Areas for the Supercollider, Berkeley (1987), p. 608, eds. R. Donaldson and M.G.D. Gilchriese, Singapore: World Scientific.
21. L. Poggioli, contribution to these Proceedings.
22. G. Drews *et al.*, Nucl. Instr. and Meth. **A290** (1990) 335.
23. C.W. Fabjan *et al.*, Nucl. Instr. and Meth. **141** (1977) 61.
24. M. de Vincenzi *et al.*, Nucl. Instr. and Meth. **A243** (1986) 348.
25. A. Beer *et al.*, Nucl. Instr. and Meth. **A224** (1984) 360.
26. T. Åkesson *et al.*, Nucl. Instr. and Meth. **A262** (1987) 243.
27. D. Hertzog *et al.*, Nucl. Instr. and Meth. **A294** (1990) 446.
28. H. Burmeister *et al.*, Nucl. Instr. and Meth. **A225** (1984) 530.
29. D.E. Groom, contribution to these Proceedings.

PHYSICS REQUIREMENTS FOR LHC/SSC CALORIMETRY

DAN GREEN
Fermi National Accelerator Laboratory
P.O. BOX 500
Batavia, Illinois 60510, USA

1. Introduction

The goal of the next generation of collider detectors is to study the origin of electroweak symmetry breaking. The mass scale for this study is roughly 1 TeV. No matter what the details of the mechanism, one can be confident that new phenomena will occur, since weak intereactions become strong, i.e. violate partial wave unitarity, at that mass scale [1]. The partial wave amplitude for ee→WW scattering is;

$$ao \sim \frac{\alpha_W}{4\pi}\left(\frac{M}{M_W}\right)^2$$
$$ao \sim 1 \text{ if } M \sim 1\,\text{TeV} \qquad (1)$$

Therefore, the detectors for LHC/SSC must be able to confront this mass scale. In particular, the electroweak dynamics is such that the study of gauge boson pairs has a high priority. Given that the simplest decay modes for gauge bosons are into leptons, the new detectors will naturally tend to optimize performance for leptons.

2. Physics Processes and Rates

Calorimetry will be the basis for the new detectors. It plays a role in identifying or measuring all the components of the Standard Model. Leptons appear as missing energy or electromagnetic (EM) showers, or minimum ionizing towers, or narrow jets for neutrinos, electrons, muons, or taus respectively. The quarks will appear as jets of particles. The gauge bosons will appear as EM showers, jets, or lepton pairs for photons, gluons, and W/Z respectively. The calorimeter measures the magnitude, position, and time of energy deposition over some angular range. The nonuniform response of the calorimeter to different particle species means that it also assists in particle identification. In addition, calorimetry is the technology which improves, or stays fractionally constant, with energy. Therefore, high energies are the natural province of calorimeters.

The weakness of the Higgs coupling to ordinary matter implies that the LHC/SSC detectors must operate at very high luminosities. The event rates will be in the range 100 to 1000 Mhz. The natural time scale for the calorimetry to respond is the rf bunch spacing of ~ 16 nsec. Some representative rates are given in Table 2.1. Much more detail may be found in the recent SDC study [2].

The rates are given as rough functions of Pt and integrated over Pt for all Pt > 20 GeV. The general angular coverage and Pt trigger threshold requirements can be seen from inspection of Fig. 1. Clearly a coverage in pseudorapidity, η, of $|\eta| < 3.0$ with a trigger cut of Pt > 20 GeV is appropriate for good Higgs detection efficiency. Typical hard scattering rates, if dominated by dynamics and not quark/gluon source distributions, will fall as the inverse third power of Pt.

The singles rates in Table 2.1 vary from Jets at ~ 1 Mhz, to single pions at ~ 10 Khz, to gauge bosons at ~ 40 Hz. Note that the copious rate of Z --> 2e allows one a source of tagged e for calibration of the calorimeters. The pairs are produced at reduced rates, the most copious being photon pairs. Pairs of W bosons come from top pair decays, while Z pairs are directly produced. The rates for gauge pairs are shown in Fig. 2.

From Table 2.1 one can immediately make some conclusions. The rate for ZZ with 1 Z decaying into ll and the other into νν is (Pt > 20 GeV) ~ 0.0008 Hz. The rate for Z + J is ~ 1 Hz. Therefore, one needs to find jets, J, with an efficiency of better than 1 part in 250. Comparing the JJ and $Q\bar{Q}$ rates, one sees that one must contain a jet to 1 part in 1/1000 in order that jet leakage not dominate heavy flavor sources of ν (= missing Pt). The relative rates for ZJ and ZZ lead to the requirements on coverage shown in Fig. 3. In order that the jet be tagged, coverage (with efficiency > 0.995) out to $|\eta|$ ~ 5 is needed, if Z --> νν decay modes are to be used in the physics analysis. Therefore, angular coverage to $|\eta|$ = 3 is for triggering and reconstruction of leptons with good efficiency in H --> ZZ and --> 2 gamma decays. Missing Pt in Z --> νν requires coverage out to $|\eta|$=5.

The simulation of the physics processes of interest and the associated backgrounds leads to a set of requirements for the calorimeters, their measurements of energy, position, time, and particle ID properties [2]. A partial list of the salient requirements discussed in Ref. 2 is shown in Table 2.2. The parameter, the physics process, and the requirement on the parameter are shown in the Table. The issues of hermiticity, calibration, angular coverage, and trigger coverage have already been briefly raised. In what follows, details of the other entries in Table 2.2 will be given.

3. Hadronic Size and Segmentation

This section discusses the physics processes which are studied in making decisions about the size, longitudinal and transverse segmentation of the hadronic calorimeter.

3.1 Total Depth : Dijet Mass and Missing Pt

The depth of the calorimetry is a cost driver since it defines the total volume of material. The requirements have been extensively studied [3]. The physics driver is to have unimpaired dijet mass resolution at the highest accessible mass, about 10 TeV. That requires a total depth of about 10 absorption lengths in order to limit leakage fluctuations. A plot of mass resolution as a function of calorimeter total depth, D, in absorption lengths is shown in Fig. 4. Other ,subsidiary, requirements are discussed in detail in Ref. 4. Suffice it to say that most of them cluster in the region of 10 absorption lengths. A scheme to weight the last sampling layer [5] allows one to shrink the depth by about 1, while retaining roughly constant resolution.

3.2 *Weighting and Depth Reduction*

In Fig. 5 is shown the contained energy for 0.45 TeV incident beam. One compares a depth of 10 for an unweighted calorimeter to a depth of 8.7 for a depth segmented weighted calorimeter. Clearly the rms in the two cases is very similiar. The Gaussian part of the energy distribution is slightly wider, but a glance at Fig. 4 shows that the degradation is acceptable. Thus, in terms of resolution, a calorimeter with D = 9 is (barely) acceptable.

Nonlinearity is potentially an issue for a too thin calorimeter. The unweighted D = 10.1 version is linear up to 0.2 TeV and then leakage causes a ~ 1% nonlinearity at 0.45 TeV. The D = 8.7 weighted option, with weight = 3, overcorrects by 1% for E < 0.3 TeV, followed by 1% leakage at 0.45 TeV [3]. Therefore, the nonlinearity induced by weighting is acceptable given the required energy resolution.

3.3 *Longitudinal Segmentation and Depth Reduction*

The total depth was set by the Gaussian part of the dijet mass resolution. However, there are fluctuations in the hadronic shower development which lead to long non-Gaussian tails in the energy resolution. These may be controlled by segmenting the hadronic absorber into 2 compartments [4]. A plot of the containment energy fraction at a depth of 10.1 with and without longitudinal segmentation, at 0.45 TeV, is shown in Fig. 6. Clearly, control is needed in order that these tails not be the dominant contribution to the missing energy [6]. The optimal split is set by the desire to tag energy leaking out the back, taking into account that some hadronic showers have low energy deposition gaps in their development. A rear section of 3 to 4 absorption lengths appears to be optimal [4].

3.4 *Transverse Segmentation : Dijet Mass and e ID*

A study was made for the impact of Higgs bosons decaying into ZZ which subsequently decay into jets. The mass resolution for a Z from 0.8 TeV Higgs bosons is degraded for segmentation > 0.05 [3]. The desire is to preserve Z dijet mass resolution for boosted Zs. A plot of dijet mass resolution as a function of segmentation is shown in Fig. 7. Note that the effect in going from segmentation of 0.05 to 0.1 is not very large.

Given the importance of gauge boson pairs, and thus e, the impact of hadron segmentation on electrons must be kept in mind. For electron ID, isolation is a crucial cut variable. Segmentation defines how near a jet electrons may be defined. The hadronic segmentation of the first compartment may be chosen to match that of the EM compartment(s) so as to make triggering most graceful. This tower matching makes "EM/HAD" and "isolation" easy to impose and insures that electrons may be identified as close to jets as the EM segmentation allows. A recent study [7] indicates a loss in efficiency of ~ 20% in going from 0.05 hadron segmentation to 0.10 at design luminosity.

Finally, it must be kept in mind that the SSC can reach 10x of the design luminosity fairly easily. The energy deposited in a LHC tower goes as the area, and fluctuations in the pileup "noise" thus go as the segmentation. It seems prudent to allow for increases in luminosity by maintaining fine hadronic segmentation.

4. EM Size and Segmentation

The EM calorimeter parameters are set by the need to make a precise measurement of EM energy and by the necessity of good e ID [8]. The needs of particle identification largely define the EM segmentation.

4.1 Total Depth and Leakage

The depth requirement is set by the need to contain the hightest Pt e without inducing a "leakage error" which exceeds the intrinsic EM resolution. For 99% containment at 0.1 TeV, a ~ 25 radiation length, t, EM calorimeter is needed. In addition, since the fine segmentation is expensive, one can consider weighting the last layers of the EM calorimetry in exactly the same fashion as was discussed for the hadronic sections. Note that a very deep EM compartment would lose hadronic rejection and thus is not desirable in any case.

4.2 Weighting and Depth Reduction

An exit weighting scheme gives essentially the same resolution for a $t = 22$ calorimeter as a $t = 50$ EM calorimeter. Therefore, at least from the point of view of EM resolution and containment, weighting the last EM samples is acceptable. However, the impact on hadronic resolution of showers developing early in the EM section has not yet been evaluated. This weighting scheme is still under active study.

4.3 Longitudinal Segmentation and Radiation Damage

The electron compartment(s) naturally divide along the "shower maximum" boundary at about 6 radiation lengths. It is not deemed necessary to have longitudinal segmentation in order to do electron ID. The use of E/p, isolation, EM/HAD, transverse shower shapes, and shower max matching are sufficient [9]. The shower maximum detector serves as a second, redundant, measurement of EM energy which will allow for a check against grossly mismeasured energies (e.g. "sparks").

There is another virtue in longitudinal segmentation. In the endcap EM sections, the radiation field is intense. The implementation of 2 EM longitudinal compartments allows one to make an event by event measurement of shower initiation points, and hence to correct even severe damage to a level allowed by precision EM calorimetry specifications [10]. In Fig. 8 is shown the "induced" constant term for damage coefficients up to 50% for uncorrected and corrected (2 compartments) operation. Clearly, the method allows one to operate at up to 50% damage for isolated electrons since the error due to damage then is less than the resolution.

4.4 Transverse Segmentation and b Decays

One potential scale for EM transverse size is the Moliere radius. At a 2 m distance, this scale means segmentation of roughly 0.01. That size can be accommodated, in 1 dimension, by an array of crossed "shower maximum" strips for an affordable strip count. There are no other natural scales to point to. In Fig. 9 we show for $t \to W + b$, $b \to c + e + \nu$ events the ratio of electron energy to all energy, on average, in a cone of radius R as a function of R. The isolation falls as $1/R^2$. A 5% isolation, comparable to the e energy resolution in these events, implies a segmentation of roughly 0.035. Note that this estimate ignores the possibility of high luminosity "minbias" pileup from overlaping events. This smooth backgound should be subtractable on average, leaving a fluctuation scaling as 1/R.

4.5 Transverse Segmentation and e Track Matching

The shower centroid can be found accurately if the strips have a width of about the size of the Moliere radius. Therefore, crossed strip detectors with a width of 0.01 in azimuthal angle and rapidity are plausible.

What determines the strip length? Clearly, the centroid measurement will be degraded by pileup fluctuations. The centroid error should scale as the square root of the length. In Fig. 10 is shown the mean deviation in ϕ and η as a function of strip length for t --> W + b, b --> c + e + v events. This error will define the cut limits for track matching with the SM centroid. The solid curve corresponds to square root (length), while the dashed curve is a smooth representation of the field on data. Setting a design criterion that the pileup error be < 1/2 the strip width, one finds that lengths of 0.5 are acceptable.

4.6 EM Segmentation and Jet Rejection

The JJ cross section far exceeds the cross section for e, as seen in Table 2.1. Typically an electron trigger requires Et > threshold in EM, EM/HAD > 10, and EM/(E in "isolation" ring) > 4. These cuts are roughly 95% efficient [7]. A jet will pass the same set of cuts, ~1/2000 of the time. The rejection factor for jets scales roughly as the inverse of the EM linear segmentation. This scaling makes jet rejection by the use of finer segmentation fiscally difficult.

5. Hadronic Energy Resolution

5.1 Z Dijets and Resolution

As discussed above, some of the fundamental entities of the Standard Model will appear in the calorimeters as jets. What energy measurement accuracy is required? A study [11] of Z_l and 1 TeV Z' decaying into dijets at both high and low Pt was made to address this question. A summary of the results is shown in Fig. 11. Clearly, the major effects come about due to clustering algorithims and the underlying event. The effects of any reasonable HAD calorimeter resolution are minimal. Therefore, one concludes that precise hadronic energy resolution is not required in the hadronic compartment of the calorimeters.

5.2 Dijets and e/h

A continuation of the study refered to in section 5.1 was made to explore the e/h requirements in more detail [12]. The results of that study are summarized in Fig. 12. The fractional dijet mass resolution at masses up to 10 TeV is dominated by fluctuations in the fragmentation and the underlying event. Calorimeter resolution is less important. Neutrino fluctuations from heavy flavor decay within the jet begin to become more important than the resolution at the highest accessible masses.

Clearly, the difficulty in precisely defining a jet alleviates the strict requirements one might place on hadronic calorimetry. In Fig. 13 is shown the fractional dijet mass resolution as a function of mass. The comparison is between e/h=1, e/h=1.3 with the EM and HAD compartments calibrated to equal response at all energies, and only at asymptotic energies. Obviously, e/h < 1.3 does not make a major impact on jet resolution.

5.3 Speed, Overlaps and Z Dijets

The weak coupling of the Higgs means that high luminosity machines are needed to confront Higgs physics. Thus, the issue of event pileup is of interest. A study has been made of event pileup for dijet decays of boosted Z [13]. Results for dijet mass resolution as a function of cone size, R , are shown in Fig. 14 for up to 16 overlapped minbias events. Without overlaps R ~ 0.7 is optimal and leads to minor inceases in resolution due to fluctuations. The overlap of 16 events is partially controlled by placing a tower threshold of 2 GeV in Pt. However, the dijet mass resolution is still degraded by a factor ~ 2.5 with respect to the case with no overlapping events. Therefore, jet spectroscopy will be badly compromised at high luminosity. The resolution should worsen at the square root of the luminosity.

5.4 Uniformity Tolerance in Sampling Calorimeters

The specification on the tolerance for both the absorber and the active medium is driven by the resolution specification. For hadrons, a study [11] showed that for constant terms < 3%, no adverse effects on jet masses were seen. The conversion factor from random plate variations to hadronic constant term was found using Lab E test beam data [3]. Fig. 15 shows the constant term as a function of rms light yield errors for several energies. A 2% constant term corresponds to 5% rms error.

6. EM Energy Resolution and e ID

There is a natural scale for electron energy resolution. The natural width of the Z is set by the electroweak coupling constant which is about 0.03.

$$\Gamma_Z / M_Z \sim \alpha_W \sim 1/30 \qquad (2)$$

Therefore, the narrow Z width implies that the electron energy should be measured to better than 1% in order that the calorimeter not degrade Γ_Z.

The implication for a sampling calorimeter is that the absorber should be made as thin as is economically feasible. A thickness of 1/2 radiation length of Pb leads to a "stochastic term" coefficient of 12%. A match to this resolution would be to control all nonuniformities such that the constant term is < 1%. Thus, for energies above 0.1 TeV, the constant term will dominate the energy resolution. The plate thickness is roughly proportional to the energy at which the resolution becomes constant.

6.1 Material Budget and Conversions

It makes little sense to construct a precise calorimeter and yet allow sufficient upstream material that the performance is degraded. This perception forces one to consider a maximum material "budget" upstream of the calorimeters. A glance at Table 2.1 shows that the rate of γ + J above Pt of 20 GeV is ~ 40 Hz. If the γ converts upstream of the EM calorimeter, it looks like an isolated e. A 10% conversion probability gives a rate of 4 Hz for isolated e above 20 GeV, which swamps the isolated electron trigger. Monte Carlo results [14] are given in Fig. 16. Indeed, photon conversions dominate all trigger sources for Pt < 25 GeV. This observation sets the conversion limit given in Table 2.2.

The issues for material budget also include e triggering and e ID. For example, the radiating e will deposit energy in several EM towers, effectively lowering the Et below threshold for any single trigger tower. One also loses "isolation" as a criterion for e ID. The E/p cut may be compromised since the radiated energy is subtracted from p to form E. Finally, the match of the track to the SM centroid will be compromised. A study was made of the losses due to bremmstrahlung [14]. There are 2 effects; a failure to find the e track due to "kinks" and the missmatch of the track to the SM centroid. For 10% material, the e efficiency is reduced to ~ 85%.

6.2 Massless Gaps and Upstream Material

The effect of inert material, ti, placed directly in front of the EM calorimeter may be characterized as an additional term in the resolution.

$$dE/E=\left(a/\sqrt{E}\right)\left(e^{ti/tc}\right) \tag{3}$$

where a = 0.019 and the characteristic scale is tc = 1.58 [15]. Thus, a scale is set for the material directly in front of the EM calorimeter, e.g. a magnet coil or a liquid argon cryostat. The "budget" for material distributed throughout, or upstream of, the tracking volume is considerably less as discussed in section 6.1.

In an attempt to alleviate that problem, one can sample the EM cascade directly behind the material using a "massless gap". Since the sampling fraction of the inert material is different from the EM calorimeter, one must weight this layer differently. For ti = 1, the resolution for different energies and angles of incidence are shown in Fig. 17. Clearly, this amount of material is tolerable in that the resolution is not badly degraded. Note also that the optimal weight of the layer is not strongly energy or angle dependent. The flat plate geometry means that dE/E scales as $a/\sqrt{ET}$.

6.3 Higgs --> 2 γ and Resolution

The most stringent limit on resolution is set by the desire to study the 2 γ decay mode of a light Higgs. The natural width is effectively zero. Therefore, since there are considerable backgrounds, the maximization of the S/N ratio argues for the best possible resolution. A plot of the mass distribution for a typical sampling calorimeter is shown in Fig. 18.

Note that the desire to observe the 2 γ decay mode implies a very good photon ID cleanliness. For example, a glance at Table 2.1 shows that the JJ rate is ~ 1 Mhz, while the 2 γ rate is ~ 0.4 Hz. Therefore, a J/γ rejection of ~ 1000 is implied if the 2 γ mode is to be cleanly identified. The desire for excellent resolution must be moderated by the realization that there are systematic effects. The calorimeter is an ensemble of towers which must be kept uniform. It is assumed that the "constant term" will be kept to < 1% throughout this ensemble.

6.4 Uniformity Tolerance in Sampling Calorimeters

An example of the type of errors which must be kept under control is the variation in plate thickness of a sampling type calorimeter. A study [3] was made of the relationship between the EM resolution and the rms plate thickness variation. As seen in Fig. 19, a plate thickness tolerance rms of 2% induce a constant term of 1/2%. The effect is again essentially energy independent and linear in the rms light yield error.

Note that this specification on thickness tolerance assumes that each tower is not separately calibrated. If it were, then one has effectively measured the plate thicknesses near shower maximum, and the tolerance can be relaxed. Monte Carlo studies show that the limit can be relaxed by a factor ~ 2/3. Note also that thinner plates make one less sensitive to plate variations since the shower is spread over several plates, and individual plate errors are then "washed out".

6.5 Speed and Pileup, High Luminosity

The issue of pileup is perhaps most severe for the EM calorimetry since the e ID requires a variety of cuts. For example, the rate for JJ and $Q\bar{Q}$ with Q --> qev is in the ratio of 1,000,000 for M ~ 100 GeV. Therefore, J/e rejection factors of that order are desired.

The e ID cuts are typically EM/HAD, "isolation", and SM track match. For a density, ρ , of 10 particles per unit of rapidity, each with <Pt> ~ 0.7 GeV, there is ~ 1 GeV of Et per unit of area, A, in η-ϕ space. For the SSC with luminosity, L , of 10**34/(cm*cm*sec) there are 16 interactions per bunch. For a calorimeter with 20% stochastic term and 1% constant term, studying e from Z over a 2x2 set of towers (for transverse containment), one needs towers < 0.16 not to degrade resolution with pileup for the fastest calorimetry imaginable.

Resolution is not the entire story. Studies [16], have shown that, for a 50 GeV Et e from Z --> ee decays, EM/HAD > 10 and (isolation E)/EM < 1/10 will, at a luminosity of L = 10**34/(cm*cm*sec), and 0.05 segmentation lead to a 8% e efficiency loss. In addition, the track match cut will suffer with increased luminosity. The results in Fig. 11 may be crudely scaled with 10x luminosity increase by the square root. This means that, for the same criteria, the SM strip length cannot exceed ~ 0.15.

7. Summary

Various parameters having to do with the position, energy, and time measurements of calorimeters are defined with respect to physical processes of interest at SSC/LHC energies. Design criteria are thus defined.

References

1. E. Eichten, et al., *Rev. Mod. Phys.* 56 4, 579 (1984).

2. *SDC Calorimeter Design Requirements*, J. Siegrist ed. August 5, 1991.

3. *Calorimeter Conceptual Design - Tile/Fiber Scintillator Option*, ed. SDC Tile/Fiber Calorimeter Group, September 3, 1991.

4. D. Green, et al., *Depth Requirements in SSC Calorimeters*, Fermilab-FN-570, August 1991.

5. L. Jones, *Reduction of Calorimeter Depth by Weighting Last Layer*, UM HE 91-27.

6. J. Hauptman, *Fake Missing ET Trigger Rates Due to Non-Hermeticity, Finite Energy Resolution and Finite Depth*, SDC Note.

   M. Barnett and Ian Hinchliffe, *The Impact of Resolution, Cracks and Beam Holes on Detection of Processes with Missing Energy*, SSC-SDE-29, LBL-28773, March 28, 1990.

7. A.B. Wicklund, *b-Jet Tagging Using Electrons at SDC*, August 14, 1991, SDC-91-0051.

8. *Letter of Intent* by the Solenoid Detector Collaboration, November 30, 1990, SDC-90-00151.

9. T. Kirk, A.B. Wicklund, SDC-91-00018, April 11, 1991.

10. D. Green, A. Para, *Radiation Damage, Calibration and Depth Segmentation in Calorimeters*, Fermilab-FN-565, May 1991.

11. A. Para, et al., *Jet Energy Resolution of the SDC Detector*, Fermilab-TM-1270.

12. D. Green, *Dijet Mass Resolution and Compensating Calorimetry*, Fermilab-TM-1740.

13. D. Green, *Dijet Spectroscopy at High Luminosity*, Fermilab - Conf 90/151.

14. Report of the Task Force on Impact of Material in Tracking Volume, SDC.

15. D. Green, *Effect of Inert Material on Z1 Mass Resolution for* $H \to ZZ \to eee$, Fermilab-TM-1736.

16. J. Colas, et al., *Calorimetry at the SSC*, DESY 90-136.

TABLE 2.1 RATES AT SSC FOR DESIGN LUMINOSITY

PROCESS	DIAGRAM	$\frac{dR}{d\eta dPt}$ (Hz/GeV)	$\frac{dR}{d\eta}$ (Pt>20GeV)
JETS	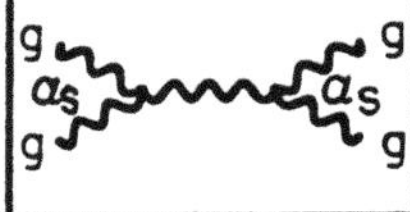	$25{,}000\ (20/Pt)^3$	1,000,000
π^o		$100\ (20/Pt)^3$	4000
W,Z	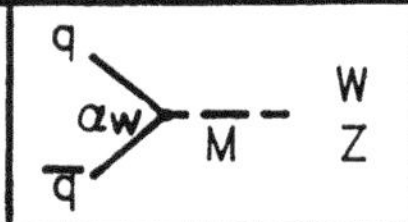	Pt~M/2	40(W) 12(Z)
γ	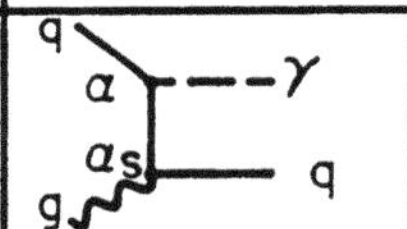	$1.0(20/Pt)^3$	40
$\gamma\gamma$	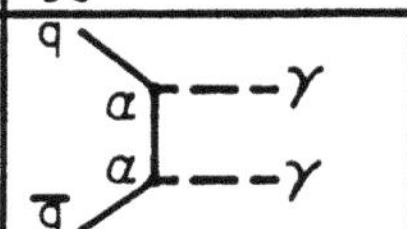	$0.01(20/Pt)^3$	0.4
$Q\overline{Q}$	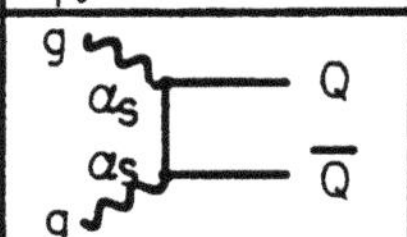	$\frac{d\sigma}{dPt^2} \sim \frac{\sigma/MQ^2}{[1+(\frac{Pt}{MQ})^2]^2}$	$8\ (\frac{100}{MQ})^3$
WW	FROM t $\overline{t}$	$\frac{d\sigma}{dPt^2} \sim \frac{\sigma/MQ^2}{[1+(\frac{Pt}{MQ})^2]^2}$	$8\ (\frac{100}{MQ})^3$
ZZ	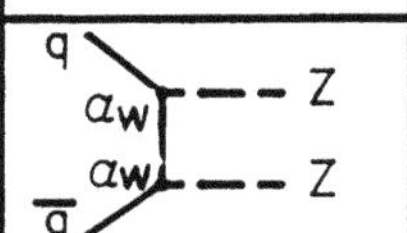		4×10^{-3}
H(1 TeV)	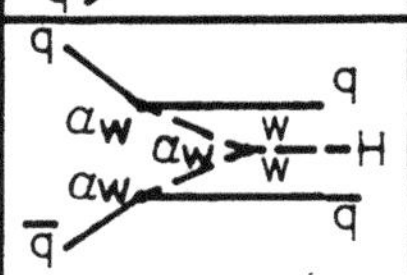		2×10^{-4}

TABLE 2.2 PHYSICS REQUIREMENTS FOR CALORIMETER PARAMETERS

PARAMETER	PROCESS	ALLOWED RANGE
HERMITICITY	JJ	J MISIDENTIFICATION $<10^{-3}$
CALIBRATION	Z	$(Pt)_e \gtrsim 20$GeV TRIGGER
UPSTREAM(COIL) MATERIAL(TRACK)	Z, Γ_Z ISOLATED e TRIGGER	$\lesssim 1\ X_0$ FOR COIL $\lesssim 0.1\ X_0$ FOR TRACKING
SPEED	$Z \rightarrow JJ$ $Z \rightarrow ee$, PILEUP	$\lesssim 1$ BUNCH CROSSING
ANGULAR COVERAGE	ZZ $\hookrightarrow \nu\bar{\nu}$	$\|\eta\|<3, \epsilon_e$ $\|\eta\|<5$, Z+J
TRIGGER COVERAGE	ZZ	$(Pt)_e \gtrsim 20$GeV
TOTAL DEPTH	$M_{JJ}=10$TeV	$D \gtrsim 10$
LONGITUDINAL h SEGMENTATION	$\not{p}_T$ IN JJ	D1, D2 FOR h
LONGITUDINAL EM SEGMENTATION	$Z \rightarrow ee$	RADIATION DAMAGE 2 DEPTH SEGMENTS FOR EM AT LARGE $\|\eta\|$
EM RESOLUTION HADRON RESOLUTION	$Z \rightarrow ee, \Gamma_Z/M_Z$ $(\alpha_W/\pi = 0.01)$ M_{JJ}	$0.2/\sqrt{E} \oplus .01$ OR BETTER $0.7/\sqrt{E} \oplus .03$ OR BETTER
COMPENSATION	M_{JJ}	$\|e/h-1\|<0.3$
TRANSVERSE h SEGMENTATION	$H \rightarrow ZZ$ $\hookrightarrow JJ$ $t \rightarrow Wb$ $\hookrightarrow ce\nu$	$\Delta\eta_h<0.05$ $\Delta\eta_h<0.05$, EM/HAD ISOLATION
TRANSVERSE EM SEGMENTATION	$H \rightarrow ZZ$ $\hookrightarrow JJ$ $t \rightarrow Wb$ $\hookrightarrow ce\nu$	$\Delta\eta_e<0.05$ J REJECTION e EFFICENCY

Figure Captions

1.a Acceptance for Higgs bosons of mass M = 0.2, 0.4, 0.6 and 0.8 TeV as a function of the rapidity coverage. All 4 leptons in the H --> ZZ --> 4e decay chain must have η < the cut.

1.b As in a, except the cut variable is the minimum Pt which each e must have. The pseudorapidity is $|\eta| < 2.5$.

2. Cross sections for WW production via $t\bar{t}$ formation and cross sections for direct production of WW and WZ pairs.

3. Relative cross sections for ZZ and ZJ with J falling outside the fiducial volume of the calorimeters.

4. Fractional dijet mass resolution for 10 TeV dijets as a function of depth, D. The points correspond to different parametrizations of data, CCFR data, • , WA1 data, o, and Lab E data, [] .

5. Measured energy distribution for 0.45 TeV beam incident on a calorimeter.
   a. D = 10.1, unweighted
   b. D = 8.70, unweighted
   c. D = 10.1, weighted by 3
   d. D = 8.70, weighted by 3

6. Distribution of fraction of contained energy, f, for 0.45 TeV test beam data.
   a. All events with D = 10.1
   b. Longitudinal segmentation with D2 = 3.5, D1 = 6.6 and ratio of energy in D2 to total less than 0.18.

7. Fractional mass resolution for Z --> JJ for low Pt Z's , ⊙ , and boosted Z with Pt = 0.55 TeV , [] , as a function of tower transverse segmentation.

8. Magnitude of the "induced" constant term due to damage coefficient of 0.1, 0.3, and 0.5 as a function of electron energy.
   a. No correction.
   b. Corrected using simple energy independent form derived from 2 longitudinal EM compartments.

9. Study of isolation for $t\bar{t}$ events with subsequent t --> Wb, b --> cev decays. The ratio of average electron energy to total EM + h energy in a cone of radius R as a function of R.

10. The mean deviation between the SM centroid and the exact tracking position as a function of strip length for t --> Wb --> cev events. Strip width is 0.01. Points are , • , field on and , o , field off x 1/10. The solid curve is square root (length). The dashed curve is to guide the eye through the data. The shaded portion corresponds to the region where the matching error is > 1/2 the strip width.
    a. azimuthal strips
    b. strips in rapidity

11. The fractional mass resolution for Z and 1 TeV Z' decays into dijets.

a. Perfect detector, no underlying event, finite cone R = 0.7.

b. Perfect detector, underlying events

c. As in b but with $0.15/\sqrt{E} \oplus 0.01$ for EM and $0.3/\sqrt{E} \oplus 0.02$ for HAD

d. As in c but with $0.5/\sqrt{E} \oplus 0.03$ for HAD

e. As in c but with $0.7/\sqrt{E} \oplus 0.04$ for HAD

f. As in d but with $0.25/\sqrt{E} \oplus 0.02$ for EM

g. As in d but with e/h = 1.3 for HAD

12. Dijet mass resolution as a function of mass. The "data points" refer to the total error. The points, • , refer to fragmentation and underlying event fluctuations, o , refer to calorimeter resolution, and ∇ refer to neutrino fluctuations.

13. Dijet mass resolution as a function of mass. The data points are for the "baseline" SDC calorimeter, • , a calibrated calorimeter with e/h = 1.3, o, and a calorimeter with e/h = 1.3 which is only calibrated at asymptotic energies, ∇.

14.a Plot of standard deviation of δ as a function of R for H(800) events keeping only tracks with kt > 1.0 GeV with no additional overlap events (solid), 8 overlap events (dashed), and 16 overlap events (dot-dashed). The results using only W fragments are shown as o symbols.

14.b As in Fig. 15a except that only tracks with kt > 2.0 GeV are allowed.

15. Induced constant term as a function of rms light yield for a hadron calorimeter. Test beam data from the "Lab E" spectrometer were used at beam energies of 0.1, 0.2, and 0.45 TeV.

16. Differential cross section as a function of Pt for leptonic W decays, semileptonic t decays, and photon conversions.

17. Fractional energy resolution for an EM calorimeter with t = 1/2 samples proceeded by ti = 1. The results are for , • , 25 GeV, $\theta = 90^\circ$; , o , 25 GeV, $\theta = 30^\circ$; ,[] , 50 GeV, $\theta = 90^\circ$.

18. Two γ mass distribution for W + H events with H --> 2 γ for the parameters of the SDC detector.

19. Induced constant term as a function of rms light yield for a EM calorimeter. EGS Monte Carlo "data" were used at incident energies of 10 and 50 GeV.

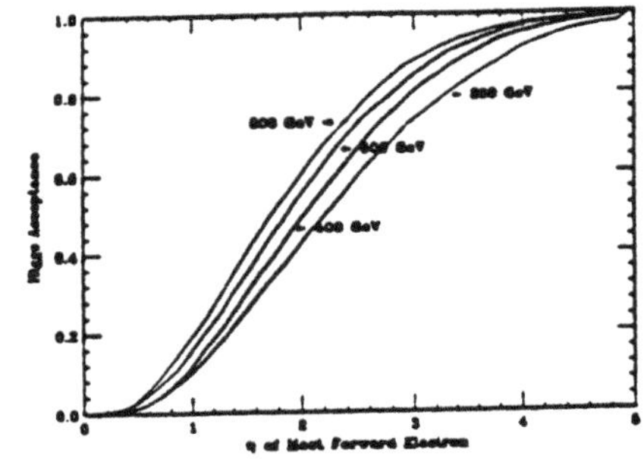

FIGURE 1.a

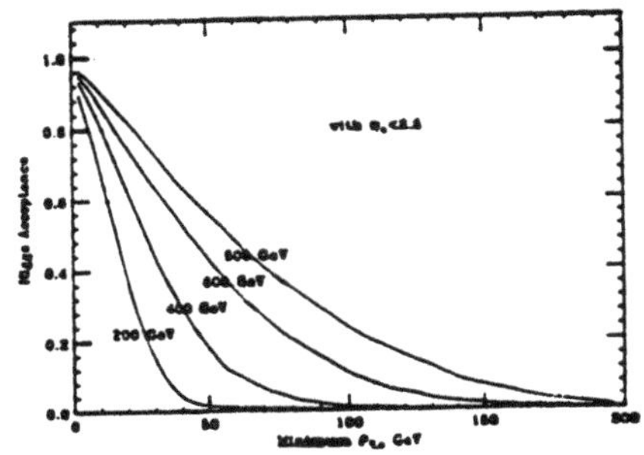

FIGURE 1.b

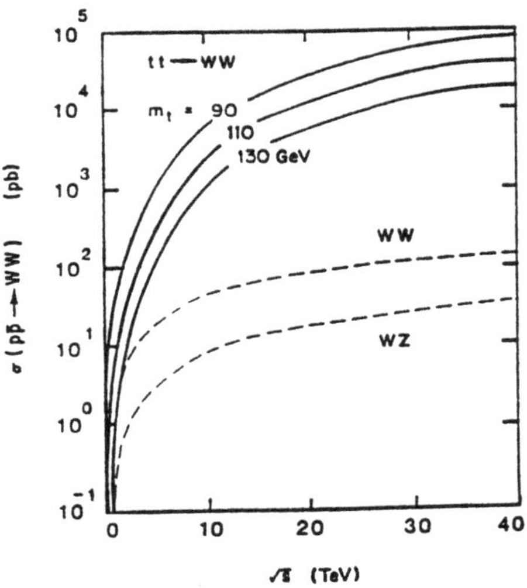

FIGURE 2

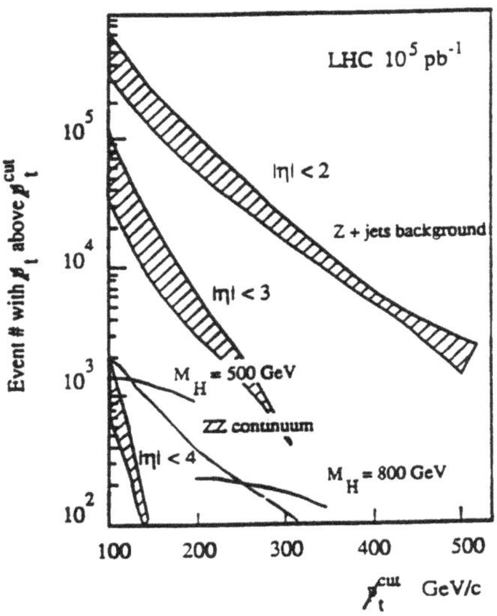

FIGURE 3

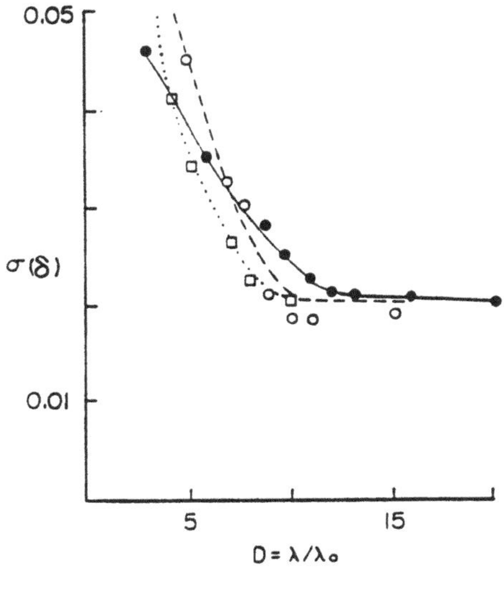

FIGURE 4

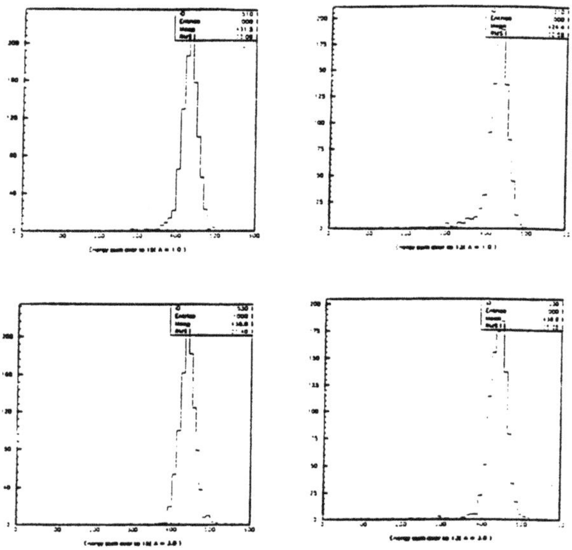

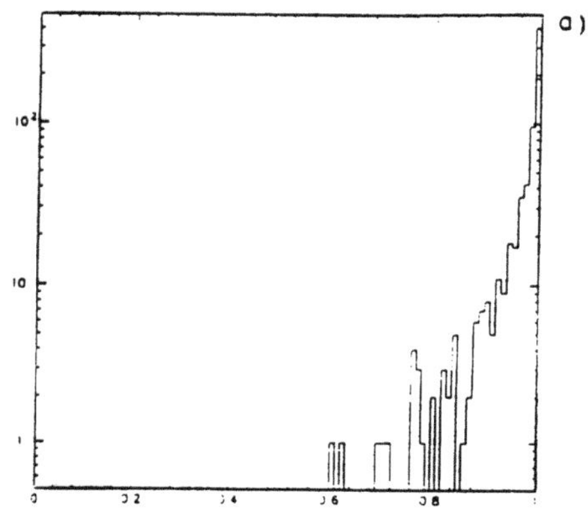

FIGURE 5

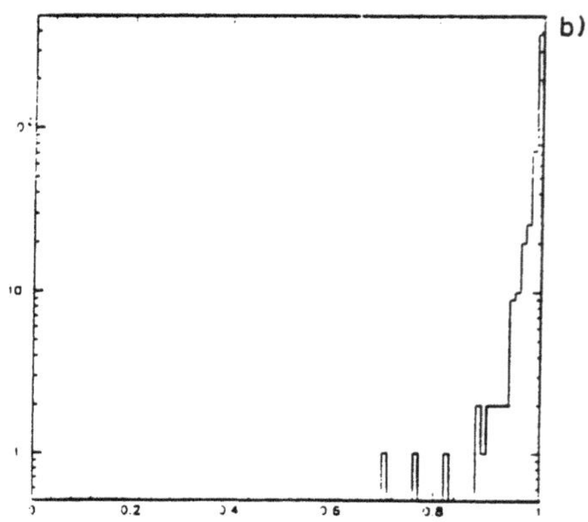

f

FIGURE 6

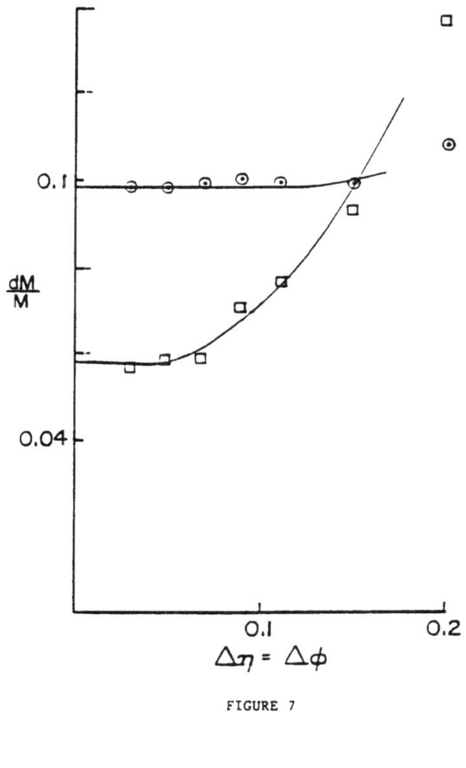

FIGURE 7

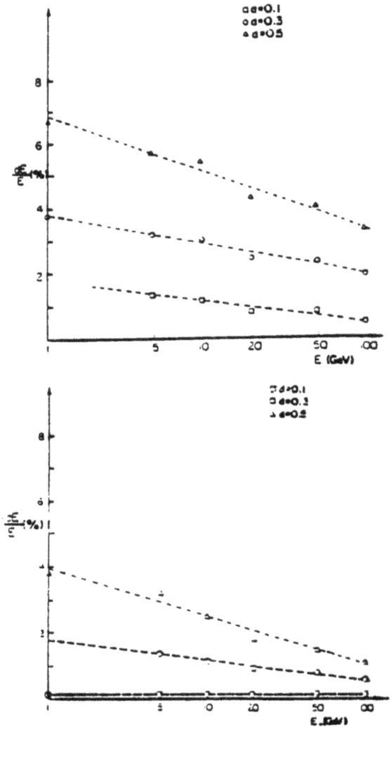

FIGURE 8

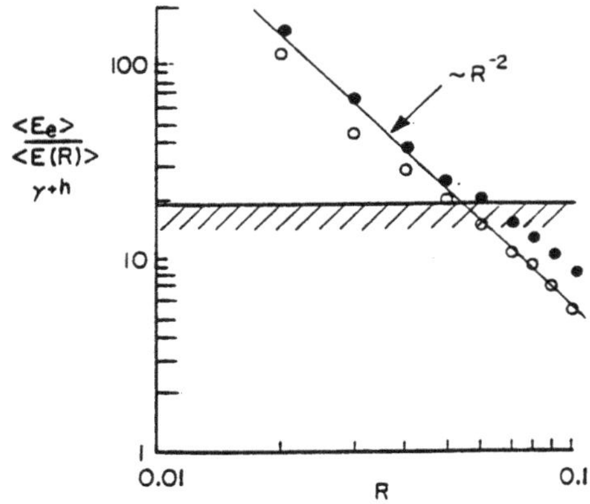

FIGURE 9

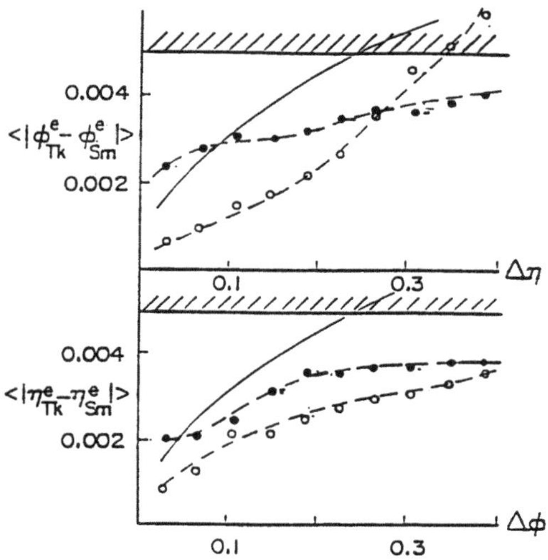

FIGURE 10

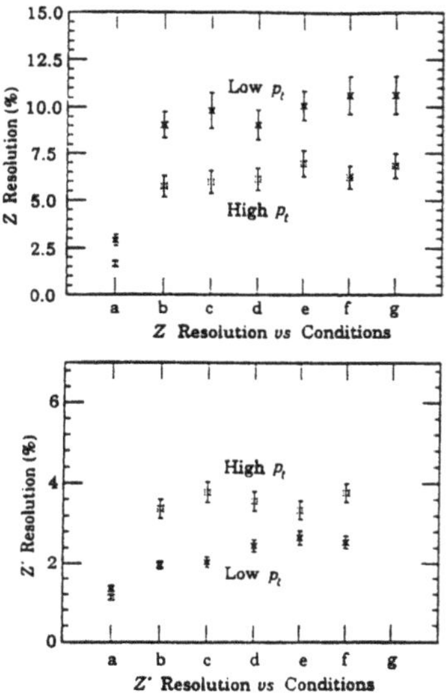

FIGURE 11

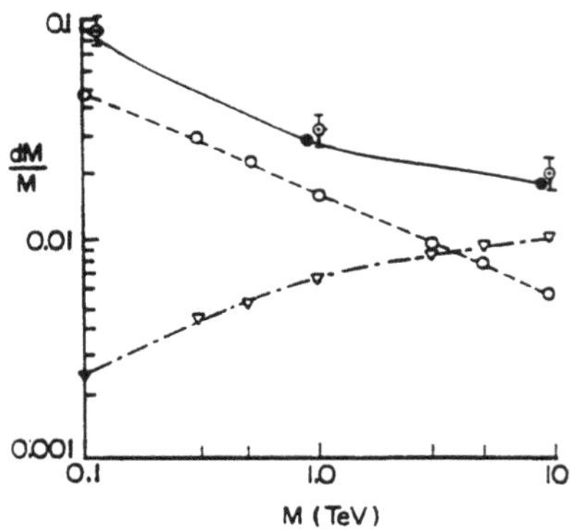

FIGURE 12

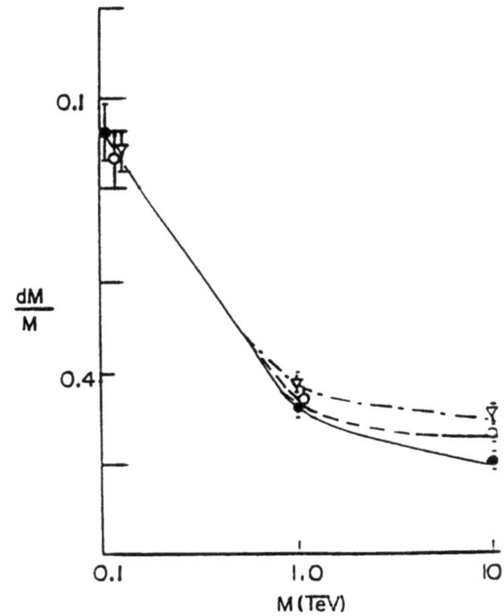

FIGURE 13

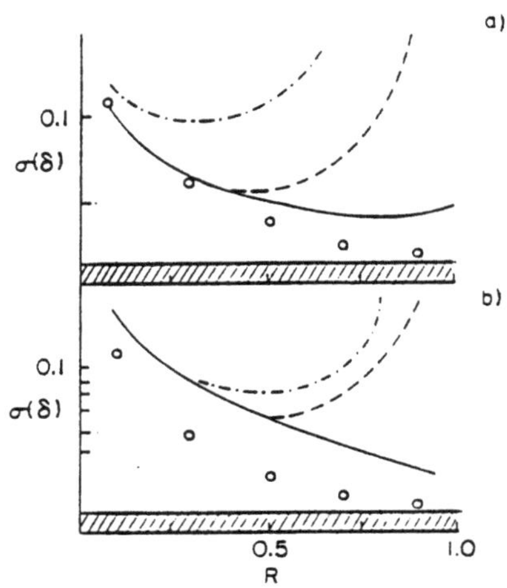

FIGURE 14

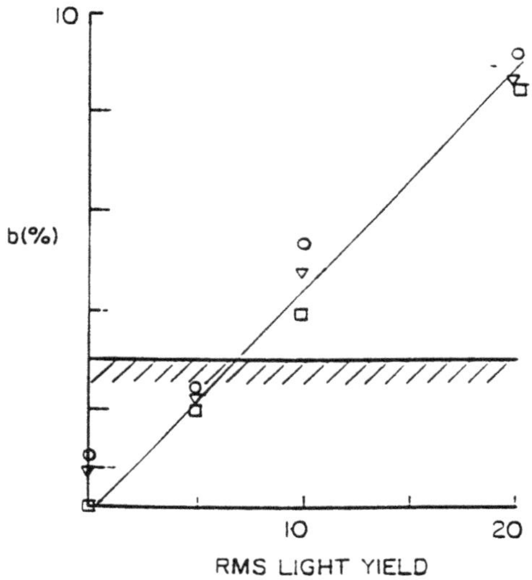

FIGURE 15

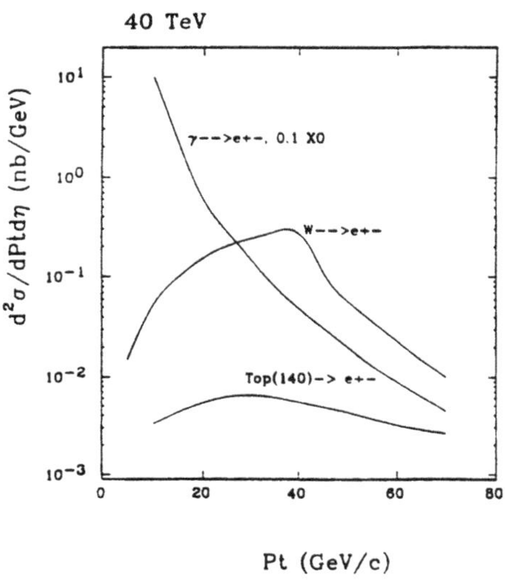

FIGURE 16

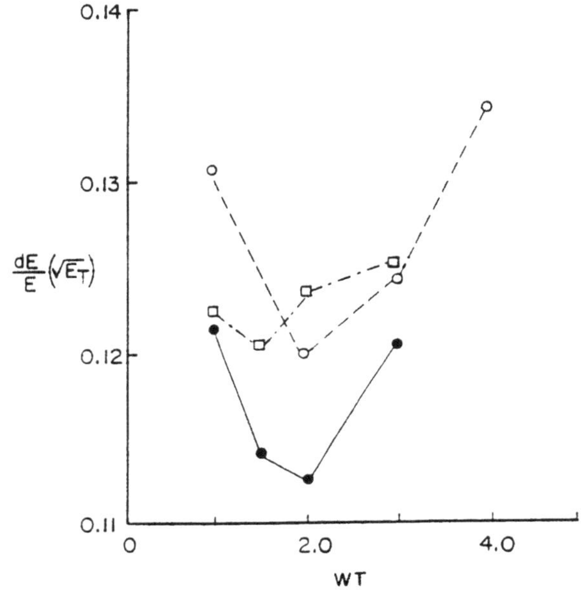

FIGURE 17

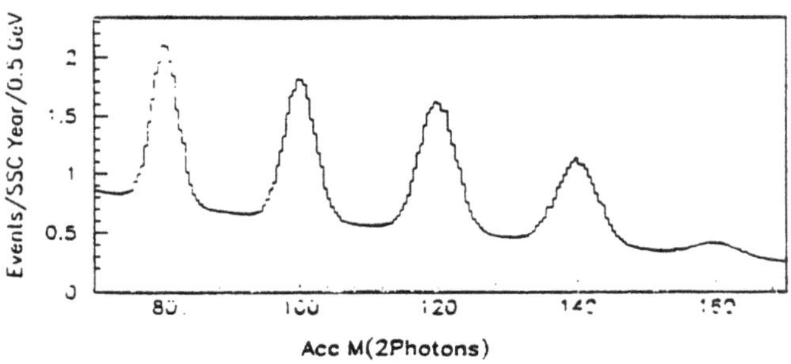

FIGURE 18

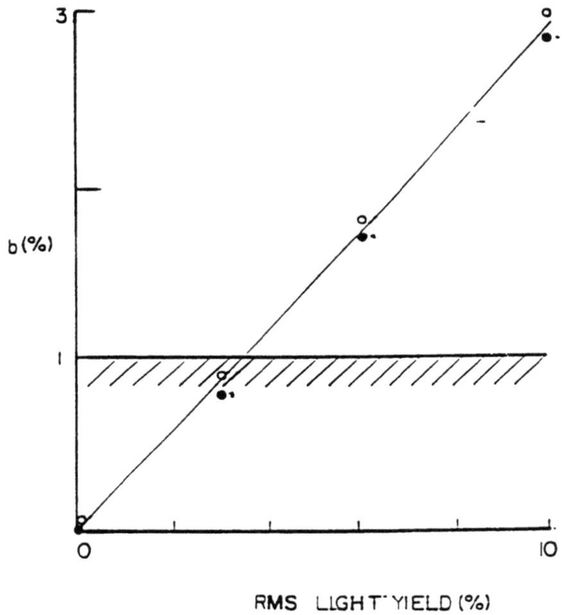

FIGURE 19

ENGINEERING ISSUES OF LARGE DETECTORS FOR THE SSC AND THE LHC

J. Pohlen
Science Systems, Martin Marietta Astronautics, Denver, CO 80201
and
T. Adams, G. Guenterberg, K. Killian
Science Systems, Martin Marietta Astronautics, Denver, CO 80201

ABSTRACT

In the design and fabrication of complex electro-mechanical systems, large or small, it is important to implement early a requirements definition and allocation process. Top level system performance requirements remain the controlling parameters while derived requirements are varied in the continuous search for an allocation that minimizes cost from design through system commissioning. Examples are given from the development of SSC detector concepts, and the tools used to perform this process are discussed.

1. Introduction

There is general recognition that engineering issues will drive detector construction at the next generation of hadron colliders. A system level engineering design approach seeks to minimize cost and maximize performance by carefully defining the requirements that each subsystem must meet and then allowing the subsystem designer to flow those requirements down to the component level. Thus the engineering issues and their solutions can be traced through a project by the requirements, tests, and analyses they generated.

As an example of the engineering design of a large detector, we will use the EMPACT configuration[1,2] proposed to the SSC. Although this proposal was not accepted for construction, many features and elements of approach are being used by the Gamma's, Electrons and Muons (GEM) collaboration. The two hermetic calorimeters proposed for EMPACT –Liquid Argon (LAC) and Spaghetti (SPACAL)– are good examples of challenges in detector design and engineering.

Engineering design for the EMPACT detector involved the construction, by a team of physicists and engineers, of three-dimensional (3-D) models of the detector systems. The models, arranged as hierarchies of components and structures, were critical for communication of the conceptual design. These models were refined and used directly to perform the required mechanical, thermal, and magnetic design analyses, particle tracking and physics performance evaluations, and interference checks and assembly studies.

2. Requirements

For a detector, the driving considerations are the overall physics signatures and goals. The physics and the detection philosophy set the top level performance requirements for individual detector components. The subsystem design then seeks to meet the top level requirements while identifying component requirements and inconsistencies or conflicts in the system level requirements.

The physics goals and the associated detector performance requirements[2] for the EMPACT calorimetry are summarized in Table 1. The physics goals were input to the physical requirements for the EMPACT LAC^3 option summarized in the following sections. Note that these requirements include both detector performance requirements and engineering issues such as safety, transportability, construction, and maintenance.

Table 1. EMPACT/TEXAS Calorimetry Specifications

Attribute	Physics Signatures/Goals	Performance Requirements
Survivability	$10^{34} cm^{-2} sec^{-1}$ /year	
EM		< 10 Mrad
HAD		< 1 Mrad
Forward		< 3 Grad
Dynamic Range		
EM	Low - 50 MeV Shower Tails High- 5 TeV Z'	10^5
HAD	Low - 3.5 GeV μ High- 15 TeV Jets	10^4
Energy Resolution		
EM	$H \rightarrow \gamma\gamma$	$\frac{7.5\%}{\sqrt{E}} \oplus 0.5\%$
	$\Gamma_{Z'}$	$\frac{15\%}{\sqrt{E}} \oplus 1\%$
HAD	$Z \rightarrow jj$	$\frac{50\%}{\sqrt{E}} \oplus 2\%$
	Compositeness	$\frac{50\%}{\sqrt{E}} \oplus 2\%$
Time Resolution		
EM	e, γ, μ	< 16 ns
HAD	jet, μ	< 16 ns
Noise EM	e/γ Isolation	$\Sigma E_T < 5$ GeV in $\Delta R = 0.2$
Depth		
EM	Resolution	25-30 Xo
HAD	Punchthrough	11-15 λ
Hermeticity	$H \rightarrow ll\nu\nu$	$\Delta p_T <$ υ Bkgd $\eta \rightarrow 5.5$
Segmentation		
EM	$\pi^{\pm}/\gamma$ Overlap	$\Delta x \sim 1$ mm, $\Delta\eta = \Delta\varphi \leq 0.05$
HAD	$Z \rightarrow jj$	$\Delta\eta = \Delta\varphi \leq 0.1$

2.1 *EMPACT Calorimetry Requirements*

A) **Performance**—Maximize the calorimeter depth and hermeticity for pseudorapidities $|\eta| < 3$. The calorimetry depth (T), in absorption lengths (λ), shall be defined by:

$T = 11 \quad [0 < \eta < 1.5]$

$T = (4/3)\eta + 10 \quad [1.5 < \eta < 3.0]$

The EM depth shall be ~25 radiation lengths with a minimum thickness of ~15 radiation lengths. Segmentation in η and φ shall be 0.025 x 0.025 for the EM and 0.05 x 0.05 for the Hadronic, with a minimum cell size of 2 x 2 cm.

B) **Safety**—Pressure vessels must be designed to meet the intent of ASME Division II.

C) **Accessibility**—Provide maintenance access to the central tracking system without major disassembly.

D) **Configuration**—The calorimeter must fit within the region specified in the EMPACT baseline EOI design, i.e., inner radius of 1.0 m, outer radius of 3.50 m, and a total length of 14.0 m. The support structure for the calorimeter must not penetrate or pass loads through the central or end toroids. The volume reserved for central tracking is 1 m radius x 5 m minimum length with a room temperature and pressure operating environment. Assume a central tracking drift chamber weight of 2.0 metric tons to be supported by the barrel calorimeter. (See Figure 1)

E) **Services**—Pass services and cabling out to the calorimeter support regions. Provide a total of one square meter of exit space for tracking signal cables and utilities.

F) **Assembly**—The structure must provide a simple and reliable overall structure assembly, module installation, maintenance, and signal cable routing.

G) **Buildability and Transportability**—Individual calorimeter modules must be self-supporting and their masses must not exceed 10 metric tons.

2.2 *EMPACT LAC Derived Requirements*

From these requirements one can then derive most of the detailed (or sub-system) requirements for the calorimeter options. Some of these requirements are common between the SPACAL[4] and LAC options and some are design choices made during the preliminary design. Some of the derived requirements for the liquid argon option are shown below:

1. The toroid, muon chambers, and LAC radial layout is depicted in Figure 1.
2. Access to the central tracker will be ensured by providing separate argon and vacuum vessels for the endcaps and barrel. The endcaps shall be mobile as required to allow access to and/or removal of the central tracking assembly and electronics. Assume one significant maintenance event per year requiring access to the central tracking assembly. During maintenance the barrel calorimeter shall remain self sufficient and cold independent of the end calorimeters.
3. Argon and vacuum vessels must be designed to the intent of ASME code Division II. The cryosystem shall be designed so that no two independent faults will result in a catastrophic failure. The construction materials to be considered are:

Calorimeter Vessels:	Aluminum (5000 series)
Vacuum vessels:	Aluminum (5000 series),
Support Structure:	Stainless steel (300 series) Aluminum (5000 series) Titanium (TBD series) Composites

4. Pressure requirements.
   The argon vessels are to be designed to 55 psid internal allocated as follows:
   15 psid due to vacuum vessel
   25 psid head
   10 psid operating over-pressure
   5 psid relief valve activation pressure
   A 15 psid internal vacuum (positive external pressure) occurs prior to filling, during purging/decontamination .
   Vacuum vessel design pressures are 12 psid internal resulting from a potential leak of the argon vessel and the 15 psid external pressure occurring during normal operation.
5. The endcap design shall allow for the routing of central barrel power/instrumentation cabling and cryogenic piping. The central barrel calorimeter endplates provide equally spaced instrumentation cabling feed-through ports as required, 6 high voltage feed-throughs, and 6 cryogenic line feed-throughs.
6. The LAC electronics constitute a 10 to 50 milliwatt/channel thermal load.
7. The calorimeter shall be symmetric wherever possible.

8. The barrel calorimeter centerline (beam line) maximum deflections shall not exceed 0.635 cm.
9. Assume average calorimetry densities of 7.91 gm/cc for the EM and 10.67 g/cc for the hadronic elements, with corresponding radiation lengths of 0.829 and 0.591 cm and absorption lengths of 28.5 and 17.6 cm.
10. Longitudinal segmentation for the EM shall be 3 layers with two sets of 4 layers in the hadronic elements for redundancy.

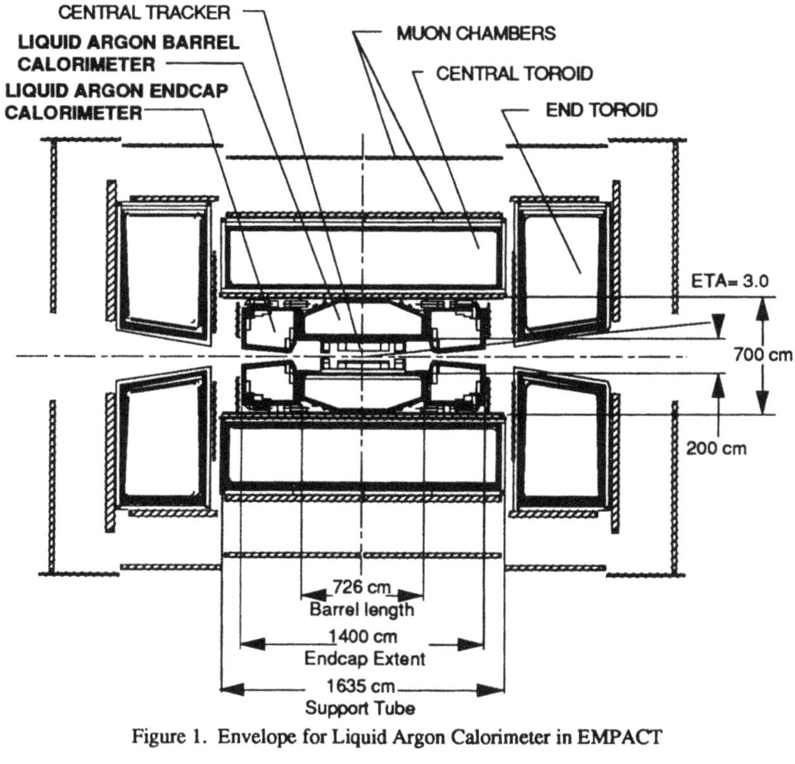

Figure 1. Envelope for Liquid Argon Calorimeter in EMPACT

2.3 *Configuration Baseline Features*

The features of the baseline LAC, the initial configuration that meets requirements, are summarized below:

a) Independent Barrel and Endcap Calorimeters—The access requirement for the tracking system mandated separate barrel and endcap calorimeters. Totally independent endcaps with no shared pressure or vacuum vessels were judged to allow the ease and range of movement necessary for major access to the tracking module.

b) Basic Calorimeter Configuration Study—To minimize the effects of the vacuum walls on the hermeticity, three basic configurations shown in Figure 2 were modeled and investigated. One model was constructed with a plug calorimeter, a second model with a short barrel and the third model with a long barrel. The plug calorimeter had significant coverage loss at the forward corner of the plug. The short barreled calorimeter required a large diameter endcap to provide sufficient active calorimetery at the crack between the calorimeters. The 350 cm radius for both the barrel and the endcap left no space for utilities and supports. The longer barrel concept was heavier, but permitted utilities and support access for both the barrel and endcap. The study indicated that the full fourteen meters allocated was required for a calorimeter which both provided good coverage and left space for utilities and supports.

c) Shell and Washer Assemblies With Self-Supporting Modules—A Stanford Large Detector (SLD) – inspired structure of shells and washers which acts as both a pressure vessel and support structure for calorimeter modules was selected as the initial concept. This configuration was viewed as having potential for less inactive material in front of the modules or between the modules and vessel heads.

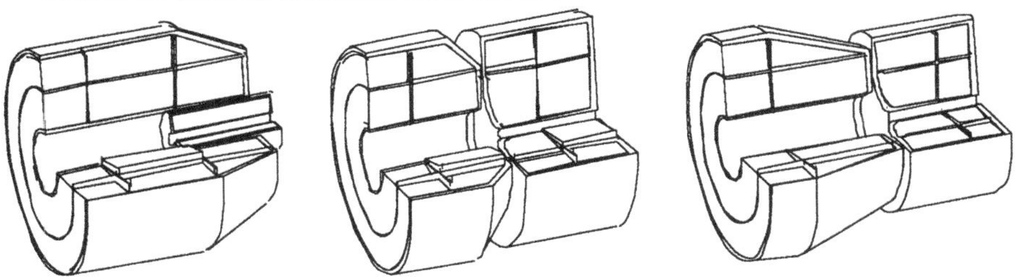

Figure 2. Calorimeter Configuration Study

d) Self-Supporting Calorimeter Modules—No specific internal structure was assumed for the calorimeter modules. The individual calorimeter modules are self-supporting, with an aluminum strong-back for handling. Once installed, the modules are supported uniformly by the washers and heads of the argon vessels.

e) Flat Heads on Barrel Vessels—To maximize hermeticity in the critical regions between the barrel and endcaps, the calorimeters must be as close together as possible. To this end, we selected flat heads for all initial barrel vessel designs. The EM in the end bay of the barrel calorimeter was terminated at a projective angle to provide a minimum calorimeter thickness of 15 cm.

f) Elliptical Heads on Endcap Vessels—Endcap heads have more area than the barrel heads thus requiring thicker walls to support the pressure loads. Analysis of several head shapes showed that a flat head with an elliptical taper near the beamline minimized the forward wall thickness. The heads of both the vacuum jacket and the argon vessel closest to the interaction point were baselined as elliptical.

g) Aluminum Vessels—We employ aluminum (5083) vessels for the baseline design, with consideration given to the use of steel or titanium in certain applications.

3. Design, Modeling and Analysis Tools

The integrated design, modeling, and analysis system we use (Fig. 3) is built around the SDRC I-DEAS[5] packages Geomod, a sophisticated 3-D modeler and database, and Supertab, a flexible analysis interface. As shown in Figure 1, a variety of structural [5,6,7,8], thermal,[9,10,11] fluid flow,[9,12] and magnetic[8,13,14] analysis codes are attached to Supertab. Most major commercial computer aided design (CAD) systems can be attached to Geomod; at present we use the CIMLINC[15] and Geodraw[5] packages.

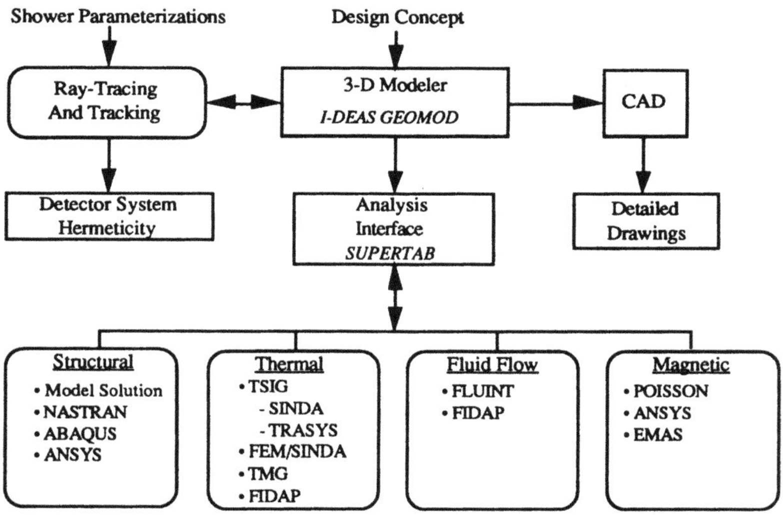

Figure 3. The Integrated Design Analysis and Modeling System.

We have also built, and connected to Geomod, a ray tracing package that tracks particles through the 3-D database used for structural, thermal, and magnetic design and analysis. This package consists of a tracing module and a post-processor. The post-processor can provide useful geometric data based on selected attributes (such as material type, depth of active material, etc). Further the post-processor provides a basic measure of the detector performance that allows us to characterize the subsytems designs.

The integrated set of tools provided a common basis for the design team to work from. All of the critical design evaluations were done in parallel from a common design database. Specific design issues were investigated, the overall design modified and the new design created within the span of a week or two. This design process evaluated a large number of options, settling on the configuration outlined in Table 2.

Table 2. EMPACT/TEXAS Specifications

Detector Elements	Detector Type	Inner Radius	Outer Radius	Length	Z min	Z max	Number of Channels	Weight
		meters	meters	meters	meters	meters	Kilochannels	Tonnes
Central Tracking								
Outer Pads	Silicon or SciFi	-	~1	6.20	-	3.1	15.0	
PreRadiator	SciFi	0.92	0.94	6.20	-	3.1	476.0	
Straw Tubes	TRD	0.35	0.92	6.00	-	3.0	290.0	
Inner Pads	Silicon or SciFi	~0.35	-	6.00	-	3.0	15.0	1
Calorimeters	LAC							
Barrel EM + Had	"	1.00	4.00	7.28	-	3.64	16.0	1306.4
Endcap EM + Had	"	~.25	~2.6	2 x 4.06	3.63	8.29	2 x 5.25 10.5	2 x 536 1072.4
Forward	"	0.07 @ Z1	1.9 @ Z2	3.10	15	18.10	2 x 2.00 4.0	2 x 225 450
	SPACAL							
Barrel EM + Had	"	1.00	4.00	7.28	-	3.64	60.8	1865
Endcap EM + Had	"	~.25	~2.6	2 x 4.06	3.63	8.29	2 x 28.6 57.2	2 x 605 1210
Forward	"	0.07 @ Z1	1.9 @ Z2	3.10	15	18.10	2 x 1 4	2 x 225 450
Toroids	Super						monitors	
Central	conducting	5.00	8.30	16.60	-	8.30		797
End	Air Core	0.82	9.30	2 x 5.1	8.7	13.80		2 x 385 770
Muon Detectors	Aluminum							
Central Barrel	extrusion							
A1 (16)	streamer	4.10	4.21	6.00	0	6.00	18.2	
A2 (16)	chambers	4.66	4.77	7.70	0	7.70	21.0	16 x 4 64
B (16)		8.35	8.86	7.95	0	7.95	78.4	
C (16)		11.30	11.61	8.25	0	8.25	53.8	16 x 19 304
Cap Barrel								
D (16)		9.30	9.60	4.94	0	4.94	63.6	
E (16)		12.52	12.67	5.63	0	5.63	43.2	
F (16)		12.94	13.09	4.37	14.02	18.39	26.8	16 x 16 256
Plug								
G (2)		1.18	3.10	0.15	7.17	7.32	9.4	
H (2)		1.42	6.23	0.15	8.52	8.67	16.5	2 x 14 28
Cap								
J (2)		1.83	10.77	0.40	14.35	14.75	35.6	
K (2)		2.65	12.73	0.20	17.85	18.05	43.5	2 x 71.5 143

4. Material Selection

Materials considered as alternatives to 5083 aluminum in the vessels were 304L stainless steel and Ti6Al4V titanium. Table 3 lists selected properties of these materials and the shell or head thickness (t), relative to aluminum, required for various loading conditions. The functional dependence of each loading condition on the working stress (S), and modulus of elasticity (E) is also shown. The relative costs for raw materials and fabrication of machined, welded, and field-assembled plates, assuming 50% of the vessels weight is in shells and 50% in heads, are shown at the bottom of Table 3. The governing factor for the cylindrical shell thickness is buckling stability. Titanium and steel are inferior in this application to aluminum when absorption and radiation lengths are

Table 3 Cryostat Material Performance, Referenced to Aluminum

Feature	Cryostat Material Performance Ratio		
	5083 Aluminum	304L Steel	Ti-6Al4V Titanium
Shells			
t/t_{Al} (Stress: $S^{1.0}$)	1	0.53	0.18
$(t/t_{Al})(\lambda_{Al}/\lambda)$	1	1.25	0.26
$(t/t_{Al})(X_{oAl}/X_o)$	1	2.69	0.46
t/t_{Al} (Stability: $E^{1/3}$)*	1	0.71	0.86
$(t/t_{Al}) \times (\lambda_{Al}/\lambda)$	1	1.66	1.24
$(t/t_{Al}) \times (X_{oAl}/X_o)$	1	3.58	2.16
Heads and Washers			
t/t_{Al} (Bending: $S^{1/2}$)*	1	0.73	0.43
$(t/t_{Al})(\lambda_{Al}/\lambda)$	1	1.71	0.61
$(t/t_{Al})(X_{oAl}/X_o)$	1	3.69	1.07
t/t_{Al} (Stability: $E^{1/3}$)	1	0.60	0.80
$(t/t_{Al})(\lambda_{Al}/\lambda)$	1	1.40	1.15
$(t/t_{Al})(X_{oAl}/X_o)$	1	3.02	2.01
Material Properties			
ρ (g/cm^3)	2.70	7.87	4.54
τ (cm)	39.40	16.80	27.50
X_o (cm)	8.90	1.76	3.56
S (10^3 psi)	10.0	18.75	55.0
E (10^6 psi)	10.0	28.0	15.5
Relative Costs			
Raw Material	1.00	1.72	9.65
Fabricated	1.00	1.06	3.42

* Controlling Feature

considered. The governing factor for head construction is bending, while for washers, stability is most important. For heads, titanium offers a significant advantage in terms of decreased absorption length. Steel is, again, inferior to aluminum.

It is clear that, in general, aluminum is preferred to steel. The use of titanium in specific applications such as vacuum vessels, where the welding and cutting difficulties might be solvable, should be considered if material of minimum absorption is desired.

5. Physics Performance

It is critical to be able to estimate the physics performance of the detector elements during engineering design and analysis. These indicators provide a consistent method of evaluating candidate designs for compliance with the physics requirements. Second, if the physics performance indicators are properly chosen, they provide a method of modifying and optimizing the response during the very flexible, early design stage and reduce the chance that detailed simulations will uncover significant weaknesses in the detector.

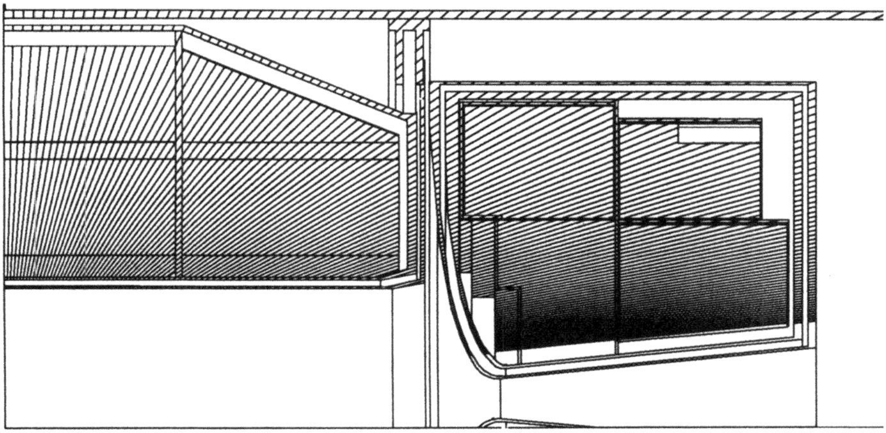

Figure 4 A Profile of the LAC With Rays Traced Radially, Starting From the Collision Point, Over the Interval $0 < \eta < 3.00$, at Intervals of $\Delta\eta = 0.025$

We considered a hermetic calorimeter to be one in which performance (e.g. single particle or jet resolution) approaches everywhere that of the pure calorimeter. To evaluate the performance we use the ray tracing capability of our engineering analysis tools to sample the engineering model. Figure 4 depicts rays traced through the baseline elliptical head LAC geometry at pseudorapidity (η) intervals of 0.025 out to $\eta = 3.00$. This sampling corresponded to the EM segmentation proposed for the EMPACT LAC.

A first indication of hermeticity is the fraction of dead material (i.e. vessels, supports, etc.) between the collision point and the last active calorimeter layer. The cryostats in a LAC are known to represent a fairly large amount of dead material, and in certain regions up to 18% of the LAC was inactive. However, total amount of the dead material is not the best measure of calorimeter performance, as it does not consider a shower's energy deposition profile. Further, an estimate of the lost energy is necessary but not sufficient to provide an adequate performance measure because the lost energy can, in principle, be compensated for during data analysis. Energy resolution was chosen as our performance measure since it is a limiting factor in the reconstruction of the underlying event.

We estimated the calorimeter energy resolution using analytic shower deposition models coupled to an appropriate estimate of the effect of lost energy on resolution as described in Reference 16. The data shown in Figure 5 were generated by applying the hadronic and EM resolution estimates to the baseline LAC. The peak in the hadronic resolution curve occurs in the region encompassing the transition from barrel to endcap and is due largely to the argon and vacuum vessel head walls. It is important to note that the calorimetry group considered this resolution acceptable and encouraged further LAC development.

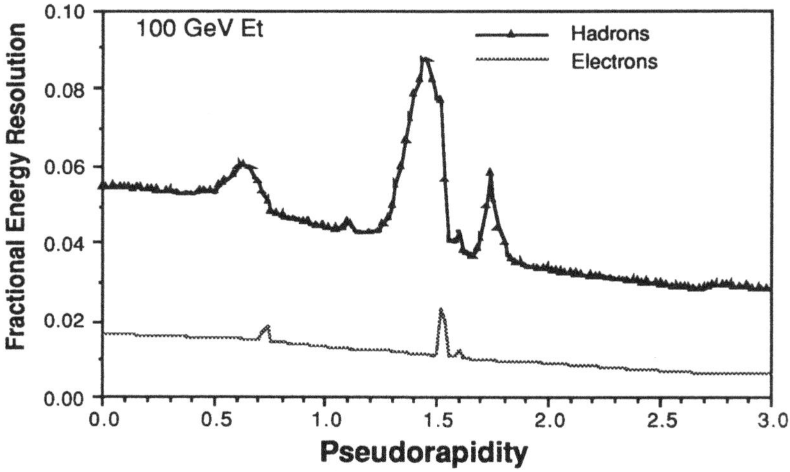

Figure 5. EM and Hadronic Resolution for the Baseline LAC

A similar type of analysis applied to the SPACAL option was used to identify and correct an initial design flaw. Ray traces generated to evaluate the hermiticity of the baseline SPACAL are shown in Figure 6. The final hermeticity is excellent, although the initial design exhibited a large spike at $|\eta| \approx 1.50$ due to locally projective barrel structure. Refinement of the corner to eliminate the projectivity resulted in the energy resolution shown in Figure 7. The perturbation in resolution at $|\eta| \approx 1.50$ is due to the now non-projective structure, and the signal cables from the preconverter and central tracker.

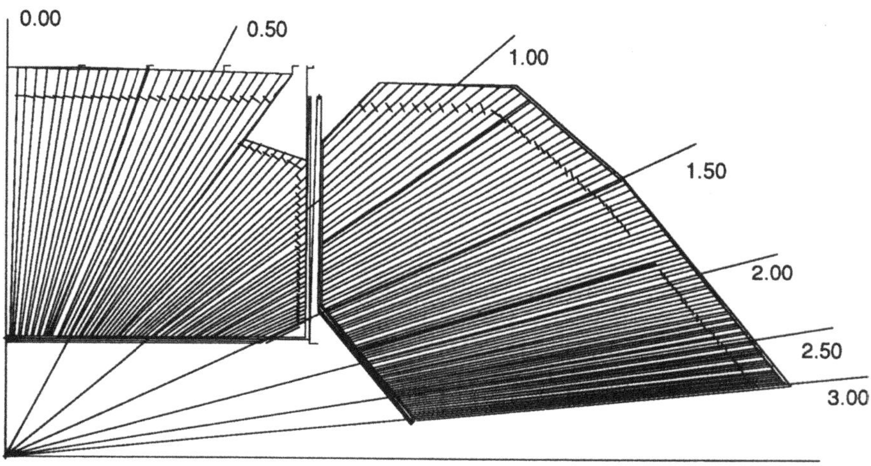

Figure 6. A Profile of the SPACAL With Rays Traced Radially

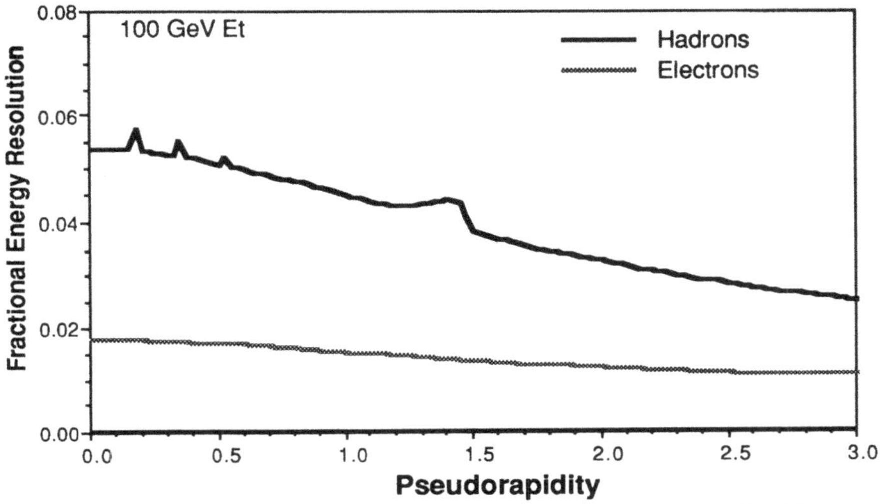

Figure 7. Fractional Energy Resolution of the Revised Spaghetti Calorimeter for 100 GeV Transverse Energy Pions and Electrons.

6. Resource Allocation Modeling

A method for deriving some design requirements has been developed. It is a systematic process using flexible computer tools for efficiently allocating and scheduling program resources (equipment, facilities, and personnel) to achieve program objectives. This process levies requirements on the working hardware and on all the supporting hardware to minimize total cost. A simplified flow diagram depicts the process in Figure 8.

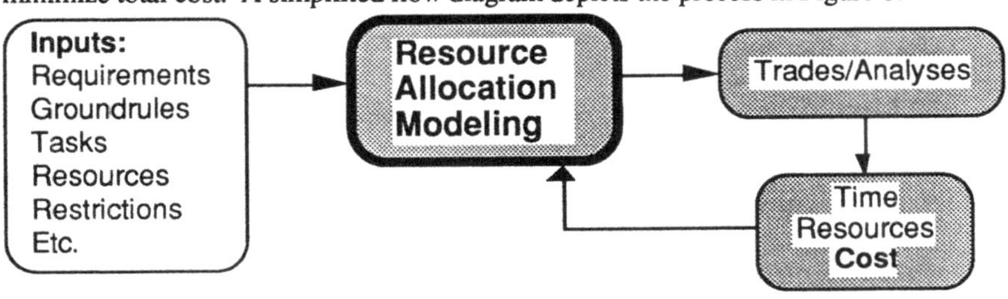

Figure 8. Resource Allocation Modeling Flow.

Table 4 lists some factors that influence the resource allocation study results. This tool can also be used to formulate a recovery strategy if an item on the critical path falls behind schedule.

Table 4. Factors that Influence Resource Allocation

• Facility Limitation	• Schedules
• Work Shift Restriction	• Milestone Schedules
• Resource Restrictions	• Subassembly Availability
• Number of Personnel	• Pass-Fail Criteria
• Learning Curve	• Total Resource Loading
• Work Time Availability	• Resource Operational Impacts
• Rework Percentage	- Availability Percentage
• Rework Time	• Safety Requirements
• Holidays	• Weather

7. Summary

Many factors must be considered to arrive at an "optimum" design for a major detector subsystem such as a calorimeter. Although, meeting the physics performance requirements must be foremost in this process, good engineering design techniques and requirements allocation can speed the design and reduce the cost of large detectors. A technical staff of physicists and engineers working with state–of–the–art engineering tools can evaluate options and perform the trade studies which grade alternative approaches to the top level goals while minimizing costs. These trade studies weigh design options for their performance, assembly and test requirements, support equipment, and personnel requirements early enough in the design cycle to minimize the risks involved with large projects. Further a complete engineering analysis will not only define the cost of the detector itself, but the costs involved with the supporting equipment such as transportation and handling gear, above ground storage and assembly space, and teams of people that need to be trained to assemble the components of detector subsystems. This engineering process was critical in helping a rather small EMPACT collaboration meet the EOI and LOI deadlines with a detector that met the collaboration's physics goals.

7. References

[1] M. Marx,*"Expression of Interest/Electrons Muons Partons with Air Core Toroids (EMPACT)"*, SSCL-SR-1161 (1990)

[2] EMPACT,*"The EMPACT/TEXAS Letter of Intent for the Superconducting Super Collider"*, SSCL-SR-1155 (1990)

[3] T. Adams, et al., *"EMPACT Technical Note 218: Status of EMPACT Liquid Argon Calorimeter (LAC) Engineering Design"*, . (1990).

[4] T. Adams, et al., *"EMPACT Technical Note 219: Status of EMPACT SPACAL Calorimeter Engineering Design"*, . (1990).

[5] I-DEAS, Structural Dynamics Research Corporation, Milford, OH (1987).

[6] MSC NASTRAN, MacNeal-Schwendler Corp., Los Angeles, CA (1987).

[7] ABAQUS, Hibbit, Karlsson and Sorenson Co., Providence, RI

[8] ANSYS , Swanson Analysis Systems Inc., Houston, PA (1987).

[9] TSIG/TRASYS/SINDA/FLUINT, Martin Marietta Corp., Denver, CO (1986).

[10] FEM/SINDA, Martin Marietta Corp., Orlando, FL (1988).
[11] TMG, Structural Dynamics Research Corp., Milford, OH (1989).
[12] FIDAP, Fluid Dynamics International, Evanston, IL (1988).
[13] POISSON, Los Alamos National Laboratory, AT-6, Los Alamos, NM (1987).
[14] MAGNETIC/MAGNUM, Mac Neal-Schwendler Corp., Milwaukee, WI (1988).
[15] CIMLINC, CIMLINC Inc., Itasca, IL (1980).
[16] T. Adams, et al., *"An engineering design study of a liquid argon calorimeter for SSC"*, Nuclear Instruments and Methods, **A306** (1991) 128-138.

Simulation of showers and detectors

René Brun Federico Carminati Isabel Josa
Kati Lassila

European Laboratory for Particle Physics
CERN Geneva, Switzerland

Abstract

We review the status of the various simulation programs currently in use in HEP. Comparisons between different programs and between programs and experimental data are presented for electromagnetic and hadronic showers. Our current strategy for the development of GEANT in view of the new experiments is also discussed.

1 Introduction

Simulation of particle detectors is assuming a completely new role in High Energy Physics. In experiments before the LEP era, simulation was mainly used to understand the behaviour of an existing detector. To perform this task, the features of the detector response were derived from data coming either from test-beam or from the experiment itself, and then fed back into the simulation program in the form of parametrizations. The simulation program was a bare framework handling the succession of the events and some service task like I/O and histogramming.

The availability of accurate simulation programs for electromagnetic showers like EGS [1] has allowed in many cases to replace the parametrization based on real data of the electromagnetic component of showers, by full shower simulation. Parametrizations based on the predictions of simulation (for instance the boostrap method [33]) have also been developed. These have the advantage, over the use of real data, that results can be obtained at any energy and for any kinematic configuration, with virtually infinite precision and no systematic errors, at the cost of enough CPU time.

The full simulation of hadronic together with electromagnetic showers is complicated by our poor understanding of hadronic interactions in the non-perturbative region, and by the intrinsic complexity of the processes to be simulated which involve intranuclear cascades and many-body final states. One of the most successful attempts to provide a full shower simulation program including both hadronic and electromagnetic (EGS) showers has been the GHEISHA [3] program, developed in the context of the MARK-J experiment at DESY. The latest version of the hadronic shower simulator of this program is now interfaced with GEANT and it has provided most of the hadronic shower simulation for LEP.

When the LEP detectors were initially designed, the need for a realistic and precise prediction of the behaviour of the detectors, independent from the data, became more apparent. The LEP experiments have been built to do high statistic and high precision physics in an extremely competitive environment. One of the main components of the success of LEP has been the careful tuning of the software chain from data acquisition to analysis. This has required a detailed simulation of the detector response to given input conditions, which could be achieved via a simulation *ab initio*, starting from the final state of the primary interaction and using full shower development. To this effect it has been necessary to built integrated simulation tools similar to those which had been developed at DESY around the GHEISHA program, but adding sophisticated geometrical modellers able to describe precisely complex geometries. Examples of such programs are the GEANT and DELSIM codes. Three out of four LEP experiments are now using GEANT, and future experiments both at LHC and SSC have expressed their plans to use GEANT as their standard simulation tool. The HERA experiments at DESY (H1 and ZEUS) feature very similar, or even greater, requirements. GEANT is also widely used to simulate HERA detectors.

A small number of codes exist today which are similar to GEANT insofar they perform full shower development for both hadron and lepton initiated cascades starting from *first principles*. Usually these programs have also a geometrical modeller associated or at least interfaced with them. The most common approach is combinatorial geometry with the notable exception of the present version of GEANT. Work is now under course to combine the advantages of the current GEANT geometrical description with Combinatorial Geometry algorithms.

Given the complexity of the subject and the number of the different fields to be mastered, the autonomous development of a general purpose simulation code has proved extremely difficult. The general trend in the last decade has been to combine the best codes available to simulate different reaction channels at different energy ranges to obtain a MonteCarlo system capable to respond to the needs of the physicists. In the vast majority of cases EGS (version 4) has been used to provide the simulation of the electromagnetic showers. The only notable exception here is GEANT, where the simulation of the electromagnetic processes has been part of the GEANT code since the beginning of the project.

One of the most remarkable examples of this technique is the HERMES code [4]. Presented as a collection of state-of-the-art physics programs for calorimeter simulation, this system emphasizes modularity. The different modules of HERMES perform different and largely independent step of the simulation process. This approach is indeed very interesting and in principle it allows to replace one module with a better code without having to change the structure of the program. Even more interesting is the possibility to test different models for a certain region of the energy spectrum and for a certain class of particles without changing the treatment of the rest of the shower. The HERMES code provides one of the frameworks for the simulation of the ZEUS detector at HERA, the other one being GEANT.

Very similar to HERMES insofar it uses partly the same modules, is the CALOR [11] code. Developed at the Oak Ridge National Laboratory it also uses EGS version 4 for electrons and photons while HETC [12] is used to transport high energy hadron showers, and MORSE [13] simulates the low energy neutron and photon component of hadron showers.

Another code in use in the HEP community is FLUKA [10]. FLUKA makes use of EGS version 4 to develop electromagnetic showers, but it is otherwise the result of a largely autonomous development for the hadronic shower generation code. FLUKA has now been interfaced with GEANT version 3.15 with the exception of the treatment of low energy neutrons.

In Figure 1 we have tried to summarize the situation with respect to the packages cited above with respect to the energy range and the processes simulated. The thickness of the lines reflects the authors' opinion on the level of detail provided by the package based on their present knowledge. This part of the table is clearly debatable and the authors will be happy to aknowledge any major omission from the table.

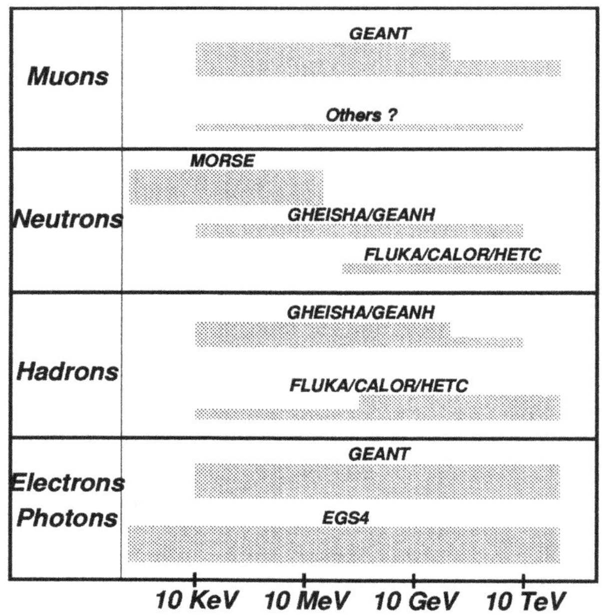

Figure 1: MonteCarlo in High Energy Physics

With the increase in the cost and complexity of detectors, simulation is now assuming a completely new dimension in High Energy Physics. Simulation is now instrumental not only because it provides the necessary test data to tune the reconstruction and analysis software, but also as a design tool to optimise the future detectors on the drawing board. In the new colliders which are now either under construction or in advanced design phase, luminosity will increase by some orders of magnitude and the cross section of the *interesting* events will decrease with respect to the present machines. As a result, the requirements on the precision of the simulation are becoming more and more stringent.

The use of the simulation program as a design tool introduces the need for high-level user interfaces and advanced graphics tools. As simulation is effectively used as a CAD tool to design and optimise a detector, a user interface of the same quality as the one currently offered by CAD programs is almost a requirement. But this is indeed a formidable requirement in itself, as no CAD tool exists today, able to handle an assembly composed by more than one million objects at a time, as is the case for the detectors described by GEANT.

Parametrizations have been developed for the LEP detectors, but the increase in speed of the computers has allowed to make a large use of full shower development and parametrization has played a somewhat secondary role at LEP. The time to perform a full shower simulation for the new LHC/SSC detectors has increased by a factor 1000 at least compared to LEP, and a general and flexible parametrization scheme is now becoming necessary for first order simulations which may possibly be complemented by a more precise full shower simulation for finer studies.

The extension of the energy range is extremely important at both ends of the scale. At the high end it is imposed by the increase in the energy of the primary interaction. At the low end of the scale the so called *pile-up* effect and the introduction of extremely small detectors, makes it necessary to follow the particles longer in time and to simulate the finest details of the energy flux, which is dominated by tails of the shower.

In the light of all these requirements the present paper will make a short review of the efforts going on at CERN to provide a coherent and efficient framework for detector simulation in the 90's.

2 Electromagnetic showers

The electromagnetic shower simulation package EGS [1, 2] has been and still is the reference model. In the past few years, the EGS code has been essentialy extended to cover the very low energy part (in the eV range). A recent survey of the use of EGS shows that 60 per cent of the applications are in the medical physics area. The GEANT electromagnetic package has been largely developed on the EGS model. GEANT does not use the EGS code directly. In the past years, many improvements have been made to the two systems independently. Many comparisons between these two codes and real data have been published in the literature. The agreement is in general pretty good. See for example [6, 7, 8]. The Figure 2 shows one comparison between EGS4 and GEANT. A source of 9 MeV gamma rays is placed at the centre of a cylinder filled with C6F6. Only a small percentage of gammas interact with the sensitive medium. The highest peak in the energy deposition correspond to the case of a positron annihilation where the two photons are leaving the medium without interacting (missing 1.2 MeV). The second peak corresponds to the same situation with one photon only interacting. This setup has been designed in order to make the best possible comparison between the two packages. We should nevertheless mention that there are still a few areas requiring major developments, in particular in the case of fine grain calorimeters.

- For very thin layers, the ESTEPE parameter of EGS has to be tuned. Large variations may be observed in the energy response (up to 20 per cent). This problem was partially solved in GEANT314 with the AUTO mode which makes the best possible automatic computation of the tracking parameters.
- The generation of delta rays and the Landau fluctuations still require further developments in the extreme cases of very thin layers or gases.

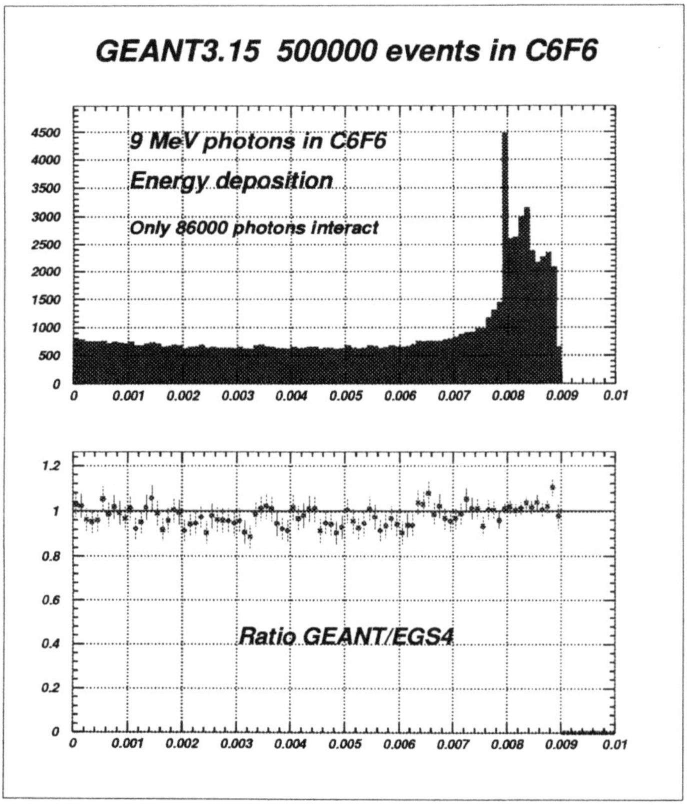

Figure 2: Comparison between GEANT315 and EGS4

3 Hadronic showers

Because of the complexity of the hadronic processes and the lack of calculation methods that can cover all cases, it is desirable to be able to use different approaches when simulating the hadronic physics. This will allow to compare the results, verify or reject the models and get a better understanding of the physics involved. With version 3.15, two interfaces with hadronic shower generators will be supported in GEANT, one with GHEISHA and another with FLUKA. The GEANT team is interested in interfacing also other hadronic shower programs — such are for example DTUNUC and CALOR.

As a module of GEANT we do not need the whole hadronic simulation program, which normally includes also the routines for the geometry definition, the electromagnetic interactions and the tracking. Calculation of the cross sections for different hadronic processes and generation of secondary particles from elastic and inelastic collisions are the only parts used.

3.1 GHEISHA in GEANT

The physics is described in GHEISHA via parametrizations of experimental data. A problem is the lack of experimental data in some energy regions. The collision model is based on the independent fragmentation model of Feynman and Field [14], but it doesn't follow the parton formalism and it assumes that the fragmenting object can be a hadron or even a nucleus.

The first interaction on a proton or on a neutron inside the nucleus is simulated according to the data of the scattering on free nucleons. The development of the hadronic shower is governed by so called "cluster invariant quantities" which remain unchanged during the development of the shower. Examples of such quantities are the energy and the quantum number flow in a certain angular region. Nuclear scattering data are used to parametrize the needed quantities.

The hadronic process is chosen according to the total cross section calculated for the possible channels: elastic interaction, inelastic interaction, neutron capture and nuclear fission.

Before passing to the cascade routines, the energy of the incoming particle is reduced by the amount of energy used in the nuclear evaporation. For the low energy neutrons there is a special tracking routine. Above the GEANT cutoff energy, the particles are passed to the corresponding cascade routines, except π^0, Σ^0 and $\bar{\Sigma}^0$, which do not interact in hadronic processes and are passed forward to GEANT decay routines.

In the cascade routine the particle interacts with a nucleon within the nucleus inelastically or elastically. In the inelastic case it is checked, if it is energetically possible to produces pions or kaons. If it is not, nuclear excitation is assumed and the energy of the incoming particle is reduced. The particle will continue and have a quasielastic interaction, and nucleons and gammas are emitted from the nuclear evaporation. If particle production is possible, the number of secondary pions, protons and neutrons is interpolated from the data tables [15, 16]. Some produced pions or nucleons are replaced by kaon or strange baryons. For the antiparticles, the annihilation channels are taken into account. To take into account the intranuclear cascade particles, the *backward* produced secondaries undergo further interaction with average multiplicity parametrized as a function of the atomic weight of the target material and energy.

For stopped particles all kinetic energy will be deposited. Negative pions and kaons are absorbed by nucleus and antiprotons and antineutrons annihilate. The other particles decay and they are handled by the GEANT decay routine.

3.2 FLUKA in GEANT

The event generator of FLUKA consists of two models. For a projectile with the momentum below 5 GeV/c a resonance production model is used [17, 18]. The hadronic shower caused by a projectile with the momentum above 5 GeV/c is generated according to the multi-chain fragmentation model [19, 20, 21]. The fragmentation of the generated parton shower is done according to Monte Carlo chain decay model developed by J. Ranft and S. Ritter [22, 23, 24].

The elastic or inelastic interaction is chosen in FLUKA by sampling the respective total cross sections, and the particle is passed to the interaction routines. In the current implementation the neutrons below 50 MeV kinetic energy are passed to GHEISHA (GNSLOW routine). The hadronic interactions of all the other particles are handled by FLUKA. Fission and neutron capture are not yet implemented.

If elastic interaction is chosen, the target is hydrogen and the projectile is a nucleon with the kinetic energy below 3.5 GeV, it will undergo a coherent scattering and the new direction is sampled according to the Hetkfa2/Bertini parametrization [25] taking into account the two-body kinematics and proton recoil. In other cases different parametrizations for the differential elastic cross sections are used. In addition, the cross sections for heavy ions can be calculated. After the collision the recoil energy of the target is calculated and it will be deposited.

If hadronic inelastic interaction is chosen and the momentum of the incoming particle is below 5 GeV/c, the particle will be treated according to the low energy model. The number of interactions inside the nucleus is sampled. A neutron or a proton is chosen as a target. If the kinetic energy is below a certain threshold, π^- and K^- will be stopped in the nucleus and the total energy is used for cascade and excitation (nuclear evaporation and gamma production) and antibaryons will annihilate.

Before the primary hadron-nucleon collision is simulated, the energy left for protons and neutrons of the intranuclear cascade and the excitation energy are calculated. Some of the collisions are peripheral. In particle production low energy model uses known cross sections to construct resonances which decay through known decay channels. All the cross sections that contribute more than 2% to the total cross section and all the decay channels that contribute more than 1% are taken into account.

The energy left after hadron-nucleon collision is used in the intranuclear cascade and excitation. In the cascade, nucleons are emitted till all the energy is used. The remaining energy, which is not enough for nuclear emission, is given to the excitation processes.

The incoming particle with momentum above 5 GeV/c is treated according to the high-energy model. First, the number of intranuclear interactions is sampled for materials heavier than hydrogen. The amount of energy needed for the intranuclear cascade is decided before the primary interaction, and intranuclear interactions are performed and the cascade secondaries are generated. In the high-energy part diffractive events are also possible forming some 20% of the high energy events.

If there is still enough energy left after the intranuclear cascade calculation, the particle has a high energy interaction with the nucleus, according to the multi-chain fragmentation model. A two-chain parton system is formed for each interaction between quarks of the projectile and the target nucleus. The first collision occurs between the valence quarks, the others between quark-antiquark pairs of the parton sea of the projectile and valence quarks of target nucleons. The chains are hadronized accroding to the Monte Carlo chain-decay model. Each chain conserves quantum numbers, energy and momentum.

The excitation energy of the residual nucleus is calculated during the process. This energy is used for nuclear evaporation and gamma production.

3.3 Comparisons between FLUKA and GHEISHA

Results obtained with the hadronic parts of FLUKA and GHEISHA agree fairly well in general; within the limits of statistical errors they are in most cases almost equal [9]. Some tendencies can however be distinguished. The main differences have been found with several test runs. The most obvious ones are the following:

- FLUKA produces considerably larger amount of photons.
- The multiplicity of FLUKA is higher.
- GHEISHA produces more events per incident particle, but with lower energy.
- GHEISHA deposits the energy faster, especially at high energies.

The higher photon production in FLUKA is due to the sophisticated de-excitation gamma production used which produces several photons per event, wherear GHEISHA produces only one or two. The higher multiplicity of FLUKA is partly a consequence of the higher photon production. The average multiplicities and secondary particle types for incoming π^-'s and protons for several energies are shown in table 1. The higher multiplicity of FLUKA is compensated by the fact that GHEISHA produces more events with less energy for incoming particle. The energy deposit and particle flow through a simple uniform block of matter are shown in figures 4 and 3.

The cross sections have been compared with experimental data. FLUKA shows in general better agreement, although total cross sections agree fairly well for both programs in the central energy region. Total cross sections for neutrons on several materials are shown in figure 5.

3.4 DTUNUC

DTUNUC [27] — still in the development phase — has been planned to be interfaced with GEANT. At its present state DTUNUC needs a large external file for each material which contains data for intranuclear cascade formation. This poses some problems that have to be solved before interfacing the program with GEANT. Certainly, for future high energy experiments an option for DTUNUC will be most welcome.

Table 1: The average multiplicities and average percentages of secondary particle types

The particle type	*The percentage of the total multiplicity*			
	GHEISHA		FLUKA	
	π^--beam	p-beam	π^--beam	p-beam
Projectile momentum 10 GeV/c				
Photons	13.56	13.86	31.99	30.86
Neutral hadrons	60.34	59.70	48.96	47.56
Charged hadrons	28.68	29.03	23.18	23.72
Average multiplicity	4.274	4.349	5.838	5.562
Projectile momentum 50 GeV/c				
Photons	14.30	14.05	31.74	31.79
Neutral hadrons	59.49	52.98	47.14	47.35
Charged hadrons	28.79	29.26	23.24	23.03
Average multiplicity	4.594	4.600	5.864	5.772
Projectile momentum 100 GeV/c				
Photons	13.96	13.98	32.30	32.10
Neutral hadrons	59.77	59.50	47.01	46.56
Charged hadrons	28.86	29.10	22.85	23.47
Average multiplicity	4.563	4.576	5.798	6.120

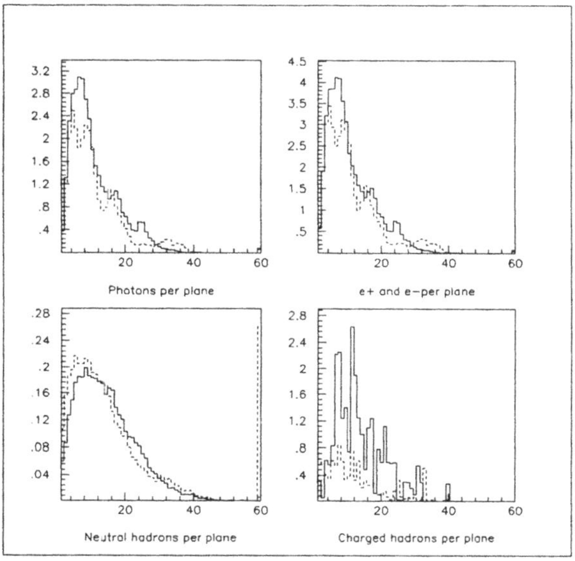

Figure 3: Flow of particles for incoming π^-'s with momentum of 50 GeV/c in copper. The continuous line shows the results of FLUKA, the dashed line those of GHEISHA.

DTUNUC produces the fast hadrons according to Glauber model. The formation of intranuclear cascades in DTUNUC is based on a theoretical model, whereas in GHEISHA and in FLUKA it is done via parametrizations. Fermi motion and Pauli's principle are taken into account when generating low-energy nucleons.

The Dual Parton Model can be extrapolated to very high energies which will give a theoretical base also for the high energy simulation for which not much experimental data is not yet available. There is also a possibility to introduce new effects, such as formation of quark-gluon plasma. Results obtained for nucleus-nucleus collisions are presented [28].

3.5 CALOR

The CALOR package has been developed in Oakridge National Laboratory for calorimeter design and analysis. The hadronic part at its present state consists of HETC code [29] for intranuclear cascades and evaporation, FLUKA87 for high energy interactions and MORSE or MICAP packages [30] for low-energy neutron treatment.

3.6 Low-energy neutrons

One of the main problems in the simulation studies for the LHC detectors will be the neutron pile-up. As inter-bunch time is very short and luminosity very high there will always be a large number of neutrons from previous interactions. Therefore neutron tracking should be done very carefully down to few eV energies. Now, in both cases, with

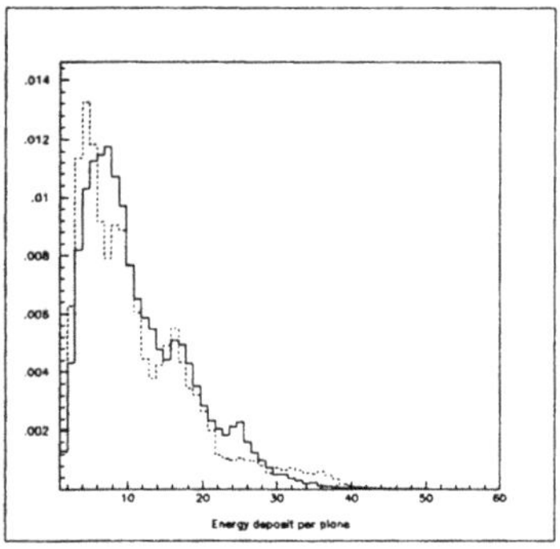

Figure 4: Energy deposit for incoming π^-'s with momentum of 50 GeV/c in copper. The continuous line shows the results of FLUKA, the dashed line those of GHEISHA.

FLUKA and with GHEISHA, low energy neutrons (kinetic energy below 50 MeV) are passed to GHEISHA's low energy routine. There they are tracked down to 100 eV.

To pass the particles from one program to the other is not a consistent solution. One possible solution already prepared for the complete FLUKA is to have an access to the nuclear reactor cross section data. The cross section tables are very large and therefore they would have to be placed as external files and only the data needed for a specific setup would be picked up for each run.

4 Parametrization

4.1 Introduction

Particle and energy showers in calorimeters are simulated by tracking all particles down to a minimum energy considering all possible physical processes the particles may undergo. The secondary particles created in these processes are also tracked in the same way. In the energy range of the coming accelerators the development of showers with increasing number of tracks makes the detailed simulation a prohibitive task in terms of CPU time. However for many purposes a very detailed simulation is not needed and several techniques such as shower libraries, parametrization of showers, particles terminators etc., can be used to speed up the simulation, generally to the cost of lossing some precission.

4.2 Shower libraries – The bootstrap method

The idea is to store a number of fully generated showers in a direct access file. The showers are generated for different particles and several values in energy in a large volume of material, divided into cells of small enough dimensions. In the simulation stage the primary particles are tracked down to a cut-off energy and then replaced by a randomly fetched shower of the library. Each primary shower is then built by a superposition of several showers coming from different files and each "frozen" shower is combined with different

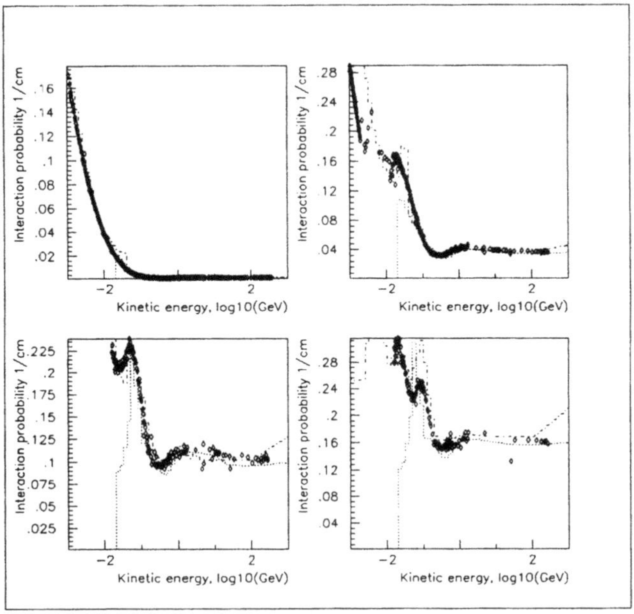

Figure 5: Interaction probabilities for a neutron in a) hydrogen, b) carbon, c) copper and d) uranium. The lozenges show the experimental points, the dashed line shows the parametrization used in FLUKA, the dotted line that of GHEISHA.

partners in different combinations. In this method shower fluctuations and correlations are still taken into account by the detailed tracking above the cut-off.

This method has been successfully used by the L3 experiment both for the electromagnetic (homogeneous media) [31] and hadron calorimeter (sampling media) [32] obtaining an overall speeding factor of about one order of magnitude. This method was also used by the OPAL experiment [33] gaining a factor of 3 in the CPU time but it is now abandonned as the I/O demands were excessive and the shower libraries had to be regenerated for each new upgrade in the showering/tracking in GEANT.

4.3 Parametrization of showers

Parametrization means to replace (totally or partially) the full shower development by a parametrized expression of the spatial energy distribution (longitudinal and transversal).

It is well known [34] that the longitudinal development of electromagnetic shower can be expressed with an expression of the type:

$$\frac{1}{E}\frac{dE}{dz} = \frac{\beta}{\Gamma(\alpha)} \cdot (\beta \cdot z)^{\alpha-1} \exp(-\beta \cdot z)$$

where the depth, z, is expressed in terms of radiation lenghts (χ_0), and E is the incident energy. α and β are dimensionless coefficients that carry the energy dependance of the shape of the shower. The position, in radiation lenghts, of the shower maximum is given by $t_{max} = (\alpha - 1)/\beta$. Both, α and t_{max} have a logaritmic dependance with energy.

For the lateral shower profile several expressions have been proposed:

$$\frac{1}{E}\frac{dE}{dr} = (a_1 \exp(-b_1 r) + a_2 \exp(-b_2 r))r \qquad \frac{1}{E}\frac{dE}{dr} = \frac{2rR_{50}^2}{(r^2 + R_{50}^2)}$$

The hadronic shower profiles are usually described by a superposition of two gamma functions [35], one describing the electromagnetic component and the other the pure hadronic component.

However the previous expressions only describe the average profiles forgetting about shower to shower fluctuations and what is more important the sampling fluctuations in the case of sampling calorimeters. All theses effects were reproduced in the past by fluctuating the parameters in an "ad hoc" way tuned to meet the experimental results.

Probably the most advanced parametrization algorithms now available are the fast simulation packages designed at DESY for HERA. Both experiments, H1 and ZEUS handle in a similar way the fluctuations of the individual shower profiles and the correlation among the parameters although they use different expressions for the transversal profiles. ZEUS in addition has included a simulation of transverse deformation of the showers whereas H1 assumes the showers to be symmetric. A description of the ZEUS algorithm can be found in [36]. GFLASH [37] was developed at HERA for the fast simulation of electromagnetic and hadronic showers in H1 experiment. It generates e.m. and hadronic showers and computes the visible energy fraction in a geometry defined by the user in the GEANT framework.

The spatial distribution of the energy deposited by an e.m. shower is parametrized according to the following expression

$$dE_{dp}(r) = E_{dp} f(z) dz f(r) dr f(\phi) d\phi$$

The longitudinal profile of individual showers are parametrized according to

$$f_i(z) = \frac{x^{\alpha_i - 1}\exp(-x)}{\Gamma(\alpha_i)}; \qquad x = \beta_i z$$

with the shower parameters (α, β) being normally distributed with means μ_α, μ_β, and fluctuations σ_α and σ_β and correlation coefficient ρ. The dependance of μ_α, μ_β, σ_α, σ_β with energy is also calculated. The lateral energy profile is represented by the expression

$$f(r) = \frac{2rR_{50}^2}{(r^2 + R_{50}^2)}$$

with no azimuthal dependance. The mean and variance of the R_{50} parameter is also parametrized as a function of energy and the shower depth.

A sampling calorimeter module is described by a single effective medium and sampling fluctuations are taken into account in the conversion from the deposited energy to the fraction which is visible in the active part of the calorimeter. Lateral energy spots are deposited which visible energy is computed using the measured sampling fractions for minimum ionizing particles and the relative sampling fractions for electrons and hadrons. To simulate the fluctuations, a Poisson-distributed number of energy spots are deposited per longitudinal integration step, according to the radial probability function.

GFLASH tracks the particles with the GEANT routines down to the first inelastic interaction. The particle is then replaced by a pseudo-shower-particle and the tracking of the original track is stopped. This pseudo-shower particle is also tracked with GEANT and an individual shower profile is generated by sorting a correlated pair of parameters (α, β). It is integrated in small steps and for each step, energy spots are deposited according to the fluctuated lateral profile and the sampling fluctuations for the volume the spot is in.

The same technique is extended for the hadronic showers. The longitudinal profile is described by three terms, each of them of the form of the e.m. one. One of them describes the purely hadronic component of the shower and the two other terms represent the π^0 components, the one generated in the first inelastic interaction and the π^0 fraction that is produced in the late development of the shower.

At CERN we are currently working on the implementation of a general purpose parametrization scheme [38]. Our idea is to track the particles in the detailed geometry down to a user defined energy threshold and then substitute them by a shower of particles (Energy-rays) that will be tracked in a simplified description of the detector (SCAN geometry) depositing the remaining energy. It has then the advantage that the energy deposition mechanism is sensitive to the cracks of the detector as the primary particles are still tracked in the real detector. In this scheme, all secondaries are replaced by the same type of particles, the energy-rays (probably generated with different parameters depending on the type of particle to be parametrized), and this allows a high degree of compacticity in the tracking code.

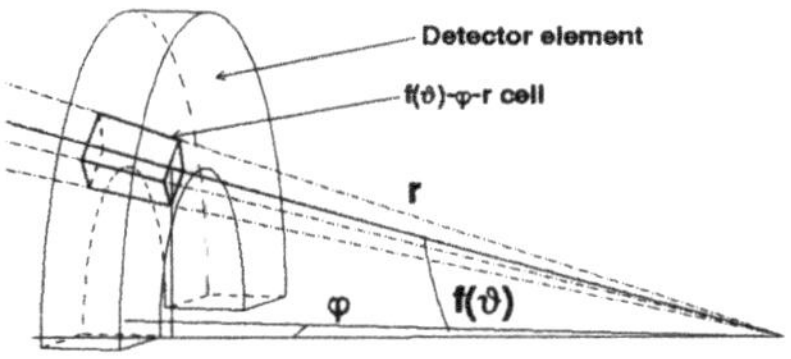

Figure 1: Scan Geometry

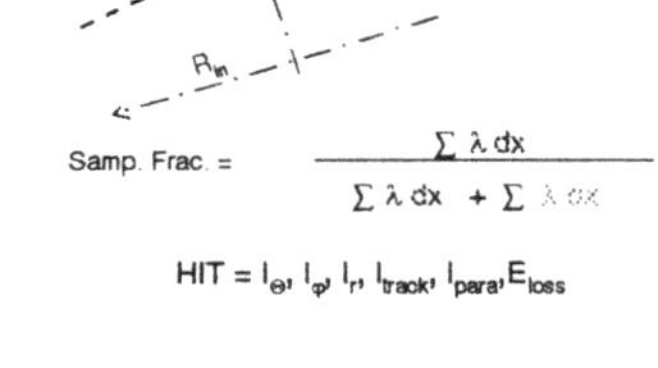

Figure 2: Energy deposition

The SCAN geometry consists in the description of the experimental set-up in terms of

$f(\theta) - \phi - r$ cells. The user defines the granularity in $(f(\theta), \phi)$ and the list of volumes to be SCANned via data cards and the program provides the routines to scan the detector and compute automatically divisions in r every time a boundary of a selected volume is found (Figure 1).

For each of these divisions several quantities such as the total number of radiation and absorption lengths up to the current cell boundary, and the electromagnetic and hadronic sampling fractions are stored in a very compact way in the general GEANT data structure.

At tracking time, if the parametrization flag is turned on and the energy of the particle falls below the threshold specified by the PCUT card, the routine GUPARA is called by GTRACK and the tracking of the particle is abandonned. At this point the user may call his/her own parametrization routines or, in the default case, let GEANT store the track for later parametrization. The energy deposited in the SCAN cells by the Energy-rays is stored in an extension of the actual HITS structure (Figure 2). The new version GEANT3.15 will include the tools to define and create the SCAN geometry, a simple shower generation model (spherical shower) and a dedicated set of routines for a fast tracking of the energy-rays inside the SCAN geometry.

For the next version GEANT3.16 we want to provide a more realistic shower generation based on a possible extension of the GFLASH algorithm. We are also trying a slight different parametrization where the shower profile is expressed according to

$$dE_{dp}(r) = E_{dp} f(\theta) d\theta f(l) dl f(\phi) d\phi$$

and

$$\frac{1}{E}\frac{\partial E}{\partial \theta} = [\exp(p_1 + p_2 \cdot \theta) + \exp(p_3 + p_4 \cdot \theta)] \times \theta^{p_5}$$

and for each θ slice the energy profile is given by

$$\frac{1}{E}\frac{\partial E}{\partial l} = \frac{\beta}{\Gamma(\alpha)} \cdot (\beta l)^{\alpha-1} \exp(-\beta \cdot l)$$

being θ the angle with the energy axis and l the distance from the shower origin along a given θ direction.

5 Graphics

The need of a graphics interface to the simulation program became evident with the simulation of the LEP detectors. When modelling a complex geometry via a procedural language, it is extremely painful and inefficient to debug the code without a graphical representation of the object described.

With the advent of interactive computing and graphics, the power of the visualization tools has been greatly enhanced, but so have been the requirements on the software. Facilities like interactive *zoom* or the possibility to measure a dimension on the screen have become almost mandatory. Some of these facilities had been provided in GEANT since an early stage of its development, but they are really exploited only now, when, with

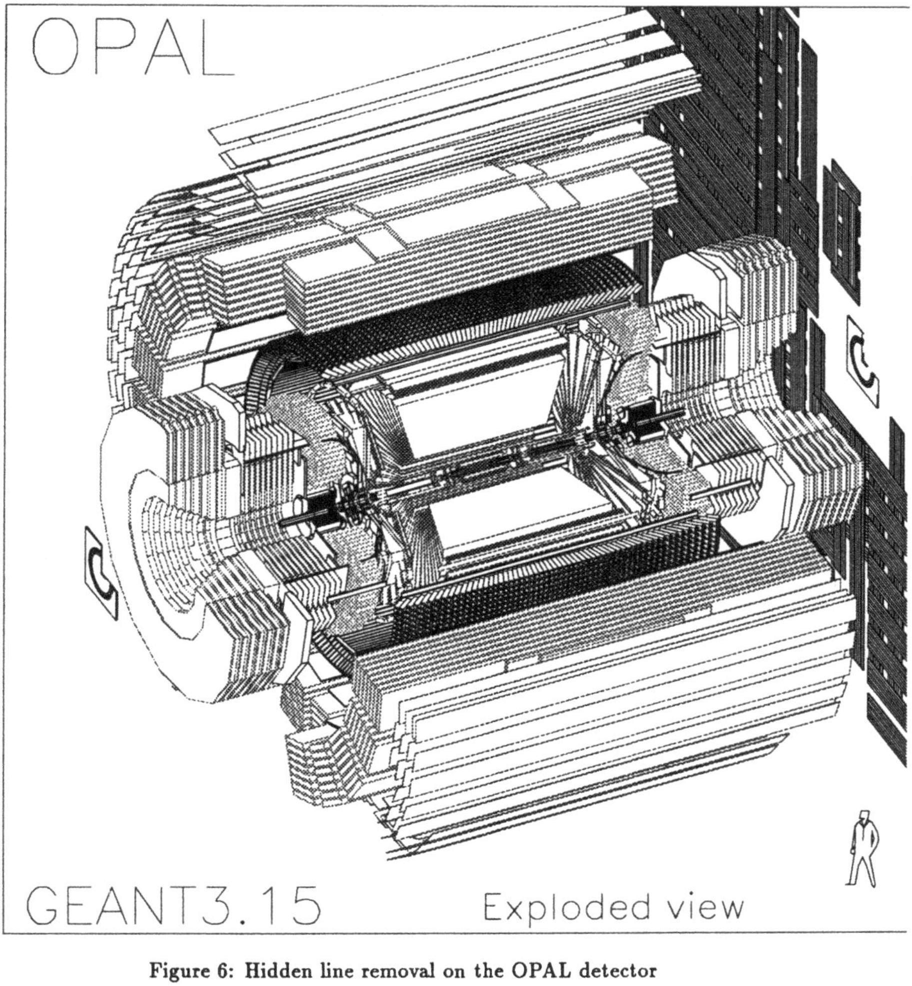

Figure 6: Hidden line removal on the OPAL detector

the advent of the RISC based personal workstations, very fast CPUs, advanced interactive environments and high-quality graphics devices are combined together.

With version 3.15 of GEANT, the interface with the Combinatorial Geometry (CG) package of E.Chernyaev (Serpukhov) has been completely rewritten, with a gain of a factor 30 in the time and 50 in the space needed. This package allows to draw a complex collection of objects limited by plane faces whilst removing the lines of an object which are hidden to the observer either by the object itself or by the neighbouring objects. The system is very powerful as it allows to draw detectors with more than one million objects like L3 on an ordinary personal workstation.

A completely new development of the CG package has allowed the introduction of a surface shading tool, which, combined with the hidden line removal capabilities, allows the realistic representation of detectors.

Already GEANT in version 3.14 the CG package offered the possibility to *cut* a detector with a box to reveal the internal details. The new interface allows to perform the cut with new shapes like spheres, cylinders and cones. Each volume can be cut up to two times with different cutting shapes. To allow a better understanding of the structure of a detector, the above techniques can be combined with the movement of parts of the detector from their original position. This can be done for a single element along any axis with the `SHIFT` tool, or, with the `BOMB` tool, for all the elements, along lines which go from the centre of the detector to the centre of gravity of each element, see Figure 6.

The introduction of these tools has allowed to find small imperfections in almost all the descriptions of large detectors now in use, which had not been detected in years of careful debugging.

A prototype user interface is now available using the tools provided by the MOTIF window manager to interact with the user via panels, sliders, pull-down menus and buttons. Further developments are foreseen in this area including the capability to interactively design and modify an existing detector and the interface with 3-D graphic packages which will allow to use the advanced hardware graphics features when they are available.

References

[1] R.L.Ford and W.R.Nelson: Stanford University report SLAC-0210 (1978)

[2] W.R.Nelson, H.Hirayama and D.W.Rogers: *EGS4 Code System*, SLAC 265 (1985)

[3] H.Fesefeldt, PYTHA 86/05, February 1986

[4] P.Cloth et al.: *HERMES - A Monte Carlo Program System for Beam-Materials Interaction Studies* KFA Julich, Report Jul-2203 (1988)

[5] K. Hanssgen, J. Ranft: *NUCRIN* Comp. Phys. Comm. 39, 53 (1986)

[6] D.R.Ward *The OPAL Monte Carlo program - GOPAL* Proceedings MC91, Amsterdam (1991)

[7] H.S.Chen: *The Detector Simulation of L3* Proceedings MC91, Amsterdam (1991)

[8] A.Givernaud: *Simulation of a warm liquid calorimeter with GEANT314* Proceedings MC91, Amsterdam (1991)

[9] J.Gayler: *Simulation of H1 Calorimeter Test Data with GHEISHA and FLUKA* Proceedings MC91, Amsterdam (1991)

[10] P.A.Aarnio et al.: *FLUKA User's guide* CERN TIS-RP (1990): TIS-RP-190(1987)

[11] T.Handler, et al,: CALOR89. Calorimeter Simulations, Benchmarking and Design Calculations Proc. of Symposium on Detector Research and Development for SSC, Ft. Worth, TX (October 1990)

[12] T.W.Armstrong and K.C.Chandler: *HETC - A High Energy Transport Code* Nucl.Sci. and Eng.,49,110 (1972)

[13] M.B.Emmett and N.M.Greene: *Morse Code System* ORNL/TM-3706 (1973)

[14] R. D. Field and R. P. Feynman: *A Parametrization of the Properties of Quark Jets* Nucl. Phys B136, 1 (1978)

[15] H. Fesefeldt: *GHEISHA The Simulation of Hadronic Showers - Physics and Applications (2.6.12)* PITHA 85/02

[16] H. Fesefeldt, Th. Hamacher and J. Schug: *Tests of Punch-through Simulations* Nucl. Instr. and Method A292, 279 (1990)

[17] K. Haenssgen and J. Ranft: *Hadronic Event Generation for Hadron Cascade Calculations and Detector Simulation I, Inelastic Hadron-Nucleon Collisions at Energies below 5 GeV* KMU-HEP 82-11 (1982), Nucl. Sci. Eng. 88, 537 (1984)

[18] K. Haenssgen, H-J. Moehring and J. Ranft: *Hadronic Event Generation for Hadron Cascade Calculations and Detector Simulation II, Inelastic Hadron-Nucleon Collisions at Energies below 5 GeV* KMU-HEP 83-6 (1983), Nucl. Sci. Eng. 88, 551 (1984)

[19] A. Capella and J. Tran Tranh Van: *A New Parton Model Description of Soft Hadron-Nucleus Collisions* Phys. Lett. 93B, 146 (1980)

[20] A. Capella and J. Tran Tranh Van: *Hadron-Nucleus Interactions and the Leading Particle Effect in a Dual-Parton Model* Z. Phys. C10, 249 (1981)

[21] J. Ranft and S. Ritter: *The Monte Carlo Codes NUCEVT and HADEVT to Simulate Hadron Production in Hadron-Nucleus and Hadron-Hadron Collisions* TIS-RP/IR/83-23, CERN (1983)

[22] S. Ritter and J. Ranft: *Simulation of Quark Jet Fragmentation into Mesons and Baryons on the Basis of a Chain Decay Model* Acta Phys. Pol. B11, 259 (1980)

[23] S. Ritter: *QCD Jets in* e^+-e^-*-Annihilation and the Transition Into Hadrons* Z. Phys. C16, 27 (1982)

[24] S. Ritter: *Monte Carlo Code BAMJET to Simulate the Fragmentation of Quark and Diquark Jets* Comput. Phys. Commun. 31, 393 (1984)

[25] H-J. Moehring, KMU-HEP Internal note 88-01 (1988)

[26] Private communication, Alberto Fasso, CERN-TIS

[27] H. J. Möhring and J. Ranft: *Dual Parton Model with Complete Formation Zone Intranuclear Cascade for the Description of Particle Production in Hadron-Nucleus and Nucleus-Nucleus Interactions* unpublished (1991)

[28] J. Ranft: *Hadron Production in Nucleus-Nucleus Collisions at LHC Energies in the Dual Parton Monte-Carlo Model DTUNUC* CERN 90-10, ECFA 90-133, 1127 (1991)

[29] RSIC Computer Code Collection *HETC Monte Carlo High Energy Nucleon-Mesin Transport Code, User's Manual* CCC-178C

[30] RSIC Computer Code Collection *MORSE-CG General Purpose Monte Carlo Multi-group Neutron and Gamma-Ray Transport Code with Combinatorial Geometry* CCC-203

[31] E.Longo and L.Luminari, Fast electromagnetic shower simulation, NIM A 239 (1985) 506

[32] H.S.Chen, Fast simulation of showers in inhomogeneous media, Computer Physics Communications 57 (1989) 375

[33] D.R.Ward, "Bootstrap" method for simulation of the OPAL electromagnetic endcap in GOPAL, OPAL Internal note, February 1987

[34] E.Longo and I.Sestili, Monte Carlo calculation of photon initiated electromagnetic showers in lead glass, NIM 128 (1975) 283

[35] R.K.Bock et al., Parametrization of the longitudinal development of hadronic showers in sampling calorimeters, NIM 186 (1981) 533

[36] J. del Peso and E.Ros, Fast simulation of electromagnetic showers in the ZEUS calorimeter, DESY 91-005

[37] G.Grindhammer, M.Rudowicz and S.Peters, The fast simulation of electromagnetic and hadronic showers, NIM A 290 (1990) 469

[38] R.Brun and F.Carminati, Detector simulation and software tools, Summary talk at the ECFA–LHC workshop, Aachen 4-9 October 1990

CAN WE PREDICT RADIATION LEVELS IN CALORIMETERS?

A. Ferrari, P.R. Sala
Istituto Nazionale di Fisica Nucleare INFN,
Sez. di Milano, Via Celoria 16, 20133 Milano
and
A. Fassò, G.R. Stevenson
CERN, Geneva, Switzerland

ABSTRACT

A description is given of radiation level simulations for LHC and SSC detectors. The various sources of uncertainty are analyzed and discussed. It is shown that relative comparison can be made at the 10 % level provided suitable simulation tools are used: however absolute values can be trusted only within a factor 2 due to systematic errors deriving mainly from the source term. Results for neutron fluxes and doses for geometries inspired to the EAGLE detector for LHC and to SDC for SSC are presented.

1. Introduction

The construction of the next generation of colliders and of the associated detectors will require large efforts in many fields. The high intensity required by the physical goals will put severe constraints both on machine and detector components. Radiation damage can greatly influence any technical choice and therefore reliable and detailed predictions about radiation levels are mandatory.

In the following sections the simulation tools used for computing the radiation levels expected for the LHC and SSC detectors are briefly analyzed: an indication of their accuracy and reliability is given. The source data used for p-p inelastic interactions at $\sqrt{s} = 16, 40$ TeV and their implications on the final results are discussed in section 3.

The last sections present detailed estimates both of neutron fluxes and doses arising from ionizing particles in realistic detector geometries for LHC and SSC. The influence of many parameters on these quantities and the possibility of minimizing the radiation levels with a careful choice of some of them are also discussed.

2. The FLUKA code

The FLUKA code has been widely used at CERN for shielding and radiation protection calculations. It has recently undergone substantial improvements as described in table 1. All the calculations presented in the following have been performed using the last version (FLUKAN) of the code. Details can be found in refs.[1,2] for FLUKA86, and in refs.[3–6] for the improvements. We discuss only the low energy neutron transport part, since it is the most relevant for the present calculations.

Table 1
Short description of the original FLUKA code and its present version FLUKAN

Topics	FLUKA86	FLUKAN
Hadron- nucleus interactions, 5-10 GeV/c $<$ p $< \approx$ 10TeV , primary interactions	Dual Parton Model (NUCEVT, HADEVT, BAMJET)	Removed known bugs and approximations
Hadron- nucleus interactions, 50 MeV $<$ E $<$ 5GeV , primary interactions	NUCRIN: resonances production (HADRIN) and their decay (DECAY)	Modified versions of HADRIN/NUCRIN, preequilibrium reactions below 200 MeV for p and n
Cascade following primary interactions	Parametrized	Strongly modified: correlation of cascade with primary int., nuclear effects taken into account
Nuclear de-excitation	None	Particle evaporation with a modified version of EVAP-5 followed by γ de-excitation
Stopping $\bar{p}$, $\bar{n}$, π^-	On spot energy deposition or decay	$\bar{p}$, $\bar{n}$ treated by NUCRIN,π^- by preequilibrium
Hadrons below 50 MeV	On spot energy deposition or decay	Protons: nuclear reactions plus dE/dx ;other charged hadrons: dE/dx; neutrons: reactions above 20 MeV, specialized treatment below; other neutrals: transported till decay.
Electromagnetic shower	EGS4	Electron transport completely modified. New treatment of photoelectric effect and of bremsstrahlung (LPM effect included)
Geometry	Combinatorial	New bodies, optimization with strong speed up factors. Transport in magnetic field
Biasing	Leading particle biasing	Leading particle biasing, region importance biasing, russian roulette and splitting for secondaries production. Biased downscattering and non-analogue absorption for low-energy neutrons. All region dependent
Time analysis	None	Complete

2.1 Transport of Low Energy Neutrons ($E_n < 20\ MeV$)

The treatment of neutron interactions in this energy range is based on data sets containing all relevant information: this approach is possible since there are large and accurate compilations of nuclear data for neutrons. This highly specialized part of FLUKA is similar to the transport part of the MORSE code[7]: however so many modifications were required that a set of new routines have been written instead of using the original ones from MORSE. It is sufficient to say that energy and direction of neutrons are selected according to appropriate downscattering matrices taking into account all relevant reaction channels.

The most important items are the cross section data sets. Data compilations must be processed to produce suitable input files for transport codes: generally available processed data sets are usually limited to a few materials and often they do not contain all informations. To overcome these problems a collaboration was started with the ENEA laboratory specialized in nuclear data processing for reactor and fusion applications. This led to the production of a data set especially tuned for detector and machine application for high energy colliders[8]. It contains data for about 40 elements using 72 neutron energy groups: whenever available gamma ray generation data have been included. Partial reaction cross sections and kerma factors are also included. The Doppler broadening for cross section resonances has been taken into account processing a few materials not only at room temperature, but also at 87° K (Liquid Argon temperature) and at 0° K (that means no Doppler broadening), which is suitable for liquid helium cooled apparatus.

2.2 Code Validation

Many tests of the new version of the code have been performed. Only a few of them will be discussed here: they mainly concern the hadronic part since for the present problem the modifications made to EGS4 are not believed to be important.

The inelastic event generator has been tested comparing its predictions against experimental data for rapidity distributions, shower, grey and black particle average multiplicities and multiplicity distributions over a wide range of energies up to 800 GeV and for a large variety of target nuclei. The agreement with the experimental data is generally good: the results obtained for grey and black particles made us confident that the modifications to the cascade and evaporation parts are correct and reliable. An example is shown in fig.1.

Much care has been put in testing the new module for nucleon-nucleus interaction below 200 MeV. The agreement with experiments is satisfactory and it is apparent that this module represents a substantial improvement in this energy range: these tests are very important for neutron transport calculations due to the large number of medium-low energy neutrons in a hadronic cascade. An example is presented in fig.2.

A very important test which can directly give informations on code reliability in predicting hadronic cascades can be found in ref.[11]. It was shown that the new version of FLUKA is usually in good agreement with experimental data in predicting neutron fluxes produced by 24 and 200 GeV proton beams on Iron and

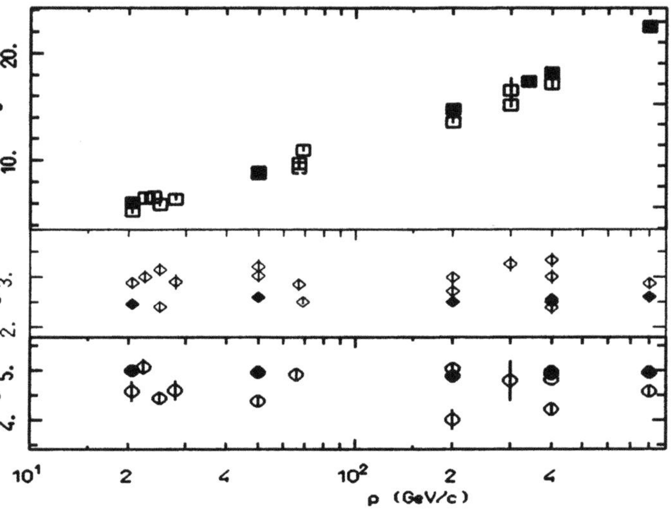

Fig. 1 : Average multiplicities of charged black(N_b), grey (N_g) and shower(N_s) particles produced by protons of different energies in emulsions. Full symbols are FLUKAN results, open symbols are experimental data[9]

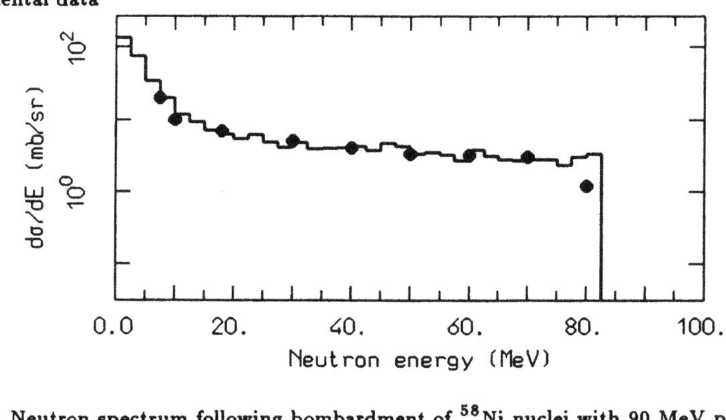

Fig. 2 : Neutron spectrum following bombardment of ^{58}Ni nuclei with 90 MeV protons. Points are experimental data[10]

Lead dumps[12]. The code successfully predicts both the radial and longitudinal distribution of the flux measured by different activation detectors with threshold as low as 1 MeV: the agreement for albedo neutrons is also good. Preliminary comparisons[11] show also nice agreement with doses measured with RPL glasses in the same experiments. These benchmarks make us confident that FLUKAN predictions can usually be trusted within the uncertainties of neutron dosimetry (which can be estimated of the order of 10-20 %).

3. The Source Term

Any calculation on the radiation levels in future detectors requires specific assumptions on the number, identity and spectra of particles emitted in minimum bias events. However due to the lack of experimental data, it is necessary to use physical models whose reliability at the energy of interest can be a major source of uncertainty in the final result. For all calculations presented in the next sections primary p-p events were generated using the DTUJET90 code[13]. A full description of the physics model underlying this code can be found in the original papers: we want only to stress that the code includes all processes relevant for the present task, and it is in nice agreement with experimental data up to the Tevatron energy. The properties of the events generated by DTUJET are summarized in table 2, for $\sqrt{s}$ = 16 and 40 TeV. However the reader must keep in mind that these events were generated using a given set of structure functions: other sets can give different results. The behaviour of the gluon and sea quark structure functions has not yet been measured down to the small x values which are relevant for LHC and SSC. The uncertainty introduced can be quite large: very preliminary results[14] show that structure functions behaving like $x^{-1.5}$ near zero can result in an increase of the multiplicity as large as 50 % when compared with the data presented in table 2 and generated according to a x^{-1} dependence. Unfortunately no discrimination can be made on the basis of available experimental data since both sets go into the same results approaching the present collider energies. A few exploratory runs were also performed using a different code[15] for generating primary events, however the resulting differences in the final results are well within the uncertainties discussed above and therefore they do not justify any further effort in this direction.

Finally it must be mentioned that, for a given luminosity, a 10-20 % scale error derives from the uncertainties on the inelastic cross section adopted for p-p collisions.

4. Radiation Environment for a Realistic LHC Detector

In the following the results of simulations carried out for a realistic LHC detector configuration are presented. The quantities scored are neutron spectra and doses due to ionizing particles in the various parts of the detector.

4.1 General Considerations and Assumptions

The detector geometry used for these calculations is presented in figure 3: the surfaces and volumes used for scoring neutron fluxes are also indicated. The main features of this geometry, taken from a simple sketch of the EAGLE collaboration, are presented in table 3. The Barrel Calorimeter covers down to η = 2, while the End Cap one goes down to η = 3. The materials of the Electromagnetic and Hadronic Calorimeters have been treated like homogeneous mixtures of the proper amounts of absorber and active media. This approximation does not cause any significant consequence on neutron fluxes calculations, since neutron mean free

Table 2
Mean properties of secondaries generated in p-p events by the DTUJET code

LHC										
	Multiplicity					Mean p_T (GeV/c)				
η	Tot.	Ch.	π^0	γ	Str.	Tot.	Ch.	π^0	γ	Str.
0.- 2.	45.	25.	13.	1.1	5.9	0.46	0.46	0.39	0.36	0.67
2.-4.	50.	28.	13.	1.1	8.0	0.43	0.42	0.36	0.38	0.57
2.-6.	39.	22.	10.	0.8	6.5	0.41	0.40	0.36	0.37	0.53
2.-8.	23.	13.	6.1	0.5	3.8	0.38	0.37	0.33	0.32	0.48
2.-10.	8.0	4.7	1.9	0.1	1.2	0.30	0.29	0.25	0.28	0.36
10.-∞	0.7	0.4	0.1	0.01	0.08	0.14	0.14	0.11	0.14	0.14
0.-∞	165.	93.	45.	3.7	26.	0.42	0.41	0.36	0.36	0.56

SSC										
	Multiplicity					Mean p_T (GeV/c)				
η	Tot.	Ch.	π^0	γ	Str.	Tot.	Ch.	π^0	γ	Str.
0.- 2.	48.	27.	14.	1.3	6.4	0.47	0.46	0.40	0.36	0.67
2.-4.	52.	29.	14.	1.2	8.7	0.43	0.42	0.37	0.38	0.56
2.-6.	43.	24.	11.	0.9	7.2	0.42	0.42	0.36	0.35	0.54
2.-8.	29.	16.	7.6	0.6	4.8	0.41	0.41	0.35	0.39	0.53
2.-10.	14.	7.9	3.5	0.2	2.1	0.35	0.35	0.29	0.33	0.43
10.-∞	2.8	1.7	0.6	0.03	0.5	0.22	0.22	0.19	0.15	0.26
0.-∞	189.	106.	51.	4.2	30.	0.42	0.42	0.36	0.36	0.56

paths are larger than the typical dimensions of calorimeter cells. On the contrary this is surely an approximation for doses: the proper dose to an active material should be computed multiplying for the ratio between its mass stopping power and that of the homogeneous medium. Moreover, such a procedure can be incorrect if the material considered significantly perturbs the radiation spectrum. Therefore, a correct computation can be made only knowing both the geometry and composition of the zone of interest and the surrounding media (which is not trivial for example for a preamplifier). For this reason, and since mass stopping power ratios are always not much different from one , in the following we report doses to the "average" materials without any corrections.

The results of the calculations have been converted to annual doses and fluxes assuming an inelastic cross section $\sigma_{in} = 60$ mb and an integrated luminosity $L = 10^{41} cm^{-2}$ corresponding to a luminosity of $10^{34} cm^{-2} s^{-1}$ and a beam time of $10^7 s$.

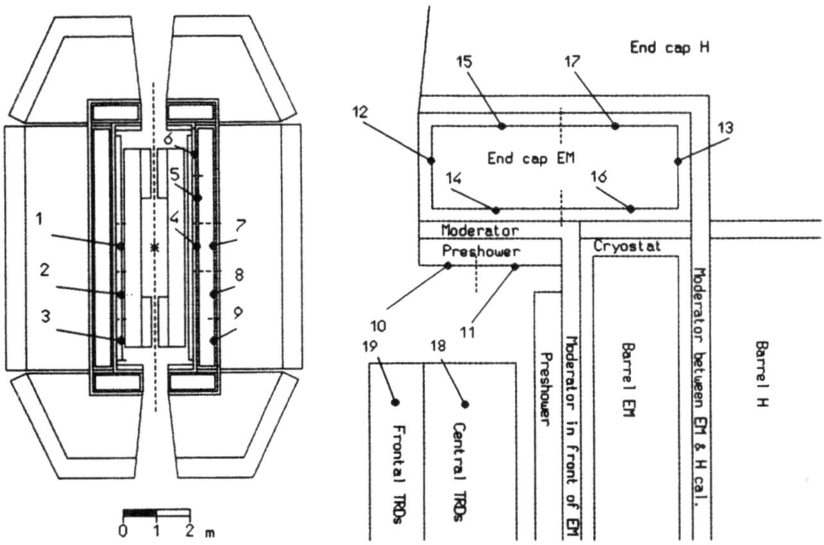

Fig. 3 : Geometry and scoring surfaces used in calculations for a LHC detector: Horizontal section (left) and magnification of the End Cap zone (right)

The cross section does not include single diffractive events which have been removed from the file generated with DTUJET90. Due to the considerations outlined in section 3, there is a systematic uncertainty arising from the source term alone which is at least of the order of 50 %. The systematic errors introduced by the whole cascading process are estimated to be of the order of 20 % for neutrons, according to the discussion of section 2.2, while are likely to be smaller for doses. Bearing in mind these systematic effects and taking into account that they are rather independent from the considered configuration, the results presented in the following can be compared at the 5-10 % level corresponding to the typical statistical errors of the calculations. Every configuration has been tested with 300 primary p-p events randomly chosen out of a set of 850. These figures should be large enough to avoid any selection effect. To save CPU time during neutron flux calculations, electrons, positrons, gammas and π^o have been immediately discarded, so neglecting the small contribution deriving from photohadron interactions.

4.2 Neutron Fluxes

The effects of the magnetic field, of different materials for the EM and Hadronic Calorimeters, of moderating materials inside the central cavity and of different forward cone apertures on neutron fluxes have been investigated and are presented in the following.

Table 3

Possible configurations for a LHC detector considered in this work

Config.	Description
1	Reference Configuration. Barrel: Lead-LAr EM cal. (R_{min}=125cm, -2 < η <2), Iron-LAr H cal. . End Cap: Lead-LAr EM cal. (-3 < η <3), Iron-LAr H cal. . Magnetic Field: 2 T solenoidal field (-2 < η <2, R <115 cm). No moderator, no TRD in the central cavity.
2	Same as 1 but without magnetic field
3	Same as 1 but with a Lead-LAr hadronic calorimeter instead of the Iron-LAr one
4	Same as 1 but with an Iron-Scintillator hadronic calorimeter instead of the Iron-LAr one
5	Same as 1 but with an Iron-Scintillator hadronic calorimeter instead of the Iron-LAr one and a Lead-Scintillator EM calorimeter instead of the Lead-LAr one.
6	Same as 1 but with a 10 cm polyethylene layer all around the EM calorimeter entrance face, both on the Barrel and the End Cap.
7	Same as 1 but with a 10 cm polyethylene layer between the EM calorimeter and the H one.
8	Same as 1 but with Transition Radiation Detectors (TRD) in the central cavity: the assemblies of both lateral and frontal TRDs are simulated as a proper amount of hydrogenated material uniformly distributed over the volume occupied by TRDs (see fig. 3).
9	Same as 1 but with a 10 cm polyethylene layer in front of the End Cap EM calorimeter only.
10	Same as 1 but with a 10 cm polyethylene layer in front of the End Cap EM calorimeter only and the TRDs in the central cavity (it is a combination of configurations 8 and 9).
11	Same as 1 but with the End Cap covering now down to η=4.
12	Same as 1 but with the End Cap covering now down to η=4 and with TRDs in the central cavity (Configurations 8 and 11).

It is customary to present neutron fluxes with a threshold set at 100 keV, which roughly corresponds to the threshold for damaging silicon based semiconductors[16] or also to normalize the fluxes assuming a unit damage for 1 MeV neutrons and folding the spectrum with the silicon damage function. However there is a growing interest in the use of GaAs based devices such as preamplifiers, both for their radiation hardness and intrinsic electronic properties: the damage function of GaAs[17] is quite different from that of silicon, and presents a non negligible sensitivity down to thermal energies. We decided therefore to present the data on neutron fluxes both with and without the 100 keV threshold, however the discussion in the following paragraphs and the comparisons between different configurations are restricted to fluxes with E_n > 100 keV to avoid confusion. Most of the conclusions are at least qualitatively the same also for the total flux.

4.2.1 Reference Configuration

To discuss the effect of the various modifications, the configuration described in table 3, point 1, has been assumed as the reference one. It is a possible configuration

Table 4
Neutron fluxes in EAGLE (in unit of 10^{13} neutrons/cm^2/yr). First line: flux with $E_n > 100$ keV. Second line: total flux. Values with statistical errors greater than 20% are marked with an asterisk

Posit.	Configurations											
	1	2	3	4	5	6	7	8	9	10	11	12
1	1.7	1.9	2.0	1.5	.79	.17	1.5	.40	.90	.33	3.2	.51
	7.2	8.1	9.0	5.5	1.8	.35	4.4	1.5	2.9	.96	16.	1.8
2	2.0	2.2	2.2	1.7	.96	.21	1.8	.55	1.0	.39	3.8	.82
	7.8	8.4	9.9	6.0	2.1	.50	4.8	2.1	2.9	1.1	16.	3.7
3	3.1	3.2	3.6	2.7	1.6	.39	2.7	2.0	1.2	.67	6.2	4.0
	10.	10.	14.	7.9	3.4	.86	6.6	6.2	3.2	1.7	23.	14.
4	1.1	1.5	1.2	1.0	.37	.41	1.1	.65	.93	.60	1.8	.71
	6.1	8.2	8.3	4.8	.95	.71	3.7	1.6	3.1	1.3	13.	1.9
5	1.4	1.7	1.5	1.2	.46	.45	1.3	.79	.96	.66	2.1	1.0
	6.8	7.6	8.9	5.1	1.2	.80	4.1	2.1	3.0	1.5	14.	3.6
6	2.0	2.2	2.3	1.7	.69	.61	1.8	1.6	1.0	.85	3.4	2.7
	8.1	8.7	10.	6.3	1.9	1.2	5.1	5.2	2.7	2.0	17.	11.
7	.57	.78	.72	.50	.03	.30	.43	.44	.54	.40	.80	.49
	3.1	3.8	4.5	2.3	.07	.69	1.3	1.2	1.8	1.2	6.4	1.4
8	.69	.85	.81	.55	.03	.36	.50	.51	.53	.43	.95	.63
	3.6	4.7	5.2	2.5	.06	.88	1.4	1.5	2.0	1.1	8.2	2.3
9	.84	.99	1.1	.70	.04	.43	.62	.78	.51	.48	1.3	1.2
	4.2	5.1	6.3	2.9	.09	1.3	1.7	3.1	1.8	1.6*	8.4	5.5
10	5.6	5.3	7.1	5.1	4.0	1.5	5.0	4.8	2.5	1.6	18.	16.
	15.	15.	20.	12.	7.4	2.8	10.	11.	5.3	3.2	41.	34.
11	3.2	3.1	3.7	2.9	1.6	.28	3.0	2.6	.80	.57	6.2	5.2
	9.8	9.6	13.	7.7	3.5	.68	6.6	7.4	2.4	1.4	22.	16.
12	14.	13.	18.	12.	11.	14.	12.	14.	13.	12.	62.	62.
	23.	23.	32.	20.	22.	21.	18.	21.	22.	18.	98.	102.
13	1.1	1.1	1.8	.79	.05	.92	.67	1.0	.79	.71	1.6	1.6
	5.3	5.8	7.1	3.2	.14*	2.7	1.6	4.6	2.4	2.0	9.0	8.2
14	8.0	7.6	9.5	7.2	4.0	7.5	7.3	7.9	6.8	6.2	43.	43.
	16.	16.	21.	13.	8.5	12.	12.	14.	12.	11.	69.	70.
15	9.3	9.0	14.	7.5	2.7	9.4	6.4	8.9	8.6	8.0	61.	59.
	19.	22.*	29.	14.	6.4	16.	10.	16.	15.	14.	110.	101.
16	2.3	2.4	3.1	2.0	.44	1.9	1.9	2.3	1.5	1.6	3.7	3.6
	7.8	8.4	11.	5.9	.97	4.4	4.5	6.7	3.4	3.3	16.	12.
17	2.4	2.4	4.2	1.9	.29	2.3	1.5	2.3	2.3	2.0	3.7	3.8
	7.2	8.9	13.	5.3	.69	5.4	2.8	7.5	6.3	4.7	18.	14.
18	2.1	2.2	2.4	1.8	1.0	.24	1.9	.53	1.1	.32	4.1	.93
	7.7	8.5	10.	6.0	2.2	.54	5.0	3.0	2.9	1.3	17.	4.6
19	2.4	2.6	2.8	2.2	1.3	.38	2.2	.84	1.3	.48	5.0	1.6
	8.8	9.0	11.	6.5	2.7	.73	5.4	3.0	3.2	1.3	19.	7.3

for the EAGLE detector without any moderating material and with a 2 T solenoidal magnetic field inside the central cavity (no field in the EM and Hadronic calorimeters). The EM calorimeter is made of Lead-Liquid Argon (LAr), and the hadronic one of Iron-LAr, with appropriate partial densities. Preshower detectors are placed in front of both the Barrel and the End Cap EM Calorimeters, and an aluminium cryostat, which can also mimic the effect of the coils, is considered. The results of table 4, 5 show that neutron fluxes will range roughly from $2x10^{13} n/cm^2/yr$ at the barrel entrance to 10^{14} in the most exposed End Cap positions for such a configuration.

4.2.2 Modified Configurations

The results for a few configurations both for the flux above 100 keV and the total flux are presented in table 4. The positions where neutrons have been scored can be seen in fig. 3. It must be noted that in the configurations 11 and 12 the scoring surfaces 12,14,15 are actually at inner positions since they have been moved down to $\eta = 4$.

Table 5 presents a direct comparison among all configurations for a few selected points: only neutrons with $E_n > 100\ keV$ have been considered here. Differences smaller than 10 % have been omitted since they are not statistically significant, of course caution must be used in interpreting variations close to that limit. In the last years many estimates of the expected radiation environment for the LHC detectors have been published in the literature[18]. Most of them combined simple analytical models with Monte Carlo simulations in idealized geometries to derive the quantities of interest. A comparison of our results with these estimates is rather difficult and position and configuration dependent. Such a comparison in a simpler geometry has been done in ref.[19]. Here the reader can find both a detailed discussion and the hint that there are positions/configurations where significant differences can arise.

4.2.3 Effect of the Magnetic Field

The 2 Tesla magnetic field has only a minor effect on the neutron fluxes: this is not unexpected since such a field with a Barrel radius of 1.2 m cannot bend enough charged particles to prevent them from impinging on the Barrel.

4.2.4 Effect of Different Materials for the Calorimeters

The replacement of iron with lead as absorber in the hadronic calorimeter causes an increase in the neutron fluxes, ranging from 20-30% on the preshower up to 60 % at the transition between the EM and H calorimeters. Only a slight reduction arises from the use of an Iron-Scintillator hadronic calorimeter despite its larger moderating power with respect to Fe-LAr. Larger effects are obtained using scintillators as active media also for the EM calorimeter. They are larger inside the barrel than on its internal surface, showing that neutron fluxes on the Barrel are dominated by low energy neutrons multiscattered in the central cavity and not by neutrons originating from interactions in the Barrel itself (see also ref.19 about this problem).

Table 5

Neutron fluxes in the configurations listed in table 3: The first line shows the fluxes concerning the reference configuration, while in the following lines the percentage difference with respect to line 1 is reported for a selected set of positions (see fig.3). Differences smaller than 10% have been omitted. Only neutrons with E_n > 100 keV have been included.

Config.	Positions						
	Barrel			End Cap			η_{max}
	1 p.s.	4 EM in	7 EM/H	10 p.s.	14 EM in	15 EM/H	12 EM in
1	1.7E+13	1.1E+13	5.7E+12	5.6E+13	8.0E+13	9.3E+13	1.4E+14
2	+ 15 %	+ 33 %	+ 37 %	=	=	=	=
3	+ 16 %	=	+ 25 %	+ 27 %	+ 19 %	+ 56 %	+ 31 %
4	=	=	- 13 %	=	=	- 19 %	=
5	- 53 %	- 67 %	- 95 %	- 29 %	- 50 %	- 71 %	- 18 %
6	- 90 %	- 64 %	- 47 %	- 74 %	=	=	=
7	=	=	- 25 %	- 11 %	=	- 31 %	=
8	- 76 %	- 43 %	- 23 %	- 15 %	=	=	=
9	- 47 %	- 18 %	=	- 55 %	- 16 %	=	=
10	- 80 %	- 47 %	- 29 %	- 72 %	- 22 %	- 14 %	- 14 %
11	+ 89 %	+ 56 %	+ 39 %	+220 %	+435 %	+553 %	+353 %
12	- 70 %	- 38 %	- 13 %	+192 %	+438 %	+538 %	+354 %

4.2.5 Effect of Moderators

The above conclusion is strongly supported by the results of the run with a 10 cm polyethylene layer in front of the EM calorimeter. It is apparent that all preshower and Barrel positions benefit from that layer up to a factor of ten, which could be explained only if fluxes were mainly due to neutrons multiscattered in the central cavity, rather than locally produced by inelastic interactions. The relative insensitivity of position 14,12 can be explained remarking that at high rapidities the contribution of locally produced neutrons is surely more important and also that no polyethylene was placed on the End Cap side directly facing the central cavity (surface 12 of fig. 3): the possibility of placing such a layer must be further investigated since it is likely that the benefits can be more than compensated by the unwanted extra amount of interactions due to the restriction of the forward cone. It must be stressed that the presence of such layer in front of the Barrel EM has so many drawbacks on its performances that it can be excluded a priori at least in this crude form. Configuration 6 must therefore be regarded as an useful exercise rather than a realistic ansatz to reduce the neutron problem. A further attempt can be to move the moderating layer beyond the EM calorimeter: this was done with Config. 7, however the resulting reductions are very discouraging. Unlike the polyethylene layer of Config. 6 the situation described in Config. 8 is realistic and shows how also minor structures can have important effects. Since it has been proposed to use assemblies of Transition Radiation Detectors (TRD) and they would contain a certain amount of hydrogenated material, their presence was schematized as a uniform region of polyethylene of appropriate density (regions 18,19 in fig. 3) in the central cavity. Rather surprinsingly, despite the integrated radial thickness is only about 3 cm, they resulted in a significant reduction of the neutron flux on the Barrel, maybe because of the large solid angle covered by such assembly and the long paths that neutrons coming from the End Cap regions must cross in it. The next step still exploits a polyethylene layer but this time limited to the End Cap region only, where it could be accepted by physicists. The goal is to try to decouple neutrons coming from the large η regions from the Barrel. A certain reduction is achieved but of course it is smaller than the one of Config. 6. Combining such a limited polyethylene shield with the TRDs gives the results shown for Config. 10. It can be seen that significant reductions can be achieved both on the Barrel and on the End Cap preshowers: however no significant reduction occurs in the End Cap EM part most exposed to radiation.

4.2.6 Effect of the Forward Cone Aperture

A test has been made with the same geometry but with an $\eta = 4$ forward cone. The flux increase is quite large, especially on the End Cap as expected, but also the Barrel regions are affected. However if we introduce the TRDs (config. 12) we can reach again the same fluxes on the Barrel EM and preshowers as with the $\eta = 3$ End Cap. This result strongly suggests that if a suitable technology will be available to build an End Cap Calorimeter covering down to $\eta = 4$ which can survive to these radiation levels, no significant consequence will occur at the Barrel Calorimeters provided an appropriate amount of hydrogenated material is present in any form in the central cavity.

4.3 Dose rates

Dose rates are mainly dominated by the electromagnetic component of the cascades. As a consequence they are much less dependent on details like the presence of moderating materials or the exact composition of the EM and H Calorimeter, provided the absorber material in the EM Calorimeter is the same.

Due to the reasons explained above, only few configuration have been run for doses, checking mainly the effect of the magnetic field, of TRDs and of the forward cone extended to $\eta = 4$. The configuration numbers refer to table 3.

4.3.1 Results

The doses at cascade maximum in the EM calorimeter as a function of η are reported in table 6 . Computed doses range from 300 Gy/yr near $\eta = 0$ up to 10^5 for $\eta = 4$. Unlike neutron fluxes they depend only on η without any significant influence of the forward cone aperture on values for small rapidities. The effect both of the magnetic field and of the presence of TRDs in the central cavity is small as expected. However it must be pointed out that these data refer to regions were no field was present, while in reality the stray field from the central solenoid can increase such estimates, especially if no iron return yoke is foreseen. These data are in reasonable agreement with previously published estimates[18] of doses in a spherical geometry once the η dependent distance to the EM calorimeter and the geometrical peculiarity of the corner between the Barrel and the End Cap have been taken into account. This conclusion suggests that unlike neutron fluxes which have been shown to be sensitive also to details, doses can reliably be derived from simple models without the need for complex and CPU expensive simulations in realistic geometries.

Table 6

Dose values (kGy/yr) as a function of rapidity in different LHC detector configurations, at cascade maximum in the EM calorimeter.

Config.	η							
	0.-0.5	0.5-1.	1.-1.5	1.5-2.	2.-2.5	2.5-3.	3.-3.5	3.5-4.
1	.33	.37	.43	.87	4.0	11.	0.	0.
2	.45	.48	.60	1.1	3.8	12.	0.	0.
8	.34	.37	.44	.80	3.6	11.	0.	0.
11	.36	.38	.46	.94	3.6	14.	52.	88.
12	.33	.38	.47	.86	3.8	14.	53.	92.

5. Radiation Environment for a realistic SSC detector

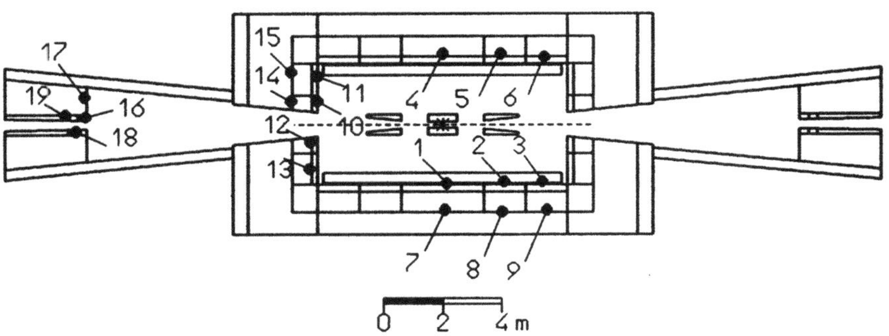

Fig. 4 : Geometry and scoring surfaces used in the calculations for SDC. (horizontal section)

The simplified geometry of the SDC detector used for these calculations is presented in figure 4: the surfaces used for scoring neutron fluxes are also indicated.

The Barrel and the End Cap cover the intervals $|\eta| < 1.44$, $1.44 < |\eta| < 3$, starting at R = 210 cm and Z = 420 cm respectively. The Forward Calorimeter goes down to $\eta = 5$ and starts at Z = 1200 cm. Lead-Scintillator EM and Iron-Scintillator Hadronic Calorimeters have been adopted both for the Barrel and the End Cap. Iron-LAr has been used for the forward Calorimeter. The results of the calculations have been converted to annual doses and fluxes assuming an inelastic cross section $\sigma_{in} = 80\ mb$ and a luminosity of $10^{33} cm^{-2} s^{-1}$ and the same beam time of $10^7\ s$ as for LHC. Systematic uncertainties are similar to the ones presented in section 4.1, but the error associated with the source term should be further increased due to the larger energy and consequent larger uncertainties in the physical model. The results presented in the following have typical statistical errors of 5-10 %. No effort has been made to run a set of configurations like the one shown in table 3, since we believe that most of the conclusions of Section 4 can directly be applied also to the SDC case. Estimates of neutron fluxes and doses for SDC Barrel and End Cap Calorimeters have been already reported and can be compared with the data presented in tables 7 and 8. What is new to our knowledge is the attempt made here to compute fluxes and doses for the Forward Calorimeter.

5.1 Neutron Fluxes

Table 7

Neutron fluxes in SDC(in unit of 10^{12} neutrons/cm^2/yr). Values with statistical errors greater than 20% are marked with an asterisk

Position	1	2	3	4	5	6	7	8	9	10
Flux, E>100 keV	0.24	0.31	0.38	0.06	0.07	0.08	0.003*	0.004*	0.002*	3.1
Flux, total	1.4	1.6	1.8	0.23	0.31	0.35	0.008*	0.02*	0.005*	6.8

Position	11	12	13	14	15	16	17	18	19
Flux, E>100 keV	0.74	7.5	0.88	2.3	0.26	35.	8.9	197.	182.
Flux, total	2.4	25.	3.7	15.	0.98	67.	25.	332.	338.

Table 7 presents the annual fluxes computed at the positions shown in fig. 4: runs have been made both with and without the forward Calorimeter, but no detectable difference was found in the results for positions 1 through 15. Neutron fluxes ($E_n > 100$ keV) are well below $10^{12}\ n/cm^2yr$ in all Barrel positions, reaching 10^{13} roughly in the most exposed position of the End Cap. On the contrary fluxes in the Forward Calorimeter are large and show a strong buildup with depth, going up to $2\text{x}10^{14}$.

5.2 Dose Rates

Annual Dose rates at cascade maximum are presented in Table 8 as a function of η. Few tens of kGy/yr can be reached in the Forward Calorimeter: this value is about 20 times larger than the maximum dose in the End Cap despite the much larger distance from the interaction point.

Table 8
Dose values (kGy/yr) as a function of rapidity for the SDC detector, at cascade maximum.

	Barrel			End Cap			Forward Calorimeter			
η	0.-0.5	0.5-1.	1.-1.5	1.5-2.	2.-2.5	2.5-3.	3.-3.5	3.5-4.	4.-4.5	4.5-5.
Dose	.016	.019	.020	.11	.50	1.7	.72	2.7	9.8	37.

6. Conclusions

Large systematic errors cannot be avoided in estimating radiation levels for detectors at the future colliders. While the uncertainty on the inelastic cross section is merely a scale factor, more fundamental sources of errors come from the questions still opened about minimum bias events spectra and multiplicities and, with minor importance, the reliability of hadronic cascade simulations. The resulting systematic uncertainty is at least a factor 2: however it has been shown that useful comparisons can be made since most of the systematic effects are the same in every calculation.

Apart from systematic effects, doses can be reliably computed and do not depend too much on details about geometry and composition of the detectors. On the contrary neutron fluxes can be strongly influenced by these factors and require dedicated realistic simulations: there is also evidence that results obtained with simple models can in some cases significantly underestimate the fluxes[19]. The much lower radiation levels of SDC compared with the EAGLE reference Configuration can be only partially explained by the lower luminosity (anyway in part compensated by the higher energy): important effects arise also from the larger Barrel radius, and, for neutrons only, from the presence of moderating materials in both the EM and Hadronic Calorimeters (to appreciate this last effect look at the difference between EAGLE Config. 1 and 5 in table 5).

7. Acknowledgements

The authors gratefully aknowledge the help given by Prof. J. Ranft, both in running the DTUJET code and with many valuable suggestions and discussions about the modifications of the hadronic part of FLUKA. A special acknowledgement must be given to Mr G. Panini of ENEA-Bologna for the processing and production of the nuclear data sets used for neutron transport, without which the present work would not have been possible. The help given by J. Zazula both in supplying unpublished data and in developing and debugging the FLUKA code is also acknowledged.

8. References

1 P.A. Aarnio, A.Fassò, H.J. Möhring, J. Ranft and G.R. Stevenson, *FLUKA86 user's guide*, CERN TIS-RP/168 (1986), and references therein
2 W.R. Nelson, H. Hirayama and D.W.O. Rogers, SLAC-265 (1985)
3 A. Ferrari, P.R. Sala, R. Guaraldi and F. Padoani *An improved multiple scattering model for charged particle transport*, submitted to *NIM*
4 A. Ferrari and P.R. Sala, *A New MonteCarlo code for inelastic hadron nucleus interactions below 1 GeV*, to be published
5 A. Ferrari and P.R. Sala, *Improvements to the Electromagnetic part of the FLUKA code*, to be published
6 P. Cloth et al., *HERMES user's guide*, KFA-IRE-E (1988) AN/12/88
7 M.B. Emmett, ORNL-4972 (1975)
8 E. Cuccoli, A. Ferrari and G.C. Panini, JEF-DOC-340 (1991)
9 A. Tufail et al., *Phys. Rev.* **D42** (1990) 2187 ; S. Fredriksson et al., *Phys. Rep.* **144** (1987) 187
10 A.M. Kalend et al., *Phys. Rev.* **C28** (1983) 105
11 J. Ranft, A. Ferrari, P.R. Sala, A. Fassò, G.R. Stevenson and J.M. Zazula, CERN TIS divisional report (1991), to be published
12 A. Fassò et al., CERN TIS-RP/90-19 (1990); J.S. Russ et al., CERN TIS-RP/89-02 (1989)
13 P. Aurenche et al.,*Multiparticle Production in a two component dual parton model*, *Phys. Rev.* **D44** in press ; F.W. Bopp et al., *Z Phys.* **C51** (1991) 99
14 J. Ranft private communication; F.W. Bopp et al., to be published
15 UA5 Collaboration, G.J. Alner et.al. , CERN-EP/86-213 (1986)
16 J.E. Gover, J.R. Srour, Sandia Report SAND85-0776 (1986)
17 A.M. Ougouag, *IEEE Trans. Nuc. Sci.* **37** (1990) 2219
18 G.R. Stevenson, *proceedings of the Large Hadron Collider Workshop*, CERN 90-10 (1990), vol III pag 566
19 G.R. Stevenson et al., *proceedings of this conference*

CALORIMETERS FOR ASTROPARTICLE PHYSICS

Michel SPIRO
Commissariat à l'Energie Atomique, DAPNIA/SPP
CE-SACLAY - 91191 - GIF-sur-YVETTE Cédex - France

ABSTRACT

There is a wide variety of techniques based on calorimetry which are used in particle Astrophysics. This review briefly covers low temperature calorimetry for dark matter detection, deep underground imaging water Cerenkov detector for proton life time, supernovae and solar neutrinos, air shower calorimetry for high energy cosmic ray study and ultrahigh Neutrino Cosmic Ray Detectors.

1. Low temperature calorimetry for dark matter detection

The flat rotation curve of spiral galaxies (fig. 1) cannot be accounted for by the distribution of known matter (stars) in those galaxies. It strongly suggests that spiral galaxies (and thus our galaxy) are embedded in large spherical dark matter halos. The nature of these halos is still at stake. One polular scenario is that it could be made of WIMPs (Weakly Interacting Massive Particles). The local density of our halo is .4 GeV/cm^3. The direct detection of WIMPs is based on the elastic scattering of a WIMP on a target nucleus. The nucleus recoil energy is in the 10 keV range for WIMPs with masses greater than few GeV.

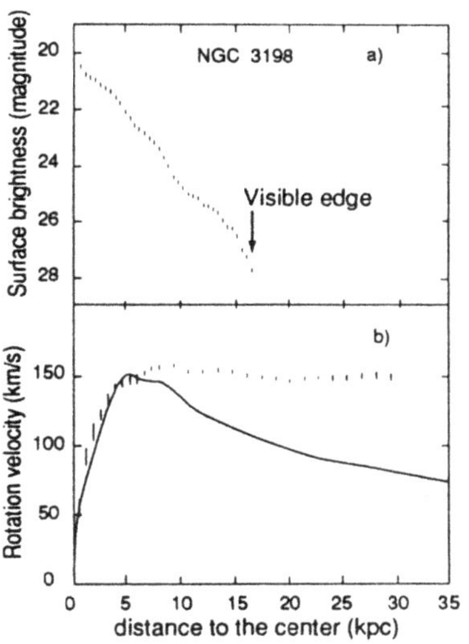

Fig. 1 - Rotation curve measurements compared to expectation (continuous curve) from visible mass distribution.

The best limits on Weakly Interacting Massive Particles as the constituents of our galactic halo, have been obtained so far from solid state Ge and Si diodes (fig.2), réf.[1]. These detectors are deep underground and well shielded from cosmic-rays and surrounding radioactivities. They are also cooled down to liquid nitrogen temperature to improve the energy resolution and lower the threshold (low noise).

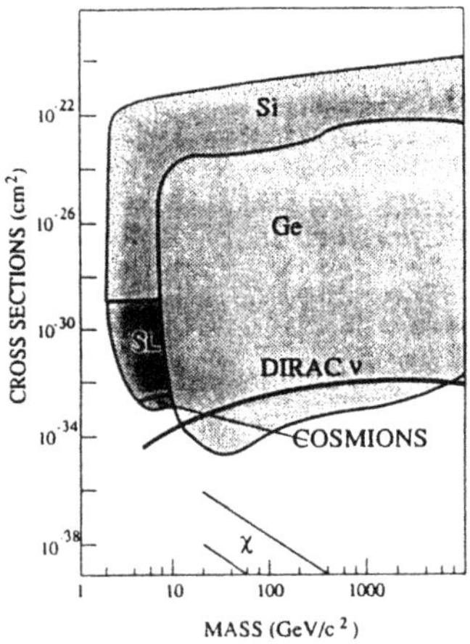

Fig. 2 - Exclusion zones for dark matter particles.

However, while they perform sufficiently well to reject heavy Dirac neutrinos and cosmion candidates, they have still a counting rate few orders of magnitude higher that one would require for the detection of the most popular candidate, the photino or more generally the neutralino.

Many groups are now studying a more direct approach of low temperature calorimetry. The idea is to measure not only the ionisation induced by a nuclear recoil due to the scattering of a WIMP, but also the small temperature rise resulting from individual nuclear recoils in low heat capacity material. For a general review on the detection of nuclear recoil energy by low temperature calorimetry, see [2]. The principle is based on the rapid reduction in specific heat with temperature in specific material :

$$C \alpha (T/T_D)^3$$

where T_D is the Debye temperature of the material. A small energy deposit at very low temperatures will then induce an important rise of the temperature.

Great technical progress has been achieved. Table 1 shows the results obtained with relatively large detectors and sensitivity close to that required.

	Mass	r.m.s. base	FWHM 60 keV
Coron et al. (Paris)	25 g Al_2O_3	260 eV	3.7 keV
Munich	19 g Si		1.2 keV
Berkeley	60 g Ge		
ionization		460 eV	2.4 keV
phonons		900 eV	2.8 keV

Table 1

Furthermore the possibility to identify a nuclear recoil and therefore get a strong rejection power against the surrounding γ activity has been demonstrated by the Berkeley Group [3] by measuring with a 20 mK, 60 g germanium crystal with an applied .4 volt bias both the heat and ionisation signal with a NTD Ge sensor. Both signals are measured with the same channel and are differentiated by their time duration (fews μs for ionisation, few ms for phonon detection). Using γ ray sources and neutron sources, they showed that the correlation between heat and ionisation is completely different for a nucleon recoil (little ionisation) than for a compton electron (fig. 3). Similar behaviour has also been seen with a 1/4 g Si crystal [UK group : Rutherford and Imperial College]. The search for neutralinos could then start by using $\sim$1 kg of ^{73}Ge enriched crystal.

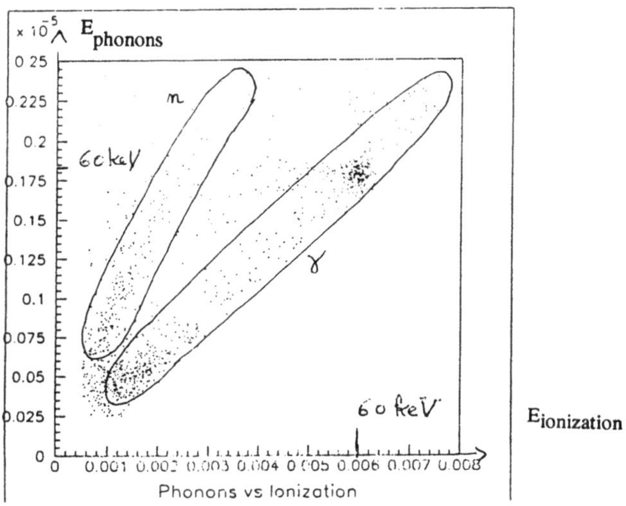

Fig. 3 - Heat vs ionization

2. Deep underground imaging water Cerenkov detector for proton life time, supernovae and solar neutrinos.

The Kamiokande detector is a 4000 t pure water Cerenkov detector equipped with 780 20" diameter photomultipliers. This detector was originally designed for searching for proton decay (with a fiducial volume of 2000 t). In 1986, it has been upgraded to measure the solar neutrino flux through the elastic scattering of solar neutrinos on electrons. The threshold on electron detection is $\sim$9 MeV with an energy resolution of 20 % and an angular resolution of 28°. The number of PM hits per MeV electron energy deposit is 2.3. By applying a series of cuts (fiducial cut 680 t, spallation cut, γ ray, etc.) they successfully demonstrated that the detector was up to the task of detecting the high energy (^8B) solar neutrino component (fig. 4).

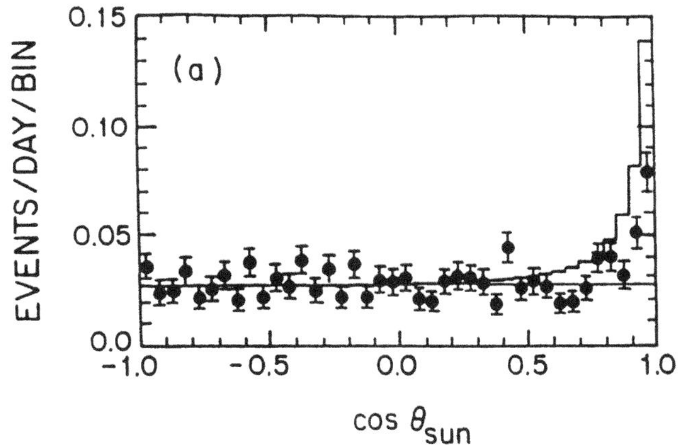

Fig. 4 - Plot of the cosine of the angle between the electron direction and a radius vector from the Sun showing the signal from the Sun plus an isotropic background. This plot is for E_c 9.3 MeV, and the time period January, 1987 through April, 1990, a total of 1040 live-dectector days.

The recently approved Superkamiokande detector is an upscaled version of the Kamiokande detector (Table 2). It should be able to push the sensitivity to the proton life time from 10^{32} to 10^{34} yrs (fig. 5), which is particularly fascinating in view of the recent results on the Grand Unification Scale and the Minimal Supersymmetric Standard Model. Table 3 and Table 4 show also the large gain one expects from Superkamiokande over Kamiokande in what concerns supernovae and specially solar neutrino detection. The fiducial volume should be $\sim$ 22000 t compared to the 680 t of Kamiokande.

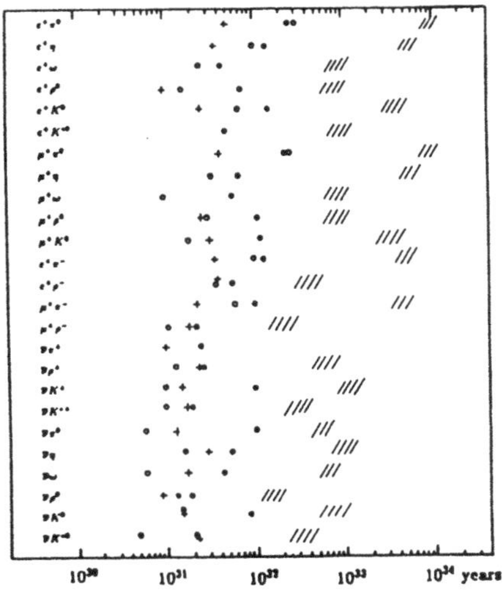

Fig. 5 - Nucleon partial lifetime limits at 90 % C.L. from Kam-I (Ë), IMB (o) and Frejus (+). The expected limits after 5 years of the Super-Kamiokande operation is also shown (////).

Parameters	Kam-II	Super-Kamiokande
Dimension(anticounter included)	16mh×19mϕ	41mh×39mϕ
Total mass	4500t	50000t
Fiducial mass	1040t	22000t
Number of PMT (inner detector)	948	11200
Coverage of photo-sensitive area	20%	40%
Energy resolution for electrons	$0.04/\sqrt{E}$ (E in GeV)	$0.03/\sqrt{E}$ (E in GeV)
	0.22 (at 10MeV)	0.16 (at 10MeV)
Vertex pos. resolution	1.7(1.1$^{(*)}$)m	0.5m
for 10MeV/c electrons		
Trigger for electrons	6.8(6.1$^{(*)}$)MeV/c	5MeV/c
Analysis for electrons	≥9.3(7.5$^{(*)}$)MeV/c	≥5MeV

(*). Preliminary values after June 1988[4].

Table 2 - Comparison of parameters between Kamiokande-II and Super-Kamiokande

Table 2 Number of neutrino events in a supernova explosion observed in Kam-II and will be observed in Super-Kamiokande.

	Kam-II SN1987A(~50kpc)	Super-Kamiokande Galactic Center(~10kpc)
$\bar{\nu}_e$	(10.5)(*)	~4000
ν_e	(0.34)(*)	~130
(Neutronization ν_e	(0.03)(*)	~13)
$\overset{(-)}{\nu}_\mu + \overset{(-)}{\nu}_\tau$	(0.21)(*)	~80
Total	11(observed)	~4200(**)

(*)Calculation based on the total observed number, 11, and on Reference 12.

(**)Estimation based on Reference 12. For the supernova neutrinos, the fiducial mass is 32000 tons.

Table 3 Number of solar neutrino events per day and the estimated number of various backgrounds in Kam-II and Super-Kamiokande for $E_e \geq 9.3$MeV. In calculating these numbers, directional correlation is not taken.

	Kam-II (/day/680ton)	Super-Kamiokande (/day/22000ton)
^{8}B solar neutrino	0.13	4.3
Total background	3.4(*)	~8
^{214}Bi decay	~2	≪1
External γ rays	~0.7±0.4	~0.4
Spallation products	~1.0±0.6	~6
U fission	$\lesssim$0.02	$\lesssim$0.6
Reactor $\bar{\nu}_e$	~0.003	~0.1
Solar $\nu(\nu_e{}^{18}O \to e^{18}F)$	~0.01	~0.3(**)

(*)After June 1988, the background number are reduced by a factor of ~3 compared with these numbers[6] (preliminary). This background reduction could be a strong support for the estimated background rate in Super-Kamiokande.

(**) Event rate is estimated based on Reference 22.

3. Air shower calorimetry for high energy cosmic ray study.

High energy cosmic rays have been studied for the past 50 years or more. The differential energy spectrum for MeV to 10^{21} eV has been measured with various techniques (fig. 6) and is shown in fig. 7. Few events have been observed with an energy greater than 10^{20} eV.

The search for possible point-like sources of high energy γ rays has raised a major interest when few groups reported evidence that Cygnus X3, a binary pulsar was may be the source of ultrahigh energy γ rays from TeV to 10^{18} eV. Even more the muon content of the showers originated from Cygnus X3 seemed to have some anomaly when compared to the expectations for γ ray originated showers (too many muons). However the signals were of marginal significance and have not been confirmed so far.

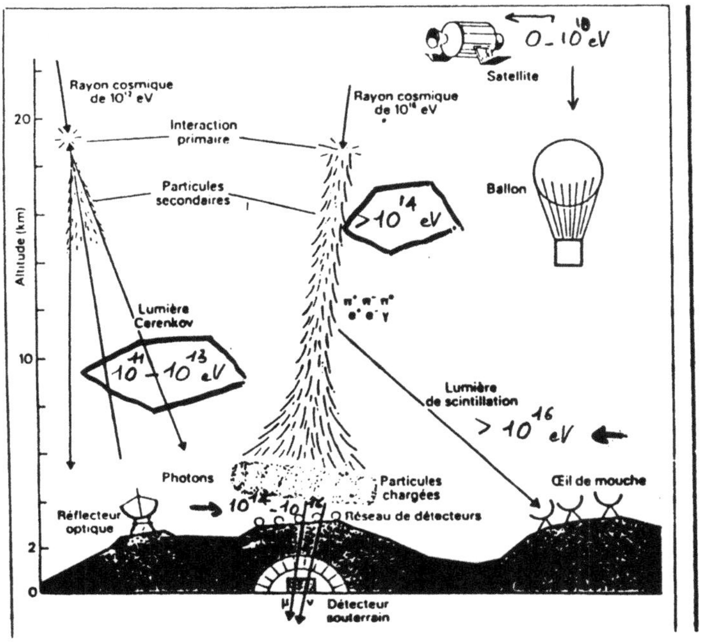

Fig. 6 - Techniques used for the cosmic ray energy spectrum measurement.

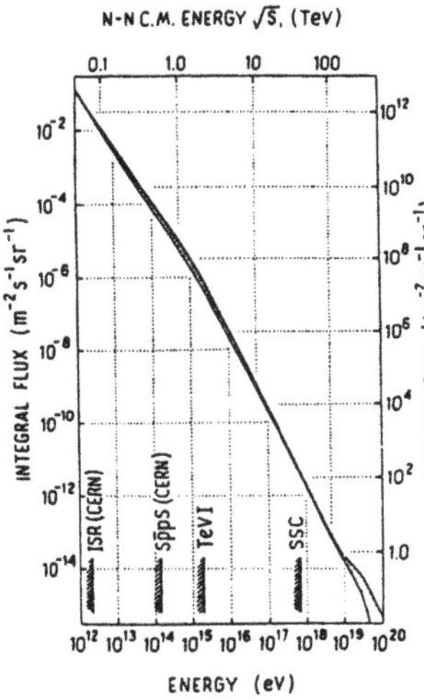

Fig. 7 - The integral spectrum of the primary cosmic-ray flux vs. energy per particle (bottom scale) and vs. the equivalent energy in the CM of a nucleon-nucleon system.

Three mains techniques are currently being used :

the air Cerenkov technique which is best illustrated by the Whipple Observatory : a compact array of PMs at the focal plane of a parabolic mirror. Each PM hit corresponds to a direction. The array samples only a fraction of the Cerenkov cone of a high energy shower [4]

For a γ ray shower (small cone aperture) which originates from an on-axis point-like source, the pattern of hits should look like an ellipse. The major axis should point towards the center of the array. To separate a point-like on center γ ray source from a diffuse proton background, the AZWIDTH (see fig. 8) parameter distribution has been found very efficient. The Crab source signal at $\sim$20 σ level [5] is shown in fig. 9. No other point-like sources have been found.

WEEKES *ET AL.*

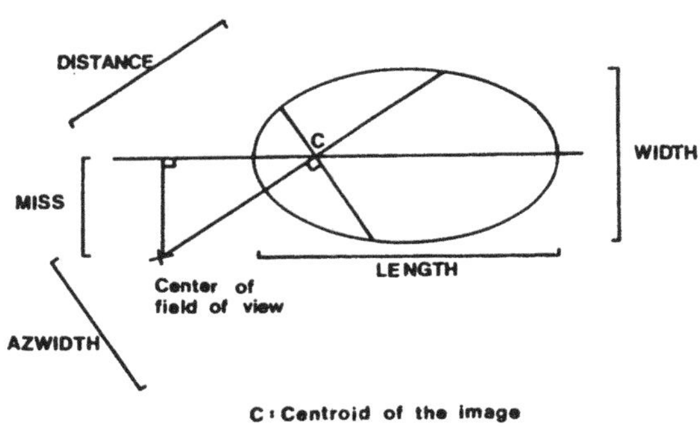

Fig. 8 - Definition of image parameters

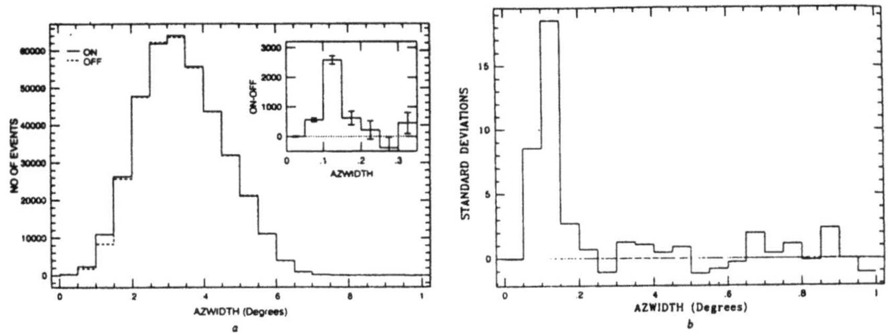

Fig. 9 - (a) Measured distributions of the Azwidth parameter for the ON and OFF per 0.05 interval of Azwidth. THe inset shows the ON-OFF distribution. - (b) on-OFF distribution plotted as number of standard deviations per 0.05 interval of Azwitdh.

The threshold for the experiment is few 100 GeV.

Another Cerenkov technique using an array of 18 parabolic mirrors (80 cm diameter) with one photomultiplier each and a 12 fold coincidence was able also to detect a signal from the Crab at 4.0 σ level using the timing of the PMs to determine the direction [6]

This technique demands showers which are extended on a larger area than the previous one and has then a higher threshold ($\sim$2 TeV). Both results are compatible (fig. 10).

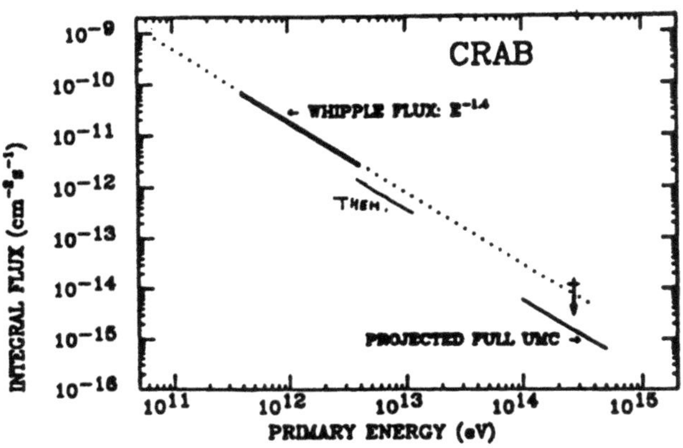

Fig. 10 - Summary of high energy measurements for the Crab.

A scintillator array demands even higher energies because the shower (particles) have to reach the ground. In the Cerenkov technique only the light has to do so. The most sophisticated detector is CASA in Utah. So far no signal from point-like sources have been found (with a threshold of $\sim$ 100 TeV).

Finally for ultrahigh energy cosmic ray showers one can use the light from N2 scintillation which reaches ground arrays of photomultipliers. The threshold of such detectors as Fly Eyes is $\sim 10^{15}$ eV. So far also, no point-like sources have been convincingly found.

4. Ultrahigh Neutrino Cosmic Ray Detectors.

Such planned detectors are based on the detection of Cerenkov light in water, using natural or artificial lakes, or even the Ocean (like Dumand in Hawaï) [7].

The hope is that Active Galactic Nuclei might be the source of very intense and energetic flux of neutrinos. These are very bright, distant and compact sources which might be fueled by the gravitational energy of matter infalling onto a supermassive ($\sim 10^8$ Mo) black hole at the AGN center.

The AGN ν energy spectrum was calculated by various authors. The flux of AGN ν is found to overcome the flux of atmospheric neutrinos (neutrinos produced in hadronic cascades in the atmosphere) above 10^{14} eV with a cut off at 10^{19} eV.

The number of events in the Dumand II detector, is shown in fig.11 with the particularly nice signature of resonant neutrino absorption in the atmosphere through the reaction $\nu_e\, e^- \rightarrow W^-$ at 6.4 10^{15} eV incident neutrinos.

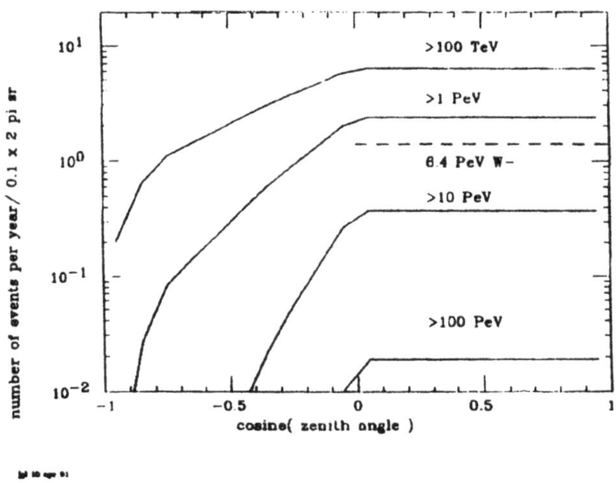

Fig. 11 - DUMAND contained nu-e, AGN model

Note also that if our galactic halo is made of light neutrinos, there should be a sharp resonant absorption line in the spectrum of ultrahigh energy intergalactic neutrinos reaching the earth (fig. 12). The detection of such a narrow line would be a proof of the neutrino halo. Its position would provide for the neutrino mass : $2 M_\nu E_\nu = (M_{Z^0})^2$. If the ν mass is ~ 10 eV, the absorption line is at $5\ 10^{20}$ eV :

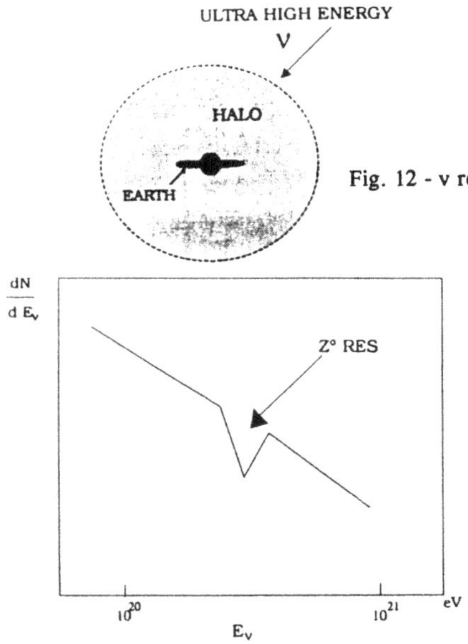

Fig. 12 - ν resonant absorption by our galactic halo.

Conclusion

Such a short review cannot be complete and cannot cover in the allocated time and space such important recent developments as illustrated by detectors or projects like :

- Superconducting granules
- Macro, Icarus, Soudan II
- Borex, SNO
- Grande, Lena, Baïkal.

References

1. G. Gerbier, ***Dark Matter, an overview***, Proceedings Conf. Physics in collision - Colmar 20-22/6/91 - to be published.
2. P.F. Smith and J.D. Lewin, ***Dark matter detection***, Phys. Reports - vol. 187 - N° 5 - march 1990 - PRPLCM 187(5) 203-280 (1990)
3. B. Sadoulet, ***Proceedings of the Texas/ESO-CERN Symposium*** - Brighton (1990)
4. T.C. Weekes et al., ***Observation of TeV gamma rays from the crab nebula using the atmospheric Cerenkov imaging technique***, The Astrophys. Journal, 342, 379-395(1989) July 1
5. G. Vacanti et al., ***Gamma-ray observations of the crab nebula at TeV energies***, The Astrophys. Journal, 377, 467-479, 1991, August 20
6. P. Baillon et al., ***Themistocle experiment***, Proceedings of the Dublin Cosmic Ray Conf. - août 91 - to be published.
7. ***Third Int. Workshop on Neutrino Telescopes***, Venezia, 26-28 Feb. 1991 edited by Milla Baldo Ceolin

Parallel Sessions

Trends in Scintillator Calorimetry

A. MENZIONE
INFN, via Livornese
Pisa, 56100 Italy

ABSTRACT
The following is an introduction to the contributed papers for the parallel session on "Scintillation Calorimetry" of this Conference. Its aim is to give an overview of the impressive work carried out in recent years to understand and develop scintillator based calorimeters.

1. Introduction

An ambitious goal for this overview would be to make a statement about whether we are ready or not to design and build calorimeters for future high energy colliders. Indeed, many contributions were based on results of R&D efforts for SSC or LHC. More realistically, at this point, we shall limit ourselves to report on the status of the art and introduce the reader to the issues and the problems which were presented and discussed.

The future p-p colliders are machines which are characterized by the extremely high energy of the secondaries [TeV] and extremely high luminosities [10^{33}~10^{34}]. This last statement implies a very short interval between bunch crossings and consequently short collection times for the signals. It appears that a scintillator based device is in principle well suited for the task. However, problems may arise in connection with hermeticity of the detector, the light readout and other geometrical constraints.

Other problems are related to the radiation resistance of the materials and front-end electronics, trigger problems etc.. These matters were discussed in other parallel sessions.

A number of new experimental results from several prototypes were presented, and also new analyses of test beam data taken in the last few years. In some cases the performance of new calorimeters installed in real experiments was presented.

With some exceptions the majority of the contributions in our session fall into two main categories: the "tile calorimeter" approach, which takes advantage of fibers doped with wave-length shifting materials (WLS); and the "spaghetti calorimeter" approach with direct collection of the scintillation light. While the first allows a natural separation between the electromagnetic (em) and the hadron (had), possibly with two different absorbing materials, the second allows a pseudo-homogeneous assembly and the separation between em and hadron showers is done using other tools.

A sizeable number of contributions were devoted to the simulation of calorimeters, both to foresee the performance of a device of novel design, and to compare the predictions with the experimental results of devices already tested.

The basic parameter which characterizes a calorimeter is its resolution as a function of energy. However, having in mind future applications at high energy, a particular emphasis was given to issues like uniformity of response and calibration methods needed to keep the non scaling term of the resolution small. These address in particular the performance of the em compartment. Parameters related to the quality of the trigger like linearity, e/π separation (does one need a preshower detector?), compensation (or, better, how large is the non-compensation factor that can be tolerated?) were also discussed by many participants. Often data were presented, or performance predicted by the simulation, on other important quality factors like position resolution, time resolution, two particle and two jet separation. The issue of the amount ofpassive material that may be tolerated in front of the em calorimeter was discussed and suggestions were made to minimize its impact. In two reports, the changes in the performance of the device by shortening the gate time were studied.

2. Tile calorimetry

A "tile calorimeter" is an array of scintillator pieces (tiles), interleaved with absorbers, which in a collider detector are organized in a more or less projective tower geometry. The light is collected and shifted in fibers of small diameter, funnelled to clear fibers which bring the light to the photocathode. Usually all fibers of a given sector a tower are connected to the same photomultiplier. This is an evolution of the technique used, in their hadron compartment, first by UA1 and later by CDF. At the time of construction of those experiments good WLS fibers were not available and light collection was based on the use of strips of WLS placed along the edges of the scintillator and connected to the photomultiplier through clear acrylic strips.

Several different approaches were presented. One was by Leslie Camilleri in which tiles were coupled to a fiber running across one diagonal. In order to minimize nonuniformity each layer was the mirror image of the following one. Another scheme, thought out only for the had compartment was presented by Otto Gildemeister. In this scheme, still under study, the scintillator is envisaged to be positioned parallel to the beam thereby solving some mechanical integrity problems of the steel when, for instance, it is used for to return the magnetic flux. The third scheme was presented by Jim Freeman, who illustrated the preliminary test results of the prototype for the new CDF plug. In the em compartment the light collection efficiency was very high and the achieved resolution as good as expected.The had resolution was not optimal, since the calorimeter had the constraint of the existing iron work of the CDF magnet. Jim also presented the readout scheme (sigma tile) foreseen for SDC calorimetry, which will hopefully allow easier construction and more uniform response. Another readout geometry, again aimed at optimizing uniformity and efficiency was presented by Armand Simon for their anticoincidence hexagonal tiles, used in an experiment in which a 13 ton spaghetti calorimeter module was used as a neutron counter (see par 5).

3. Spaghetti calorimetry

With this name one indicates a calorimeter in which scintillating fibers are inserted in an absorbing material. Usually the fibers are oriented at a small angle relative to the incoming particles, and all fibers of the same tower are connected to the same photomultiplier. The reason for the above mentioned small angle (a few degrees) is to minimize the channeling effect, namely abnormal energy deposition for particles running along a single fiber. The residual channeling contributes to the constant term of the resolution, which is relatively more important in the em compartment. It was shown by Luc Poggioli that the use of thinner fibers (.5 mm instead of 1 mm) reduces the effect, while improving the resolution by 40%. Giacomo Polesello's talk was devoted to the performance of a prototype with 1 mm fibers in lead and projective tower geometry. The approach was to group all fibers which can be reached by an em shower into a photomultiplier and all the shorter ones into another photomultiplier. By comparison of the two signals good e/π separation can be achieved. Still aimed at the design of the calorimetry for future collider experiments, Masa Mishina focused his report on geometrical and performance problems, plans and simulation of new generation prototypes. In order to reach the radiation resistance needed for devices to be used in the forward direction at LHC, Valerij Khovansky discussed the option of using liquid scintillator flowing in small diameter capillaries rather than plastic fibers.
A different application of the lead-fibers technique was presented by Lucia Votano in the framework of the experimentation at Daphne (phi factory). The idea is to take advantage of the good time resolution of the scintillator to locate the decay point of $k^o \to \pi^o$'s $\to \gamma$'s. The typical energy range of these photons is 20~200 MeV. Results of their prototype show that a resolution of ~250 psec might be achieved.

4. Simulation and data analysis

The status of the many programs that simulate particle interactions with matter was given, in the plenary session, by R. Brun. In the parallel session, Theresa Tymieniecka presented the results of an upgraded FLUKA with neutron detection, and compared them with experimental results. In particular she showed that the best energy resolution is expected at values of e/h which might be rather far from 1, especially in the low energy range [GeV]. This goes in the direction of confirming the statement that good em resolution and compensation are mutually exclusive.

One often discussed issue, with reference to spaghetti calorimeters, is the channelling effect which contributes to the non-scaling term of the resolution. From the presentation of Valerio Vercesi one can guess how large this effect is, with increasing energy. This effect depends critically on the impact point of the particle. Both test beam data and GEANT simulation are in good agreement in predicting a very small constant term if one forces the impact point to be in the lead.

Besides other parameters, the channelling effect was investigated by Carl Bromberg who simulated the behavior of a fiber calorimeter with fibers of different WLS in the em and had compartments. The two sections of the fiber are spliced together, but at the end a mirror transparent to the green (em) and reflective to the blue (had) sends separate fractions of light to separate photomultipliers.

Adele Rimoldi focused her contribution on the capability of the spaghetti calorimeter prototype to separate two nearby pions with a χ^2 algorithm based on the covariance matrix.She compared events built by superimposing two pions on the same positions to events where the two pions entered at a non-zero variable distance. A similar analysis was also performed for-two jet separation.

5. Contributions on other issues

A description of the Zeus Uranium-scintillator calorimeter, one of the few compensating systems in use, was given by Joannis Gialas. Details of its performance derived from test beam data were presented. A study of the time structure of the light collection in the same calorimeter was given by J. Parson. After a short pulse (10~20 nsec), a sizeable quantity of light is collected, for hadron showers, in the following few hundred nsec, due to thermalization neutrons. This feature can be used to give good e/π rejection.

Another report on an application of scintillator calorimetry in a real experiment came from Armand Simon. They used a spaghetti calorimeter module to measure neutrons, working out a fast em/had shower separation scheme taking advantage of the different time spread of light arrival. Christine Scheel discussed, with reference to the same apparatus, the changes in performance of the calorimeter if one decreases the charge collection time to 15~16 nsec, as one expects in future colliders.

Both in the plenary and in the parallel sessions, the issue was discussed whether it is tolerable or not to install the em calorimeter outside the coil in a collider experiment. The coil will be at least one radiation length thick and have an impact on energy resolution and on e/pi separation. Hartmute Plothow-Besch discussed the possibility of instrumenting the coil material with the insertion of a layer of glass scintillating fibers and gave figures for the improved energy collection and resolution.

The basic ingredient of a scintillator calorimeter is the scintillator material itself and in particular the fibers, either scintillating or WLS. Parameters like light yield, attenuation length, diameter tolerance, radiation resistance etc. enter into the quality factor of the final device. On these issues was focused the contribution of Charles Hurlbut, who presented the production and the plans of Bicron ltd.

In future collider experiments the calorimeters will have such a large number of towers, that is unlikely that one will have the possibility to calibrate them all in the test beam. Furthermore one will need to keep the device calibrated for many years with an accuracy of 1% or better. Virgil Barnes devoted his presentation to this problem. He presented a scheme in which a gamma source is guided through thin tubes to each individual tile in the calorimeter. The system, which is an evolution to the SSC solenoidal detector of the system implemented for CDF, is under test now. Preliminary results show that it will be suited for the task as long as the on-going reliability studies will be completed.

PRELIMINARY RESULTS FROM THE TEST OF A LEAD-SCINTILLATING FIBRE CALORIMETER PROTOTYPE WITH PROJECTIVE GEOMETRY

THE RD1 COLLABORATION
Cagliari, CERN, Clermont-Ferrand, Ecole Polytechnique, Lisbon, Marseille, Napoli, Paris VI, Pavia, Rio de Janeiro, UCSD, Weizmann

presented by

Giacomo Polesello

Istituto Nazionale di Fisica Nucleare, sezione di Pavia

1. Introduction

The RD1 collaboration aims at investigating the adaptability of lead-scintillating fibres calorimetry to the LHC environment building on the excellent results obtained in this field by an R&D effort in the framework of the LAA project[1].

A fundamental step in this investigation is the design, construction and test of a lead-scintillating fibre prototype with projective geometry. The first part of this paper describes the construction of such a prototype, which was tested during August-September 1991 on the H2 beam of the CERN SPS with electron and pion beams.

In the second part, some preliminary results from the analysis of collected data are described.

2. The calorimeter

2.1 Geometry

The basic structure of the calorimeter is based on longitudinal scintillating fibres uniformly distributed in a lead matrix in the volume ratio 1:4 as required for compensation. The fibres are 1 mm diameter blue fibres (KURARAY SCSF81 and BICRON B) arranged in such a way that each fibre is equidistant from the (six) neighbouring ones. A total of 19 modules were built, each corresponding to a section of a projective calorimeter with internal radius 130 cm, covering an angular range $\Delta\eta \times \Delta\phi \approx 0.03 \times 0.03$. Each module has the shape of one quadrant of a truncated pyramid, with a front face of 4 x 3.85 cm^2, a back face of 10.3 x 9.8 cm^2, and a length of 200 cm (Fig. 1). The 19 modules were arranged in a staggered layout, which approaches the geometry of a central module with 2 surrounding rings.

2.2 Readout segmentation

All fibres run parallel to each other in the direction of incident particles. Therefore some of them run all along the 2 meters of the whole calorimeter length, whereas the ones in the sloping part of each cell become shorter as they approach the outer edge, and start at progressively increasing distances from the front face with a pitch of $\approx$ 7 cm. This property was exploited in order to obtain a kind of longitudinal segmentation in the readout (fig. 2).

All the fibres running up to the front face of the modules, together with the first two rows of shorter fibres (a total of 399 fibres per module) are grouped together and named 'EM' section. The remaining 1947 fibres, which start at more than 20 cm from the front face (27 radiation lengths) are named 'HAD' section .

The two groups of fibres are connected to two separate tapered light guides which are read out with 1" XP2961 phototubes (EM section) and 2" XP2282 phototubes (HAD section) respectively.

2.3 Construction

The construction technique of the modules was a first attempt at a mass building procedure with industry involvement.

The basic components are extruded grooved lead plates cut at a length of 230 cm. A 5 μm tin layer is deposited electrolytically on the plates, which are then stacked in parallelepiped blocks with a $\approx$ 10 x 14 cm^2 cross-section, and brazed by heating at 245 °C for 12-14 hours. The blocks are then machined in order to obtain from each block two modules with the geometry described above.

The modules built with this procerdure, after some cleaning of partially obstructed holes, have on average 5% of the holes in the EM section unusable for fibre insertion. The percentage goes down to 1.3 % for the central octet of modules.

3. Preliminary analysis results

3.1 Energy resolution and uniformity

The energy resolution of the prototype was studied with 10 to 150 GeV electrons hitting the centre of the calorimeter with an incidence angle of 3° with respect to fibre axis in order to avoid channelling effects. The size of the beam spot was 8 mm in diameter. The energy dependence of the resolution (fig. 3) is well parametrized as:

$$\frac{\sigma(E)}{E} = \frac{(0.146 \pm 0.009)}{\sqrt{E(GeV)}} + (0.016 \pm 0.001)$$

The uniformity in response was studied with 20 and 80 GeV electrons by a position scan covering the surface of a cell in steps of 1.5 cm. Drops in response of the order of 10% were observed at the interface between two modules. Given the reproducibility of the effect across different modules, a preliminary study indicates the possibility to correct for it, once the impact point is precisely known.

3.2 Position resolution

The position resolution was studied using the data from the position scan. For 80 GeV electrons, the relationship between the true impact point in the vertical (Y)

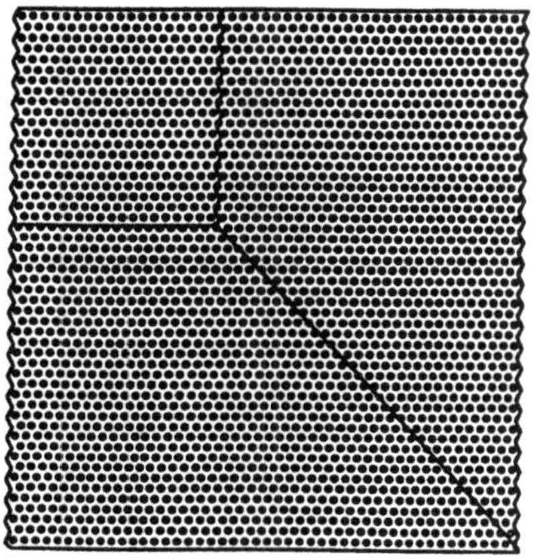

Figure 1: Front view of a calorimeter module, with the electromagnetic front face in the top left corner.

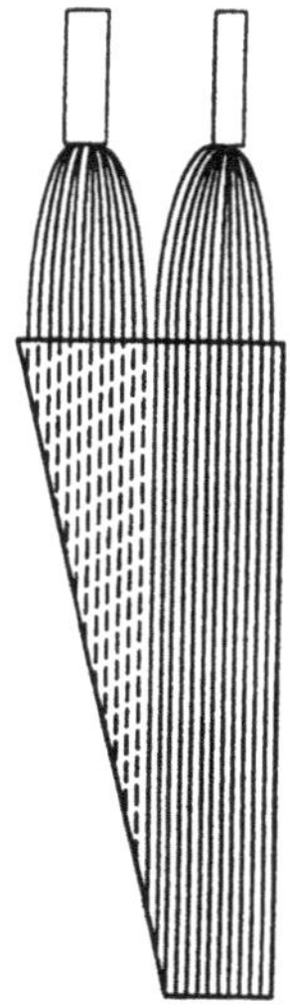

Figure 2: Side view of a calorimeter module (not to scale), illustrating the readout segmentation.

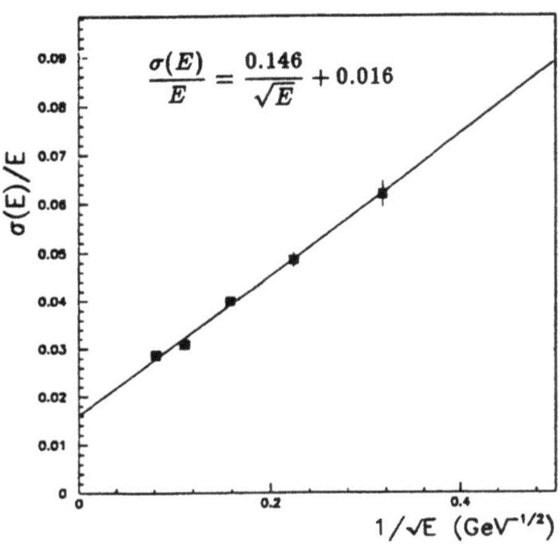

Figure 3: Energy resolution for electrons as a function of $E^{-1/2}$ as measured at cell centre at a beam impact angle of 3°.

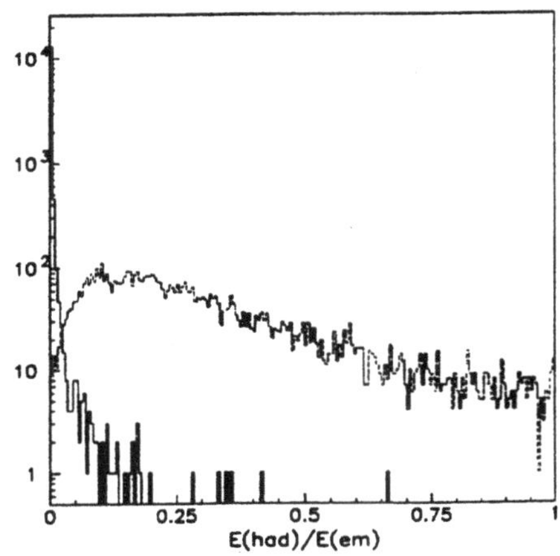

Figure 4: Ratio of HAD to EM energy deposition for electrons (full line) and pions (dashed line). Note the logarithmic scale.

coordinate and the center of gravity of the electron energy deposition (Y_{calo}) can be parametrized as :

$$Y_{corr} = 14.6 \ \arctan(0.45 \ Y_{calo})$$

After applying this correction we obtain a space resolution averaged over the surface of a cell of 0.88 mm for 80 GeV electrons. Such a resolution has been measured to depend on the impact point in the cell and ranges from $\approx$ 0.6 mm at the cell boundary to $\approx$ 1.1 mm at cell centre. The energy dependence of the position resolution was studied at cell centre with electrons from 20 to 150 GeV. After unfolding the resolution of the beam defining chambers, we find a dependence of the form:

$$\sigma(Y_{corr}) = \frac{9 \ mm}{\sqrt{E(GeV)}}$$

Similar results are found for the horizontal coordinate.

3.3 Electron-pion separation

The readout segmentation of the calorimeter can be exploited to discriminate between electrons and pions. In a preliminary way this was studied with 80 GeV electron and pion beams hitting the boundary between two cells. To improve the beam purity, a preshower counter made of 1.7 X_0 of lead, followed by a scintillator slab was used. Pions were required to give a signal smaller than 1.5 minimum ionizing particles (mip) in the preshower counters, whereas electrons were defined by requiring a signal greater than 6.7 mip.

The variable used for this study was the ratio between the energy deposited in the HAD section and the energy deposited in the EM section, both summed over 19 cells. The distribution of this variable for electrons and pions is shown in fig. 4. Requiring that the value of this variable be smaller than 0.01, 96% of the electrons are kept, whereas only 0.3% of the pions pass the cut.

The properties of cuts employing the longitudinal segmentation were also studied as a function of energy, yelding results of the same order of magnitude of the one quoted above.

3.4 Number of photoelectrons per GeV

In order to measure the number of photoelectrons per GeV, special electron runs were taken at 10 GeV with neutral attenuation filters in front of the PM. The energy resolution was studied as a function of the filter reduction factor for 3 different attenuation filters. The result of the measurement is 471 $\pm$ 23 photoelectrons per GeV deposited in the calorimeter.

References

1. D. Acosta et al. Nucl. Inst. and Meth. **A308** (1991) 481

PRELIMINARY RESULTS ON A LEAD/500 MICRONS SCINTILLATING FIBER PROTOTYPE THE RD1 COLLABORATION

Luc Poggioli, CERN, Geneva, Switzerland

1 Introduction

The RD1 collaboration aims to build a lead-scintillating fibers calorimeter with a good matching of both electromagnetic and hadronic performances to the LHC physics requirements, while keeping as much as possible all the nice features provided by this technology and observed experimentally by the SPACAL collaboration in the framework of the LAA project [1].
In that spirit the RD1 collaboration has built and tested 2 m long industrial projective modules with 1 mm diameter fibers and two read- outs providing e/π separation [2].
In the same spirit the very stringent requirements on calorimeters for LHC have led the RD1 collaboration to move to a calorimeter segmented in an electromagnetic and hadronic compartment with a different sampling fraction in each while maintaining the ratio 4:1 in volume for Pb:fibers to achieve compensation. The advantages thus obtained are substantial:

1. the electron identification becomes "natural".
2. one gets a better electromagnetic resolution and uniformity (including the possibility to tilt this part at larger angles w.r.t. the beam to decrease channeling effects).
3. Moving to 2-3 mm diameter fibers in the hadronic part should still provide good enough energy resolution for jets ($\simeq 45\%/\sqrt{E}$).
4. One thus gains in price and feasibility.

All these considerations led the RD1 collaboration to study an electromagnetic prototype using 500 μm diameter fibers with a ratio Pb:fibers of 4:1 in volume.
This prototype has been tested during August-September 1991 in the H2 beam of SPS at CERN with electron and pion beams.

2 The calorimeter

The lead structure (stiffened with 2% antimony) is made of sheets 1 mm thick in which are grooved U-shaped slits to insert the fibers. The grooves are obtained by machining. After inserting the fibers, the plates are stacked together. No glue nor soldering has been used.
The fibers are 500 μm diameter polystyrene plus 3HF and PTP used as dopants (yellow light). They are produced by KURARAY Company in Japan and are radiation hard. Each fiber is equipped on the front face with a mirror deposited by sputtering technique (reflection coefficient 85 %).
This leads to a non-projective structure of 8 x 8 x 30 cm^3 active zone (42 X_0 long).

For the read-out the fibers are bunched in 4 different groups, which represents a 4 cells equivalent prototype with a surface of 4 x 4 cm^2, corresponding to a granularity $\Delta\eta \ X \ \Delta\phi = (0.03)^2$ at a radius of 1.3 m.
Each fiber bundle is bunched in an hexagonal shape then diamond machined. The light is mixed via an hexagonal light guide which is then connected to a phototube Philips XP2961 (green extended 1" photocathode).

3 Preliminary results for electrons

3.1 Energy resolution

We investigate the energy resolution between 10 and 150 GeV for an incident angle $\theta_z = 2^0$. One gets (Fig.1): $\sigma/E = (9.3 \pm 0.3)\%/\sqrt{E} + (0.67 \pm .04)\%$.
This has to be compared to the previous results obtained with 1 mm fibers and $\theta_z = 3^0$: $13\%/\sqrt{E} + 1.2\%$.
The gain in the sampling term is $\sqrt{2}$, as expected.
The gain in the constant term comes from less contribution due to channeling effects (even smaller for γ's), and fiber-to-fiber fluctuations (divided by $\sqrt{4}$).
This resolution is found to be unchanged when using a $1.5X_0$ preshower upstream, and the constant term is expected to improve for larger incident angles ($\sim 6^0$).

3.2 Uniformity

A uniformity scan has been performed with 40 GeV electrons on the central region of the calorimeter to avoid side leakage (4X4 cm^2). The result shown in Fig. 2 exhibits clearly a pattern in the response leading to a non-uniformity in the response with $\sigma^{RMS} = 2.17\%$.
This effect is under investigation. The presence of a pattern (absent on previous prototype with 1 mm fibers) indicates that this non uniformity comes from the bunching and machining of the 500 μm fibers and the way light is collected.

3.3 Light yield

The statistical fluctuation coming from the number of photoelectrons produced shows up in the resolution as a contribution to the $1/\sqrt{E}$ term (to be added in quadrature to the sampling fluctuation contribution).
We used different filters to change the effective light yield and observed the degradation of the resolution. This gives a light yield of 500 pe/GeV.
This comfortable number (even with a small sampling fraction of 2.3%) can be increased if needed by using fibers with larger numerical aperture.

3.4 Linearity

A non linearity of 2% is observed between 10 and 150 GeV. This effect is coming from the small attenuation length $\lambda_{att} = 1.5\ m$ measured for short fibers due the extra light propagating in the cladding. This effect can be easily solved by black painting the end of the fiber to eliminate the cladding light.

3.5 Position resolution

Comparing the coordinates of the impact point defined by the beam wire chambers to the shower center of gravity defined by $x(y) = \Sigma_{i=1,4}x(y)_i E_i^\alpha / \Sigma_{i=1,4} E_i^\alpha$ with $\alpha = .3$, one gets (Fig. 3)

$$\sigma_{x,y} = 2.5/\sqrt{E}\ mm + .45\ mm$$

for electrons shot in the center of the calorimeter. The intrinsic beam chambers resolution has not been unfolded and contributes to the constant term.

3.6 Two showers separation

We investigated the possiblity to identify 2 neighboring electromagnetic showers based on the lateral energy deposited. This possibility can help in rejecting the π^0 background in the $H \to \gamma\gamma$ channel.

Comparing the lateral profile of one 80 *GeV* shower to the one obtained with 2 X 40 *GeV* showers, one gets for 70 % efficiency on one shower a misidentification of 20 % for a distance between the 2 showers of 1 *cm* (Fig. 4).

3.7 Pions showering

We investigated the effect of pions starting a shower in the EM part and then travelling through the bundle of fibers where the signal should be amplified (around 100 % sampling fraction compared to 2.3 % in the calorimeter), thus spoiling the hadronic response. The preliminary results show a very small effect.

References

[1] D. Acosta et al., CERN PPE-91-85, Submitted to NIM. references inside.
[2] G. Polesello, these proceedings.

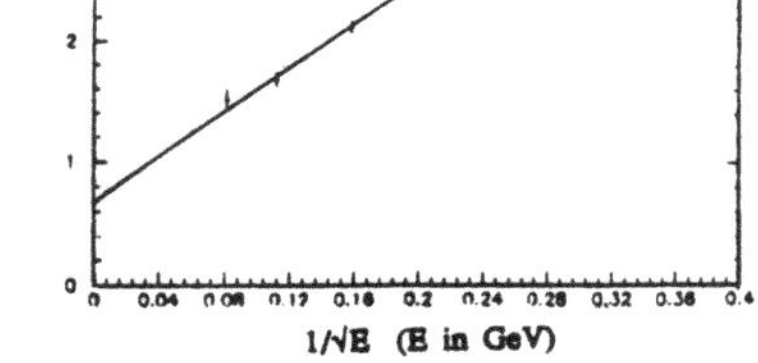

Figure 1: Energy resolution for electrons

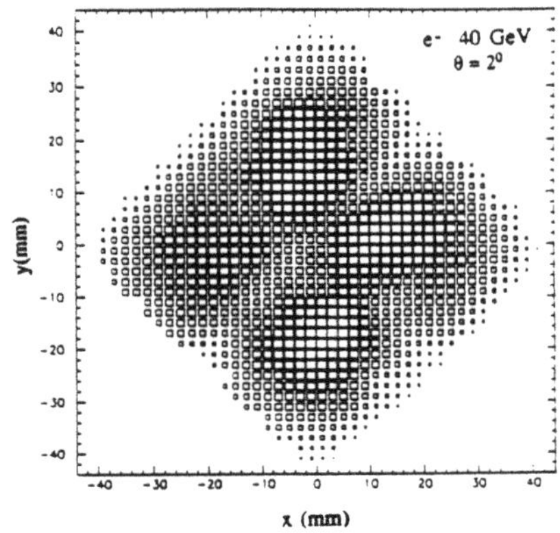

Figure 2: Uniformity scan for electrons. 1 box corresponds to 2x2 mm^2

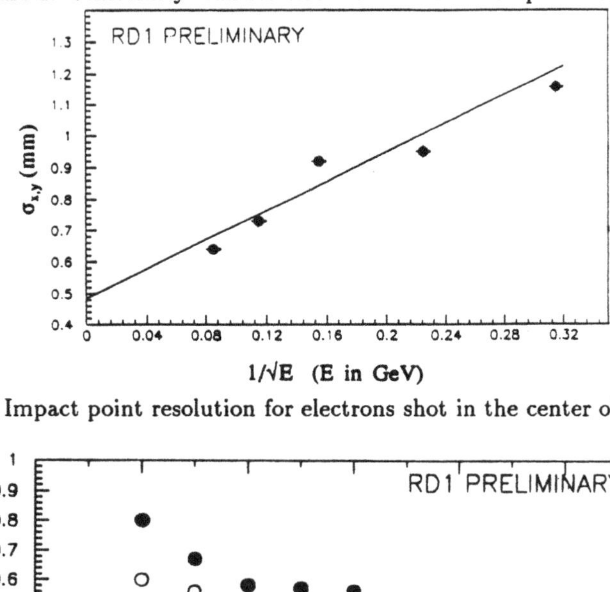

Figure 3: Impact point resolution for electrons shot in the center of the calorimeter

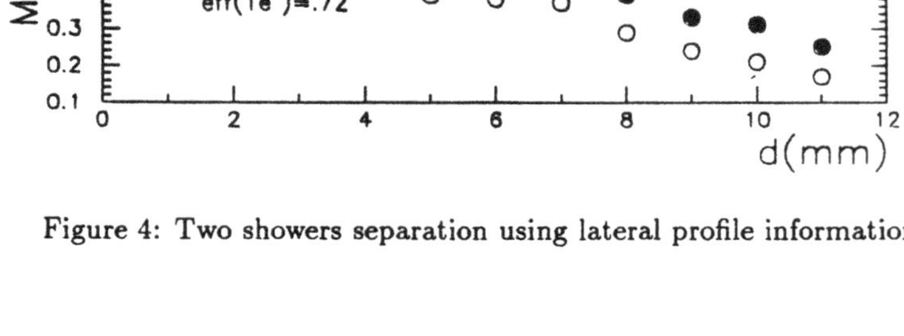

Figure 4: Two showers separation using lateral profile information

SCINTILLATION FIBER CALORIMETER PROTOTYPING

M. MISHINA
National Laboratory for High energy Physics, Tsukuba-shi, Ibaraki-ken, Japan
and
Fermilab, P.O. Box 500, Batavia, IL 60510, USA

ABSTRACT

We have made a design study of depth segmented scintillation fiber calorimeter with practicality for a large system for the future high energy hadron collider environment in our mind. We have now a comprehensive set of mechanical structure, associated fabrication technique, and assembly procedure. A preliminary study was made on the overall support structure and the cost estimate. As a result we believe this technology is rather simple compared to any other calorimeter technologies and hence the estimated cost, quite competitive with other technologies, should be reliable. Four prototype modules are under construction and will be tested at Fermilab soon.

Since the basic properties of scintillation fiber calorimeter were explored by SPACAL group, though in an idealistic configuration, our study was focussed on the practical design of the depth segmented projective tower calorimeter. The practicality has to prevail so that the overall design is adaptable to a large scale mass production.

What is apparent, as listed in Table 1, is the significant difference between the e.m. and the hadronic segments both in terms of the requirements and the physical parameters. The e.m. segment is about 1/10 of the hadronic segment in the length and about 1/20 in the volume.

Therefore by segmenting the calorimeter towers in depth, one can optimize the design of the two segments independently. The small size e.m. segment can be built with much greater care in order to achieve the quality required for e.m. signals. On the other hand large volume hadronic segment must be built economically because the looser requirement for hadronic signal measurement is not difficult to achieve.

Taking a segmentation of $\Delta\eta \times \Delta\phi = 0.05 \times \pi/64$, the design is schematically shown in Fig. 1. The fiber diameters are 1 mm for e.m. and 2 mm for hadronic segment, respectively. The scintillation fibers in the e.m. segment are spliced to clear fibers which penetrate through the hadronic segment and are connected to a phototube at the rear end of the hadronic segment. Only the e.m. segment is tilted by 6° in order to avoid channeling of electrons. The e.m. and hadronic segments are encased in steel sheathes separately and then connected together. The mechanical support is provided by a steel box which holds the steel sheath at the end of the hadronic segment. Phototubes are housed in this steel box. In order to minimize the total length, short, 6.5 cm long, phototubes are used. Due to the mesh dynode structure, these tubes have fast timing and large dynamic range.

Along with the mechanical design, we have developed the associated fabrication techniques and tools.

Table 1. Features of depth segmented scintillation fiber calorimeter.

Pb : FIBER = 4 : 1 (VOLUME)
X_0=7.4 mm , λ=20.3 cm

E.M.	+	HAD
25 ~ 30 X_0 (~ 1 λ)		10 λ
~20 cm		~2 meter
Volume = 5 %		95 %

FEATURES
- Easy-to-implement e/π separation
  - Trigger signals at earliest level
  - Higher level trigger and off-line analysis
  - *Lateral profile and timing are additional handles.
- Cleaner electron signal : No big antenna for noise
- Independent minimum ionizing signal for μ identification
- Possible weighting of E.M. and HAD for e/h compensation
- Independent optimization

E.M. - Isolated e, γ:
* Good σE/E ⇒ Fibers ≤ 1 ^{D}mm
  - Finer fibers for better homogeneity
  - Higher light output than longer attenuation length
  - Tighter dimensional tolerances
  - Fiber sorting
  - Fiber treatment...Aluminization
* Good hermeticity : tighter dimensional tolerance
* Possible smaller lateral segmentation
* Possible faster timing
* Needs to avoid channeling : Tilt of fibers > 6°
* Radiation peak
  - Radiation resistant fiber...Different fluor/doping
  - Possible special gas atmosphere
* Possible more sophisticated fabrication technique
* Possible independent support for hermeticity

HAD- Jets :
* Coarser energy resolution
  - Coarser homogeneity : Thicker fibers/Pb plates
* Behind 1 λ : Support structure
* Lower radiation level
* Good economy ⇒ Fibers ≥ 2 ^{D}mm
  - Simpler economical structure/assembly
* Compensation by neutrons : Longer integration time
* Need good projective tower for energy flow measrmnt
* No channeling : No tilt
* Long module
  - Longer attenuation length than higher light output

Diamond flycutters for cutting individual fibers and also epoxied bundle of fibers were built and has been in heavy use for the past several months. While our assembly procedure does not require finishing of the individual fiber end , single fiber cutter has been essential for any optical measurement of fibers in order to have consistent reproducibility.

Many techniques for making lead matrix have been tried . Casting eutectic lead around stainless steel pins which are extracted afterwards, casting calcium lead with stainless steel tubings,which may be left in the lead matrix or dissolved by acid, and bonding hexagonal lead tubings together with low temperature solders have been tried. All of these method worked successfully but were abandoned eventually due to the complexity of the procedure and or the additional cost for the material.

The method we have settled to is to cut grooves by a set of gang saws. Properly U-shaped saw blades in the number , about 50, corresponding to the number of grooves in a row are mounted on an arbor with precise spacings in between. The shape and the spacings of the grooves are completely controlled by the gang saw and the results have been quite satisfactory both for 1 mm grooves of e.m. segment and 2 mm grooves of hadronic segment. The operation is one-pass and the plate for hadronic segment, about 20 cm wide and 190 cm took 8 minutes without tuning. This technique is particularly suitable to calcium lead which has high strength, both in terms of hardness and in the resistance against long term creeping. The finish of the grooving is smooth and clean and even the 1.9 meter long plates is strong enough to be handled by a single person.

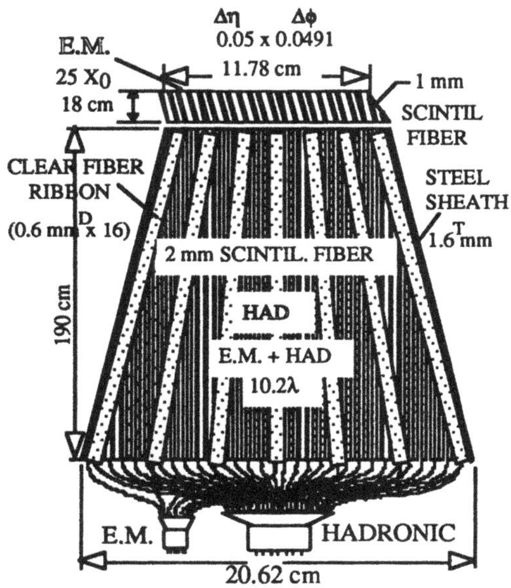

Fig. 1.: Longitudinally segmenterd scintillation fiber calorimeter.

One of the key elements in the depth segmentation design is the method of splicing fibers of e.m. segment to clear fibers for readout. Two methods have been worked out. One is to splice the fibers individually by welding them at the joint. An automatic machine has been developed and is operational. The other is a connector to splice the fibers collectively. Scintillation fibers and clear fibers faces each other being held in the same pattern by a pair of perforated plates. The accuracy of the alignment is controlled by numerically controled machine which usually have a tolerance of better than 10 μm over the necessary size, about 11 cm square, of the plates. Many such perforated plates have been made using aluminum and calcium lead. Calcium lead is used for the end of the scintillation fibers in order to maintain the same lead-fiber ratio throughout the e.m. segment.

One of the features of the scintillation fiber calorimeter is that the same fiber density can be maintained independent of the polar angles of the towers. Simply the shape of the module has to be adjusted at the front and the rear end of the e.m. segment so that the thickness of the e.m. segment stays constant. Then the angles of the holes on the perforated plate for the fiber splicing has to be inclined with the angles of the projective towers. Such plates with holes inclined by 35° corresponding to the end of barrel region have been also successfully made in the same technique as the plates for η=0. Thus we have developed all the necessary technique to fabricate depth segmented module.

An additional instrumentation is a fiber sorter to eliminate fiber to fiber variation and a semi-automatic unit is under development.

We are in the process of collecting the parts of four prototype modules based on such a design and we expect that the modules will be tested in the beam in November.

A conceptual design of how the calorimeter modules are supported has been made. Some number of modules are bundled together into a super module and connected to a supporting rib which is supported by an outer strong structure. This strong structure is a part of the solenoid flux return. Space has to be left inside for the access to the phototubes. The cost of the structure seems to be minor in the overall cost.

We now turn to the discussion on scintillation fibers. Polystyrene fiber has a sharp rise of loss below 500 nm and also the radiation damage occurs at shorter wave length. Therefore it is desirable to shift the light to green region. A possible candidate is 3HF which has a large Stokes shift and therefore does not have self absorption. Currently we are studying the effect of different fluor concentration.

Another issue is the cladding material. Ordinary fibers have PMMA cladding on polystyrene core. However PMMA is known to be weak against radiation and the effect is noticeable in long wavelength component. A possible remedy is fluorinated polymer which has been used for PMMA cored fiber. This material is much stronger against radiation, as shown in Fig.2 and there is some evidence of improvement in the attenuation length after high irradiation. A side benefit is the larger, 5 %, light capturing efficiency, compared to 3% for PMMA-Polystyrene combination due to the smaller refractive index. There is an effort by Kuraray to improve the mechanical affinity between polystyrene core and this material.

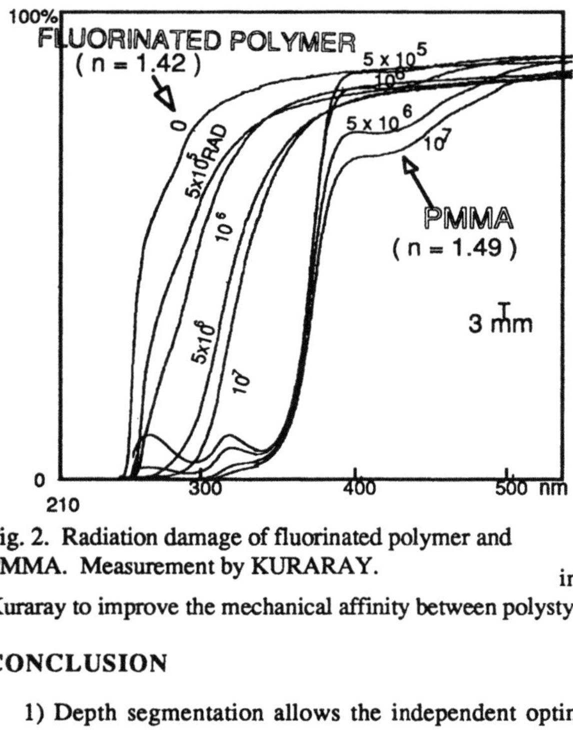

Fig. 2. Radiation damage of fluorinated polymer and PMMA. Measurement by KURARAY.

CONCLUSION

1) Depth segmentation allows the independent optimization of e.m. and hadronic segments.

2) We have developed a comprehensive set of mechanical design, associated fabrication techniques and toolings, and the assembly procedure for a depth- segmented scintillation fiber calorimeter with projective tower geometry.

3) Scintillation fiber calorimeter is a simplest form of scintillator based calorimeter and strongest against radiation damage.

4) Assembly is simple and therefore the cost is competitive.

ACKNOWLEDGEMENT

This work was supported by the funding from US Department of Energy for SSC R&D.

PERFORMANCE OF A LEAD/SCINTILLATING-FIBER CALORIMETER PROTOTYPE AT SHORT GATE WIDTHS

CHRISTINE V. SCHEEL
NIKHEF/H, Amsterdam, the Netherlands

Representing the SPACAL Collaboration

ABSTRACT

The performance of a lead/scintillating-fiber electromagnetic and hadronic calorimeter is studied using ADC gate widths down to those compatible with LHC/SSC timing. It is found that the good performance is hardly affected for gate widths down to ~40 ns. Below that, we find incomplete charge collection, an increase in energy resolution, a rise in e/π, a narrowing of the hadronic lateral shower profile, and worse e-π separation. However, even for gates as short as the LHC/SSC bunch spacing, these effects are no unacceptably large.

1. Introduction

At the future multi-TeV pp colliders LHC and SSC, event rates will be very high, with a bunch crossing every ~15 ns. The detectors at these colliders will have to be able to recognize interesting events and provide selection criteria for the trigger and data acquisition system within a short time interval.[1] In particular, the energy measurement provided by the calorimeter will be needed for the first-level trigger decisions, so the charge collection should be as fast as possible. A fast calorimeter is also needed to avoid a pile-up of unrelated events.

The spaghetti calorimeter (SPACAL), an integrated electromagnetic (e.m.) and hadronic calorimeter, is a type of calorimeter intended for use in a detector on such a collider. In order to see how the calorimeter would perform under such circumstances, a prototype was tested using various ADC gate widths. Below, the results of these studies are briefly presented, showing the relative effects of short charge collection times.[2] In Sect. 2, the calorimeter and the beam tests are described. The results are given in Sect. 3. A summary and conclusions are found in Sect. 4.

2. Description

2.1. *The Calorimeter*

The SPACAL prototype is composed of layers of extruded lead and plastic scintillating fibers running longitudinally, the latter constituting ~20% of the volume. The fibers are 1 mm in diameter and 2.2 m long and are aluminized on the front. With such a composition, the radiation length amounts to 7.2 mm, the nuclear interaction length (λ_I) to 21 cm, and the Molière radius to 2.0 cm. In full the calorimeter has a roughly cylindrical shape, with a diameter of ~1 m (4.8 λ_I) and a length of 2 m (9.5 λ_I), large enough to contain >97% of an hadronic shower. At the back, the fibers from each of the 155 cells are bunched and coupled to a light guide and a photomultiplier. One cell contains ~95% of an e.m. shower. The calorimeter is almost compensating, with $e/h = 1.15$.[3]

2.2 *The Testbeam and Data*

The prototype was tested in June 1990 in a testbeam at CERN. Measurements were done with gate widths ranging from 5 to 358 ns, where *gate width* here means the time over which the charge is collected, from the beginning of the signal to the end of the ADC gate. The normal gate width was 358 ns, enough to collect all of the charge. Beams of electrons and pions from 5 to 150 GeV were steered into the calorimeter at an angle of 3° with the fiber axis.

3. Results

The signals produced by the calorimeter depend on whether they are resulting from an e.m. or an hadronic shower.[4,5] An e.m. shower is produced through fast processes, such as ionization and Bremsstrahlung, so the corresponding signal from the scintillation light is also fast. In addition, each shower is very similar to another, producing very uniform signals. An hadronic shower, on the other hand, is much larger and contains a slow component (time constant ~ 10 ns[4]) from soft neutrons, and thus it takes longer to collect the entire signal. In addition, the production of scintillation light occurs deeper on average for hadronic showers than for e.m. ones. The light that is reflected by the the mirror then travels further before reaching the PM, contributing to the tail of the hadronic signal. If the charge collection time is shortened, then it is clear that the pulse from the hadronic shower will be cut more than that from the e.m. shower.

3.1. Charge collection

The fraction of charge collected for 80 GeV particles at various gate widths is shown in Fig. 1a. One sees that 90% of the charge is integrated after 80 ns for electron and 90 ns for pions.

With such fast charge collection, we discovered that the long coaxial signal cables that had been used for these tests lengthened the calorimeter signals considerably. The effect of the cables was studied in detail and could be well-described by a sum exponentials, with an overall rise time of ~65 ns. This could then be deconvoluted from the signal by software, which was done whenever possible for the results here in order to provide a clearer picture of the calorimeter behavior.

Fig. 1b shows the fraction of charge collected with the cables unfolded. There is a drastic improvement: 90% of the charge is collected in 37 ns for electrons and 47 ns for pions.

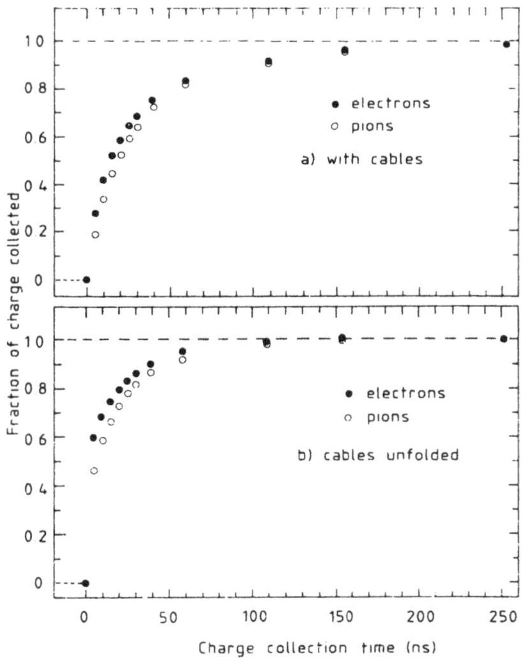

Figure 1. The fraction of charge collected as a function of the charge collection time, as measured **(a)** and with cables unfolded **(b)**, for 80 GeV particles.

3.2. Energy Resolution

Since the pulse from an e.m. shower is quite narrow, there is little or no change in the e.m. energy resolution down to gate widths of ~40 ns (see Fig. 2a). As the charge collection time is decreased and some of the signal is lost, the e.m. signal distribution becomes less Gaussian and the resolution gets worse.

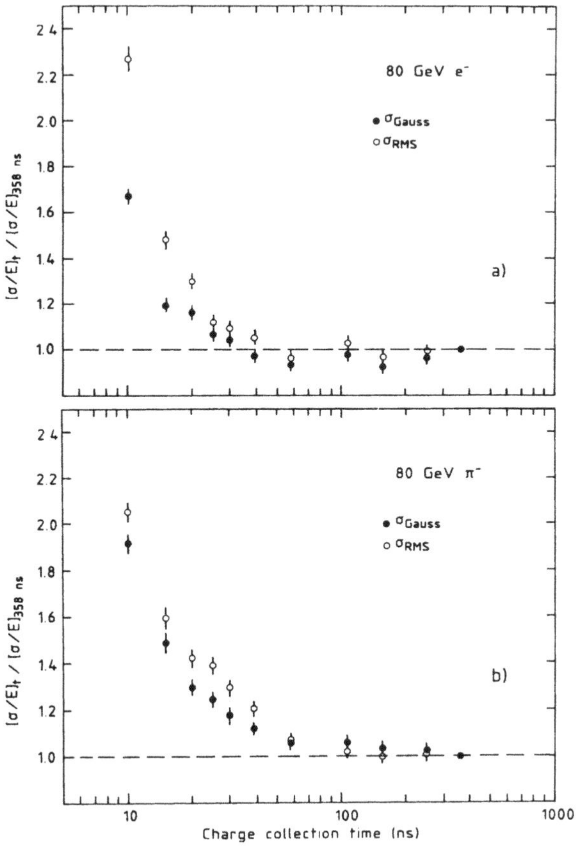

Figure 2. The e.m. (**a**) and hadronic (**b**) energy resolution, normalized to the values at the full gate width, as a function of the charge collection time, for 80 GeV particles.

The signal from an hadronic shower is longer than that from an e.m. shower and includes a slow tail, so that charge from the hadronic signal is lost at slightly longer gates. Down to ~60 ns, however, there is a less than 10% increase in the hadronic energy resolution (Fig.2b), so only the tail is truncated. As the charge collection time is shortened the distribution becomes less Gaussian, until the resolution roughly doubles at ~15 ns.

3.3. e/π Signal Ratio

Since a larger fraction of the hadronic signal (π) is lost as compared to the e.m. signal (e) for the same charge collection time, the e/π signal ratio increases at shorter times (Fig. 3). However, as seen in Fig. 1, the difference in the fraction of charge collected is not very large, and therefore the increase in e/π is not so sharp. For example, it is less than 10% larger at a 25 ns gate for 80 GeV particles. Since the calorimeter effectively becomes less compensating at short gate widths, the hadronic signal becomes less linear with respect to the energy.[2]

3.4 Hadronic Lateral Shower Development

Most of the slower components of an hadronic shower develop longitudinally deeper in the calorimeter and laterally further away from the impact point. As the gate width is decreased, the hadronic shower profile becomes narrower, with a larger decrease in signal at a larger distance from the shower center. One consequence of this is that e.m. and hadronic shower profiles start to look alike at short gate widths, making e-π separation more difficult. However, position resolution and the particle separation capability should improve.

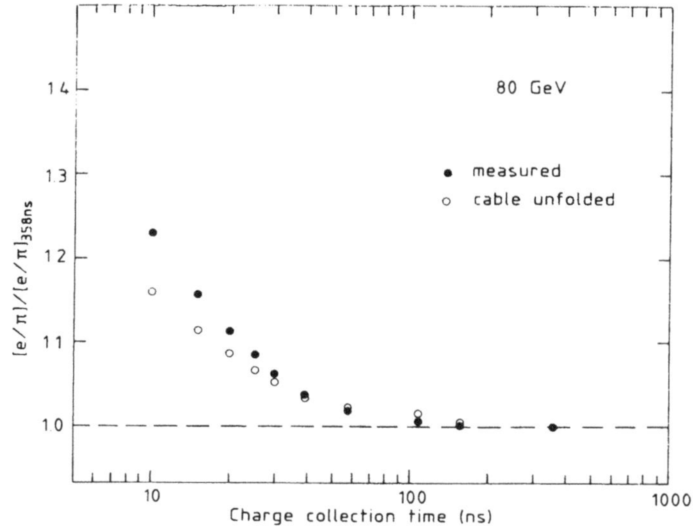

Figure 3. The e/π signal ratio, normalized to the value at the full gate width, as a function of the charge collection time.

3.5. Electron-Pion Separability

Since it has no longitudinal segmentation, the SPACAL relies on the differences in lateral shower profiles and in signal shape between electrons and pions in order to distinguish between them. For the many methods studied, pion rejection factors of 500 to 1000 were found for an

Table 1. Summary of the calorimeter performance at long and short gate widths.

At charge collection time:	**358 ns**	**15-16 ns**
e- charge collection	100%	52% measured 75% with cable unfolded
π- charge collection	100%	45% measured 65% with cable unfolded
e.m. energy resolution	$\frac{\sigma}{E} = \frac{12.9\%}{\sqrt{E}} + 1.2\%$	20% increase (Gaussian fit) 50% increase (RMS)
hadronic energy resolution	$\frac{\sigma}{E} = \frac{30.6\%}{\sqrt{E}} + 1.0\%$ without corrections: $\frac{\sigma}{E} = \frac{29.5\%}{\sqrt{E}} + 2.2\%$	without corrections: $\frac{\sigma}{E} = \frac{61.2\%}{\sqrt{E}} + 3.1\%$
$\frac{e}{\pi}$ signal ratio	$\frac{e}{h} = 1.15$	at 80 GeV $\frac{e}{\pi}$ increases 16% (measured) and $\frac{e}{\pi}$ increases 12% (cables unfolded)
hadronic lateral shower profile		effective radius reduced by 10%
e-π separability by lateral shower profile	pion rejection factors ~500-1000	pion rejection factor worsens by factor 2-5

electron efficiency of ~99%.[5,6] One method is based on finding a quantity which gives the effective shower radius, where the hadronic shower radius is distinctly bigger than the e.m. radius. The e-π separability was found to get worse at shorter gate widths, as the effective hadronic shower radius got smaller and closer to the e.m. radius.

4. Summmary and Conclusions

The performance of the SPACAL lead/scintillating-fiber calorimeter prototype is summarized in Table 1 for gate widths of 358 ns, the nominal value,[3,6] and of ~15 ns, the time between LHC/SSC bunch crossings.

After studying the performance of this calorimeter with charge collection times ranging from 5 to 358 ns, we see that it is practically unaffected down to ~40 ns, even with the effects of long signal cables. No signal shaping was done, nor even necessary. Even at a charge collection time corresponding to an LHC/SSC bunch crossing interval, the quality of the calorimeter information is still very acceptable.

References

1. S. Rescia, Readout of Calorimeters at High Luminosity Colliders, these proceedings; G. Jarlskog and D. Rein (eds.), Proc. Large Hadron Collider Workshop, Aachen 1990, Vols. I and III, CERN 90-10/ECFA 90-133.
2. A full report of these results can be found in the following preprint: D. Acosta *et al.*, The Performance of a Lead/Scintillating-Fiber Calorimeter at LHC/SSC Compatible Gate Widths, preprint CERN-PPE/91-195 (1991).
3. D. Acosta *et al.*, Nucl. Instr. and Meth. **A308** (1991) 481.
4. R. DeSalvo *et al.*, Nucl. Instr. and Meth. **A279** (1989) 467.
5. D. Acosta *et al.*, Nucl. Instr. and Meth. **A302** (1991) 36.
6. D. Acosta *et al.*, Nucl. Instr. and Meth. **A305** (1991) 55.

JET STUDIES IN SPACAL: THE COVARIANCE MATRIX APPROACH

The SpaCal Collaboration

Amsterdam - Cagliari - CERN - CERN LAA - Lisbon
Napoli - Paris VI - Pavia - Rio de Janeiro
UCSD - Weitzmann

ADELE RIMOLDI
Dipartimento di Fisica Nucleare e Teorica Universitá di Pavia
and Istituto Nazionale di Fisica Nucleare, Sezione di Pavia
ITALY

Abstract

Results of a study for improving particle separation in high energy calorimeters through a general technique based on the covariance matrix approach are presented. Using a sample of showers produced in the SpaCal calorimeter the observed pattern of energy deposition is parametrized in terms of a standard covariance matrix and error matrix formalism. A χ^2 test is used to define the goodness of the fit.An event with a pion near a jet or a jet near another jet does not fit a typical jet event and, as a consequence, it is flagged by a large χ^2. 80 GeV pions within 120 mm of similar particles are identified with 90 percent efficiency. 80 GeV jets can be identified within the same distance of 80 GeV jets with an efficiency of about 70 percent. These results are obtained with a probability smaller than 10^{-3} for misidentifying an isolated jet or pion as having a pion accompanying it. Similar results are obtained when the energy is 40 or 150 GeV both for pions and jets. Internal jet structure is also investigated and an estimation of internal multiplicity is presented as a function of the χ^2 cut.

1 Method

The high granularity of the SpaCal calorimeter is well suited for studies based on lateral shower energy deposition aimed at identifying and separating incident single particles and jets. Beams of known particle composition (electrons,pions or jets) and of given energy were used. For details on the SpaCal apparatus see ref.1 - 5 together

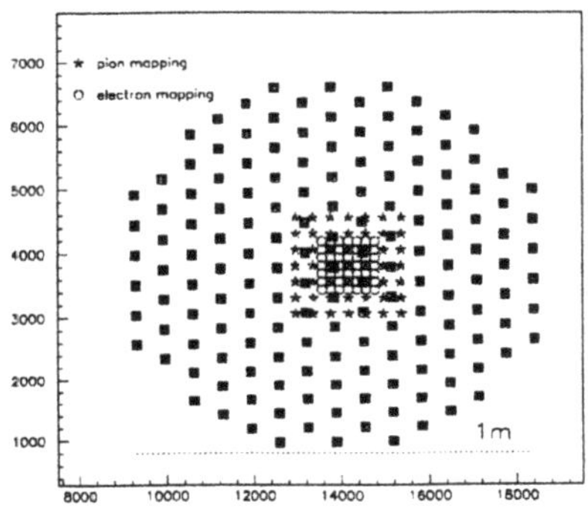

Figure 1: Diagram of the front surface of the 155 cells calorimeter prototype showing the center of the cells (black squares) and the mapping of beam impact points for this analysis(stars:pions, circles:electrons)

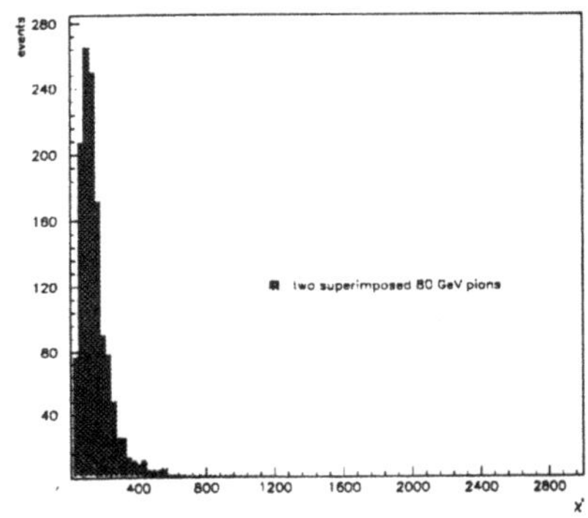

Figure 2: χ^2 distribution for two 80 GeV pions hitting the calorimeter front face at the same point

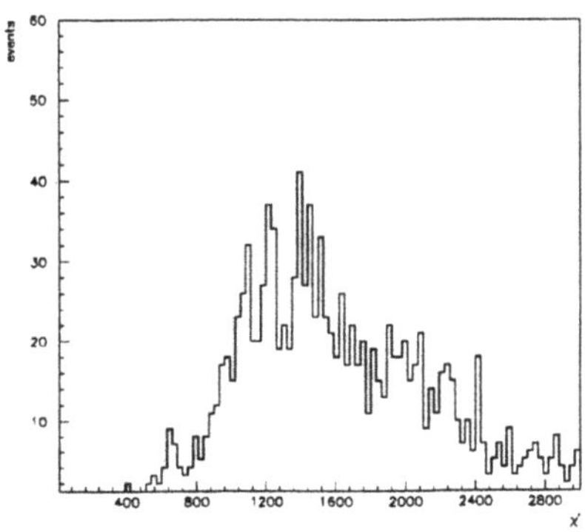

Figure 3: χ^2 distribution for two 80 GeV pions hitting the calorimeter front face at 160 mm distance

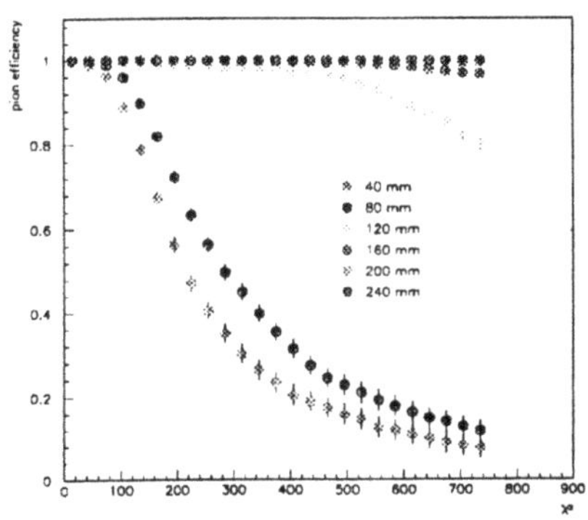

Figure 4: π efficiency for distinguish two pions as a function of the χ^2 cut when different hypotheses in the distance between pions are taken

with its preminent results. In Fig.1 the front surface of the SpaCal calorimeter is shown with its exagonal global composition. This calorimeter is segmented in 155 modules whose centers are shown in the figure together with the mapping impact points available with electron and pion beams. All the available data are taken at incident angle of 3 degrees and different energies (40, 80, 150 GeV).
From the known segmentation of the calorimeter the measured total energy E is given by the total sum over the single calorimeter elements:

$$\mathrm{E} = \Sigma_{i=1,j}\ E_i$$

(j is the number of calorimeter cells partecipating in the event). The covariance matrix $M_{i,j}$ is defined as:

$$M_{i,j} = N^{-1}\ .\ \Sigma_{n=1,N}\ (E_i^n - \overline{E}_i)(E_j^n - \overline{E}_j)$$

where N is the total number of events in the sample, i and j are the running cell labels, $\overline{E}_i$ and $\overline{E}_j$ are the average energy in the cell i and j respectively. The inverse of the covariance matrix is then the error matrix $\epsilon_{i,j}$. Given an event with a certain pattern of energy deposits in the calorimeter it is possible to calculate the related χ^2 as:

$$\chi^2 = \Sigma_{n=1,N}\ (E_i^n - \overline{E}_i)\ \epsilon_{i,j}\ \ (E_j^n - \overline{E}_j)$$

This variable allows to quantify the similarity of any given event to those in the control set.

2 Pion Separation

In order to study the problem of particle separation in our calorimeter using the covariance method, 80 GeV pion data were considered at first as a calibration. Two pions beams of known energy at different impact points were superimposed varying the distance between the two impact points. Then a covariance matrix was built using pion data at twice this energy. The hypothesis of two superimposed pions was considered and a χ^2 was calculated using this covariance matrix (Fig.2). Only one pion with a χ^2 greather than 1000 was found. A χ^2 was also calculated using the hypothesis of two pions at different impact points showing a sharp dependence on the separation between the two beams (Fig.3).It is then possible to quote the confusion probability between two pions (Fig.4) for different impact points as a function of the χ^2 cut. At least in 90 percent of the cases the minimum distance required to distinguish the two components is 120 mm for pions at 80 GeV. This situation corresponds, at present calorimeter segmentation, to a distance of more than one cell between the two pion beams.

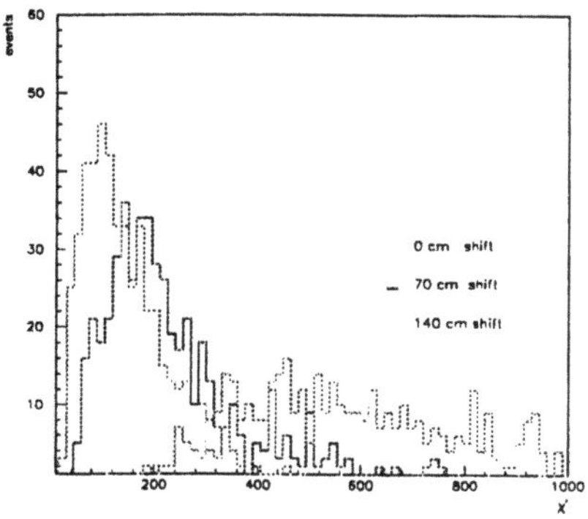

Figure 5: χ^2 **distribution for two 80 GeV jets hitting the calorimeter front face at different points.The shaded area represents the** χ^2 **distribution for two jets superimposed**

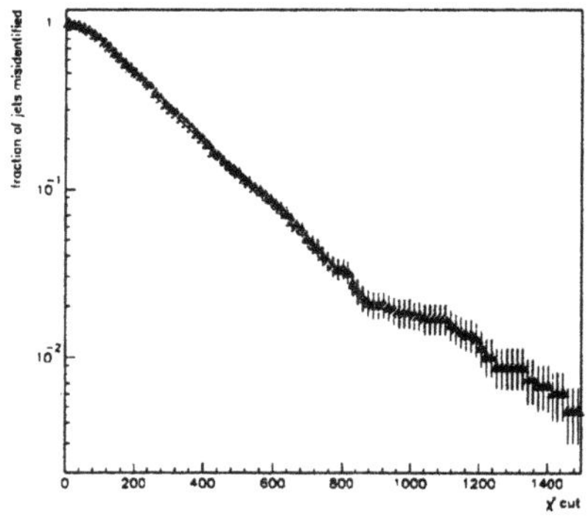

Figure 6: Fraction of misidentified 40 GeV jets as a function of the χ^2 **cut**

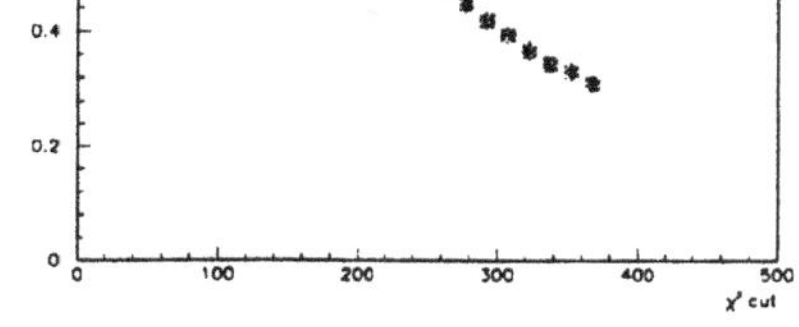

Figure 7: jet-jet identification efficiencies for a 40 GeV jet sample as a function of the χ^2 **cut**

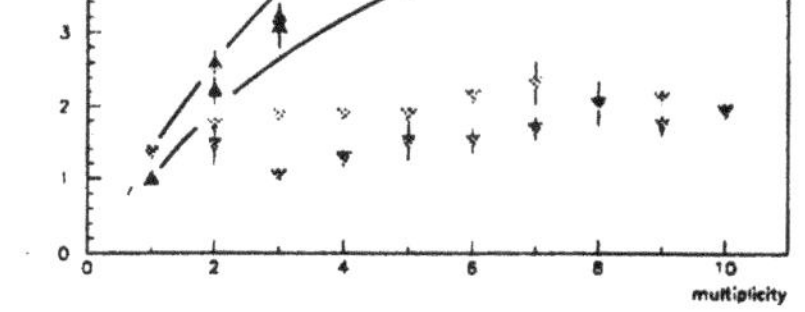

Figure 8: Distribution of the ratio between the mean χ^2 **and the number of degrees of freedom as a function of multiplicity when the far (144 cm) and the close (44 cm) target position is chosen (dashed curves are drawn to guide the eye)**

3 Jet Separation

Pions that interact in a paraffin target upstream the calorimeter are defined in this study as jets. Pions and jets produced in our test beam have very similar transverse shower developements. Fig.5 shows the χ^2 distribution for 80 GeV jets when the two impact points of the jets considered are separated by 0, 1, 2 calorimeter cells. As with pions it is then possible to show the fraction of misidentified jets as a function of the χ^2 cutoff (Fig. 6). This distribution does not change very much with incident energy and shows that this fraction falls to a few percent when the χ^2 cutoff is as low as 800. For the same value of χ^2 cut the identification efficiency is only 30 percent when the barycenters of the two jets are separated by only one calorimeter cell. This value reaches above 70 percent when the separation is greater than one cell (for 40 GeV jets) (Fig. 7).

4 Internal Jet structure

The same technique used above is also suitable for evaluating the internal composition of a jet. In this case it is possible to calculate the χ^2 distribution as a function of multiplicity for different jet samples. Among the data taken in test run period there are jet samples taken at close distance from the front face of the calorimeter prototype (44 cm) and farther upstream (144 cm) at different energies (40, 150 GeV). Very different behaviour is indeed seen when, at the same energy, jets produced at different distances are considered (Fig.8). When the close position is considered, the distribution flattens and it is impossible to distinguish high multiplicity from low multiplicity jets. In the case of far position it is possible to distinguish a χ^2 band with low associated jet multiplicity from a χ^2 region in which high jet multiplicity dominates independently of jet energy.

References

[1] D. Acosta et al. : Nucl. Instr. and Meth. **A294** (1990) 193.

[2] D. Acosta et al. : Nucl. Instr. and Meth. **A305** (1991) 55.

[3] R. De Salvo et al. : Nucl. Instr. and Meth. **A279** (1989) 467.

[4] D. Acosta et al. : CERN - PPE/91-85 (1991) to be published in N.I.M.

[5] D. Acosta et al. : CERN - PPE/91-107 (1991) to be published in N.I.M.

DETECTING AND TRIGGERING ON NEUTRONS USING THE SPACAL Pb - SCINTILLATING FIBER CALORIMETER *

A. Simon
Heidelberg University, Physikalisches Institut, D-6900 Heidelberg, FRG

Abstract

We present an application of the SPACAL spaghetti calorimeter prototype [1, 2] in the CERN hyperon beam experiment WA89. First results on the performance of the upgraded SPACAL calorimeter in a high rate experimental environment are presented. In order to reconstruct invariant masses from Σ^- hyperons in final states, the energy of the decay neutrons and their impact point have to be determined with the best possible accuracy. As a first step in the analysis, Σ^- beam particles have been easily reconstructed. For future runs we intend to trigger on the neutrons from Σ^- decays using fast calorimeter information and a scintillating tile predetector. Testbeam results on both components will be presented.

1 Motivation

The performance of the almost compensating (e/h=1.15) electromagnetic and hadronic SPACAL spaghetti calorimeter has been extensively investigated in two *testruns* in 1990, and many results on its intrinsic properties as well as on shower characteristics have already been published[2, 3, 4, 5]. The first *application* of this calorimeter prototype was recently made in the hyperon beam experiment WA89[6] at CERN aiming to measure the neutrons from Σ^- particles ($\Sigma^- \rightarrow n\pi^-$, $BR = 99.85\%$) steaming from charmed-strange baryon or "exotic" multiquark state decays. The WA89 experimental setup (fig.1) - which was up to then only sensitive to Λ hyperons in the final states (via $\Lambda \rightarrow p\pi^-$, $BR = 64.1\%$) - was extended for 2 1/2 weeks in 1991 by placing the SPACAL under an angle of 0^0 behind the Pb-glass electromagnetic calorimeter (which was moved out of the beam for 1 week). As the physical processes to be studied by the hadron calorimeter are rare processes (order of several hundreds of events per year), a trigger on final states with high energy neutrons is essential for future runs.

*This work was performed in the framework of the CERN hyperon beam experiment WA89

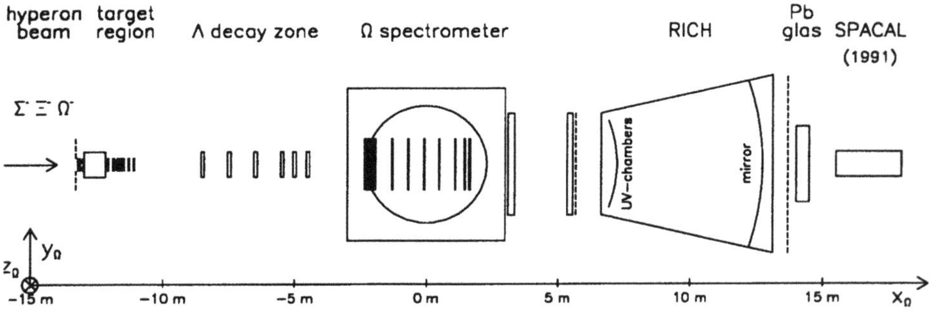

Figure 1: *WA89 experimental setup in 1990 and 1991*

2 Upgrade of the SPACAL prototype

In contrast to testbeam conditions, the calorimeter is exposed in the WA89 experiment to almost 10^6 hadronic particles ($\Sigma^- and\, \pi^-$) per spill, making some essential changes to the original prototype necessary. Each PM base was equipped with a booster system to stabilize the voltage between the last 2 dynodes; this improved the high rate capability by 2 orders of magnitude. The heat development in the closed readout section reached several *Watt* per module and required an efficient cooling system. The PM outputs were split, one branch going to a 12/15 bit ADC with automatic range switch, the second one being used for trigger electronics tests (see sect.5). Finally, a flash lamp monitor system distributing the light via clear fibers to the individual modules and - for the most irradiated modules - a LED monitoring system were installed [7]. It should be stressed that both systems only monitor the readout chain from the PM on; in order to look for possible fiber damage, the calorimeter was calibrated before and after the main run in WA89. The front face of the calorimeter was opened to air (for better scintillator recovery in case of damage) by replacing the previous black plexiglas plate by a light tight cloth.

3 Neutron reconstruction and Σ^- invariant mass determination

High energy neutrons are detected by looking for hadronic showers in SPACAL with a minimum energy of 40 *GeV* and a reconstructed shower barycenter that cannot be linked to a track found in the spectrometer. To determine the direction of the neutron, one further needs the coordinates of the Σ^- decay vertex from tracking information. This is quite straightforward to do for beam trigger events where a clean sample of Σ^- can be obtained by requiring a track in the vertex detector and a single track in the spectrometer with a minimum kink angle between both. The neutrons hit the central region of the calorimeter whereas the bending by the strong Ω magnetic field prevents more than 90 % of the decay π^- to hit the calorimeter surface (as shown in the event display in fig.2). This fraction should even be higher for the lower energy π^- originating from Σ^- decays in interaction events.

Fig.3 shows the invariant mass distribution for 330 GeV/c beam particles. The mean value of this distribution is offset from the PDG value by about 5 MeV/c^2 and the width is about 29 MeV/c^2 (still a factor 2-3 too large w.r.t. MC simulations). It should be pointed out, however, that this mass plot was obtained with online calibration constants, with no corrections whatsoever (see sect.4 for the importance of the monitoring) and with Pb-glass in front of the SPACAL calorimeter.

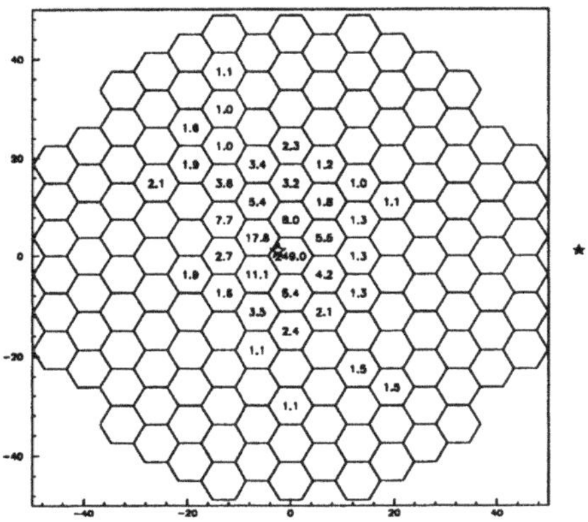

Figure 2: *Event display of the neutron impact point (open star) on SPACAL and the reconstructed π^- track (black star) (Display cut of $\geq 1\,GeV$ energy deposit / module).*

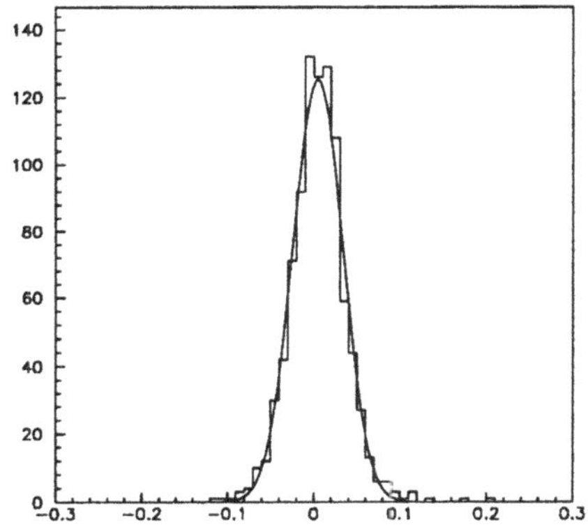

Figure 3: *Invariant mass distribution for Σ^- beam particles, plotted as the difference from the PDG mass. Preliminary result.*

4 Response stability and damage - first observations

First results from the analysis of the monitor data show two kinds of sizable corrections to be applied to the experimental data.

Long-term response variations of larger extent have been observed in the data for modules around the impact point of the non-interacting beam particles (maximal effect: 50 % over 2 1/2 weeks). The same effect, of the same size, is also observed by the offspill flash lamp monitor (covering only the readout elements behind the scintillating fibers!). Consequently, the damage has to be attributed to the photomultipliers - either to the photocathode or the dynode surfaces. Thus, within the error (actually 2 %) with which we are able to compare particle response and monitoring data, we have no indication for any scintillator damage (dose for the most irradiated module: $10 - 100\,krad$).

As for short-term effects, we observe pedestal and gain variations during the spill for the modules most exposed to particles (maximal effect for gain variations: 15 %). For these modules we also observe significant contributions from pile-up.

5 Triggering on neutrons

The main motivation for hadronic calorimetry in WA89 comes from rare physical processes with Σ^- in the final states. The concept for a trigger on the Σ^- decay neutrons exploits basically the high speed and the shower information in the shape of the SPACAL signals combined with the response from an hexagonal scintillating tile hodoscope. This trigger can be subdivided into 3 stages:

A fast distinction between *e.m.* *and hadronic* types of particles can be done using a width measurement (at 20 % of the amplitude = Full Width Fifth Maximum) of the analog calorimeter pulses ('FWFM-method', see ref.[3]). The electronics (a LATE CFD = Leading And Trailing Edge Constant Fraction Discriminator) that we designed for our 1991 testrun provides online (typically after 40-60ns) an amplitude independent shower separation with more than 99 % π-efficiency for a 99 % e-rejection (Fig.4). A hadron-trigger signal is generated by the LATE CFD if the amplitude and the width of the calorimeter signal are above preset amplitude- and FWFM-thresholds.

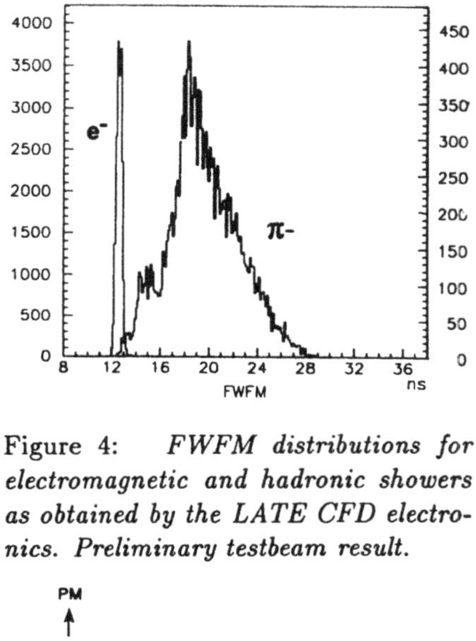

Figure 4: *FWFM distributions for electromagnetic and hadronic showers as obtained by the LATE CFD electronics. Preliminary testbeam result.*

Charged and neutral particle separation is done with a hodoscope consisting of hexagonal scintillating tiles of equal granularity as SPACAL and placed just in front of the calorimeter. The light readout is done on one surface via 5 WLS (Wave Length Shifting) fibers (Fig. 5) that are coupled directly to small size photomultipliers. Preliminary testbeam results on 19 tiles show efficiencies for each module of better than 98 % .

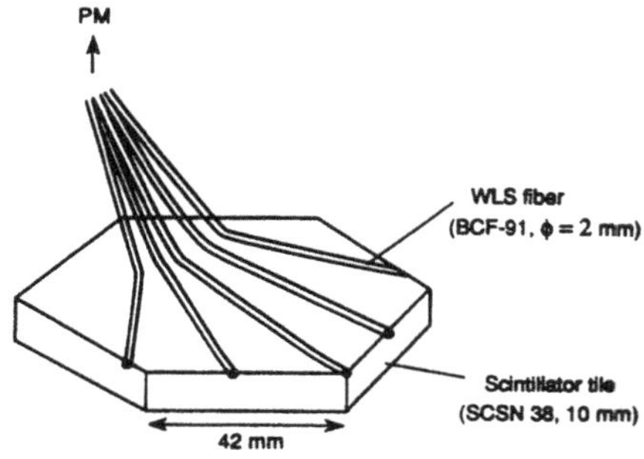

Figure 5: *An hexagonal scintillating tile with WLS fiber readout.*

The last step then consists of an online correlation between both sets of information. A *'high-energy neutron'-trigger* will be generated if a high-energy hadronic shower is observed in SPACAL but no signal is seen in the hodoscope module (and eventually its six neighbors) facing the location of the shower barycenter.

6 Summary and Outlook

The upgraded SPACAL spaghetti calorimeter was successfully operated in an experimental environment of 10^6 hadrons per spill. Σ^- beam particles were easily reconstructed and the analysis of Σ^- in final states is in progress. The most irradiated modules show no observable fiber damage (up to the level of $10 - 100\,krad$) however a considerable response decrease (up to 50 %) is observed in the PM electronics for some of these modules over 2 1/2 weeks. The additional gain variations of up to 15 % during the spill lead us to suspect that the corrections extracted from the monitoring data will play a crucial role in the quality of the neutron reconstruction. An hexagonal scintillating tile hodoscope and electronics for a fast separation between electromagnetic and hadronic showers were successfully tested in view of a future neutron trigger in WA89 using the SPACAL calorimeter information.
Beside the physics application, interesting data are available for further studies on the intrinsic properties of the SPACAL calorimeter e.g. with protons of energies up to 250 GeV/c, with high rate and background environment, and with and without Pb-glass in front of SPACAL.

References

1. A. Simon, *Results from an integrated electromagnetic and hadronic Pb - Scintillating fiber calorimeter*, Proceedings of the First International conference on CALORIMETRY IN HIGH ENERGY PHYSICS, Fermilab, october 1990, world scientific (1991) 155.

2. D. Acosta et al., *Electron, Pion and Multiparticle Detection with a Lead/Scintillating-Fiber Calorimeter*, NIM **A308** (1991) 481.

3. D. Acosta et al., *Electron-Pion Discrimination with a Scintillating Fiber Calorimeter*, NIM **A302** (1991) 36.

4. D. Acosta et al., *Localizing Particles Showering in a Spaghetti Calorimeter*, NIM **A305** (1991) 55.

5. D. Acosta et al., *On Muon Production and other Leakage Aspects of Pion Absorption in a Lead/Scintillating-Fiber Calorimeter*, NIM **A309** (1991) 143.

6. *WA89 Proposal to the SPSLC*, CERN/SPSC 87/43 (1987).

7. K.H. Brenzinger, ***diploma thesis (in German)***, Heidelberg, November 1991.

TEST BEAM CALIBRATION OF THE ZEUS BARREL CALORIMETER

Ioannis Gialas
Columbia University - Nevis Labs
136 S. Broadway, Irvington, NY 10533, USA

A. Bernstein[β], T. Bienz[γ], A. Caldwell[β], L. Chen[δ], M. Derrick[α], R. Imlay[δ], S. Kartik[δ], H.J. Kim[δ], T. Kinnel[θ], H. Kreutzmann[γ], C.G. Li[ε], J.N. Lim[ζ], R. Loveless[θ], B. Lu[η], U. Mallik[γ], K. McLean[ε], R. McNeil[δ], W. Metcalf[δ], B. Musgrave[α], B.Y. Oh[ζ], S. Park[ε], J. Parsons[β], D. Reeder[θ], J. Repond[α], S. Ritz[β], M.T.P. Roco[γ], P.H. Sandler[θ], F. Sciulli[β], W.H. Smith[θ], R.L. Talaga[α], G. Tzanakos[β1], L. Wai[β], M.Z. Wang[γ], J. Whitmore[ζ], J. Wu[γ], S.H. Yang[β]

[α]) *Argonne National Lab.*, [β]) *Columbia Un., Nevis labs,* [γ]) *Un. of Iowa,* [δ]) *Louisiana State Un.,* [ε]) *Ohio State Un.,* [ζ]) *Pennsylvania State Un.,* [η]) *Virginia Polytechnic Inst. and State Un.,* [θ]) *Un. of Wisconsin, Madison*

ABSTRACT

A fully compensating uranium-scintillator calorimeter with specially designed electronic readout was constructed for the ZEUS detector at HERA. Modules of the Barrel part of the calorimeter, built by U.S. collaborators, were subjected to beam tests at Fermilab during two running periods in the summers of 1990 and 1991. Preliminary results from these runs indicate energy resolutions of $38\%/\sqrt{E}$ for hadrons and $19\%/\sqrt{E}$ for electrons, tower-to-tower and module-to-module uniformities less than 1% , energy nonlinearity smaller that 1%, and $e/h \simeq 1.0$. Analysis of the data is continuing.

[1]On leave from University of Athens.

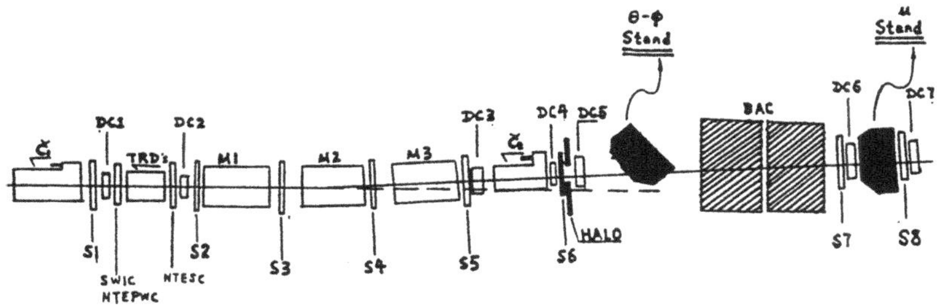

Figure 1: Experimental apparatus used to calibrate the ZEUS Barrel calorimeter modules in the Fermilab test beam.

1. Introduction

Meeting the physics goals of HERA imposes certain restrictions on the design of the ZEUS calorimeter. It must be hermetic over nearly 4π, have good hadronic resolution, and use readout electronics that will have low noise and be compatible with the 96 *nsec* HERA bunch structure. The solution chosen was a uranium-scintillator sampling calorimeter with wavelength shifter (WLS) readout designed to achieve compensation of the hadronic and electromagnetic response ($e/h = 1$), along with specially designed pipelined readout electronics[1].

The calorimeter is divided into three sections; the Forward (FCAL), Rear (RCAL), and Barrel (BCAL) calorimeters. The BCAL[2] is segmented transversely into 32 modules and longitudinally into three sections, the electromagnetic section (EMC) and two hadronic sections (HAC1 and HAC2). The EMC is divided into 53 projective towers and the HAC1 and HAC2 into 14 nonprojective towers. The EMC towers are composed of 21 uranium plates 3.3 *mm* thick interleaved with 22 scintillator plates 2.6 *mm* thick. All plates were machined with a tolerance of less than 0.2 *mm*. The HAC1 and HAC2 towers contain 49 uranium and 50 scintillator plates each[2] .

2. Experimental apparatus

The experimental setup used in the Fermilab test beam is shown in figure 1. The magnetic spectrometer was specifically calibrated to provide absolute momentum to better that 1% . An array of scintillator counters provided the beam trigger. Four drift chambers upstream of the dipoles and four downstream, each with position resolution of 200 μm, provided position measurement implying an event-to-event momentum resolution of less than 1% . Electrons were selected with an array of 6 TRD modules. To estimate the leakage energy out the back of the BCAL modules,

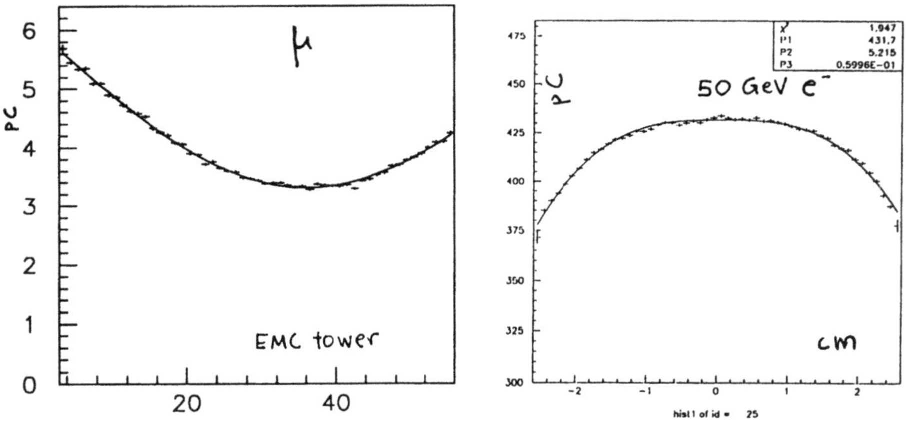

Figure 2: **a.** Collected charge for a mip versus EMC tower. **b.** Electron response versus tower dimension parallel to the BCAL module.

part of the CCFR neutrino experiment target[4] was used as a backing calorimeter (BAC).

Up to four BCAL modules could be mounted on a computer controlled fixture capable of independent rotation about the (ZEUS) θ and ϕ directions in such a way so that the beam particles entered the calorimeter modules as if they came from the HERA mean interaction point. A second fixture holding only one module and moving transversely to the beam direction was located behind the BAC, allowing the study of the nonprojective HAC towers with muons. This muon fixture was supplemented with its own trigger counters and drift chambers.

Data were taken during the summers of 1990 and 1991. In the 1990 run, six modules were tested and in the 1991 run, three modules were tested. Preliminary results from the 1990 run have been presented[5] .

3. Results

The discussion here reports preliminary conclusions primarily from the 1991 run. We expect that some of these conclusions will improve with further analysis. The short and long term stabilities of the readout electronics, and of the photomultiplier gains were extensively studied. The data were taken principally to understand the response of the calorimeter to muons, electrons, pions, and protons. Quantities that were measured included the energy calibration and resolution of the calorimeter to

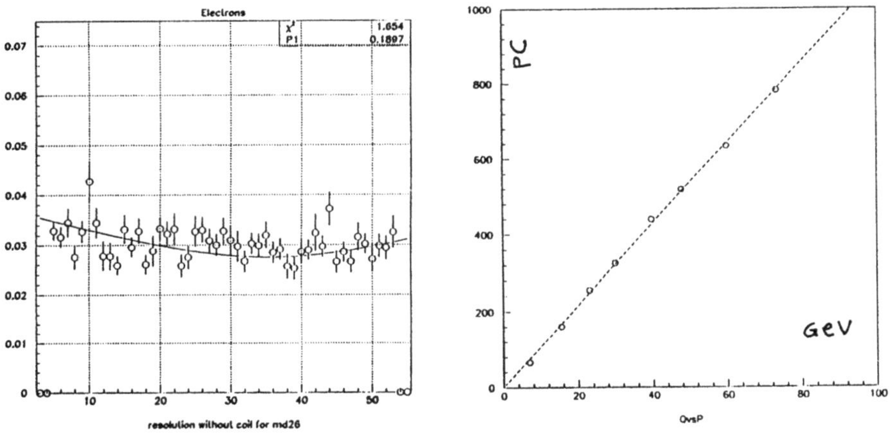

Figure 3: **a.** Electron resolution versus EMC tower number. **b.** Energy linearity for electrons.

electrons and hadrons, mip studies with muons, energy linearities, and tower-to-tower and module-to-module uniformities.

The PMT gains were chosen so that, within each tower type (ie. EMC, HAC1, HAC2), the calibration of anode charge to energy deposit would be equal for all towers. The calorimeter depleted uranium (DU) induced signal, after being integrated over 20 $msec$, generated stable anode currents (DU currents) which were continuously monitored so that changes in gains would be immediately obvious. The high voltage values were set corresponding to nominal DU currents of 133 nA, 708 nA, and 912 nA, for the EMC, HAC1, and HAC2 sections respectively[3] . If the tube gain changed between the time when the HV was set and when measurements were made, the intercalibration between towers of the same section could then be achieved by multiplying the reconstructed charge by $DU_{nominal}/DU_{measured}$.

An effort was made to determine the absolute beam momentum with precision. The calibration obtained from the tracking and magnetic field information was compared to the energy measured by the BAC which had been accurately calibrated by the CCFR collaboration[4]. The two values were consistent to 1%.

With a module in the muon fixture, beam muons entered the HAC towers parallel to the tower axis. Using the mean HAC tower responses to muons (mip), the tower-to-tower and module-to-module uniformities were obtained. Results from the HAC muon studies are presented in table 1. Each row shows the mean and rms of the mip for the 14 HAC1 and 14 HAC2 towers in a single module. The last row averages

Module	$HAC1$		$HAC2$	
	Q^{sum}_{mips} (pC)	δQ^{sum} (%)	Q^{sum}_{mips} (pC)	δQ^{sum} (%)
12	9.34	0.85	9.48	0.64
26	9.50	0.64	9.62	0.92
32	9.64	0.64	9.58	0.78
Ave.	9.49	1.47	9.56	0.99

Table 1: Results from the HAC muon studies.

over all modules. It can be seen that the tower-to-tower uniformity is less than 1%. Module-to-module uniformity is about 1%-1.5% for the three modules studied.

The θ-ϕ fixture was used to get the response from the EMC towers for muons travelling parallel to the EMC tower axes. Figure 2.a shows the collected charge from the phototubes of each EMC tower plotted versus the tower number. Most of the charge variation is a consequence of the geometry of the EMC section. Muons traverse a thicker layer of scintillator as the angle of incidence increases. The solid line is a fit to the parametrization $Q_0/\cos(90° - \theta)$ where θ is the incident angle and Q_0 the collected charge at $\theta = 90°$. The tower-to-tower uniformity in Q_0 is within $\sim 1\%$.

The calorimeter response to electrons was determined by adding the signals from the targeted EMC tower and the two neighboring towers on each side as well as the HAC towers behind. Most of these data were taken at a nominal beam momentum of 47 GeV. Typical tower response along the direction parallel to the module is shown in figure 2.b. The energy resolution versus EMC is shown in figure 3.a. If the resolution is dominated by sampling fluctuations, one expects the rms to be approximately proportional to $1/\sqrt{\cos(90° - \theta)}$. Electron energy resolutions of about $19\%/\sqrt{E}$ were obtained from this preliminary analysis. In a subsequent setup, one radiation length of aluminum was positioned just upstream of the BCAL modules to simulate the ZEUS superconducting coil. With this additional material, the measured energy resolution for electrons was about $20\%/\sqrt{E}$. Tower-to-tower EMC uniformity was measured with electrons to be within 1%. Results from two modules show the average response for each tower to be within 0.5%. Measurements taken at different energies show energy nonlinearity for electrons to be no worse than 0.5% , as can be seen in figure 3.b.

Study of the calorimeter response to hadrons in the EMC and HAC sections has not yet been completed. Preliminary results show an energy resolution of about $38\%/\sqrt{E}$. These incomplete results also indicate that the response to electrons and hadrons is nearly the same.

4. Acknowledgements

These results would not have been possible without the very strong support by the Fermilab administration. Special thanks to J. Peoples, P. Garbincius, H. Montgomery, and T. Yamanuchi. G. Koizumi was indispensable in the effort to deliver beam to the experiment and C. Hojvat made sure that needs were met in the best possible way. We owe special gratitude to K. Gray for essential support. C. Billups saw that operations in the Lab F were efficient and safe. Also thanks go to colleagues who worked to maintain modules and on the first stages of the experiment, including G. DeClute and J. Kehres from Univ. of Wisconsin; H. Abdelkader and W.G. Liou of Univ. of Iowa; R. Stanek from ANL; J. Xu, F. Xu and D. Chen from Columbia Univ.; and R. Nolty, D. Asner, W. Wen, and X. Shi from California Institute of Technology.

References

[1] W. Sippach *et al.*, *IEEE Trans. Nucl. Sci.*, **36**, 465(1989); A. Caldwell et al., Design and implementation of a high precision readout system for the ZEUS calorimeter, to be published in NIM.

[2] M. Derrick *et al.*, *Nucl. Inst. and Meth.*, **A309**, 77(1991).

[3] I. Ambats *et al.*, Cosmic ray tests of the ZEUS barrel calorimeter modules, ANL-HEP-PR-91-40 (Submitted to Nucl. Inst. and Meth.).

[4] W.K. Sakumoto, *Nucl. Inst. and Meth.*, **A294**, 179(1990).

[5] U. Mallik, Proc. International Conference on Calorimetry in High Energy Physics, FNAL, October 1990, Ed. A. Para.

A PB-SCIFI E.M. CALORIMETER FOR AN EXPERIMENT ON CP VIOLATION AT DAΦNE

A. Antonelli, S. Bertolucci, S. Bianco, C. Bloise, F. Bossi,
P. Campana, R. Casaccia, M. Cordelli, M. Curatolo, B. Esposito,
F.L. Fabbri, S. Miscetti, F. Murtas, G. Nicoletti, V. Patera,
S. Sarwar, L. Votano, A. Zallo

INFN - Laboratori Nazionali di Frascati
Via Enrico Fermi, 40 - I-00044 Frascati, Rome, Italy

A. De Martinis
Istituto Nazionale di Fisica Nucleare, Sezione di Trieste, Italy

presented by L. Votano

ABSTRACT

A proposal for a PB-SCIFI sampling calorimeter for an experiment at the Φ Factory DAΦNE is reported. The design closely matches the experiment requirements. The results on timing and energy resolutions obtained with prototypes are presented.

1 - Calorimeter Requirements

The DAΦNE[1] project of the INFN-Frascati National Laboratory concerns the construction of a e^+/e^- storage ring running at the Φ peak (1020 MeV) with very high luminosity ($L=10^{32} \rightarrow 10^{33}$ cm^{-2} sec^{-1}).

The near future experiments that search for direct CP violation in the K_S-K_L system should aim at the measurement of the $\Re e\,(\varepsilon'/\varepsilon)$ ratio with an absolute accuracy of 10^{-4} and should keep systematic errors below 5×10^{-5}, in order to represent a real step forward with respect to the latest experimental results and to contribute to the understanding of the phenomenon[2, 3].

In particular, the requirements for the calorimeter are challenging: the accuracy in reconstructing the $K^\circ \rightarrow \pi^\circ \pi^\circ$ neutral decay vertex has to be kept under 1 cm, even though the energy spectrum of the photons is very soft (20 MeV-Eγ-300 MeV) and events originated by K_L are more or less uniformly distributed over the whole detector volume.

A complete hermeticity and photon detection coverage in the full kinematic range is also required to reach a total rejection factor of 10^{-5} against K_L background decay channels. The K° neutral decay length can be reconstructed with a global fit technique[4], using the energies of the four photon showers, a three-D measurement of their conversion points, and the energy-momentum conservation on Φ decay. Thus, the energy and space resolution requirements are very severe. However, they can be relatively loose if one can also use information from the photon arrival times, which alone provide a procedure for the measurement of the decay point of K°'s produced with very low velocity ($\beta \leq .22$)[5]. A complete Montecarlo simulation has been performed[6], proving that a calorimeter measuring the photon arrival times with excellent time resolution (300 psec at 20 MeV, scaling with $1/\sqrt{(E)}$), good energy resolution ($7\%/\sqrt{(E)}$, E in GeV) and reasonable spatial resolution ($\sigma x = \sigma y$ = 1cm σz=5cm) is able to determine the K° vertex with a $\sigma \simeq .75$ cm.

The simulation also considers the machine energy spread (300 keV) and Φ localization (σx=.2 cm, σy = 20 μm, σz = 3 cm). Furthermore, if the calorimeter time resolution is assumed to be 300 psec for any photon energy, the vertex resolution is worsened by about 30%, while, for the spatial resolution, one gets only a 10% improvement, assuming 1-cm precision on the z coordinate. Finally, if the least energetic photon escapes from the calorimeter, no appreciable loss on resolution is observed.

2 - The Detector

The proposed solution is a sampling calorimeter: 1-mm scintillating fibers are embedded in grooved lead plates (. 38 mm thick, <.1 X_0.), with a lead to fiber ratio of 35:50 (15% glue), giving a X_0=1.6 cm and a sampling fraction of about 15% (Fig. 1a). The barrel calorimeter is 4 m long, with an internal radius of 2 m and a depth of 15 X_0(24 cm). The fibers are read at both ends and are segmented into a first part of square (3.3 x3.3 cm^2) elements and a second part (tail catcher) with coarser (5.x 5.cm^2) granularity. This should give a transverse resolution of 1 cm^2, while the z coordinate along the beam direction is reconstructed from the differences in arrival times (Fig. 1b).

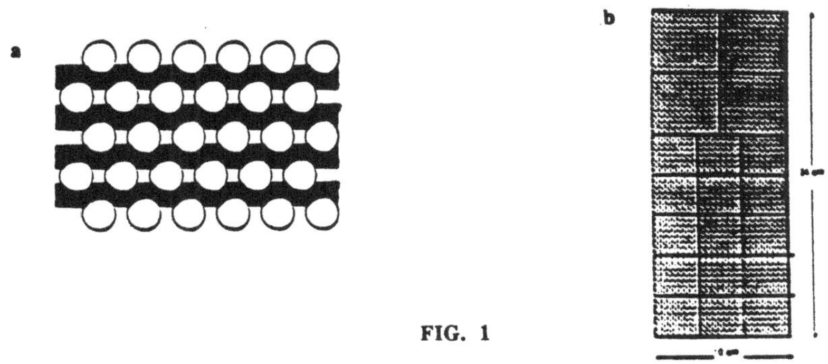

FIG. 1

Calorimetric modules employing plastic scintillating fibers and grooved lead plates in the ratio of 1:1 or more, with the fibers parallel to the impinging particles (head-on configuration) provide a well established technique for noncompensating e.m. calorimeters with very good energy resolution. They have been used in the energy range 0.1-10 GeV, where an energy resolution of $\sigma/E \simeq 6{,}3\%/\sqrt{E(GeV)}$ has been obtained[7]. Tests have been performed at Frascati to extend the measurement down to 20 MeV using the LADON tagged photon beam of ADONE. Figure 2 shows the energy resolution obtained[8] with two (9.8 x 9.8 x 22) cm^3 modules assembled with OPTECTRON S101-S 1mm fibers and different optical cement, with a fiber-lead-glue ratio 50:35:15. The modules are exposed to the beam in a head-on configuration. The photon energy is varied between 20 and 80 MeV, using different endpoint energies of the LADON beam tagged by a microstrip solid-state detector; the single strip energy resolution is ± 2% at 80 MeV. An energy resolution better than $6\%/\sqrt{E(GeV)}$ down to 20 MeV has been measured.

A calibration with cosmic muons crossing transversally the module prototype has shown that the energy released on calorimeter active medium by a 20 MeV photon corresponds to the energy released by a minimum ionizing particle in 3 cm of detector (~ 20 layers of fibers).

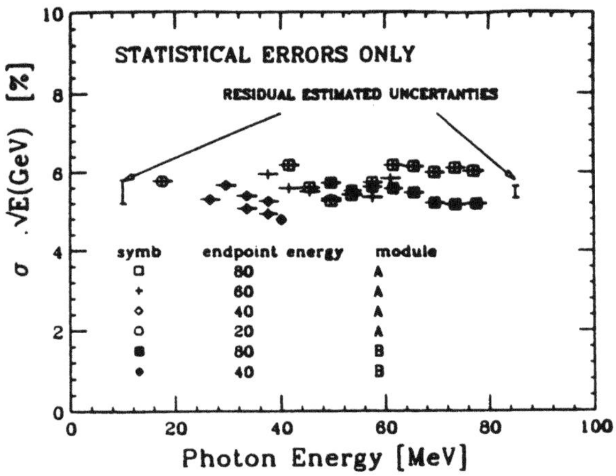

FIG. 2

In order to test the timing performances, dedicated tests have been performed at the Frascati Laboratory, with a set of counters formed of layers of 1 mm blue fibers and using minumum ionizing particles as the trigger[5].

Figure 3a shows the uncorrected timing distribution obtained with a 50-cm-long counter formed of 19 layers of fibers; if the start jitter is removed a σ=250 psec is obtained. A 200-cm-long counter consisting of 9 layers of fibers yields a σ = 390 psec (Figure 3b), confirming a $1/\sqrt{(E)}$ trend.

The result is very encouraging, since it has been obtained without any optimization either on the fibers, the PM, or the electronics; however R&D studies and tests are being carried out along these directions.

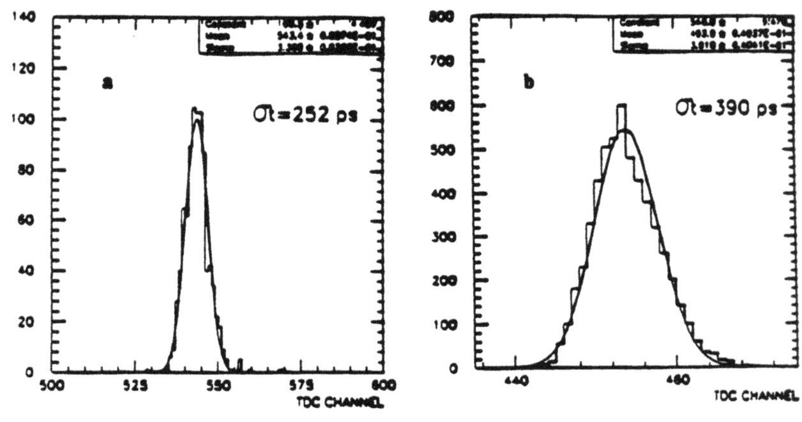

FIG. 3

3 - Conclusion

The PB-SCIFI sampling calorimeter proposed is very suitable for the experimentation at DAΦNE. The tests performed on prototypes seem to confirm that the required energy and time resolution can be achieved. A proper engineering design that can also ensure homogeneity and hermeticity is under study.

Acknowledgements

We wish to aknowledge the Organizing Committee for a perfectly organized and very interesting workshop.

References

1. Proposal for a Φ Factory, LNF - 90/031, (1990).
2. P. Franzini, Proceedings of the Workshop on Physics and Detectors for DAΦNE, Frascati, April 9-12, 1991 - p. 733.
3. M. Piccolo, Proceedings of the Workshop on Physics and Detectors for DAΦNE, Frascati, April 9-12, 1991 - p. 707.
4. G. Barbiellini and C. Santoni, CERN - PPE - 90-124 (1990).
5. A. Antonelli et al., Proceedings of the Workshop on Physics and Detectors for DAΦNE, Frascati, April 9-12, 1991 - p. 557.
6. C. Bloise, Proceedings of the Workshop on Physics and Detectors for DAΦNE, Frascati, April 9-12, 1991 - p. 495.
7. D. Hertzog et al., NIM A294, (1990).
8. D. Babusci et al., Proceedings of the Workshop on Physics and Detectors for DAΦNE, Frascati, April 9-12, 1991 - p. 581.

MONTECARLO SIMULATION STUDIES FOR SCINTILLATING FIBER CALORIMETRY

presented by Valerio Vercesi

Istituto Nazionale di Fisica Nucleare Sezione di Pavia

Via Bassi 6 - I 27100 Pavia - ITALY

1 Introduction

The scintillating fiber calorimeter technique has been developed to meet the requirements for detectors at future high energy colliders. Compensation properties can be achieved for a lead/plastic-scintillator calorimeter only if the detector has a small sampling *fraction*, a factor of 4 smaller then typical values, which implies that sampling fluctuations tend to dominate the energy resolution. This can be counterbalanced by using a high sampling *frequency*, hence the choice of fibers as active material rather then the traditionally used scintillator plates. As a consequence this calorimeter is very interesting as an object of simulation studies: its notable properties (such as good electromagnetic and hadronic resolution, compensation, linearity) address problems of fundamental physics, involving the detailed description of showers evolution. We will investigate in the following two different setups: the non-projective one built by the Spacal Collaboration, the other one projective tested by the RD1 Collaboration, both operating at CERN.

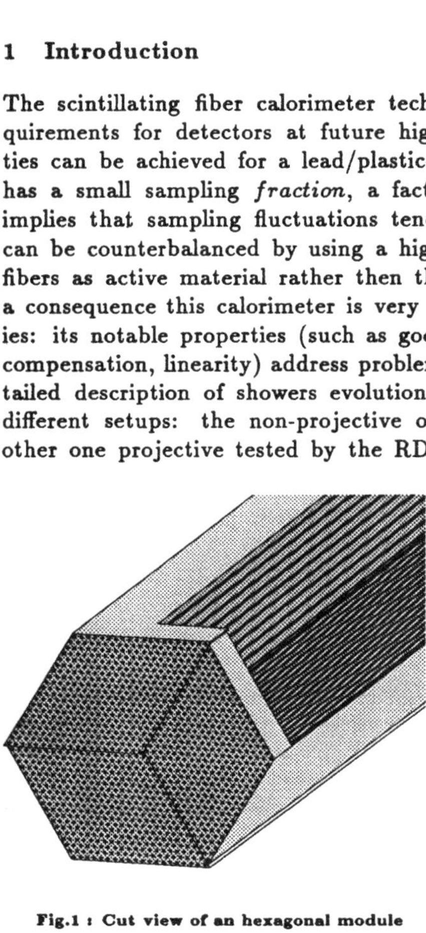

Fig.1 : Cut view of an hexagonal module

The Spacal Collaboration implemented this technique through the construction of hexagonal modules of 4.3 cm to the side and of 2 m length. Each module consists of a lead matrix with 1141 doped polystirene fibers of 1 mm diameter, assembled by stacking in layers extruded lead foils and fibers; the final setup is made by 155 modules, an EM+HAD integrated calorimeter. The RD1 Collaboration have built and tested a projective prototype made of 19 pyramidal modules arranged in a staggered geometry. Each module is 1/4 of a pyramid trunk, with R_{in}=127 cm and R_{out}=317 cm. The inner face is 3.85×4.00 cm^2 and the outer face is 9.89×10.28 cm^2.

2 Geometry description

We have used for this work the GEANT programme, version 3.1416, always with the latest corrections. This implies that part of the work had to be redone when some major bug has been discovered and that unfortunately not all the results are completely consistent with a single version. The simulation of these geometries in an efficient way is limited by the absence in the current version of GEANT of elementary shapes suited to build honeycomb structures or "pyramids with holes". To overcome this problem we have tried an approach which adopts a decompostion technique and a clonation principle, to make the widest possible use of the division structure inside GEANT. For the non-projective setup we have considered an hexagon as a combination of three slightly staggered rombs made of lead (see Fig. 1), which have been divided along the x and y axis to build the search tree. Into each division a "hole" of 2 meters length and 1.18 mm diameter is put, and in the hole a fiber of 0.471 cm radius made of polystirene, surrounded by the lucite cladding of 29 microns. The division technique in GEANT automatically replicates this structure in each partition and in this way an exact 4:1 ratio between lead and fibers is reproduced (other details on the way this geometry is realized can be found in [1]). In the end we position 155 copies of the hexagon to simulate the experimental setup, adding also a preshower detector of $1.7X_0$ which was present in the test beam.

The implementation of the projective geometry needs even more artificial solutions then the non-projective one. The basic motivation comes from the impossibility in GEANT of subdividing a pyramid (or a trapezoid) in the way necessary to position fibers perpendicularly to the bases. The way out is to approximate a pyramid by a sequence of boxes where each box starts at the z value where one more layer of fibers is needed, which is of course function of the opening angle of the pyramid. This description implies that fibers are in practice formed by segments of equal length which join at each box start. This artifice allows to use the division technique instead of positioning fibers of different length, when the oblique face is considered (see Fig. 2). The drawback is that the edges of each module become cracks because the correct lead/fiber ratio is not reproduced. This problem is also seen in reality, because of lead machining precision, but to a lesser extent.

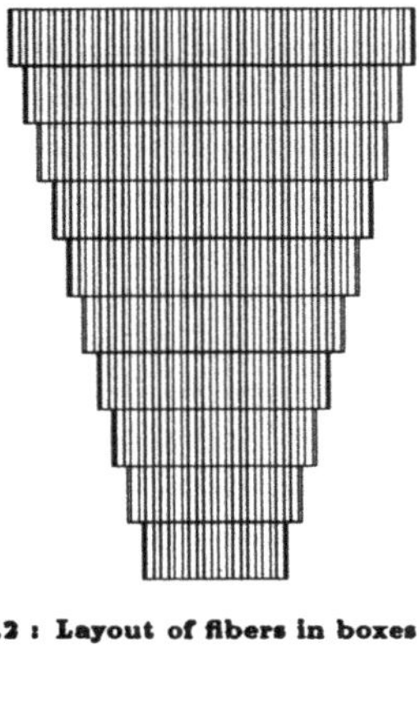

Fig.2 : Layout of fibers in boxes

With these implementations we can keep at a reasonable level the CPU time needed: electrons score is 4.5 s/GeV/ev and pions 4 s/GeV/ev, for both setups, on CRAY X/MP and HP720.

3 Electrons results

All the results refer to incident particles entering at 3°: the agreement between GEANT simulation and the non-projective Spacal data [2] is fairly good: the fit gives a resolution of $13.0\%/\sqrt{E} + 0.8$ to be compared to the $12.9\%/\sqrt{E} + 1.2$ of the experiment. In Table 1 results for different energies are summarized: in particular when selecting the impact point only in lead the simulation shows the correct trend with a constant term compatible with zero.

	Non-Projective				Projective	
E (GeV)	Data (all)	MC (all)	Data (lead)	MC (lead)	Data	MC
5	7.0	6.8		6.6		
10	5.4	5.0		4.6	6.2	5.0
20	4.0	3.7	3.6	3.0	4.9	3.7
40	3.3	3.0	2.7	2.1	4.0	3.0
80	2.7	2.5	2.0	1.7	3.1	2.5
150	2.1	1.8	1.4	1.2	2.8	1.8

Table 1: $\sigma(E)$ for electrons

This confirms the interpretation of this term as the result of the channelling of particles in the fibers at the early stages of the electromagnetic cascade. In the experimental data other sources of systematics are of course present, like the fiber-to-fiber fluctuations, and this explains why the experimental constant term is higher. As an example Fig. 3 shows the radial shower profiles for electrons: the agreement over up to seven orders of magnitude between the GEANT points and the experimental results is impressive.

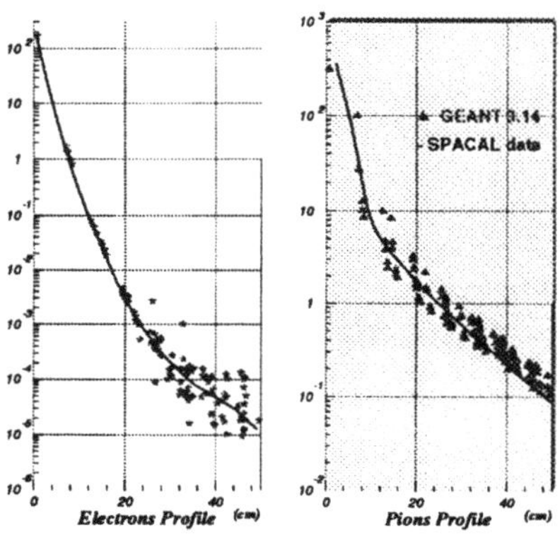

Fig.3 : Radial Shower Profiles

The projective setup is still at a preliminary stage of study, both from the experimental and the Montecarlo point of view. In Table 1 results are compared to available data [3]: the fit to data points gives $14.6\%/\sqrt{E} + 1.6$ while the simulation predicts $13.1\%/\sqrt{E} + 2.0$. Since we spread the impact point with a gaussian of $\sigma = 3$ mm (as in the testbeam) a possible origin for the increase of the constant term might be looked for in the cracks present at the modules' edges. A different approach to the geometry for avoiding this problem is currently under evaluation.

4 Pions results

The description of the hadronic showers is unfortunately still far from being complete, expecially at very low energies. We have tested both GHEISHA and NUCRIN packages and we have found the second one slightly better.

	Non-Projective				Projective
E(GeV)	Ex (λ)	MC (λ)	Ex (no λ)	MC (no λ)	MC (λ)
5	15.6	16.3	14.8	16.3	
10	11.7	12.7	11.0	11.9	12.8
20	8.2	8.6	7.5	7.4	9.9
40	6.9	6.7	5.7	5.6	7.2
80	5.6	4.8	4.4	4.1	6.0
150	4.8	3.9	3.4	3.2	4.2

Table 2: $\sigma(E)$ for pions

In the non-projective case the experimental resolution is $27.7\%/\sqrt{E}+2.5$ if the attenuation length (λ) effect in fibers is left. When we implement the λ effect in the simulation we obtain a value of $35.0\%/\sqrt{E}+1.1$. When this effect is disentangled from the data (leaving only the Birk's contribution) the fit is $30.6\%/\sqrt{E}+1.0$, to be compared to the simulated $35.0\%/\sqrt{E}+0.1$. No experimental data are yet available for the projective setup so Table 2 contains only the predicted values for this case.

We observe that the constant term variation with λ is identical in the data and in GEANT, and the simulation shows that in this case there is no channeling effect. This is understood because of the completely different developement of the shower, much longer in depth and without symmetry properties. The resolution suffers in the simulation from the lack of knowledge of details in the hadronic showers and of the dE/dx of slow massive particles: more work is ongoing to understand these problems.

As a general conclusion GEANT 3.14 proved to be a useful tool for understanding and reproducing the properties of fiber calorimeters. Other studies are ongoing to refine and complement the current activities, including muon simulations and new geometries developement.

References

[1] A.Dell'Acqua, G.Fumagalli, V.Vercesi, Proceedings of the 1st CERN-ECFA LHC Workshop, Volume 3, page 14.

[2] D. Acosta et al., NIM **A259**(1991) 481.

[3] Refer to the contribution of G. Polesello, in these Proceedings.

The FLUKA Monte Carlo Studies of Lead-Scintillator Calorimeters for Hadrons

T. TYMIENIECKA
University of Warsaw, Warsaw (Poland)

Monte Carlo study of the hadron detection is done for lead-scintillator calorimeters. The FLUKA90 code[1] is used modified as follows. A model of nuclear evaporation including the competition with high energy fission is attached[2] to the FLUKA generator. It models the emission of neutrons, protons, heavier products and gammas of nuclear origin. The low energy neutron transport is extended down to 1 keV, that is, below the 50 MeV threshold of the original FLUKA; the neutron collision subroutine is adopted from the MORSE code with the HILO86 multigroup cross section data available for 18 elements. In parallel, the ONMC neutron transport model[3] is used which provides a detailed simulation of neutron transport and neutron signal without statistical weights and in the continuous energy scale. Energy loss in scintillator for protons of energy below 10 MeV is corrected; light saturation is included. Furthermore, some improvements are introduced to the generator of the high energy inelastic processes (EVENTQ). The geometry routines are modified to describe the multilayer structures of calorimeters.

The calorimeter response for hadron is given in terms of the equivalent electron response (*the e/h ratio*) and the hadron energy resolution $\sigma_h/\sqrt{E_h}$.

The calorimeter response is estimated as follows. In the calorimeter region of interest the energy loss of hadron shower is subdivided into components characterized by the same detection feature, e.g. electromagnetic cascade or low energy neutrons. Each component is multiplied by its gaussian distributed detection efficiency to give its calorimeter signal. This method permits the fast but detailed simulation of the calorimeter response without affecting the correlation within a single hadron shower. The method is tested against the full shower simulation with included the EGS4 code and the ONMC code.

The signal of hadrons at TeV energies is estimated as follows. Hadrons of energy from 3 GeV to 200 GeV produced in interactions are not showered. Their signal is calculated from the e/h ratio smeared according to the gaussian distribution with its standard deviation equal to the hadron energy resolution $\sigma_h = C_h \cdot \sqrt{E_h}$. Both values, e/h and C_h, are energy dependent. They are taken from these FLUKA studies at the lower energies.

The Monte Carlo prediction is compared with the experimental data from a pure scintillator calorimeter[4] and from the existing 7 lead-scintillator setups[5-11] covering the wide range of ratios of the thicknesses of absorber plates to active plates d_{Pb}/d_{Sc} from 8 down to 1. The results are partially described elsewhere[12]. The conclusions are as follows.

Monte Carlo prediction for the hadron signal and its energy resolution agrees with measurement within the experimental errors for lead scintillator experimental designs as well as pure scintillator. The hadron shower features like the measured sampling fluctuations[13] are well reproduced.

This Monte Carlo code reproduces the gross features of the experimental spatial profiles of the calorimeter signal. Some discrepancies are observed. They suggest that Monte Carlo provides the slightly narrower *lateral* profiles of the energy deposition than measured in the experiment. Uncertainties in the experimental details, for example the selection criteria in ref.10, excludes the more definite conclusion.

Changes in the adopted values of the model parameters including the FLUKA ratio of π^0 production to other hadrons cause some systematical shift of the final results for the energy resolution and the e/h ratio within at most 10%. These shifts do not affect the position of the minimum of the energy resolution in the d_{Pb}/d_{Sc} scale.

The good reconstruction of the calorimeter signals proved that modelling of hadron interactions, the particle transport and conversion of energy deposition into the scintillator response are satisfactory.

The study of the infinitely large calorimeter at 3 GeV to 200 GeV suggests that the calorimeter signals of the electrons and hadrons are equal to 1 for a wide range of calorimeter structures. The range depends on the incident energy. The minimum energy resolution is situated for $d_{Pb}/d_{Sc}<4$. The deterioration of the energy resolution is only partially correlated with deviation of the e/h-ratio from 1.

For a better understanding of the e/h≠1 contribution the e/h ratio and energy resolution are studied as a function of the π^- incident energy E_{beam} for two calorimeter structures, one with the thickness of the absorber plates equal to 1 cm (experimentally proved to be a compensating one), the second with plates 0.25 cm thick (the undercompensating calorimeter). The scintillator thickness is fixed to 0.25cm. The results are as follows:

d_{Pb}	d_{Sc}	E_{beam}	5 GeV	20 GeV	200 GeV	2 TeV
1cm	0.25cm					
		e/h	1.01	0.97	0.98	0.99±0.01
		$\sigma_h/\sqrt{E_h}$	0.47	0.43	0.41	0.41±0.04
0.25cm	0.25cm					
		e/h	1.15	1.09	1.07	1.05±0.01
		$\sigma_h/\sqrt{E_h}$	0.36	0.34	0.38	0.63±0.06

where the errors at 2 TeV comes in the most part from the uncertainty of the adopted values of the energy dependent e/h ratios.

Besides the energy region below 10 GeV the e/h ratio for the full detected showers is weakly energy dependent. Similarly the energy resolution is roughly energy independent except for TeV showers in the undercompensating composition.

The good reconstruction of a series of measurements suggests a big predictive power of the FLUKA code.

Some improvement has been recently done in the FLUKA code as presented by A.Ferrari at this conference. Besides a test of the detected discrepancies, in particular, the lateral profile of hadron shower or a search for the optimal calorimeter structures, it is of interest to investigate the origin of fluctuations in hadron shower for heavy absorbers.

References

1. P.A.Aarnio, H.J.Möhring, J.Ranft, A.Fasso, G.R.Stevenson, FLUKA89 Users' Guide,*CERN* 1990.
2. J.M.Zazula, *DESY*-D3-66, 1990.
3. T.Tymieniecka, One Neutron Monte Carlo, *DESY* 91-012, 1991.
4. A.Bienvenuti et al., *Nucl. Inst. and Meth.*, **25**(1975)447.
5. O.Botner, *Physica Scripta*, **23**(1981)556.
6. M.F.Duffyet al., *Nucl. Inst. and Meth.*, **228**(1984)37.
7. E.Bernardi et al., *Nucl. Inst. and Meth.*, **A262**(1987)229.
8. G.d'Agostini et al., *Nucl. Inst. and Meth.*, **A274**(1989)134.
9. A.Tsirou, *On the Optimization of a Lead-Scintillator Compensating Calorimeter*, Thesis, University of Wisconsin-Madison, 1989.
10. E.Bernardi, *On the Optimization of the Energy Resolution of Hadron Calorimeters*, Thesis, Hamburg University, 1989.
11. D.Acosta et al., *Nucl. Inst. and Meth.*, **A294**(1990)193.
12. T.Tymieniecka, *Tests of Some Aspects of the FLUKA Hadron Shower by Calorimetry Measurement*, Proceedings of *Workshop on Detector and Event Simulation in High Energy Physics*, Amsterdam, 1991.
13. G.Drews et al., *Nucl. Inst. and Meth.*, **A290**(1990)335.

ORGANIC SCINTILLATORS FOR MODERN CALORIMETRY

CHARLES R. HURLBUT

and

MICHAEL R. KUSNER

Bicron Corporation
12345 Kinsman Road
Newbury, Ohio, USA, 44065

ABSTRACT

Scintillator development for new calorimetry needs include long castings up to 5 meters, scintillating optical fibers, formulations for increased radiation hardness in both liquids and plastics, and new, fast fluors in the entire spectrum from near-UV to the red. We present data on scintillator performance and some developmental materials.

Plastic scintillators may be manufactured by different methods to achieve the most important properties required of the particular material. Cell-cast PVT or polystyrene (PS) based scintillator, valued for their overall high performance, have been the most frequently used materials. Extruded PS scintillator has been used in selected areas during the last 10 years because of its relatively low cost. Cell-cast PMMA based scintillator has also been used occasionally because of its lower cost.

Table 1 presents properties of these four basic plastic types. The light output data was measured at Bicron on small identical discs of commercially made blue-emitting scintillator. The emission spectra are virtually identical and assure an equal match to the quantum efficiency of a bialkali PMT. Excitation was with a Cs-137 conversion electron (624 KeV) source which produces a well defined peak. The effects of the plastic base material as well as the manufacturing process are evident in the extruded scintillator data. While being the least expensive material, its use is limited to small pieces due to poor light transmission that is primarily a consequence of low surface reflectivity.

Table 1. Properties of Generic Organic Scintillators

Plastics

General Type	Density	Light Output	Atten. Length	Decay Time	X_o	Rad-Hardness
Cast PVT	1.03	64%	280 cm	2.2 ns	43 cm	10^4 Gy
Cast PS	1.05	56%	280 cm	2.2 ns	42 cm	10^4 Gy
Extruded PS	1.05	41%	70 cm	2.4 ns	42 cm	10^4 Gy
Cast PMMA	1.2	24%	250 cm	6 ns	34 cm	10^3 Gy

Liquids

General Type	Refr. Index	Light Output	Atten. Length	Decay Time	X_o	Rad-Hardness
Mineral Oil	1.48	30%	20 m	4 ns	52 cm	10^5 Gy
1 Methyl Naph.	1.61	70%	3 m	4 ns	44 cm	-
Isopropyl Biphenyl	1.58	55%	5 m	4 ns	45 cm	-

Liquid scintillators are mainly of interest for their low cost and intrinsic radiation hardness. Mineral oil based liquids with excellent clarity, low cost, and high flash point are the choice in large volumes; the data here is for BC-517P. Liquids with a high refractive index are being used in capillary systems. The high index liquid is enclosed in a lower index capillary and effectively functions as a scintillating optical fiber. There is growing interest in liquids for the very high radiation regions of collider systems where they can be circulated through chemical rejuvenation systems and reservoirs to maintain uniform long-term performance in the calorimeter.

Long Scintillator Sheets

There is growing interest in very long scintillator plates for TOF systems and calorimeters such as the Large Area Electromatic Spectrometer at CEBAF. Major facility upgrades have been implemented at Bicron to provide extra long sheets. Single strips up to five meters long are now being tested by several projects worldwide. The critical performance factor is light collection for purposes of signal uniformity and amplitude. Figure 1 plots the signal amplitude in microamps as a function of distance in a thin sheet of BC-412. The test was performed at CEBAF using an X-ray generator positioned at several points along the sheet.

As in most tests of this type, the far end of the scintillator is made non-reflective. The sharp upturn in amplitude normally seen at the end near the PMT has been avoided here by introducing a short acrylic light pipe which was partially painted black to eliminate highly scattered photons. The effect of introducing a Wratten No. 3 optical filter, which passes light above 450 nm, is evident. The improved uniformity is achieved at the expense of about 60% of the signal amplitude.

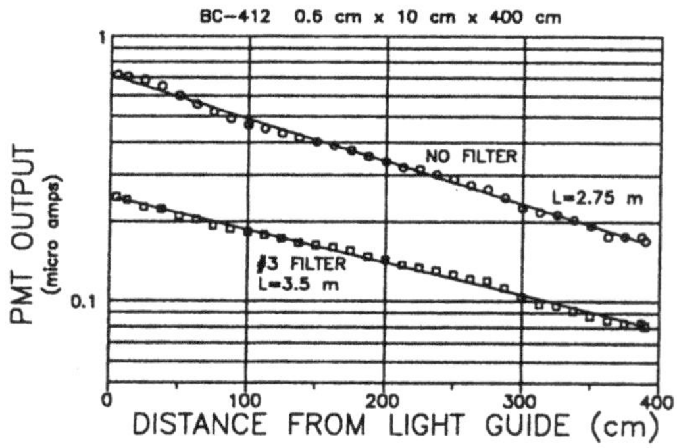

Fig. 1 Light Attenuation Plot
Effect of Green Filter
Data from C. Zorn, CEBAF

In actual use, however, scintillators are configured for maximum signal by either placing PMTs at both ends or by making the far end reflective. Figure 2 demonstrates the latter in a test performed at Bicron. In this case the signal amplitude is measured as the peak channel of a Bi-207 conversion electron spectrum (976 KeV). The 7.7 meter attenuation length applies to the data beyond 100 cm from the PMT. Note that making the far end reflective, achieved by polishing it and using a film of aluminized mylar, increased the signal even for points near the PMT. Note also that this particular scintillator is relatively narrow, and therefore the diamond milled edges make up a large part of the reflective surfaces along the length. The high attenuation length of the bar, with or without a reflective end, indicates that the diamond milled edges have a very high reflectivity.

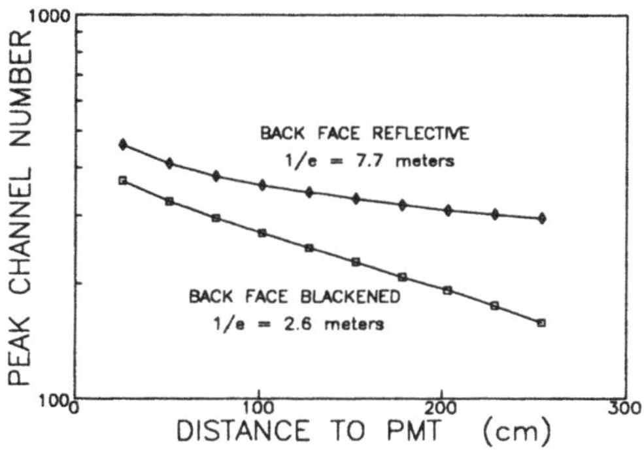

Fig. 2 Signal vs. Distance
Effect of reflective ends
BC-408 Bar, 3.8 x 7.6 x 265 cm

Timing with Long Scintillators

Additional useful information may be obtained from long scintillators by optimizing their timing resolution. This provides both position information and enhanced particle identification. Table 2 presents a number of examples recently measured by a variety of researchers and passed to the authors by conversation. In the first four cases the long scintillators were set up with PMTs at both ends, and the stated timing resolution applies to signals taken at midpoint along the sheet length.

Table 2.Timing With Long Scintillators

Scintillator/Size	Resolution	Experiment	Reference
BC-408 5 cm x 10 cm x 2.8 m 1.2 m light pipes	150 psec 1 sigma (353 FWHM) corrected	Cosmic muons midway	M.S. Alam CLEO II, SUNY - A 1969
BC-408 10 cm x 12 cm x 2.5 m no light pipes	450 psec FWHM uncorrected	250 MeV protons midway	George Lolos Univ. of Regina 1991
BC-408 3 cm x 10 cm x 2 m 35 cm light pipes	510 psec FWHM uncorrected	100 MeV protons midway	Norm Davidson Univ. of Manitoba 1991
BC-408 0.5 cm x 13 cm x 3 m short light pipes	900 psec FWHM (500 - 1600) uncorrected	Cosmic muons midway	E. Kuhlmann Univ. of Bochum 1990
BCF-10 Fiber 3.5 mm sqr. x 3.25 m	1200 psec 1 sigma uncorrected	MIP pions far end	Jim Frank Brookhaven 1991

The data for the TOF barrel bars at CLEO II is corrected for PMT jitter and electronic signal spread; all other data are uncorrected. The fourth item is unusual in that the sheet is very thin. The additional data for this strip applies for trigger detectors positioned at several spots along the 3 m sheet. A FWHM time resolution of 500 psec was achieved at the end near the PMT and 1600 psec at the far end. The best timing is achieved with thick scintillators that provide bright signals and enable triggering on the arrival of the first photons.

The last item presents data taken with 3.5 mm square fibers 3.25 m long. Ten fibers were lined up each with a PMT at one end. Minimum ionizing kaons were targetted at the far end passing sideways into the fiber array. The resulting pions from the kaon decay were measured with the indicated time spread. This is very preliminary data, and the experimenter anticipates much better results with optimized electronics.

Scintillating Fibers

Scintillating optical fibers are a relatively new medium being extensively investigated for new high resolution electromagnetic and hadronic calorimeters and for trackers. As with scintillator sheet, signal brightness and light attenuation are the most important performance characteristics. Scintillating and waveshifting fibers emitting in the blue and green dominate among the materials available for a number of practical reasons. Many blue emitting formulations abound utilizing fluor systems developed over the years in traditional organic scintillator work. These are fast and, when well made, provide good attenuation length.

Green emitting variants tend to provide slower timing properties and lower signals as a result of mismatch with most PMTs. However they provide superior attenuation length by avoiding the shorter wavelength regions where the polymers exhibit strong intrinsic optical absorption. The best transmission region is between 500 and 700 nm. Superior radiation hardness is also achieved by shifting to longer wavelengths.

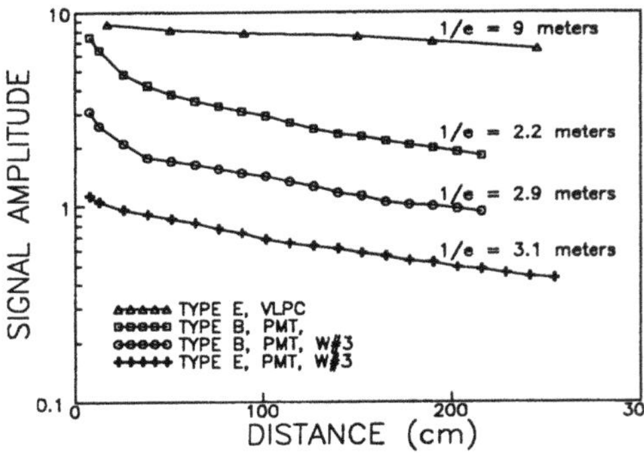

Fig. 3 Fiber Signal vs. Distance

Figure 3 shows light attenuation plots of several fibers. The lower three plots were measured using a bialkali PMT. The Type B fibers are blue emitting and 1 mm diameter, and the strong signal of the unfiltered one is equal to about seven photoelectrons at 1 meter. The bottom plot is for a green emitting fiber (530 nm peak) 780 microns in diameter. Their relative signal intensities are as indicated by their positions in the graph. The top fiber is from the same lot as the bottom one but was measured at Fermilab, by M.Atac and A.Bross, using a green sensitive solid state VPLC. The location of the plot on the graph does not indicated its amplitude relaive to the other three. However, a signal of 4-5 potoelectrons was generated from excitation at the far end and with the fiber connected to the VPLC through a 5 m long optical waveguide.

Wavelength Shifters

Waveshifting sheets and fibers are receiving extensive use in constructing highly segmented and hermetic systems. Figure 4 shows the optical spectra of a standard blue scintillator - green wavelength shifting (WLS) system. The BC-482A WLS plate material is often replaced by the WLS fiber product BCF-91A with the same spectral properties. The system exhibits moderate rad-hardness (1-2 Mrads) and a relatively slow timing property due mainly to the WLS fluor (12 nsec decay time).

The search for new fast green emitting fluors led to the very efficient G-2 compound with emission peak at 492 nm. This also has very little of the large self absorption of the WLS fluor shown above. Consequently, scintillating fibers with high attenuation

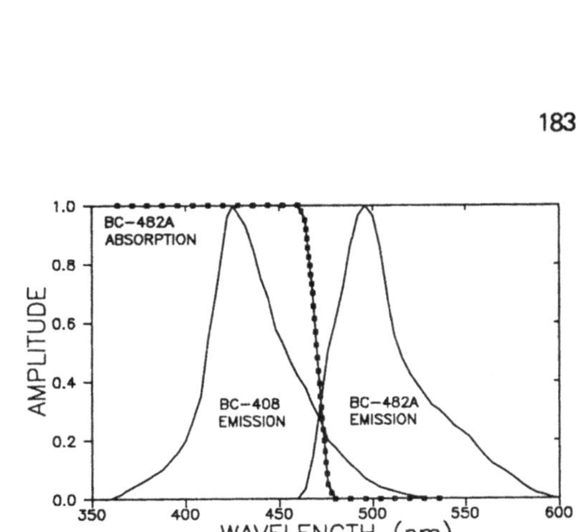

Fig. 4 BC-408 Scintillator - BC-482A WLS System

length and signal amplitude have been made. The compound is also very fast with a decay time of 2.64 nsec. This almost ideal new material suffers from relative radiation softness, and therefore is limited to experiments where radiation damage is not a threat. Its emission spectrum is shown in Figure 5.

Continued effort is underway formulating new fluorescent additives. Figure 6 shows the absorption and emission spectra of experimental green and orange compounds K-1 and K-2 in polystyrene. Both show little self-absorption and are efficient spectrum shifters. Scintillators made with them are moderately fast with fluorescence decay constants of 5.6 and 7 nsec respectively. At the of this writing, WLS fibers and plates are being prepared for optimization and characterization with regard to light transmission and rad-hardness.

Further research is underway for the development of rad-hard scintillator plates and new versions of large Stokes shift proton transfer fluors.

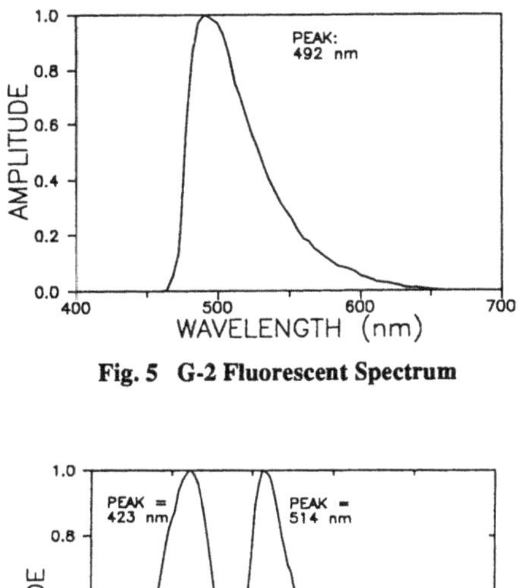

Fig. 5 G-2 Fluorescent Spectrum

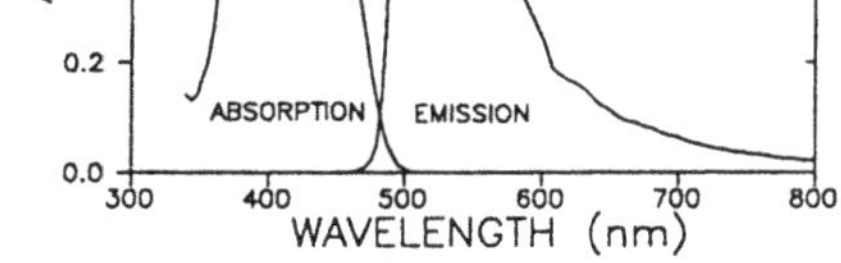

Fig. 6a K-1 Optical Spectra

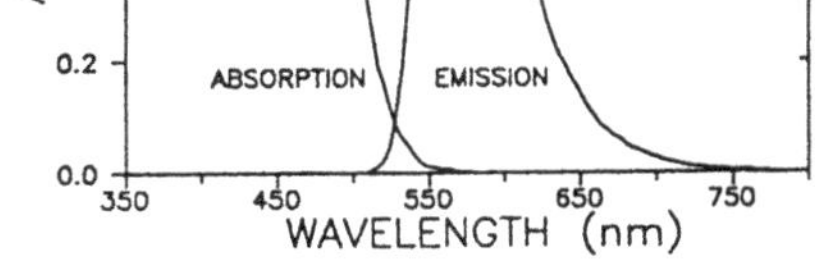

Fig. 6b K-2 Optical Spectra

MEASUREMENT OF THE TIME DEVELOPMENT OF HADRONIC SHOWERS

J. A. Parsons*,

A. Caldwell*, L. Hervas**, S. Ritz*, F. Sciulli*, W. Sippach*, L. Wai*

*Nevis Labs, Columbia University, U.S.A.
**Univ. Autonoma de Madrid, Spain

Abstract

Using custom electronics, we have performed a detailed study of the time evolution of hadronic showers in ZEUS barrel calorimeter (BCAL) modules. On-going analyses of the data should provide valuable information concerning the mechanism of compensation, the possibilities for electron-hadron separation based on the time structure of the shower, the problems of pile-up in a high rate environment, and other issues central to the design of future calorimeters.

There has been considerable progress in the understanding of hadronic showers over the last several years. Indeed, extensive test programs[1)], combined with detailed Monte Carlo programs[2)], have made possible the design and construction of compensating calorimeters, such as that of the ZEUS detector[3)]. The hadron response is expected to comprise several contributions which are collected on very different time scales, including a prompt ($\approx$ 0 ns) dE/dx and electromagnetic shower component, a delayed component from protons recoiling after a collision with a neutron and from photons from secondary fission ($\approx 0 - 100$ ns), and a late component ($\approx 10 - 1000$ ns) for gammas resulting from de-excitation of heavy nuclei after thermal neutron capture.

While the existence of these late energy deposits in hadronic showers has been verified by several groups[4)], no precise measurements are available. In order to study in detail the time development of hadronic showers, both on average and event-by-event, we designed and constructed electronics[5)] which: 1) performs a DC integration of the phototube signal; 2) samples the integrated signal at 60 MHz; 3) stores the samples in a 3μs long pipeline; and 4) digitizes the samples for a triggered event with 12 bit ADC's.

During the 1991 fixed target running period at FNAL, we took data with this electronics connected to 4 ZEUS BCAL modules stacked one on top of another. The BCAL is constructed of interleaved plates of scintillator (SCI) and stainless-steel clad depleted uranium (DU), with the SCI light transmitted via wavelength-shifter

(WLS) material to photomultiplier tubes (PMT) for conversion to an electrical signal (for details, see ref. 3). The 192 channels of analog electronics sufficed to instrument almost 6 full supertowers of each module, where a supertower has 3 longitudinal sections: the first comprises four electromagnetic cells, while the others contain 1 hadronic cell each. The transverse dimensions of the instrumented region were almost 1 m × 1 m at the front face increasing to about 1.8 m × 1.4 m at the back, with a depth slightly in excess of 5 interaction lengths. The transverse size was more than adequate to contain the showers, while the longitudinal leakage was monitored by reading out a tail catcher behind the BCAL modules.

The response of the system to a laser pulse, where the laser light was transmitted via optic fiber to the WLS immediately in front of the PMT, is shown in figure 1(a). "Differentiating" the integral with respect to time by taking successive differences yields a pulse shape which is well described by the convolution of a gaussian with an exponential with a 16 ns time constant (see figure 1(b)).

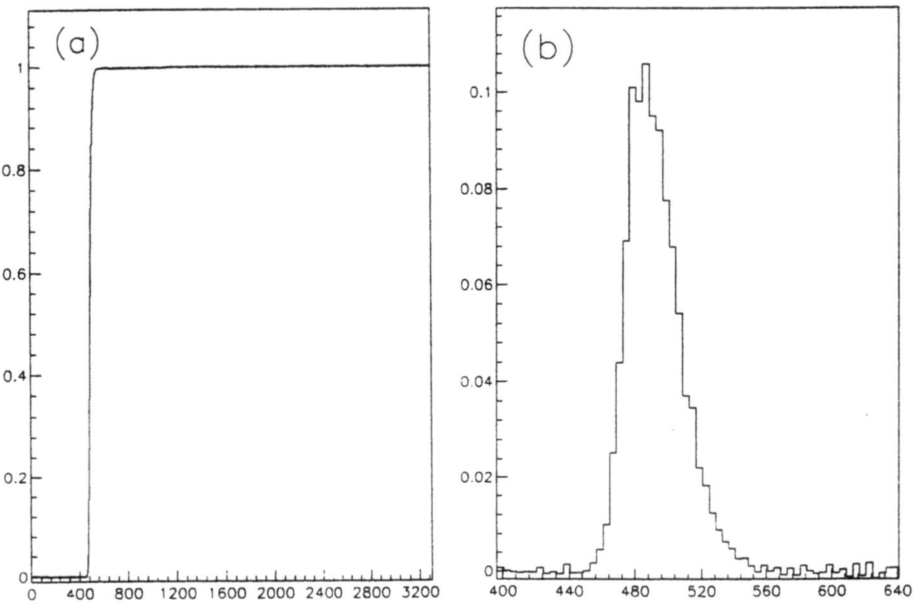

Figure 1: (a) Integrated response to a laser pulse, normalized to a total integral of 1.0. The position of t = 0 ns is arbitrary. (b) "Differentiated" pulse, in 4 ns bins.

The response of the BCAL to particle showers was studied using 20 and 50 GeV/c electron beams, and 20, 50, and 100 GeV/c pion beams. The integrated response versus time is shown for each case in figure 2, where each integral has been normalized to unity for an integration time of 100 ns. A single curve for electrons is shown since the electron response, after normalization, was seen to be independent of momentum.

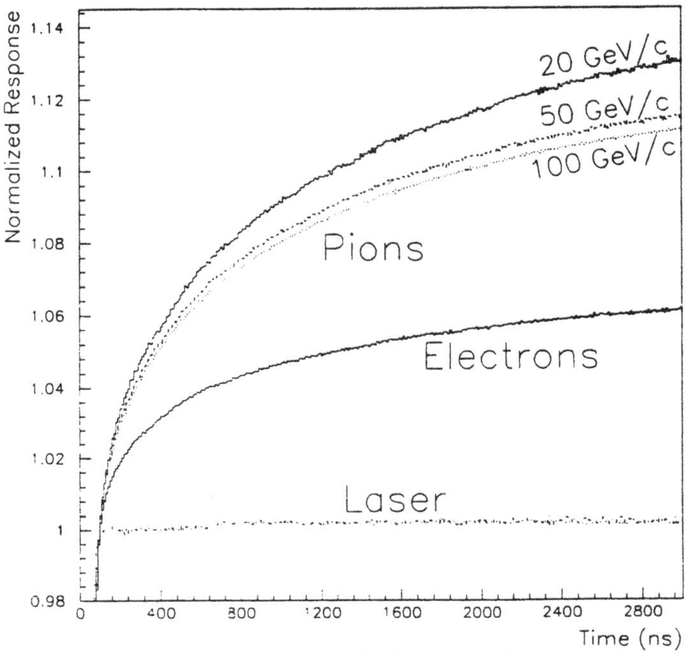

Figure 2: Integrated response (normalized to 1.0 after 100 ns) to a laser pulse, to electrons, and to pions of momentum 20 GeV/c, 50 GeV/c, and 100 GeV/c, versus time (t = 0 ns marks the beginning of the pulse). Note the suppressed zero.

From figure 2, it is apparent that the electron response is not fully saturated after 100 ns. Indeed, fitting the tail of the response shows that approximately 6% of the electron signal is collected with a time constant of about 1 μs. This surprising result, and the lack of such an effect in the case of the laser (where the light goes only through the WLS), suggests a long decay time component in the scintillator response. Of course it is possible, although unexpected, that the effect is caused by the electromagnetic shower itself, with photons perhaps exciting some long-lived states in the DU. However, 100 GeV/c muons passing through the BCAL were seen to produce the same time response as electrons. Upcoming tests, where the laser light will be directed into the scintillator, will soon settle the issue.

Comparing the pion response to that of electrons, it is immediately apparent that a substantial fraction of the pion signal is deposited late in the shower development. While the 1 μs time constant seen for the electron case explains some of the delayed energy in the pion response, there is a large excess of late energy above this contribution, which must come from energy depositions which occur some time after the shower has been begun. As discussed above, such depositions are expected due to relatively slow nuclear processes which are initiated by the hadron shower, such as the emission of photons from nuclei which have become excited after absorbing a thermalized neutron.

The fractional contribution of the electromagnetic component of the hadronic shower, which results from π^0 production, is expected to rise logarithmically with energy. This is consistent with our data where, as evident in figure 2, a comparison of 20, 50 and 100 GeV/c pions shows a clear trend for the higher energy showers to more closely resemble the electron response. However, a more detailed analysis, including leakage corrections, is needed.

The differences in the time distribution of the electron and hadron responses allow for some electron-hadron discrimination based on timing information[6]. For example, figure 3 shows the integrated signal after 1 μs integration time divided by that after 100 ns, for both electrons and 100 GeV/c pions. Clearly, placing a cut on such ratios can be used to discriminate between electrons and pions. For example, an electron efficiency of about 60% is maintained while rejecting a factor of 100 of the pions, using the ratio described above. If one instead uses the signal after 200 ns divided by the signal after 100 ns, in order to achieve a faster discrimination, the pion rejection is reduced to a factor of about 5 - 10 for the same electron efficiency.

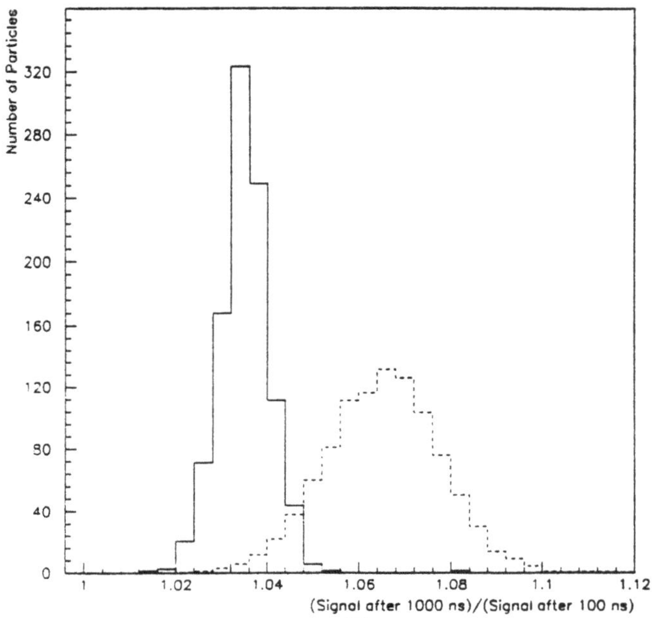

Figure 3: Response after 1 μs normalized to the response after 100 ns, for electrons (solid) and for 100 GeV/c pions (dashed). Both have been normalized to 1000 particles.

In principle the timing discrimination should work even better for lower momentum hadrons, where the difference between electrons and hadrons is more pronounced. However, for the ZEUS calorimeter, with about 100 photoelectrons/GeV, the results at 20 GeV/c are degraded by poor photostatistics and the discrimination is weaker than at 100 GeV/c.

Among the on-going analyses of the data are studies which will: (a) deconvolve the electromagnetic component of the pion response to allow a comparison of the purely hadronic contribution with MC predictions; (b) examine the "energy flow" in the shower, using the time development of the signal in each of the 112 cells of the BCAL which were read out; (c) look event-by-event in order to study the fluctuations in the hadron response, and yield a measure of the energy resolution versus integration time; and (d) study the short-time pulse shape in 0.5 ns bins.

Acknowledgements

The excellent work of Herb Cunitz and the Nevis Labs electronics shop are gratefully acknowledged.

This research has been supported in part by National Science Foundation grant PHY-89-21320, by Department of Energy grant DE-FG02-90ER40540, and by Columbia University. One of us (J.P.) would like to acknowledge financial support from the Natural Sciences and Engineering Research Council of Canada.

References

1. For example, see R. Klanner, DESY 87-058.
2. J. Brau and T. Gabriel, Nucl. Instr. and Meth. **A238** (1985) 489;
   H. Brückmann, et al., DESY 86-155; DESY 87-064;
   R. Wigmans, Nucl. Instr. and Meth. **A259** (1987) 389.
3. ZEUS Detector Technical Proposal, 1986, and ZEUS Detector Status Report, 1987, 1989.
4. M. DeVincenzi, et al., Nucl. Instr. and Meth. **A243** (1986) 348;
   M.G. Catanesi, et al., Nucl. Instr. and Meth. **A260** (1987) 43;
   T. Akesson, et al., Nucl. Instr. and Meth. **A262** (1987) 243.
5. For more details, see A. Caldwell, et al., Proc. of the 1st International Conference on Calorimetry in High Energy Physics, FNAL (1990) 243.
6. A. Caldwell and F. Sciulli, ZEUS Internal Note 88-84;
   A. Caldwell, ZEUS Internal Note 89-75;
   R. DeSalvo, et al., Nucl. Instr. and Meth. **A279** (1989) 467;
   D. Acosta, et al., Nucl. Instr. and Meth. **A302** (1991) 36.

The CDF Upgrade Calorimeter

J. Freeman[b], M. Albrow[b], G. Apollinari[g],
P.de Barbaro[f], V.E.Barnes[e], A.Bodek[f], H.S.Budd[f],
Y. Fukui[c],G. Goulianos[g], J.Hauser[a], D.H.Kim[b],
K. Kondo[h], P. J. Limon[b], M. Lindgren[a], M. Mishina[c],
S. Olsen[f], G. Pauletta[d], T.Mueller[b], W.K.Sakumoto[f],
R. Walker[f], Q. F. Wang[g], S.White[g], and B. Winer[f]

[a] *University of California at Los Angeles, Los Angeles, CA 90024*
[b] *Fermi National Accelerator Laboratory, Batavia, IL 60510*
[c] *National Laboratory for High Energy Physics (KEK), Tsukuba-gun, Ibaraki-ken 305, Japan*
[d] *Istituto Nazionale di Fisica Nucleare, University and Scuola Normale Superiore of Pisa, Pisa, Italy*
[e] *Purdue University, W. Lafayette, IN 47907*
[f] *University of Rochester, Rochester, NY 14627*
[g] *Rockefeller University, New York, NY 10021*
[h] *University of Tsukuba, Ibaraki 305, Japan*

Abstract

We report on a prototype calorimeter built for the CDF Upgrade project. Test beam results from the scintillating tile/wave-shifter fiber readout calorimeter are presented.

1 Introduction

The CDF experiment has commenced on an upgrade program that will enhance its performance for future runs of the Tevatron Collider which have higher luminosities and shorter time spacing between proton-antiproton bunch crossings. A central element of this upgrade program is a new plug calorimeter for the region $\theta < 35°$. This calorimeter replaces the existing plug and forward gas calorimeter systems. Because the new calorimeter extends its coverage to 3°, the forward calorimeters are eliminated. This allows the forward muon toroids to be moved closer to complete the CDF muon coverage.

This paper reports on the construction and test beam evaluation of a large prototype calorimeter. The technology chosen for the upgrade is scintillator plates with wave shifter (WLS) fibers embedded in the tiles. The WLS fibers are spliced to clear fibers which carry the light to phototubes in the back. Figure 1 schematically shows the optical system.

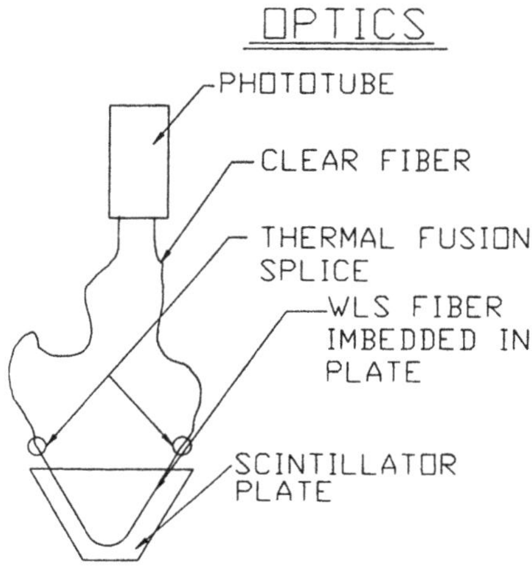

Figure 1: Optical system for the tile/fiber calorimeter.

2 Construction

The prototype was designed to have as realistic as possible geometry. We wanted the prototype to explore any potential problems with this technology, some of which are geometry dependent. The prototype had both EM and Hadron compartments. The EM (hadron) section is composed of 48 (24) projective towers. 4 EM towers map onto one hadron tower. The Em compartment is 22.5° wide in ϕ while the hadron compartment is 45° wide. The EM compartment used 4.8mm lead plates for the radiator. The hadron compartment used 2 inch iron plates. This choice of hadron plate thickness was forced by the decision to use the existing CDF endplug iron structure for the upgrade. Both compartments had 23 layers of absorber. Scintillator plastic used for the prototype was 4mm thick Bicron BC408 PVT for the EM compartment and 6mm thick Kuraray SCSN81 for the hadron compartment. The tile edges were not polished after machining. Optical white paint (Bicron BC 600) was painted on the tile edges. WLS fiber (Bicron BC91A) of 1mm diameter was placed in machined grooves in the tiles. The WLS fiber made a "U" pattern on the tile face. The WLS fibers were welded to clear fibers using a specially constructed splicer machine. Typically 10% of the light was lost passing through the splice with an rms variation of 4 %. To strengthen the splices, protective sleeves were glued around the splice region. Clear Kuraray polystyrene optical fiber was used for the readout. Layers of tiles with their associated fibers were put into "Pizza Pans" formed from sheet aluminum or steel. The tiles were not "wrapped" individually for optical isolation. Rather as pointed out earlier, white paint optically separated them. In the hadron pans, aluminized mylar covered the tile layer on top and bottom. In the EM, one side of the pan had a computer generated black/white pattern (transverse mask) that was optimized to correct for transverse non-uniformity of the tiles. (The same algorithm was used for all EM tiles.) The other side of the tile layer was covered with aluminized mylar. These pans were then installed into the absorber, and the clear fibers were dressed along the outer perimeter to phototubes located in the back. Typical clear fiber lengths were 3.5 meters for the EM and 2 meters for the hadron compartments. The Pizza Pans had steel tubes running parallel to the scintillator plane for radioactive source calibration. This scheme is described in the talk by V. Barnes in this conference. At the phototube, fibers from each tile were routed into separate holes in the phototube cookie. Neutral density filters were placed between the cookie and the phototubes. The gray scale for each tile location were determined from radioactive source measurements. These filters allowed us to correct optical irregularities in the construction process. RCA 8575 phototubes were used for the readout, into CDF RABBIT electronics with 16 bit ADC's. Figure 2 shows the layout of the prototype. Note a position detector imbedded at EM shower maximum.

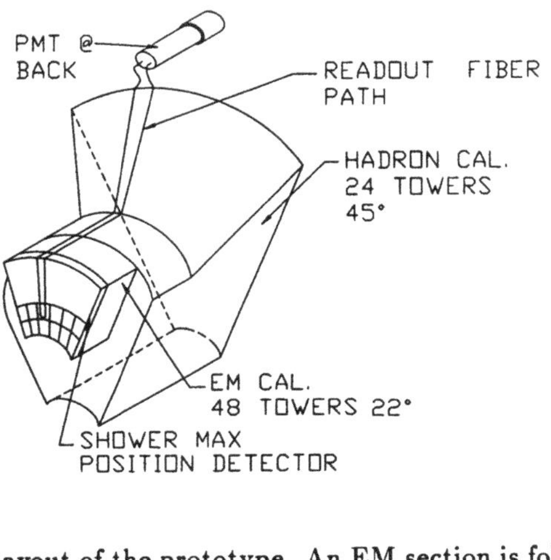

Figure 2: Schematic layout of the prototype. An EM section is followed by a hadron compartment.

3 Test Beam Results

The prototype was exposed to pions, muons, and electrons in the FNAL MTEST beamline. The light yield of the EM calorimeter was found to be 400 photoelectrons/GeV, corresponding to 4.3 pe/MIP/tile. This result was consistent with test bench light yield measurements using cosmic rays. The hadron compartment had 39 pe/GeV. Figure 3 shows the EM compartment response for 100 GeV electrons. Plotted is the energy/momentum distribution and a Gaussian fit. The Sigma of the

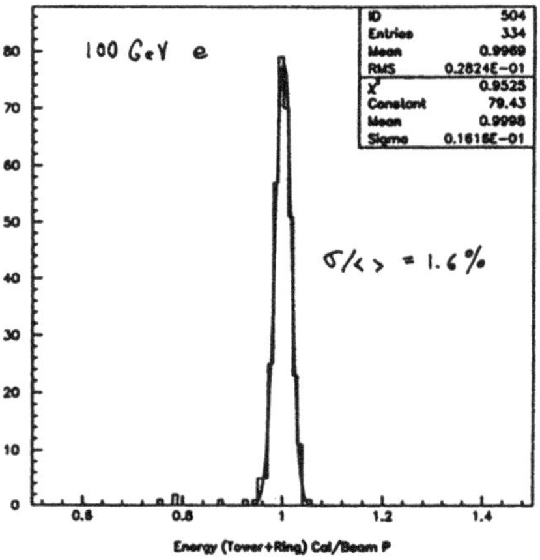

Figure 3: EM compartment E/P response for 100 GeV electrons.

fit is 1.6%. σ/E vs E for a typical tower is shown in Figure 4. Typical fits to σ/E were $0.15/\sqrt{E} + 0.01$.

The uniformity of the EM compartment was determined by scanning it with 100 GeV electrons. We found that there was about a 1% rms variation in the response scanning in either theta or phi directions. The variation was reproducible tower-to-tower, so a response map would correct it. Applying a reponse map correction we found that the energy resolution did not depend on position in the tower. (We note that the observed non-uniformity was predicted by radioactive source scans of the tiles before installation. Based on this we have constructed a 2nd generation prototype to explore an optimized fiber layout on the tile. Radioactive scans of this new pattern show it to be much more uniform, requiring no transverse mask.)

A position detector was installed at EM shower Max. This detector consisted of two planes of 2mm scintillating fibers, with the planes rotated by a stereo angle of 15° to provide both coordinates. The fibers were read out by Hamamatsu multi-anode phototubes. Accuracy of shower centroid location was found to be 0.7mm in the ϕ view, and 3mm in the θ view.

Figure 5 shows E/P for 227 GeV pions hitting a typical hadron tower. These pions were required to be minimum-ionizing thru the EM compartment.

σ/E as a function of energy is shown in Figure 6.

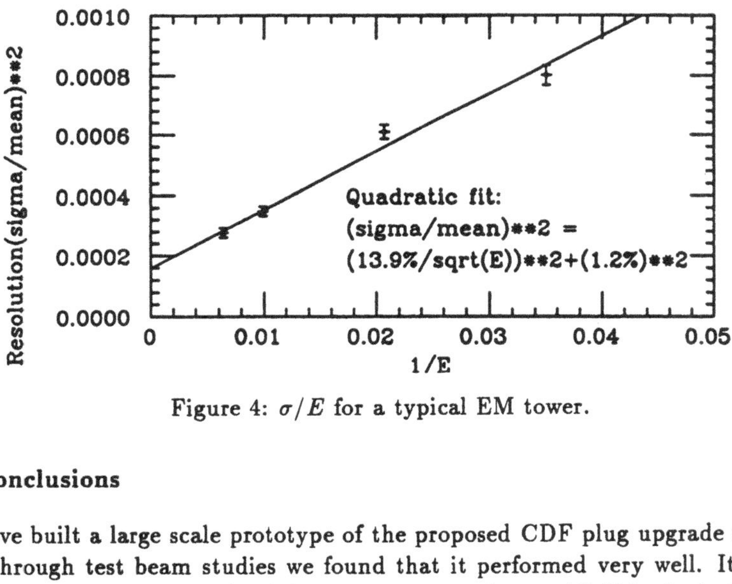

Figure 4: σ/E for a typical EM tower.

4 Conclusions

We have built a large scale prototype of the proposed CDF plug upgrade calorimeter. Through test beam studies we found that it performed very well. It had 400 photoelectrons per GeV in the EM compartment and 39 pe/GeV in the hadron compartment. Energy resolutions were as expected from EGS and scaling by absorber thickness. Transverse uniformity for electrons was adequate (however a second generation prototype is being tested). We look forward to building the real thing and installing it into CDF.

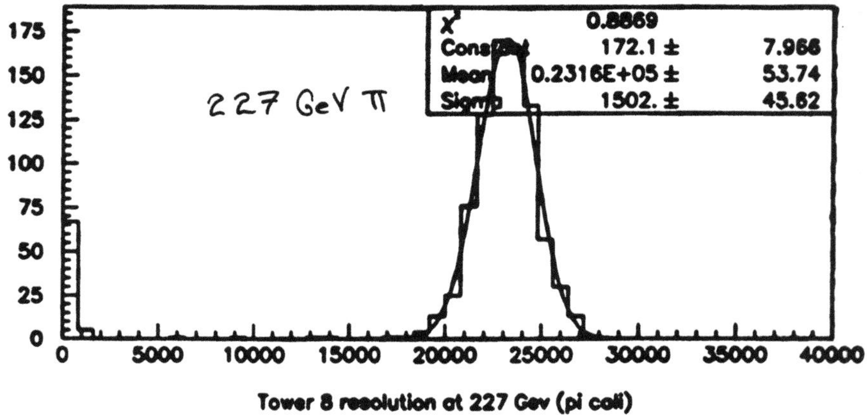

Figure 5: Hadron compartment E/P for 227 GeV pions.

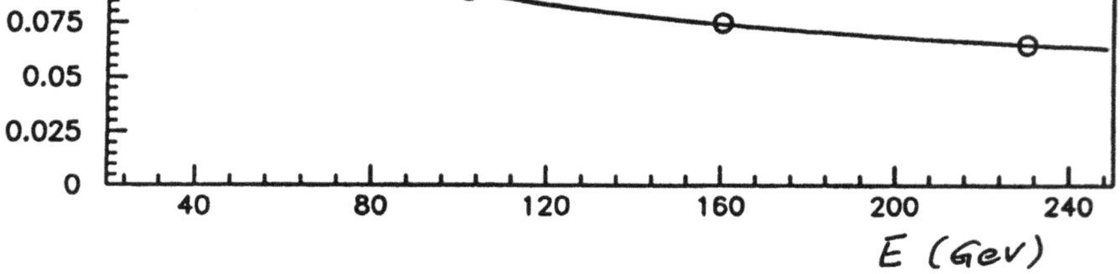

Figure 6: σ/E vs. E for a typical hadron tower for pions minimum ionizing thru the EM compartment.

CALIBRATION OF SCINTILLATING TILE CALORIMETERS USING A MOVING RADIOACTIVE SOURCE

VIRGIL E. BARNES*, ALVIN T. LAASANEN, and JAMES A. ROSS

Physics Department, Purdue University
West Lafayette, Indiana 47907 USA

ABSTRACT

Progress towards the goals of (i) equalizing the responses of the scintillating tiles within a calorimeter tower, and (ii) predicting the testbeam responses of towers, using a moving radioactive gamma source, is presented based on current measurements of the CDF Plug Upgrade EM calorimeter prototype.

1. Introduction

We are calibrating several prototype calorimeters for current Fermilab testbeams, including the CDF Plug Upgrade EM and Hadron prototypes, the SDC cast Lead EM prototype, the SDC Steel Hadron prototype, and the SDC "Hanging File" Reconfigurable Stack Calorimeter. We have also, in collaboration with Fermilab, built and provided sourcing hardware for a set of Radiation-Hardness-testing single tower calorimeters which are being irradiated at IHEP Beijing, Saclay/Orsay, and KEK in intense few-GeV electron beams.

We present here partial and preliminary results from the CDF Plug Upgrade EM prototype at the time of construction and in the early part of the testbeam run: on characterization of the modules, on adjustment of individual tile responses for uniformity, and on the accuracy and repeatability of the measurements. The present results are relevant to the proposed SDC calorimeters, as well as to the proposed CDF Plug Upgrade calorimeters. All of the above-mentioned calorimeters use scintillating tiles with imbedded wavelength shifting readout fibers.

2. Calibration Goals

One major goal is to measure the relative responses of all tiles within a tower to permit correction of: (A) manufacturing imperfections and/or (B) radiation damage if it should occur, by either (1) grey scale masking (at the "cookie" on the phototube) of the fibers from each tile to equalize the tile responses, or (2) software correction of the energy scale of each tower using weighted sums over the tile responses. If two depth compartments in the tower are read out separately, and the relative responses of all tiles are known, partial software correction of the energy resolution is also possible.

The second major goal is to maintain testbeam calibrations over the

*Talk presented by V. E. Barnes for the CDF Plug Upgrade collaboration: Argonne National Laboratory, University of Bologna, Fermilab, KEK Japan, University of Padova, University of Pisa, Purdue University, Rochester University, Rockefeller University, Texas A & M University, and University of Tsukuba.

long term, or, more ambitiously, to use the source measurements in lieu of testbeam calibrations. In practice, such source calibrations would be relative to testbeam exposures of a small subsample of typical towers (perhaps in a faithful prototype). The testbeams are needed to set the energy scale and linearity, to study cracks and transverse uniformity, to study rear and side leakage of showers, and to generally characterize typical towers. However, the hope is to avoid having to put very large calorimeter modules, or many modules, into testbeams.

3. Calibration Method

We use a new model of the source driver designed by V. E. Barnes[1], having a remotely actuated motor-driven reel which can push a 6 m long "wire" into fine channels throughout the calorimeter. The "wire" is actually a stainless steel hypodermic tube with 0.71 mm O.D. and 0.41 mm I.D. , containing a Cs-137 gamma source of ~6 mCi activity and 5 mm active length located near one end of the wire. A motor driven circular indexer can select one of 72 channels into which the one source can be driven. This number could be greater than 200 without major redesign of the system. The motors can be controlled either by switches or via a CAMAC or PC TTL interface. The source position is digitized with a 0.1 mm least count. The source and indexer counters can be read via an RS232 serial port. The source motor can be switched between two adjustable speeds, which were set to 3 cm/sec for use within the calorimeter and 6 cm/sec for travel outside the calorimeter to and from a Lead garage. Nylon tubing (1/8-inch O.D.) is used to flexibly connect the indexer disk to small plastic funnels mounted on stainless steel source tubes, or scabbards, of 1.3 mm and 1.8 mm O.D., respectively, imbedded in the EM and Hadron calorimeters. The driver automatically senses the closed end of a source tube and switches off.

The calibration signal is a DC current from the anode of the photomultiplier tube. In the present calorimeter, a source tube passes across the center of the face of every tile in every tower. The tubes are in close contact with the tiles, separated only by one thin layer of aluminized Mylar. Tiles lie perpendicular to the collider axis. There are 8 EM projective towers covering polar angles (θ) between 22° and 37.5^{o}, times 6 towers in azimuth (ϕ) covering a 21° interval. Towers at the outermost two polar angles are slightly or severely "clipped" by the cylindrical outer boundary of the CDF endplug. The severely clipped towers, which are missing over 50% of the scintillator layers, are excluded from most comparisons.

Source tubes run radially outward with closed ends at the outer radius. A typical calibration run consists of running the source on a series of complete round trips from the garage to the ends of several (or all 23) layers at a given azimuth, thus lighting up in succession all eight towers at that azimuth. The eight phototube signals at one selected azimuth, for one layer, give a set of 16 peaks for the extension and the retraction of the source. The eight phototube signals from neighboring towers 3.5^{o} away in azimuth also show "minor peaks" approximately 10% as strong as the major peaks from the direct source.

Peaks are defined to be the highest single reading at a local maximum, and are presently used in preference to polynmial fits which, while they agree reasonably well with the maxima, are slightly less robust in their present form. The extension and retraction peaks for a given tower typically agree within 1%, and are averaged to represent the response of that tile.

The source measurements are highly reproducible. The patterns of 23 tile responses in a tower measured at two different times agree with a sigma typically less than 0.65%. A few exceptions correspond to clipped towers or towers with known phototube instabilities.

Operationally, it is very important to have a flasher system, with stable or easily calibrated reference phototransducers, to inject light to each photomultiplier tube, in order to follow possible short- and intermediate-term gain drifts. Unless the phototubes are absolutely stable, such a flasher system is essential to link the source calibrations to the beam signals at other times.

4. A Systematic Tile-Size Effect With a Gamma Source

The response to a gamma source is expected to grow with tile size, since the tiles are rather transparent to the radiation. This effect must be understood and quantified if we are to make meaningful comparisons between towers of different transverse sizes. Indirect evidence for the effect can be seen in the minor peaks, and in the tails of the major and minor peaks corresponding to edge or corner neighbors of the tile seeing the source directly. The sum of the 8 tiles neighboring the sourced tile has a response roughly 50% as big as the direct tile response. A simple approximate calculation scales this 50% to a 6% growth in response of tiles from the front to the back of an EM calorimeter tower as the linear tile size grows by 13.5%.

An exact calculation[2] for a single EM calorimeter tile indicates 4.5% growth in response from front to back, and 13.8% growth in tile response going from $\theta = 22^o$ to 37.5^o. The front to back growth for a typical Hadron calorimeter tower tile is calculated to be 10.7%, in good agreement with direct measurements of about 10% for Hadron tiles using a collimated gamma source vs. a naked gamma source.

An exact calculation for a complete set of calorimeter towers is in progress, and will address not only the above tower-size corrections which must be applied to gamma source calibrations, but also the smearing of the individual tile responses caused by leakage of gammas between layers within a tower, to be discussed next.

5. Unfolding of Individual Tile Responses

Leakage to neighboring layers within a tower has been measured by masking off all but one tile in a tower and then passing the source through a series of neighboring layers. In the Hadron calorimeter (2-inch Fe plates, 6 mm tiles) the sum of adjacent layer responses is of order 8% of the direct tile response. For the EM calorimeter the adjacent tile sum lies between about 60% and 75%, depending on the lateral tower size. We denote the relative response of a tile k layers away by f_k, with $f_o = 1.00$. R_i is the individual tile response, and R^*_j is the tower response with the source in layer j. Then we have:

$$R^*_j = \Sigma f_{j-i} R_i , \quad i = 1 \text{ to } 23$$

and we can solve for the individual tile responses provided all 23 layers in the tower have source measurements.

Typical values of f_1 are 20-25%, $f_2 \simeq$ 7-8%, $f_3 \simeq$ 2-3%, and $f_4 \simeq$ 1%, for the EM calorimeter. There is some asymmetry between f_{+k} and f_{-k}, since the source tubes lie on one side of the tiles. There is some uncertainty in the rather limited measurements we were able to make of the f_k. Varying the unfolding matrix over the range of this uncertainty causes variations in the unfolded individual tile responses with a sigma of 2%.

6. Masking Equalization of Tile Responses

The unfolded tile responses were used to create masks, with 23 fields of neutral density grey filters, by laser printing a pattern of fixed-diameter dots in various densities. The first masking used randomly positioned dots and had relatively large deviations from the requested light reductions. A second masking used regular grids of dots, with ±1 dot (=1.75% reduction) jitter. The first masking also achieved only about 40% of the requested light reduction --- this reduction in dynamic range is attributed to subtle effects of the light exposure in photoreducing the laser-printed mask pages onto transparent film. With great care in controlling the exposure time, the second masking achieved substantial improvement in the RMS tile-to-tile uniformity, from 10-20% before masking to 2-10% after masking. A few tiles have half or close to zero response, attributed to possible damage of one or both of the readout fibers, and cause the worst of the above calculated RMS values.

A setup using a green readout fiber driven by a green LED and viewed by a silicon photodiode, passively read out by a Keithley 486 picoammeter, now permits rapid measurement of the masks for quality control.

7. Source Prediction of Beam Response

We routinely make weighted sums over the tile responses in each tower, using (a) uniform ("muon-like") weighting, (b) 25 GeV electron shower weighting, and (c) 150 GeV electron shower weighting. If the tiles in a tower had uniform responses the a/b/c ratios should be the same for all towers. With the equalizations achieved to date, the ratio a/c has an RMS of ≈ 2% and the b/c ratio (reflecting differences in non-linearity of energy response in different towers) has an RMS ≈ 1%.

In principle, since the tiles responses were equalized using the "raw" gamma responses, the actual intrinsic tile responses to muons or showers is expected to decrease by ≈ 4.5% from the front to the back of each tower. This has not yet been, but will soon be, incorporated into the above weighted sums.

All of the available links in the calibration chain appear to have 2% or better accuracy. Electron test beam data are described in the accompanying paper by J. Freeman. Data from the Nitrogen laser flasher system, the final essential link in the chain, will be available soon.

8. References

1. V. E. Barnes and A. T. Laasanen, Calibration of a Scintillating Fiber Calorimeter Using a Moving Radioactive Source, Proceedings of the 1990 SSC Detector Symposium, Ft. Worth, Texas, p.441 (World Scientific, 1991); V. E. Barnes and A. T. Laasanen, Proceedings of the 1990 International Calorimeter Conference, Fermilab, p. 189 (World Scientific, 1991).

2. Numerical integration over the volume of a tile, assuming a pointlike source with inverse square law intensity modified by attenuation of 663 keV gamma rays in plastic. Computer program written by Aaron Daly.

This work was supported by the U.S.Department of Energy under Contract No. DE-AC02-76ER01428.

AN ECONOMIC CONCEPT FOR A BARREL HADRON CALORIMETER WITH IRON SCINTILLATOR SAMPLING AND WLS-FIBER READOUT

O. GILDEMEISTER, F. NESSI-TEDALDI, M. NESSI
CERN, Geneva

ABSTRACT

An unconventional concept for a hadron calorimeter with scintillator plates and WLS fiber readout is discussed. It leads to a simple and low-cost design.

1. Introduction

We propose a concept for a barrel hadron calorimeter with scintillator plates positioned - inside the iron absorber structure - in R-Φ planes, instead of the usual Φ-Z orientation. Wavelengthshifting (WLS) fibers coupled to the two radial edges of the scintillator plates can thus be kept straight in radial direction. Outside the absorber structure they are grouped according to the wanted segmentation of the calorimeter and coupled to photomultiplier tubes (PMT).

This configuration can be made very hermetic and has the additional advantage that the absorber can be built up from "self supporting" plates in R-Φ planes.

For close to zero rapidities the plate structure is about parallel to the primary particles. Most of those will have interacted already in the EM-calorimeter in front, to be about one interaction length in depth. To prevent high energy particles from travelling over long distances in the scintillator the individual plates are kept short and are staggered with respect to each other (Fig. 1). (The proposed structure would not be adequate for an EM-calorimeter).

2. GEANT simulations

GEANT simulations have been made to check the effect of this particular geometry on the energy resolution. They used a calorimeter model consisting of a conventional liquid argon EM compartment, 25 X_0 in depth, with 3 mm lead plates and 4 mm gaps, in front of the hadron compartment of Fig. 1, with ℓ = 10 cm, a = 1.4 cm, d = 0.6 cm and a depth of 9.5 interaction lengths. Jets from 25 to 300 GeV were generated at 1.2 m distance from the EM-compartment and at different impact points.

The results of the energy resolutions lead to a fitted parametrization of

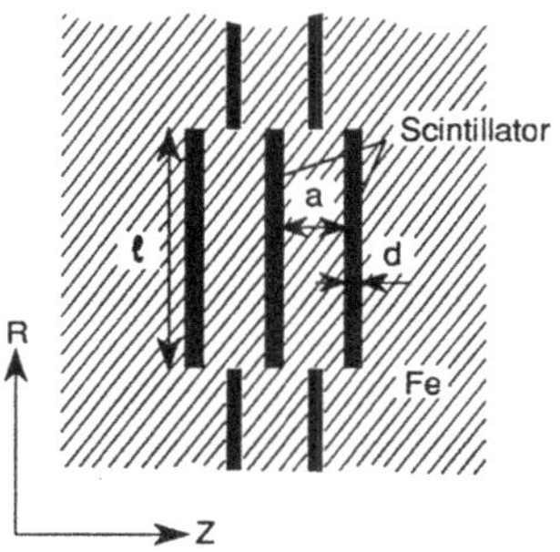

Fig. 1 Basic structure

$$\frac{\sigma_E}{E} = \frac{41\%}{\sqrt{E}} + 2\% \text{ (E in GeV)}$$

A higher statistics run with 150 GeV jets (using the parameters ℓ = 10 cm, a = 1.4 cm, d = 0.3 cm) gave

$$\frac{\sigma_E}{E} = (5 \pm 0.2)\%$$

Replacing the hadron compartment by a liquid argon structure (10 cm Pb plates, 4 mm Ar gaps) or by a standard geometry scintillator structure (10 cm Pb plates, 2.5 mm scintillators) results within the statistical errors to similar energy resolutions.

A scan with 50 GeV jets across the calorimeter front face did not show any evidence of non-uniform response beyond a 1% statistical error.

These simulations did not include any instrumental imperfections like light attenuation or fluctuations of the light coupling efficiency.

3. Scintillator plate - WLS fiber coupling

For an efficient optical coupling of 3 mm scintillator plates to 1 mm round or squared read-out fibers the scheme of Fig. 2 has been tested. The fiber is placed into a 1 mm groove cut into the narrow face of the plate, the adjacent edges of which are cut at 45 degrees and polished. A comparative measurement gave 70% light yield with respect to a standard 2 mm WLS plate (as used in the UA2 endcap calorimeter, K-27 doping) coupled to a flat edge of the same scintillator plate.

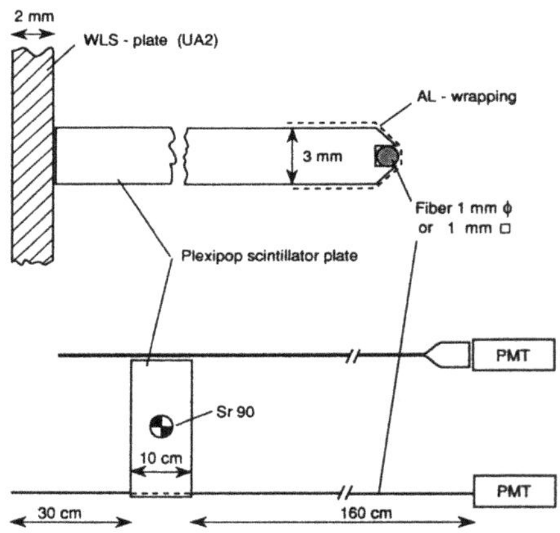

Fig. 2 Test of optical coupling

The distance between scintillator and PMT was 170 cm in both cases. The rear end of the 2 m long fiber was Aluminium plated, while the WLS plate had a white paint on the rear edge.

Another comparative measurement gave 80% light yield with one squared fiber in the groove, with respect to 3 squared fibers coupled side by side to the flat edge of the scintillator.

4. Conceptual design of a barrel hadron calorimeter for LHC

Typically a LHC barrel hadron calorimeter will be about 8 m long with an inner radius of about 2 m. A possible design concept is based on a granularity of $\Delta\Phi \times \Delta\eta = 0.1 \times 0.1$ with a fourfold radial segmentation. The barrel would be subdivided into

64 azimuthal sector modules, each containing 96 cells in R-Z (Fig. 3). The individual cells are not pointing to the vertex, but the centers of the 4 cells forming a tower are pointing.

The whole absorber of a module (Fig. 4) is laminated in the Z-direction into 1.7 mm sheets, which can easily be punched and would not need any machining. The stacked absorber is compressed by a series of rods (~ 8 mm diam.) passing through holes already punched into the sheets and drilled in the centre of each scintillator plate. A beam supporting the PMT's is welded to the outside face of the absorber stack and a 10 mm plate to its entrance face over the whole length of the module.

The 3 mm thick polystyrene scintillator plates are slid into gaps left free in the absorber stacks. WLS fibers are mounted from both sides and fit into radial slots formed by the stack structure, each slot containing - with increasing radius - 1 to 4 fibers.

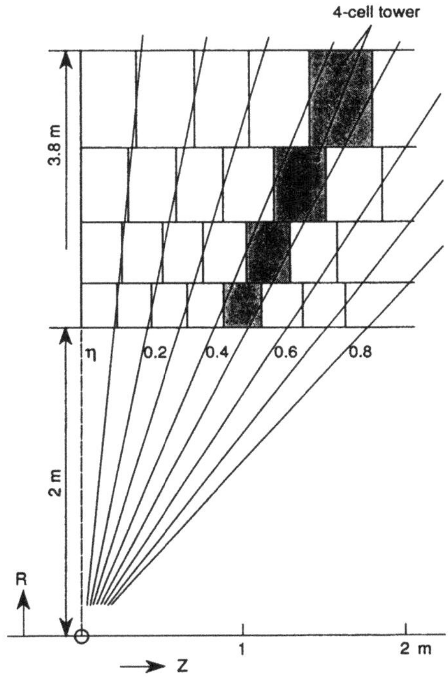

Fig. 3 Cell structure within a module

The absorber thickness between adjacent scintillator plates is $8 \times 1.7 = 13.6$ mm and the pitch between staggered plates is $5 \times 1.7 = 8.5$ mm. The length of a scintillator plate in the radial direction is 10 cm in the most inner cells and might be increased with the radial position of the cells.

The module parameters are :

inner radius	2 m
outer active radius	3.8 m
length	8 m
active iron weight	25.4 t
scintillator weight	0.72 t
iron volume	80%
scintillator volume	18%
depth at zero rapidity	9 int. lengths
number of cells	96
number of PM channels	192

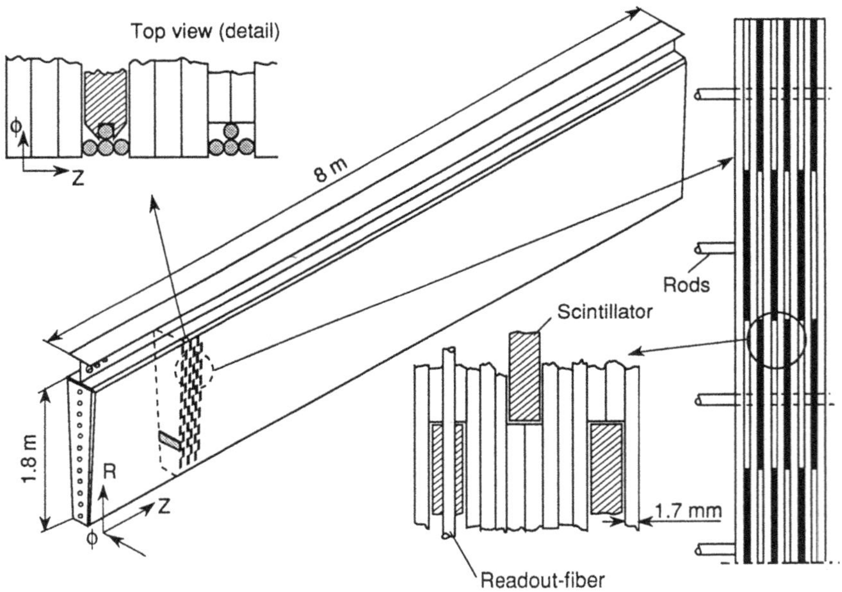

Fig. 4 Conceptual design

Since the absorber structure of the module does not need any machining and is simple and easy to mount this design concept leads to substantial cost savings compared to more conventional constructions.

Acknowledgements

We thank Dr. P. Jenni for many fruitful discussions and suggestions. The company BICRON kindly provided the WLS-fibers for the tests.

A LEAD-SCINTILLATOR ELECTROMAGNETIC CALORIMETER USING WAVE LENGTH SHIFTING OPTICAL FIBRE READ OUT

G. Ballocchi[1], L. Camilleri[1], L. Di Lella[1], J.-P. Mendiburu[2], P. Nédeléc[2], M. Rebut[1], G. Snow[3], M. Stipcevic[1], M.-T. Tran[4], F. Vannucci[5]

1) CERN, Geneva, Switzerland
2) LPC, Collège de France, Paris, France
3) University of Michigan, Ann Arbor, MI, U.S.A.
4) Université de Lausanne, Lausanne, Switzerland
5) LPNHE, Univ. de Paris VI et VII, Paris, France

presented by
Leslie Camilleri

A prototype for electromagnetic calorimeter to be used in a recently approved neutrino oscillation experiment, WA-96 referred to as NOMAD[1] (Neutrino Oscillation with a Magnetic Detector) was built and tested in a CERN SPS electron and pion beam. The calorimeter was of the lead-scintillator type with 3 mm thick lead plates alternating with 10 mm thick NE-110 scintillator[2] planes (fig. 1). Each scintillator plane consisted of nine 10 x 10 cm^2 scintillator tiles arranged in a 3 x 3 matrix. Each scintillator tile was read by a 2 mm diameter KURARAY polystyrene base fibre[3] with PMMA cladding and Y7 doping glued with EPO-TEK 305 epoxy[4] in a 2.38 mm deep groove machined along the diagonal of the tile. The last 15 mm of the groove were ramped up such that the fibre emerged from the tile before reaching the corner. The WLS fibre was fused to a 4 m long clear fibre 20 mm after emerging from the tile surface. The tiles were individually wrapped with aluminized mylar and black tape. The nine tiles belonging to a plane were stuck to a 1 mm thick stainless steel plate using double faced sticky tape. The planes of tiles were hung from two steel rods as shown in fig. 1. The fibres from the 40 tiles directly behind one another were grouped together and glued into a hollow plexiglas cylinder. The face of the cylinder and fibres were then machined flat and a 1 cm thick cookie of Röhm GS 2000 plexiglas[5] was glued to it. The whole assembly was then pressure jointed to a Philips XP2012 10 stage 11/2 inch photomultiplier[6] operated at about 1250 volts. The prototype was thus divided into 9 towers. In order to improve the uniformity of response the orientation of the WLS fibres in the odd numbered planes was rotated by 90^0 with respect to the fibres in the even numbered planes[7] (fig. 1).
The response of the calorimeter to electrons of 3, 10, 50, 80 and 90 GeV was studied. The response of the calorimeter to a given electron was taken to be the sum of the pulse height of the nine towers. At each energy a Gaussian fit to the summed pulse height distribution was made. The means of the Gaussians were then fit to a straight line through the origin. The deviations from linearity are shown in fig. 2. They are less than 1% everywhere except at 3 GeV where the larger deviation is likely to be due to the

uncertainty in the beam setting at such low energy. The light yield of the prototype was found to be > 1.8 photo-electrons per plane for a minimum ionizing particle and 155 photo-electrons/GeV for electrons.

The energy resolution of the calorimeter, $\sigma(E) / E$, obtained from the standard deviations of the Gaussian fits is shown in figure 3 plotted as a function of $1/\sqrt{E}$ together with a straight line fit. The resolution is found to be

$$\sigma(E) / E = 0.095 / \sqrt{E} + 0.004.$$

The uniformity of response over the face of a tile was tested by sending a 50 GeV electron beam at several points over the face of the central tile. The pulse height dropped by 3.5% 2.5 cm from the centre of the tile and by 6.5% 4 cm from the centre of the tile (1 cm from the edge of the tile). The same effect was seen with muons, indicating that it is due to a light collection effect rather than due to part of the shower being lost in the cracks between adjacent tiles. The results of a scan across one such crack is shown in fig. 4. The impact point of each individual electron was determined using wire chambers in the beam. The same effect as mentioned above can be seen but, in addition, within ± 5 mm of a crack an additional drop in pulse height is observed; probably due to particles being lost in the crack. It is hoped to correct this non uniformity of response by using masks over the face of the tile and by better packing of the tiles to reduce the crack dimensions.

The pion/electron discrimination of the calorimeter was also studied. The response of the calorimeter to 30 GeV pions and 30 GeV electrons is contrasted in fig. 5. It can be seen that very few pions deposit the same amount of energy as electrons of the same energy. In order to improve the pion rejection a pre-shower consisting of 8 mm of lead followed by a 1 cm thick scintillator was installed in front of the calorimeter. By rejecting events having less pulse height than 2.5 times the pulse height of a minimum ionizing particle and depositing little energy in the calorimeter hadron rejections in excess of 500 were obtained between 3 and 30 GeV. Electron efficiencies remained above 95% for these cuts.

References

1. Search for the oscillation $\nu_\mu \rightarrow \nu_\tau$ CERN SPSLC/91-21, SPSC/P261.
2. NE Technology Sighthill Industrial Estate, Edinburgh, EH11 4BY
3. KURARAY Co. Ltd., Shuwa Higashiyaesu Bldg. 2-9-1Hatchobori-ku, Tokyo, Japan.
4. EPOXY Technology Inc., 14 Fortune Drive, Billerca, MA 01821, USA.
5. Röhm, Kirchenallee, D-6100 Darmstadt, Germany.
6. Philips Nederland B.V., Marktgroep Philips Components Postbus 90050, 5600 PB NL-Eindhoven.,The Netherlands.
7. M.G. Albrow et al., NIM A256 (1987) 23.

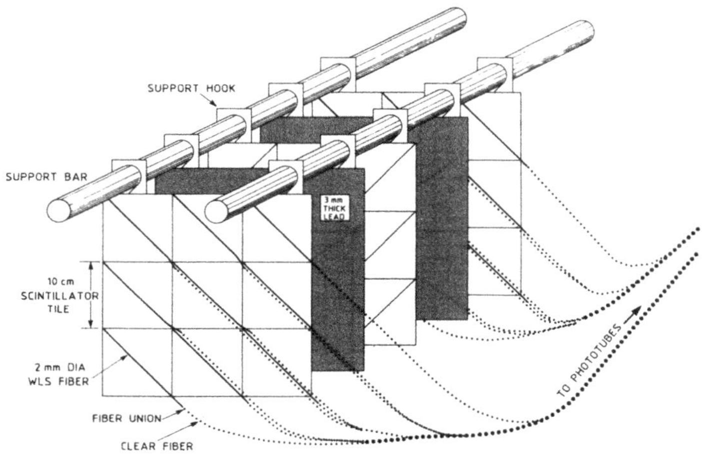

Fig. 1 The calorimeter prototype

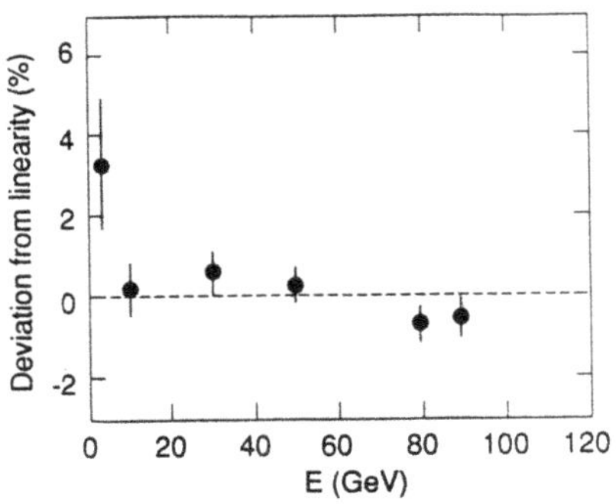

Fig. 2 Deviations from linearity as a function of energy

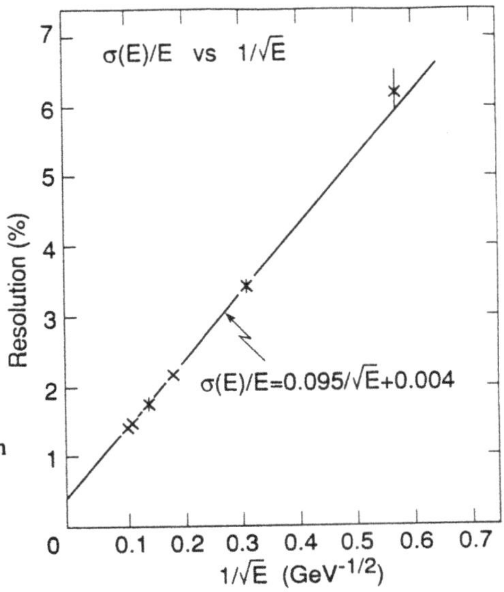

Fig. 3 The resolution as a function of $1/\sqrt{E}$

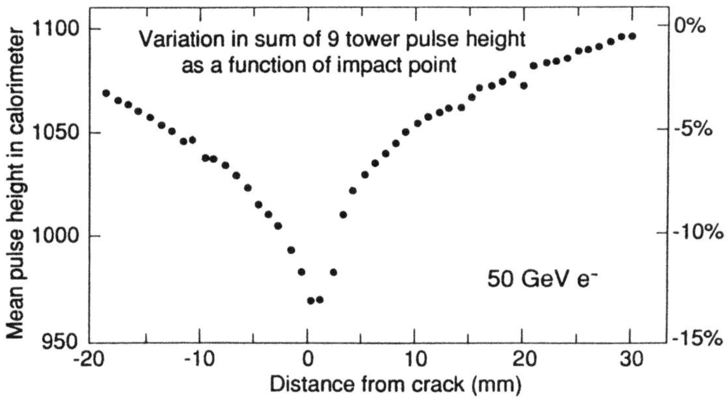

Fig. 4 Response of the Calorimeter as a function of impact point

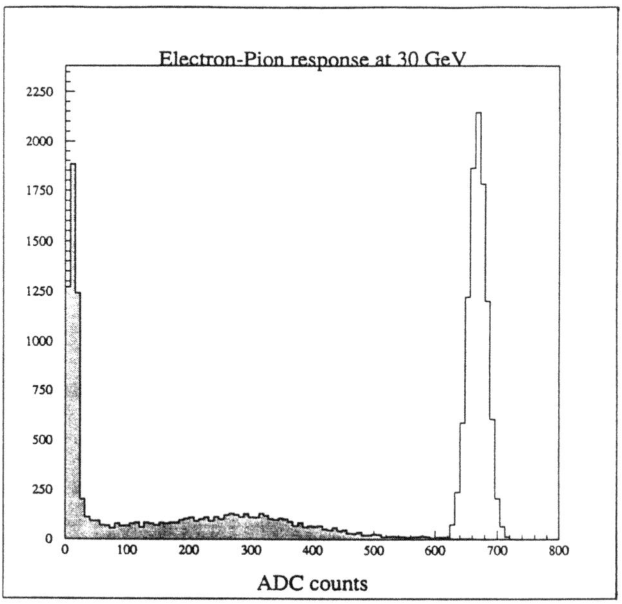

Fig. 5 Response of the calorimeter to electrons (white histogram) and to pions (grey histogram) of 30 GeV

VERY FORWARD LEAD-LIQUID CALORIMETER FOR LHC

A.ARTAMONOV, V.EPSTEIN, P.GORBUNOV, V.KHOVANSKY,
S.KRUCHININ, A.MASLENNIKOV, I.ZUCKERMAN
Institute of Theoretical and Experimental Physics (ITEP), Moscow.

and

A.EREDITATO, F.RICCARDI
Università e Istituto Nazionale di Fisica Nucleare (INFN), Naples.

(presented by V.Khovansky)

ABSTRACT

A conceptual design of a "very forward" calorimeter for an experiment at LHC is described.

1. Physics Motivation

The forward calorimeter is an essential component of a collider detector at LHC/SSC, with the main function to complement the measurement of missing E_T. The examples of important physics signatures are the decay $H \rightarrow ll\nu\nu$ of a heavy Higgs (m_H > 0.5 TeV) and decays of supersymmetric particles with missing energy. Another purpose is the jet tagging to identfy WW/ZZ fusion processes which are of great interest even in a scenario with no Higgs in the LHC/SSC discovery range.

2. Design considerations

A forward calorimeter is supposed to cover the preudorapidity range of $3 < |\eta| < 5$. This range spans a relatively small angular interval, $5.7° > \theta_z > 0.77°$, where the detector occupancy grows very rapidly with pseudorapidity, as illustrated by Fig.1. The most difficult region is $4 < |\eta| < 5$ ($2.1° > \theta_z > 0.77°$), where the energy flow is about 3.5 TeV per bunch crossing and the occupancy is almost 100%. The detector must withstand here extremely high radiation doses (up to 100 Mrad/year at $|\eta| = 5$ for a luminocity of $2 \cdot 10^{34}$ $cm^{-2} \cdot s^{-1}$). A use of *radiation hard materials* and appropriate protection of read-out electronics is, therefore, essential. The detector will become radioactive and might turn out to be inaccesible – hence, a *long-term reliability* is required. High occupancy implies also that the detector must be *fast* to minimize a "pile-up" effect.

In this paper we describe a *liquid scintillator filled tubes* approach which is being developed within the framework of a LoI preparation by the EAGLE proto-Collaboration for LHC.

3. The Proposed Solution and its Feasibilty

Basic concept. The proposed solution is a combination of a spaghetti calorimeter geometry and a liquid scintillator active medium confined to thin tubes. Thus, the advantages of a spaghetti type calorimeter, namely, very fast responce (rise and decay times of typically 2 ns) and high hermeticity and uniformity, will be complemented by the possibility to replace the active medium in order to keep the light yield constant with time. Liquid scintillators can stand the doses of $1 \div 10$ Mrad and can be recycled.

The tubes will be aligned in parallel to the beam line and exposed to incoming particles at angles from ~0.7 to ~7 degrees, which makes the calorimeter not exactly projective. On the other hand, this reduces *channeling effects* that spoil the calorimeter electromagnetic energy resolution. We have chosen a tube diameter of about 3 mm. Increasing the diameter would lead to a rapid deterioration of the elecromagnetic energy resolution: firstly, because the average distance between the tubes will become comparable with the Moliere radius ($R_M \approx 20$ mm in this case) and, secondly, because of the increasing importance of channeling effects.

The absorber material is lead, which is readily available, relatively cheap and allows an extrusion as a basic manufacturing technology. Some basic features of the proposed calorimeter are: lead/hole volume ratio 4/1; sampling fraction $1.5 \div 1.9\%$ (for a tube internal diameter from 2.7 to 3.0 mm); average radiation length $X_0 = 7.5\ mm$; average nuclear interaction length $\lambda_{nucl} = 21$ cm; absorber depth L=2m (~9.5 λ_{nucl}); average density $\rho = 9\ g/cm^3$.

Two options are presently being considered for the tubes:

Plastic tubes with inner optical cladding. A wide spectrum of very transparent mineral oil based liquid scintillators with an index of refraction n_{ref}~1.47 is available from industry. They emit predominantly in the blue. To provide a sufficient trapping of the scintillation light in such liquids the tubes should have $n_{ref} < 1.40$, which excludes glass. Meanwhile, *plastic* light guides with inner cladding having even lower index of refraction (down to $n_{ref} = 1.35$) are known to be used for medical applications. This option would feature a high light output at $L = 2\ m$ (for example, ~3000 photons/GeV per module of 90 tubes filled with BICRON BC517P liquid scintillator [1]), but the resistance of optical and mechanical properties of plastic tubes to a high dose radiation is still to be investigated.

Glass capillaries. An advantage of this option is the well known radiation hardness of glass. A drawback is that the capillaries are very fragile and difficult to handle. A more fundamental problem, though, is related to the light transmission. Commercially available glass capillaries have, typically, n_{ref}~1.49, hence the choice of liquids is restricted to the ones with a high refraction index ($n_{ref} > 1.55$). Very few commercial liquid scintillators of this kind are known. They emit predominantly in the green and exhibit a poor transparence to their own fluorescent light. As a consequence, a relatively low light output at $L = 2\,m$ is expected (for example, ~300 photons/GeV for a module of 90 tubes, filled with NE 209B [2] liquid scintillator having an attenuation length of ~80 cm around the emission peak).

4. A Mechanical Design

The elementary *module* of the proposed calorimeter is shown on Fig. 2. It consists of a 2 m long extruded lead absorber with effective cross-section of 55.5×57.6 mm^2 containing 90 longitudinal holes for the tubes, and two end pieces equipped with inlet and outlet pipes. The tubes come out from the absorber as a bundle which is supported inside a back cap by a shaper and held together by a clamp. A light from the tubes is collected and transported to the photodetector by a light guide sealed into a photodetector housing attached to the back cap by a flange. The entire volume is filled with a circulating liquid scintillator.

Upper and lower surfaces of the module have a zig-zag profile which allowes to keep uniformity of the lateral lead/liquid structure and also to bind layers of modules stacked on top of each other with a half a module width shift. The latter ensures the absence of cracks and thus a good hermeticity of the calorimeter.

The choice of the lateral size of the module is governed by design and reliability considerations. A MC study shows that below 20×20 cm^2 the calorimeter E_T resolution becomes insensitive to a lateral granularity. The described module weighs about 60 kg and is relatively easy to handle.

The proposed calorimeter consists of about 3100 such modules, packed within a cylindrical steel support, 368 cm in diameter, having a step-like structure on its inner surface. The central hole is formed by a hollow steel insert with inner diameter of 30 cm. The overall sensitive area covers a pseudorapidity range of $2.8 \leq |\eta| \leq 5.0$ (assuming a distance of 15 m from the beam crossing point). Equipped and filled, the calorimeter weighs about 200 t. The total volume of liquid scintillator in the calorimeter is about 4.7 m^3.

5. The Calorimeter Performance, Monitoring and Read-Out

The expected energy resolution of the calorimeter is $\sim 60\%/\sqrt{E}$ for hadrons and $\sim 30\%/\sqrt{E}$ for e.m. particles, with a constant term of a few percent. These estimates are based on the extrapolation from SpaCal results [3,4].

The calorimeter signal will be sampled every bunch-crossing, therefore the integration time should not exceed the inter-bunch interval of 15 ns, which implies that the proposed calorimeter will not be compensating. Assuming the e/π ratio dependence on the gate width observed in SpaCal [4], we expect $e/\pi \approx 1/0.8$ for 15 ns integration gate. Theoretically, a better compensation can be achieved by a proper choice of the lead/scintillator volume ratio [5], given the scintillator's hydrogen content.

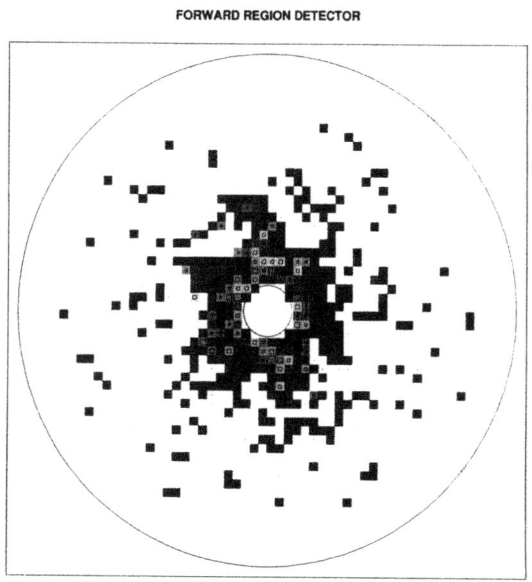

Fig. 1. The distribution of the energy deposited in the forward calorimeter located at 15 m from the interaction point after a single bunch crossing with 20 overlayed minimum-bias events (simulation by PYTHIA + GEANT). One cell corresponds to 6×6 cm^2.

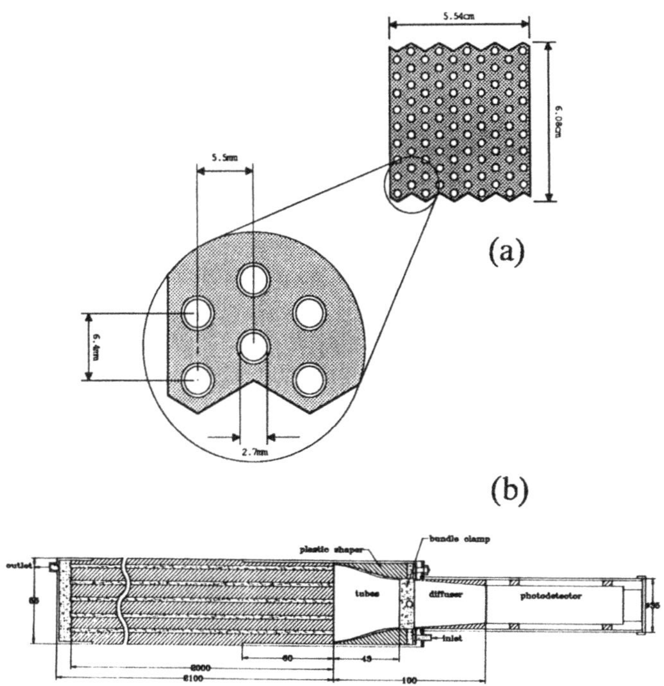

Fig. 2: The calorimeter module: (a) lateral cut; (b) longitudinal cut.

The photodetector should provide a dynamic range of 10^5 (which corresponds to signals from a few MeV to hundreds of GeV) and a stability at the percent level. This rules out the use of conventional PMTs. Taking into account the relatively high expected light output from the calorimeter module, a hybrid photomultiplier [6] with a gain of a few times 10^3 could be a promising candidate.

Several methods are forseen for the calorimeter response monitoring: (i) by recording upright cosmic muons selected by a special trigger (an estimated yield from a minimum ioizing vertical muon is ~100 photoelectrons per module, with plastic tubes); (ii) by recording signals from a radioctive source dissolved in the liquid scintillator with a concentration sufficient to obtain a few kHz count rate; and (iii) by use of LEDs fixed at the photodetector (to check the electronic read-out chain).

6. The Research and Development Activities

At present, we are making two 0.5 m long test modules, one with plastic tubes and the other with glass capillaries, to be exposed to cosmic rays and to test beams of electrons and muons. In the framework of the R&D program we are going to investigate the following main issues: liquid-tight module design; radiation hardness of the plastic tubes, long-term reliability and radiation hardness of the sealants, the calorimeter monitoring; the choice of photodetectors; the signal shaping and the calorimeter compensation.

References

1. BICRON Corporation, Newbury, Ohio 44065, USA
2. Nuclear Enterprise, Sighthill, Edinburg, Scotland EH114BY
3. D.Acosta et al., *NIM* **A294** (1990) 193–210
4. *Scintillating fiber calorimetry at the LHC* CERN/DRDC/90-23 DRDC/P1
5. R.Wigmans, *NIM* **A259** (1987) 389–429
6. See, for example, R.DeSalvo et al., Preprint CERN-PPE/91-142 and references therein.

A NEW IDEA: AN ACTIVE SUPERCONDUCTING COIL USING SCINTILLATING FIBERS

H. PLOTHOW-BESCH
CERN - PPE, CH - 1211 Genève 23, Switzerland

ABSTRACT

A new idea of an active superconducting coil using scintillating glass fibers as active medium inside the vacuum vessel is discussed. The properties of solid-state photomultipliers are very well adapted to be used as read-out system of the fibers in the proposed scheme. Such a design would allow to correct the deterioration of the energy resolution of the electromagnetic calorimeter caused by a magnetic coil in front of it.

1. Introduction

Any dead material such as i.e. a magnetic coil or a cryostat in front of an electromagnetic (EM) calorimeter results in a deterioration of the energy resolution[1], σ/E. This is especially true for a high resolution EM calorimeter, such as a lead/liquid-argon or a lead/scintillating-fiber calorimeter with resolutions $\sigma/E \leq 10\%/\sqrt{E}$, where E is the energy of a showering particle. The deterioration becomes worse in the forward region due to the increasing amount of dead material from the coil or cryostat walls with polar angle. This deterioration of the energy resolution could be corrected for by using an active medium right after or close to the dead material.

In the following we describe our idea to equip a superconducting coil with an active medium after the conductor inside the vacuum vessel using scintillating fibers to make the coil behaving like a mini-calorimeter. We also describe the read-out system envisaged for those fibers. In addition, the result of a study using such a devise to recover the deterioration of the energy resolution is presented.

2. The Idea

The dead material of a small superconducting coil in front of a high resolution electromagnetic calorimeter - i.e. like in the ZEUS experiment[2] - amounts to ~0.9 radiation lengths (X_0) at a polar angle of $\theta = 90^{\circ}$ (pseudorapidity $\eta = 0$) and increases by a factor of two at $\theta = 30^{\circ}$ ($\eta = 1.3$). If the EM calorimeter is a lead/liquid-argon (Pb-LAr) calorimeter the energy resolution[1] in this example is $\sigma/E = \sim 10.6\%/\sqrt{E}$ at $\theta = 90^{\circ}$ ($\eta = 0$) and deteriorates to $\sigma/E = \sim 16.1\%/\sqrt{E}$ at $\theta = 37^{\circ}$ ($\eta = 1.1$) using a cone size of 3x3 cells which corresponds to a cluster size of 10.5x10.5 cm^2. To correct for this deterioration of the energy resolution we propose to make the superconducting coil acting as a mini-calorimeter by using scintillating glass fibers[3] (GF-PS) as active sampling medium inside the vacuum vessel of the coil. A schematic view of the layout is shown in Fig. 1. The basic idea is to have the active medium as close as possible to the major amount of dead material. Installation of the active medium right after the coil could be another possibility. As read-out system for these glass fibers we propose to use solid state photomultipliers (SSPM[4]). The use of scintillating fibers as active medium has the advantage that the detector can also be used as position device. This device can even be completed to become a real preshower detector if additional converter material is used before the active medium which might be an advantage to recognize photons from any source.

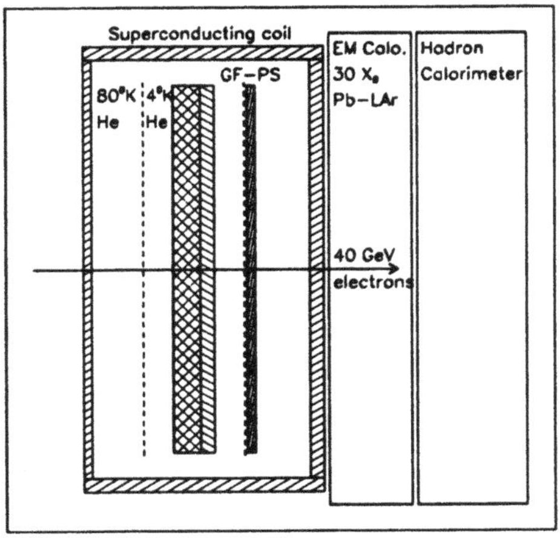

Figure 1
Schematic view of a superconducting coil using scintillating glass fibers (GF-PS) as active sampling medium
(Dimensions are not on scale)

3. Scintillating glass fibers

Only recently scintillating glass fibers[3] have been evaluated seriously as new materials for tracking applications in high energy physics. Locally emitting scintillators with high light yield, ns-decay times and good attenuation properties have become available, i.e. Cerium-doped silica (N327)[5]. This multicomponent glass exhibits a large Stokes-shift paired with a reasonable energy conversion efficiency which has been measured to 2.1 photons/keV. Measurements of the light intensity as a function of the pathlength, integrated over the emission spectrum, give an attenuation length of ~2m at 20cm pathlength.

The advantages of using scintillating glass fibers as active sampling medium in our proposal compared to scintillating plastic fibers are manifold:

- any detector put in position inside the vacuum vessel of a superconducting coil has to be solid. Access in case of failure is practically impossible. A fiber detector (glass or plastic) fulfils this condition.
- the fiber detector has to work in cold environment which might be as low as 4°K. This should not be a problem for glass fibers (tests[6] have been performed down to 80°K). In case of scintillating plastic fibers the cladding might get separated from the core and therefore affect the light yield substantially.
- the heat leakage can be kept minimal in case of glass fibers because glass is a good insulator whereas plastic is not. The heat dissipation at 4°K for 6 layers of glass fibers of 500μm diameter (Δr = 3mm) at a radius of 1.3m using 30cm for insulation going from

4°K to room temperature ($\Delta T = 224°K$) amounts to only ~33W at both ends.
- The shrinkage in length which have to be taken into account when going from room temperature to 4°K can easily be adopted by very small bends with practically no light yield reduction in case of glass fibers.

It should be pointed out that the existing Ce-doped silica show an afterglow which might make them unsuitable for high rate applications. This problem might be overcome with fluorid glasses.

4. The SSPM as read-out system for scintillating glass fibers

The solid-state photomultipliers (SSPM[4]) developed at Rockwell International provide the means of detecting a small number of photo-electrons with a high detection rate and very good time resolution which makes them very suited for fiber read-out systems in the proposed scheme. Such a device consists in principle of four different operating layers:

- a silicon substrate (SS) doped above the metal-insulator transition,
- an infrared active layer (IRL) strongly doped with As to provide a reasonable mobility for both carrier types (electrons and holes). This is the photo-sensitive part,
- a blocking layer (BL) with a doping concentration of As being 2000 times smaller than the IRL
- a transparent collection electrode.

The advantages of such a device are: a very high quantum-efficiency ($Q_{eff} \gtrsim 85\%$), a homogeneous gain and good resolution because the absorption region and the gain region are well separated, a possible high gain of $4 \cdot 10^4$ at very low operating voltages (6.5V) and good linearity for 'moderate' photon fluxes. The only disadvantage this device has for usual applications is that is has to be cooled to very low temperatures (~ 6°K) which implies cryogenics. For the application we propose (usage inside the superconducting coil) cryogenics is needed already for the coil itself and therefore available. The radiation hardness of the SSPM has to be measured.

5. Correction of the energy resolution using scintillating glass fibers

In this chapter we report on a simulation study to use the scintillating glass fiber preshower detector (GF-PS) to improve the energy resolution of a lead/liquid argon electromagnetic calorimeter. A full GEANT[7] shower simulation of electrons of 40 GeV energy at 90° incident angle has been performed. All materials such as lead, liquid-argon, steel and/or aluminium of the cryostats of the Pb-LAr calorimeter and the magnet coil and the aluminium of the coil conductor and its support have been taken into account in the simulation of the energy loss. To simulate the scintillating glass fibers a composite material of 55%SiO_2 + 24% MgO + 11%AlO_3 + 6%LiO_2 +4%Ce_2O_3 resulting in a radiation length of 9.77cm and a density of $r = 2.64 g/cm^3$ has been used. The event structure has been generated in the SLUG[8] environment, and for the geometry the DICE program as developed by Ref. 9 has been used. Effects from electronics noise, attenuation length effects, light yield effects and pile-up effects have been ignored so far in the study.

A 40 GeV electron deposits in 3mm scintillating glass fibers (six layers of 500µm diameter each) on average $\bar{E}_{GF\text{-}PS} = 38$ MeV which has to be compared to $\bar{E}_{Calo} = 5.9$ GeV in a 30 X_o deep Pb-LAr calorimeter. The energy resolution of the calorimeter has been calibrated by optimizing the relation $R = \Delta E/(E_{Calo} + a_1 \cdot E_{GF\text{-}PS})$. The result of the optimization is shown in Fig. 2 where the energy resolution $\sigma/E \cdot \sqrt{E}$ is plotted as a function of the calibration constant a_1. A sharp minimum at $a_1 = 8.5$ is visible.

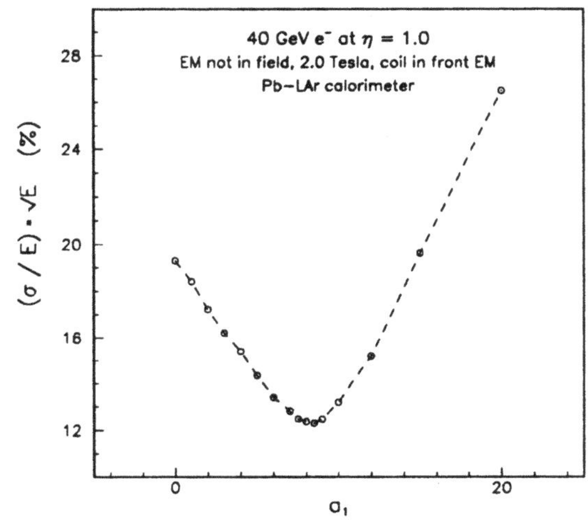

Figure 2
The energy resolution, σ/E·√E, as a function of the calibration parameter, a_1

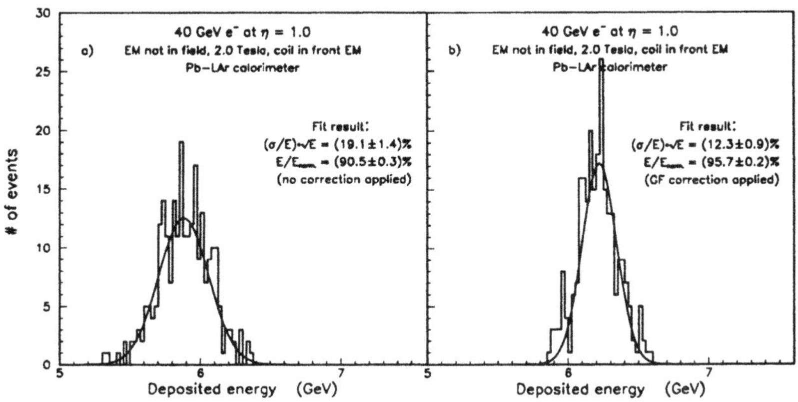

Figure 3
Deposited energy
a) for the Pb-LAr calorimeter alone b) after correction using the GF-PS detector
The curves are the result from a gaussian fit

The distribution of the deposited energy, E_{Calo}, in the calorimeter is shown in Fig. 3a before correction, and in Fig. 3b after the energy correction with the glass fiber detector has been applied. At $\theta = 40^\circ$ ($\eta = 1.0$) the energy resolution is $\sigma/E \cdot \sqrt{E} = (19.1 \pm 1.4)\%$ before and $\sigma/E \cdot \sqrt{E} = (12.3 \pm 0.9)\%$ after correction. The ratio of the deposited energy increases from $\bar{E}_{Calo}/\bar{E}_0 = (90.5 \pm 0.3)\%$ to $\bar{E}_{Calo}/\bar{E}_0 = (95.7 \pm 0.2)\%$ with $\bar{E}_0$ being the energy deposited on average at $\theta = 90^\circ$.

6. Conclusions

The idea of making a superconducting coil acting like a mini-calorimeter seems very attractive. A small superconducting coil with active material inside the vacuum vessel seems feasible. A first iteration engineers design is existing with reasonable cost estimates[10]. Using glass fibers as active material inside vacuum vessel could be a possible solution to correct for the deterioration of the energy resolution caused by the dead material from the coil conductor, support and cryostat. Results from simulations look very promising.

References

1. H. Plothow-Besch, *Energy Resolution Studies of Different Types of Electromagnetic Calorimeters as a Function of Radiation Lengths and Magnetic Field Effects*, these proccedings.
2. The ZEUS Detector, Technical Proposal, DESY, Hamburg.
3. M. Hemig et al., Workshop on Appl. of Scintillating Fibers in Particle Physics, Blossin, September 1990.
4. M.D. Petroff, M.G. Stapelbroek, W.A. Kleinhans, Appl. Phys. Lett. **51** No. 6 (1987) 406.
5. M. Hemig et al., Proceedings of the ECFA Study Week on Istrumentation Technology for High Luminosity Hadron Colliders, Barcelona (1989), p. 231.
6. P. Naß, FA. Schott, Mainz, FRG, private comunication.
7. R. Brun, F. Bryant, M. Maire, A.C. McPherson, P. Zanarini, *GEANT3*, DD/EE/84-1 (1987)
8. R.S. DeWolf, *Simulations for LHC: SLUG, GENZ and related packages*, Proc. of the 3rd Workshop on Detector and Event Simulation in High Energy Physics, Amsterdam (1991).
9. A. Dell'Acqua and M. Nessi, *the DICE program*, private comunication.
10. E. Baynham, RAL, Chilton, Didcot, Oxfordshire, OX140QX, private comunication.

IONIZATION CALORIMETERS

SESSION SUMMARY

JACQUES COLAS

L.A.P.P. , Annecy-le-Vieux, France

1. Introduction

The future large hadron colliders LHC/SSC with their high design luminosities and small time interval between bunch crossings, poses new experimental challenges to calorimetry. Not only the chosen technique should allow for the construction of very large systems, but calorimeters will have to face new difficulties :

- luminosities in excess of 10^{34} cm^2 s^{-1} require radiation hardness far beyond what has so far been achieved,
- to limit pileup to a level where it does not hinder the physics potential of these new colliders, calorimeters have to be fast and highly granular,
- the standard model Higg particle places very stringent requirements on the performances of EM calorimeters,
- measurement of jets with energy up to several TeV and detection of processes with apparent missing transverse momentum, require powerful and hermetic hadronic calorimetry down to very forward angles.

In this session summary, I have chosen to emphasized and put in perspective new experimental results which address some of the above items. Much more was said during the session and can be found in the author's papers.

2. Ionization Calorimeters and Large Systems

A number of large liquid argon devices have been built and done physics in the past years. Several new, large, collider based, detectors are presently coming into operation : SLD at SLAC, D0 at Fermilab, H1 at HERA. The last two have reported results at this conference.

The D0 calorimeter features a uniform, fine grained uranium liquid argon calorimeter which provides good energy resolution and hermetic coverage for electrons, photons and jets. It is constructed as three separate detectors : central and two endcaps. Each unit weights ~ 500 tons and is finely subdivided in ~ 16000 readout channels [1]. During the spring of 1991, D0 had a two month cosmic ray run to test its major subsystems, including the central calorimeter. Good performances were measured and reported at this workshop [2]. The endcaps are now installed and physics data taking should start in the first part of 1992.

The H1 HERA detector uses cold and warm sampling calorimeters. The central and forward regions are covered by a large (~ 400 tons, ~ 50000 channels) EM and hadronic liquid argon calorimeter [3]. The calorimeter was cooled down beginning of 1991 and is now ready to take data [4].

These two examples show that large liquid argon calorimeters can be built. With others, they constitute an invaluable base of experience upon which to design detectors for operation at LHC/SSC. However, the amount of work required to get these detectors off the ground should not be underestimated : more than 100 people are working in each of the two teams to build, measure and understand the calorimeter performances.

Warm liquid calorimetry does not have yet the same past large system experience.

3. Ionization Calorimeters and Radiation Damage

Calorimeters will have to survive a level of radiation and a neutron fluence which are orders of magnitudes above what is experienced by today's detectors (see report by A. Ferrari at this conference [5]). This is even more true at LHC with a design luminosity in excess of 10^{34} cm^2 s^{-1}.

Cells filled with TMP or TMS were irradiated by gammas from a Co^{60} radioactive source at two different dose rates (0.09 and 1.50 kGy/h) [6]. The electron lifetime (τ) was monitored as function of the irradiation level. As shown in figure 4 of reference 6, the lifetime decreases steadily as the integrated dose increases, irrespective of dose rate. Starting from a very pure liquid ($\tau \sim 1$ ms), lifetime values below 100 ns, in the case of TMS, were measured for a 10 kGy total dose.

This slow degradation of warm liquids due to radiation may not be too important for a swimming pool calorimeter design since the liquid can be recirculated and repurified. Nonetheless, radiation damage of the other materials in the calorimeter may release contaminants into the liquid which may perhaps not easily get flushed out.

The same comment apply to liquid argon calorimetry and there is a need in both cases for tests of radiation resistance of entire calorimeter prototypes.

4. Pileup, Detector Speed and Granularity.

At a luminosity of 2.10^{34} cm^2 s^{-1} there are ~ 20 inelastic pp interactions in each bunch crossing. Hence even an ideally fast calorimeter will suffer pileup. The vast majority of hadronic collisions which constitute the pileup produces particles with low transverse momenta (≈ 600 MeV) and with an average multiplicity of 10 charged and neutral particles per unit of rapidity.

To limit pileup effects, the effective detector sensitive time must be kept short not to integrate too many bunch crossings. Fine granularity in the EM calorimeter which decreases the cell occupancy also helps due to the small lateral size of EM showers [8]. Requirements are slightly less stringent for SSC with a design luminosity an order of magnitude smaller.

4.1 Detector speed and electronics noise

Although liquid ionization calorimeters have been traditionally used with slow readout, their signals have a fast risetime, which can be exploited to obtain fast pulses [9]. Two groups have recently obtained experimental results.

The SSC liquid argon collaboration has built a lead liquid argon calorimeter to measure its response to pions and electrons with a 50ns shaping time suited for experimentation at SSC [10]. Trying to design a detector for the higher luminosity expected at the LHC, the RD3 "Accordion" collaboration, pushed the shaping time down to 20ns [7,11]. Figure 1 shows a 60 GeV electron pulse from their prototype EM calorimeter. Electron energy is proportional to peak voltage. The 400 ns electron drift time in the 2mm liquid argon gap is clearly visible. This pulse corresponds to an effective integration gate of only 20ns and for a 15 ns bunch spacing, the effective pileup noise is only 1.6 times larger than the pileup from a single crossing.

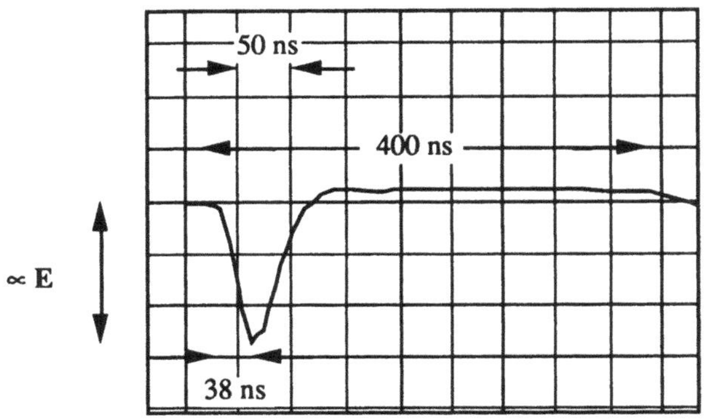

Figure 1 A 60 GeV electron pulse in the "accordion" EM calorimeter. The electron energy can be measured after 38ns. This corresponds to a 20 ns long effective integration gate (see reference 11).

The price to pay for such a fast shaping is an increase in electronics noise. However, as shown by G. Costa in reference 11, the signal from minimum ionizing particle is still visible above noise.

More work is still to be done to define the optimum signal processing as function of the machine luminosity but the above results already show that ionization calorimeters can meet the speed requirement for calorimetry at the LHC/SSC.

4.2 Detector granularity

Easy segmentation is one of the nice properties often quoted for ionization calorimetry. However, the ganging together of the signal in depth can introduce dead area or cracks. Two examples, the D0 end electromagnetic calorimeter module [12] and the "accordion" prototype [11], were presented at this conference and show that pad sizes down to ~ 2 x 2 cm^2 can be achieved without spoiling other calorimeter properties like space uniformity, hermeticity, signal speed. As discussed below, this fine segmentation results in a very good space resolution.

5. Performances for EM Showers

5.1 Space and angular resolution

There are several physics advantage to accurately measure EM showers position with calorimeters. This is the only way to get some information on photons and, for instance to select the vertex within a crossing with multiple interactions. If precise enough, angular measurement would allow to compute the $\gamma\gamma$ mass in a H --> $\gamma\gamma$ search. For electrons, a precise matching in space between a charged track measured before the calorimeter and the EM shower has proven to be very powerful to reject background coming from charged track, π^0 overlaps.

Results from the D0 electromagnetic end calorimeter module [12] and the "accordion" prototype [11] presented at this conference, show that space resolution below 1mm can be achieved for high energy showers (360 μm at 100 GeV in the accordion prototype). The "accordion" team has also measured an angular resolution using the segmentation in depth of their calorimeter which can roughly be parametrized as σ_θ = 80 mrad / $\sqrt{E}$ (GeV).

5.2 Space uniformity

All ionization chamber calorimeters can have in principle a uniform response when scanning the beam across the detector surface. However, to achieve the type of performance demanded for the discovery of a low mass H --> $\gamma\gamma$ (< 0.5%), all mechanical tolerances of the fabrication need to be controlled to a very high level. N.A. Roe presented measurements done on the D0 electromagnetic end calorimeter module [12] . The mean response of the calorimeter to 100 GeV electrons is constant over the area scanned to within 0.5% (rms)

5.3 Stability with time

Once a large liquid argon system is cold and operating, there are minimal changes in the response. This was already shown in the past for instance by the NA31 collaboration

(< 0.2% over months) [13]. The monitoring of the liquid is typically done with a few radioactive sources. D0 and H1 collaborations have presented evidence of signal stability better than 0.3%/year (Fig. 2) [2,4].

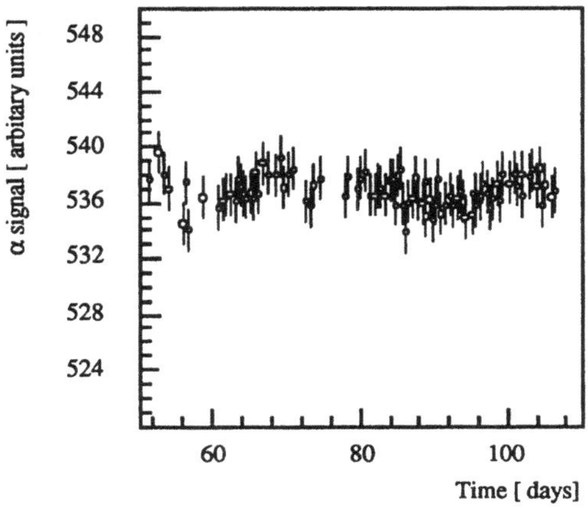

Figure 2 Liquid argon purity stability during three months of operation of the D0 central calorimeter.

Although not monitored with the same accuracy, the same stability of the response with time (< 2%) was observed by the WALIC collaboration on his TMP test calorimeter [14].

5.4 Linearity and Energy resolution

Test beam results reported at this workshop exhibit quite good linearity from a few GeV up to ~ 200 GeV [11, 12]. This confirms earlier measurements.

Energy resolution is usually parametrized as the quadratic sum of three terms :

$$\frac{\sigma}{E} = a \oplus \frac{b}{\sqrt{E}} \oplus \frac{c}{E} \quad (\text{E in GeV})$$

In ionization calorimeters, the constant term a is experimentally measured to be very small (< 0.5%)[11, 12]. However, with very fast shaping, a 1 ns timing jitter may very easily contributes 0.5% to the error on the energy measurement. Ways of controlling effects like this one should be incorporated in the design of the calorimeter readout.

The term b typical of sampling calorimetry depends on absorber material characteristics. Typical values are in the range 10-20%. As shown by the RD3 collaboration at this workshop, this value does not depend on the readout speed [11].

The last term describes various sources of noise : electronics noise, pileup noise. As discussed in 4.1, the electronics noise increases rapidly for fast readout. A value of 300

MeV for a nonet of cells large enough to contain an electromagnetic shower was reported at 20ns shaping time[11]. This contribution to the noise becomes negligible when the shower energy increases and, in the region of interest to LHC/SSC physics, is not too harmful.

6. Ionization Calorimeters and Hadronic Calorimetry

A lot of efforts has been devoted in the past few years to understand the difference in response between electrons and pions and to improve energy resolution for hadronic showers. The presence of hydrogen in the detecting medium was felt important to detect low energy neutrons present in showers, by elastic scattering on free protons[14].

The WALIC collaboration, using TMP as detecting medium, has launched a systematic study varying absorber types and thicknesses and hydrogen content of the detector. Results obtained with a lead/TMP calorimeter exposed to 10 to 150 GeV electron and pion beams, show that details of signal formation must be taken into account to understand the measurements [15]. Signal saturation for densely ionizing particles was measured by this collaboration a few years ago as function of the electric field intensity and angle with the ionizing track [16]. According to simulation, this effect decreases the response of 25 GeV hadronic showers by ~ 35% .

Results from a similar systematic study, using TMS ionization chambers, were also reported by a Karlsruhe, ITEP collaboration [17]. The beam energy was in that case much lower (2 to 6 GeV). Again, agreement is found with Monte Carlo simulations if signal saturation effects are taken into account.

Signal saturation for highly ionizing particles is due to recombination of electrons with positive ions. Using the Kramers Jaffe columnar recombination model included in the Geant simulation package, A. Givernaud showed in the case of TMP, that he can reproduce with a single parameter a wide variety of results : response to α particle as a function of electric field, saturation properties as a function of electric field intensity and angle with respect to the ionizing track, hadronic resolution of the very complicated UA1 U/TMP calorimeter [6]. The same model also reproduces the smaller saturation effects observed in liquid argon ionization chambers.

Ionization calorimeters using liquid argon as detecting medium do not contain hydrogen and compensation should be more difficult to achieve. Using uranium as absorber, the HELIOS collaboration has shown e/π to be in the neighborhood of 1.1 for 20 GeV incident particles [18]. More convenient, 12mm thick, lead absorber plates were used by the SSC liquid argon collaboration which measures e/π as $1.17 \pm .01$. Within errors, this ratio do not depend on the shaping time (50ns, 100ns, 200ns) [10].

The H1 calorimeter uses iron as absorber and is far from compensation ($e/\pi = 1.26$ at 20 GeV). A lot of beam test studies were performed paralleled by a large simulation effort and as shown at this workshop [4], a very good agreement between simulation and measurement is observed. This gave confidence in an innovative weighting method which takes advantage of the very fine granularity of the H1 calorimeter [19]. Hadronic resolution $\frac{\sigma}{E} = 1\% \oplus \frac{44\%}{\sqrt{E}}$ is achieved in test beam measurements. Very good linearity and energy resolution are also expected for jets (figure 3).

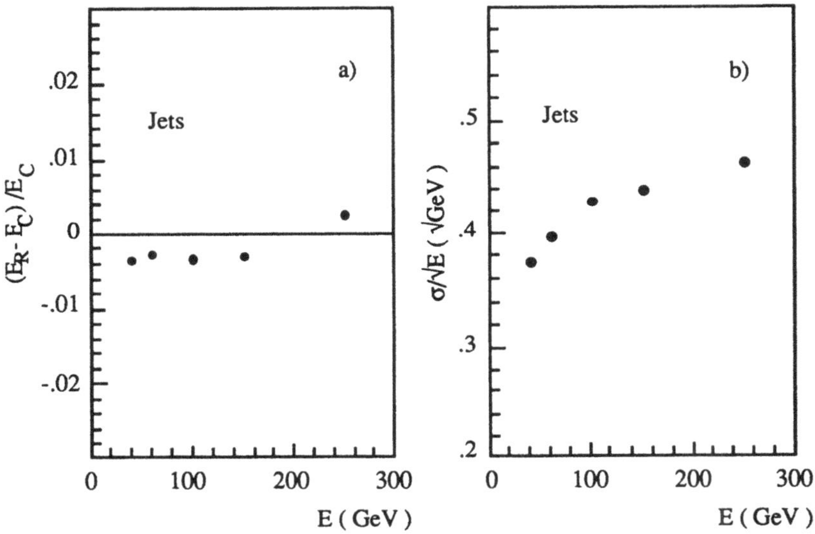

Figure 3 H1 test : Linearity of jet response and the energy resolution for jets using the H1 weighting algorithm.

A lot of progress has been made in understanding the complex phenomena which take place in hadronic calorimetry. Simulation tools, able to reproduce measurements to a few % when the details of signal formation are taken into account, seems to be now available. It is quite interesting to note that this deeper understanding has lead to two very different ways to improve hadronic calorimeter resolution : compensation using signals from the hadronic shower neutrons, software weighting using the high granularity achievable in ionization calorimeters.

8. Conclusions

Considerable progress has been made in the understanding of calorimetric measurements in the LHC/SSC environment. Performances measured on new, large liquid argon calorimeters and first results obtained by the R&D program indicate that ionization calorimeter detectors should be viable at LHC/SSC. Nonetheless a lot of work has still to be done. Just to list a few lines of required action :

- radiation damage studies of all detector components and of a complete prototype,

- engineering studies of a full size 4π detector. Can cracks and dead zones due to cryogenics be kept at a reasonable level in a large liquid argon detector ?
- fast and low noise electronics for a finely segmented calorimeter with a large number of channels (~ 200K channels ?)
- ...

There is clearly matter for discussions at an other workshop in a year time !

References

1. Design Report,
   The D0 Experiment at the Fermilab Antiproton-Proton Collider, November 1984.

2. S. Feher, *These proceedings*,
   Cosmic Ray Performance of the D0 Central Calorimeter

3. V. Korbel, *Nucl. Inst. and Meth.* **A263** (1988), 70

   H1 Collaboration, Technical Progress Reports, DESY 1987, 1988, 1989, 1990

4. H. Oberlack, *These proceedings*,

5. A. Ferrari, *These proceedings*,
   Can We Predict Radiation Levels In Calorimeters ?

6. A. Givernaud, *These proceedings*,
   Monte Carlo Simulation Of Calorimeters Based On Liquid Ionization Chambers

7. B. Aubert et al., *CERN/DRDC/90-31*
   Liquid Argon Calorimetry with LHC Performance Specifications

   B. Aubert et al, *CERN/DRDC/91-21*
   Hadronic and Electromanetic Liquid Argon LHC Prototype Calorimeter

8. J. Colas, R. Klanner, T. S. Virdee, *Large Hadron Collider Workshop*,
   Aachen 4-9 October 1990, Calorimetry At The LHC

9. V. Radeka, S. Rescia, *Nucl. Inst. and Meth.* **A265** (1988), 228
   Speed and Noise Limits in Ionization Chamber Calorimeters

10. E. Kistenev et al., BNL AGS users Newsletter Oct-Dec 91,
    AGS Test Beam Results for Liquid Argon Calorimetry with Fast Shaping

11. G. Costa, *These proceedings*,
Liquid Argon Calorimetry with an Accordion Geometry for the LHC

12. N. A. Roe, *These proceedings*,
Test Beam Results from the D0 End Electromagnetic Calorimeter

13. H. Burkhardt et al., *Nucl. Inst. and Meth.* **A268** (1988), 116

14. R. Wigmans, *Nucl. Inst. and Meth.* **A259** (1987), 389

15. B. Mansoulie, *These proceedings*,
Compensation Studies of a Lead-TMP Calorimeter

16. B. Aubert et al., *Nucl. Inst. and Meth.* **A286** (1990), 583
Saturation of Ionization Signals in TMP and TMS

17. J. Engler et al., *These proceedings*,
Calorimeter Tests with Liquid Ionization Chambers

18. HELIOS collaboration

19. W. Braunschweig et al., *Nucl. Inst. and Meth.* **A275** (1989), 246

20. H1 Calorimeter group, XXV International Conference on High Energy Physics Singapore,1990
Study of Software Compensation for Single Particles and Jets in the H1 Calorimeter

TEST BEAM RESULTS FROM THE D0 END ELECTROMAGNETIC CALORIMETER

NATALIE A. ROE†
Lawrence Berkeley Laboratory
1 Cyclotron Road, Berkeley, California 94720
Representing the D0 Collaboration

ABSTRACT

Test beam results are presented for the D0 end electromagnetic calorimeter. Data were taken with electrons and pions ranging in energy from 5 GeV to 150 GeV. Results from the analysis of the test beam data are presented on energy resolution and linearity, stability and uniformity of response, position resolution and electron-pion separation.

1. Introduction

D0 is a large multi-purpose detector located at Fermi National Accelerator Laboratory, on the Tevatron $p\bar{p}$ Collider. The uranium–liquid argon calorimeter system of the D0 detector is comprised of two end calorimeters and a central calorimeter. Each of these three calorimeters contains an electromagnetic section with thin (3 or 4 mm) uranium plates, a fine hadronic section with thicker (6 mm) uranium plates, and a coarse hadronic section with thick (46 mm) copper or stainless steel plates. This contribution describes test beam measurements of an electromagnetic end calorimeter module (ECEM)[1] from the D0 detector.

1.1 Calorimeter Construction

In order to minimize energy losses and non-uniformities which arise in inter-modular cracks, the end electromagnetic calorimeters in the D0 detector feature a monolithic design in which the signal boards and uranium absorber plates are large disks of approximately 1 m radius. The basic sampling cell in the ECEM module consists of a 4 mm depleted uranium absorber plate, a 2.3 mm liquid argon gap, a 4.3 mm NEMA G-10 multilayer printed circuit signal board, and a second 2.3 mm liquid argon gap. The signal disks are constructed from five-layer printed circuit boards laminated with facesheets coated with a high resistivity epoxy. The resistive coat is maintained at positive high voltage to provide the drift field and the ionization charge drifting in the gap induces a signal on the pads.

† This work was supported by the Director, Office of Energy Research of the U.S. Department of Energy under Contract No. DE-AC03-76SF000098

The ECEM contains 18 sampling cells in depth, for a total of 20.1 radiation lengths (X_0) of material at normal incidence. The signals from the 18 cells are ganged into four separate longitudinal depth segments consisting of 2, 2, 6 and 8 cells each. The longitudinal segmentation of the ECEM provides information about the longitudinal development of the electromagnetic shower and is valuable in distinguishing electrons from charged pions. The ECEM module is followed by a fine end calorimeter hadronic module[2] (ECIH) which contains a total of 7.3 interaction lengths, divided into five longitudinal sections.

Transverse segmentation in both modules is provided by the segmented copper readout pads on the signal boards, with each tower corresponding to an $\eta \times \phi$ interval of 0.1×0.1.† In the third longitudinal section of the ECEM, which contains typically 65% of the electromagnetic shower energy, the transverse segmentation is doubled in both directions to 0.05×0.05 in order to provide better shower position resolution for electrons. The segmented pads in both the ECEM and the ECIH are arranged in a semi-projective tower geometry.

1.2 Test Beam Setup

The D0 end calorimeter configuration was tested in the Neutrino-West beamline at Fermilab. The ECEM module was located in front of the ECIH module, and both modules were immersed in liquid argon inside a testbeam cryostat which had a thin window consisting of two 1.6 mm steel plates. A liquid argon excluder located between the cryostat window and the ECEM module was constructed of Rohacell slabs; it contained a 2.5 cm thick stainless steel plate, to simulate the D0 endcap cryostat wall, and a 4.4 cm thick aluminum plate at low angles, to simulate the material in the D0 vertex detector. The entire cryostat was located on a transporter under computer control, allowing the calorimeter response to be studied over the full range in η and over $\pm 15°$ in azimuthal angle.

The calorimeters were instrumented with the D0 electronic readout system[3] over a region of $\approx 35°$ in azimuth and over the full region in η; in addition, a border region was instrumented in which groups of channels were ganged together. This allowed the energy flow out of the fully instrumented region to be monitored, while keeping the total channel count low.

Scintillators in the beam line provided the trigger, and PWC's located before and after the final bend magnet measured particle momenta (to 0.2%) and trajectories. Both pions and electrons were used, with momenta ranging from 5 to 150 GeV/c and a typical momentum spread of 1.5%.

2. End Electromagnetic Calorimeter Results

The analysis of the ECEM data proceeded by first correcting for electronic pedestals and gains, which were measured once every eight hours and once per day, respectively. Next, the response was corrected for the temperature dependence of

† The pseudo-rapidity variable, η, is defined by $\eta = -\ln \tan \theta/2$, where θ is the polar angle measured from the beamline, and ϕ is the azimuthal angle.

the electronics (0.08%/°C), for runs where this information was available.† The spread in the beam momentum was then corrected for event-by-event using the momentum measured by the PWC system. The energy of electromagnetic showers was reconstructed by summing all four longitudinal sections in the ECEM and the first section of the ECIH. The number of towers summed to contain an electromagnetic shower varied with η, corresponding to the variation of tower size with η. For example, for η=1.95 with towers of 5.2 cm square, 99.5% containment was achieved by summing an array of 5 by 5 towers, while at $\eta = 2.5$ with towers of 2.8 cm square, an array of 9 by 9 towers was needed.

2.1 Energy Resolution and Linearity

To measure the total energy of an electromagnetic shower in the ECEM, relative weights for the signals from each longitudinal section must be determined. These weights correspond to the inverse of the relative sampling fractions (SF's), and were obtained by simultaneously minimizing the energy resolution and the deviation from linearity for electrons ranging from 10 to 150 GeV/c. The data for this analysis were all taken at $\eta = 1.95$, at the so-called benchmark spot. A single set of optimized SF's was found for all electron energies; these values are very similar to the relative SF's calculated from minimum ionizing dE/dX energy deposition. The energy resolution obtained with these weights is given by

$$\frac{\sigma_E}{E} = (0.0034 \pm .0019) \oplus \frac{(0.157 \pm 0.005)}{\sqrt{E}} \oplus \frac{(0.330 \pm 0.030)}{E}, \quad (1)$$

where the three contributions are, respectively, the constant (or calibration) term, the sampling term, and the noise term (which is consistent with the measured noise), and all three terms are to be added in quadrature. The ECEM response was linear to within 0.2% over the energy range of 10 to 150 GeV.

2.2 Stability and Uniformity of Response

The temporal stability of the ECEM response was monitored by taking data with 100 GeV electrons at the benchmark spot (η=1.95) at approximately daily intervals during the test beam run. Only data for which the temperature of the readout electronics was monitored were included in the analysis. A fit to the data showed a change in response of (- 0.12± 0.04)% /40 days. If we assume this decrease is due to increasing oxygen contamination in the liquid argon, this would correspond to an increase of 0.013 ppm / 40 days, [4] to be compared to the measured effective oxygen content of 0.45 ppm.[5] The oxygen contamination of the liquid argon was independently monitored at the test beam with α and β sources;[6] the results of those measurements are also consistent with a small increase in oxygen contamination during the test beam run.

To study the spatial uniformity of the ECEM response, we performed azimuthal scans of the module at $\eta = 1.85$ and 1.95 with 100 GeV electrons. The mean pulse height as a function of azimuth is shown in Fig. 1 for η=1.85, where the azimuth

† The electronics temperature was monitored only during the last 40 days of the run.

is given as the arc length, s = radius×ϕ. The mean response is constant to within 0.5% (rms). Similar results were obtained at $\eta = 1.95$ (0.5% rms). Small regions of non-uniform response near structural elements were also studied; these amount to less than 3.4% of the total area, and the maximum correction required is 15%.

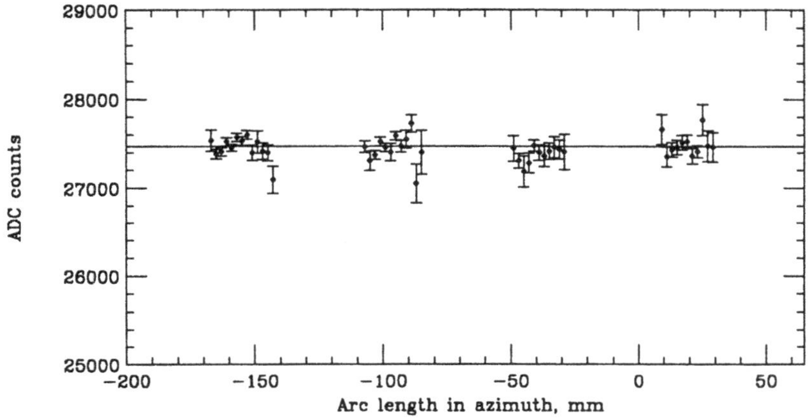

Fig. 1. ECEM response in ADC counts for azimuthal scan at η=1.85; the rms is 0.5%.

2.3 Position Resolution

The centroid of an electromagnetic shower can be determined to much better accuracy than one would naively calculate based on the calorimeter segmentation, due to the transverse spread of the shower which results in sharing of the energy among several towers. In the ECEM, the shower position can be most accurately determined in the third longitudinal section (EM3), which has a transverse segmentation of .05×.05 in $\eta \times \phi$. We have studied the position resolution in EM3 using the shower impact position extrapolated from the PWC's in the beamline. Several techniques are available to extract the impact point of an electromagnetic shower; we have studied two algorithms, both of which give very satisfactory results.

In the first method, the transverse shower shape is fit to the sum of two exponentials and a ratio of observed energies is constructed, from which the shower impact position can be iteratively determined.[1,7] In the second method,[8] a first estimate of the position is made by calculating the energy weighted center-of-gravity, $X_{cog} = \sum X_i E_i / \sum E_i$, and then correcting it with the following algorithm:

$$X_{ccog} = X_m + B \cdot \sinh^{-1}\{\frac{(X_{cog} - X_m)}{\Delta} \cdot \sinh(\frac{\Delta}{B})\}. \quad (2)$$

In this expression, X_m is the position of the EM3 tower with the maximum signal, Δ is the pad half-width and B characterizes the transverse shower shape assuming a single exponential transverse shape.

The position resolution as determined by these two techniques was very similar. In both cases, the best resolution is obtained when the shower impacts near a tower edge, and gets worse as the square of the distance from the tower edge. In addition, due to the inherently statistical nature of shower development we expect the position resoluton to vary approximately as $1/\sqrt{E}$. In Fig. 2 the position resolution obtained using the corrected center-of-gravity technique is shown as a function of energy. Both the resolution at the tower edge, and the resolution integrated over the tower width at η=1.95 (where the pad size in EM3 is 2.6 cm) are shown. The former does vary approximately as $1/\sqrt{E}$ while the average resolution shows a slightly steeper dependence. The average resolution is approximately 2 mm at 25 GeV and 1 mm at 75 GeV.

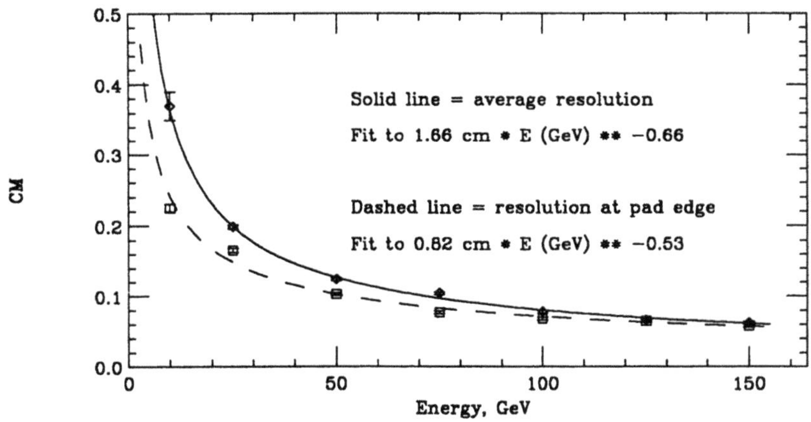

Fig. 2. Position resolution vs. energy for electrons incident at η=1.95.

2.4 Electron-Pion Separation

The development of electron and pion showers is quite different, and the resulting difference in shower shape can be exploited to identify electrons and reject the larger background from pions. Both longitudinal and transverse shower shape variables can be used; in addition there are correlations which should be taken into account to maximize the pion rejection while retaining good electron efficiency.

We have used an H-matrix technique[9] to perform e/π separation. In this technique, a training sample of electrons is used to compute the mean energy deposited in each element along with the correlations to every other element. The inverse of this matrix is then used to compute an effective χ^2, which we call ζ; a cut is then placed on the value of ζ to separate electrons and pions. Two different H-matrices have been used: in the first, all signals in each longitudinal section of the ECEM and the first ECIH layer are individually summed to create a 5×5 matrix; in the second, all channels in each layer within a 0.3× 0.3 tower are used to create a 72× 72 matrix. The 5× 5 H-matrix only takes into account the difference in the longitudinal shower development while the 72× 72 H-matrix includes both longitudinal and transverse shower characteristics.

To obtain the best e/π separation, both H-matrices were applied sequentially. On an independent sample of electrons the ζ cuts were chosen for an overall electron efficiency of 95%. The pion rejection was then determined by applying the same cuts to pion test beam data; the measurement is statistically limited so we quote a 95% C.L. lower limit on the pion rejection factor, of 700 at 50 GeV and 2×10^3 at 100 GeV.

4. Conclusions

Analysis of testbeam data from the D0 ECEM calorimeter module has been presented. The module has good energy resolution with a small constant term; is very linear, stable and uniform in response; has position resolution better than 1 mm for high energy electrons; and has an e/π rejection factor greater than 700 at 95% C.L., for energies greater than 50 GeV and an electron efficiency of 95%.

5. Acknowledgements

We would like to thank the Fermilab Accelerator Division and the staffs of the collaborating institutions on D0 for their contributions to the success of this project. This work was supported by the Director, Office of Energy Research of the U.S. Department of Energy under Contract No. DE-AC03-76SF00098.

6. References

1. See also H. Aihara, *et.al.*, *The Design, Construction and Performance of the D0 End Electromagnetic Calorimeter Module*, LBL-31378, to be submitted to NIM; H. Aihara, IEEE Trans. on Nuc. Sci **38** (1991) 398; and A.L. Spadafora, in *Proc. of the 1st Int'l Conf. on Calorimetry in High Energy Physics*, ed. Anderson *et.al.* (World Scientific, Singapore, 1991) p. 77.
2. N. A. Amos, in *Proc of the 1st Int'l Conf. on Calorimetry in High Energy Physics*, ed. Anderson *et.al.* (World Scientific, Singapore, 1991) p. 84.
3. M. Demarteau, in *Proc of the 1st Int'l Conf. on Calorimetry in High Energy Physics*, ed. Anderson *et.al.* (World Scientific, Singapore, 1991) p. 91.
4. H. Aihara and A. Allen, D0 Note # 1262, Oct. 1991 (unpublished).
5. A. Allen and A.L. Spadafora, D0 Note # 1212, Sept. 1991 (unpublished).
6. G. Blazey, in *Proc of the 1st Int'l Conf. on Calorimetry in High Energy Physics*, ed. Anderson *et.al.* (World Scientific, Singapore, 1991) p. 101.
7. M. Demarteau, D0 Note #842, May 1989 (unpublished).
8. G.A. Akopdjanov, *et.al.*, NIM **140** (1977) 441, and A. De Angelis and F. Mazzone, NIM **A287** (1990) 397.
9. R. Englemann *et.al.*, NIM **216** (1983) 45.

COSMIC RAY PERFORMANCE OF THE DØ CENTRAL CALORIMETER

SANDOR FEHER
State University of New York at Stony Brook
Stony Brook, New York 11794-3800, USA

THE DØ COLLABORATION†

ABSTRACT

Cosmic ray muon signals recorded with the DØ central calorimeter are discussed. We present studies of the electronic noise levels (both random and coherent), studies of gain and pedestal stabilities, and studies of the stability of the liquid argon purity.

1. Introduction

During the Spring of 1991 DØ had a cosmic ray run to test its major subsystems. The systems included were: central tracking, central calorimeter, central muon chambers, data acquisition and controls. The normal configuration[1,2,3] of the DØ detector was very slightly modified to take cosmic ray data by adding three scintillators to serve as trigger counters for cosmic muons. The locations of the trigger counters were: (1) on top of the muon chambers, (2) outboard of the forward drift Chambers and (3) in the center of the central tracking chambers in lieu of the beam pipe. The major goal of the run was to integrate the individual detector elements into a single system.

The 1991 run was the first test of the fully assembled DØ central calorimeter (CC) in a LAr environment. About 400,000 events were collected during a two month running period.

This paper will concentrate on the performance of the DØ central calorimeter which is the largest uranium LAr calorimeter in the world with 5,000 gallons of LAr, 200 tons of uranium, 150 tons of copper and 17,000 readout channels.

2. Noise studies

2.1 Random Noise

The performance of a system with a very large number of electronic channels can be severely affected by sources of noise. We determined the noise by studying pedestals of the DØ CC electronics[4,5]. Table 1 summarizes the results of the random noise study.

The values shown in Table 1 are well below the level generated by a minimum ionizing particle (MIP) which passes through a calorimeter cell as is shown later

† The DØ collaboration includes members from University of Arizona, Brookhaven National Laboratory, Brown University, University of California - Riverside, Columbia University, Fermi National Accelerator Laboratory, Florida State University, University of Florida, University of Hawaii, Indiana University, Lawrence Berkeley Laboratory, University of Maryland, University of Michigan, Michigan State University, New York University, Northern Illinois University, Northwestern University, University of Rochester, CEN Saclay France, Institute for High Energy Physics Protvino USSR, State University of New York - Stony Brook, Tata Institute of Fundamental Research Bombay India, Texas A&M University and Yale University.

Table 1.
Random noise level in the average electromagnetic (EM), fine hadronic (FH) and coarse hadronic (CH) channels in terms of ADC counts, energy, and equivalent electrons at the input of the preamplifier.

	σ					
	ADC Counts		MeV		Electrons	
	Electronic	Uranium	Electronic	Uranium	Electronic	Uranium
<EM Layer 1 and 2>	1.9	2.8	6.6	9.8	19.8 K	29.4 K
<EM Layer 3>	1.1	1.7	7.6	12.0	22.8 K	36.0 K
<EM Layer 4>	1.8	4.0	12.6	28.0	37.8 K	84.0 K
<FH>	4.2	11.0	14.7	38.5	44.1 K	115.5 K
<CH>	3.7	3.7	13.0	13.0	39.0 K	39.0 K

(Fig 4). Figure 1 shows the distribution of the sigmas of the pedestals versus the detector capacitance with the HV off. The linear dependence of the capacitance, as expected[6], can be clearly seen. The larger sigma with HV on, which is apparent in Fig 2, is due to uranium noise. The pedestal widths of channels that are the same with HV on or off comes from the CH modules which have copper absorber plates (i.e. no uranium noise).

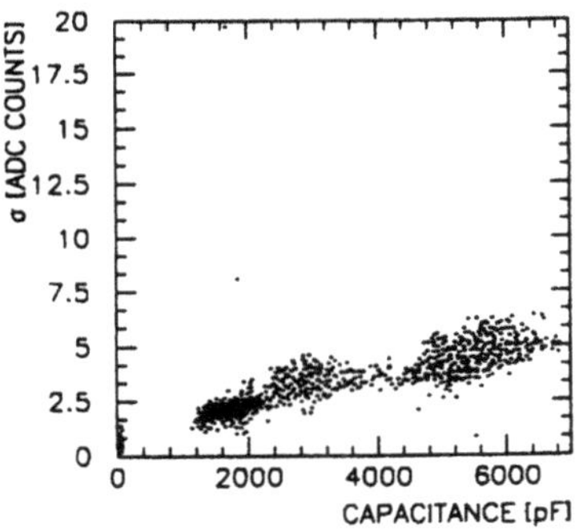

Fig 1
Widths of the pedestals versus capacitance when the high voltage is off. No contribution from uranium noise.

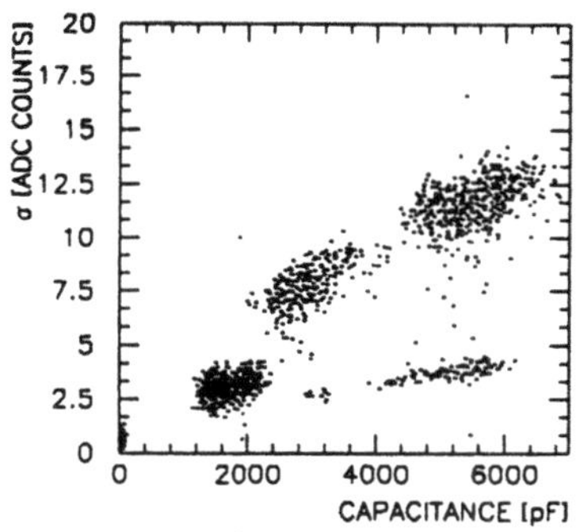

Fig 2
Widths of the pedestals versus capacitance when the high voltage is on. With contribution from uranium noise.

2.2 Coherent Noise

To measure the energy of particles or jets one must sum the signals from many channels. If the system has coherent noise, the energy resolution can degrade significantly. We determined the effect of coherent noise by measuring the variance of the calorimeter signal sum of $N = 4608$ channels (one quadrant of the CC). This variance is composed of two terms. The first is a sum over the independent random contributions of the N channels and the second is a double sum that measures the correlations of the ith and j^{th} channels. Since the first term increases like N and the second like N^2, there will be some number of channels for which random and coherent contributions are equal. For the CC the number is more than 5000 channels. This demonstrates the coherent electronic noise is indeed very low.

3. Gain and pedestal stability

Stability of electronic gains depends mainly on the temperature of the preamplifiers and the sample and hold circuits that in DØ are called Base Line Subtractors (BLS). The temperature dependence of the preamplifiers and BLSs was measured by varying the temperatures in the preamp-boxes and the BLS crates. The relative gain dependence was 0.05 %/°C for the preamplifiers and 0.03 %/°C for the BLS. The relative pedestal dependence was 0.02%/°C. The temperature was measured to ± 0.2°C. During the cosmic ray run the temperatures of the preamplifier-boxes and the BLS crates varied less than 2°C. This resulted in relative gain changes of less than 0.3 % and relative pedestal changes of less than 0.1%.

4. HV status

Since resistive coating which is the HV plane, covers the signal boards that are placed between the U or Cu calorimeter plates, a short between the grounded metal plates and the resistive coating on the signal board will locally degrade the voltage and therefore the signal over a fraction of a board. Each board covers 0.2 (0.4) rad in ϕ and several units of η for the electromagnetic (hadronic) modules. Faced with this potential problem we decided to extensively test the CC modules at 3,000 V in dry air to find and eliminate shorts ($I > 250$nA). Two shifts per day for almost 13 months were required to complete the testing of the 64 modules in the CC. This scrutiny paid off as there was only one noisy readout channel out of 17,000 that was attributed to a HV problem.

5. LAr purity

To provide information about variation in the response due to contamination of the LAr, each calorimeter was instrumented with a LAr purity monitoring system[7]. These monitors directly measured the mobility of electrons in the LAr. Three LAr monitors were placed in three different locations within the CC. Each monitor contained two sensing channels with the same HV gaps as exist in the CC modules, i.e. 2.3 mm. Two different calibrations were done with (1) Am^{241} as an α (heavily ionizing) source and (2)Ru^{106} as a β (minimum ionizing) source. Both types of cells showed that the LAr purity was very stable during the cosmic ray run (Fig 3). The relative change in response was less than 0.1 %.

Although the purity was stable it was not ideal because of nitrogen contamination which, as shown later, did not significantly affect the muon signals (see Fig 4). The nitrogen contamination resulted from a leaky Ar relief valve that communicated with the nitrogen vent line. The valve was repaired at the beginning of the cosmic ray run which accounts for the stable signals shown in Figure 3.

From β cell measurements the oxygen-equivalent contamination level of the LAr was 0.6 ppm. The nitrogen contamination, which was of order 250 ppm, only affected the α source signal. This was verified later in independent tests[8].

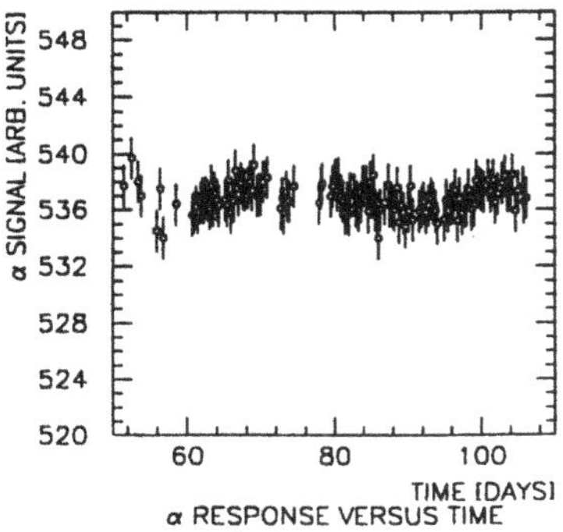

Fig 3
Liquid argon purity stability during three month of operation of the central calorimeter.

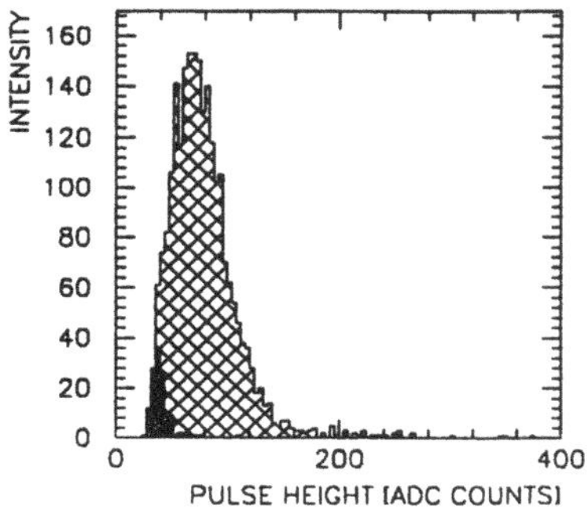

Fig 4
Distribution of the energy deposited by cosmic ray muons in layer 1 of fine hadronic CC modules.

6. Cosmic ray muons

The MIP signal from muons in the calorimeter is much smaller than signals generated by showers during a collider run. Consequently their signal to noise ratio is minimal and provides a stringent test for the DØ calorimeter electronics. With the small amount of random and coherent noise quoted earlier, e.g. 11 ADC counts (equivalent to 2.85 MeV deposited in LAr) for the first section of the CC fine hadronic module (FH1) and the expected energy deposited in the 21 double gaps of the FH1 module of about 20 MeV it should be possible to see a MIP peak for muons. Fig 4 shows a histogram of ADC counts for those calorimeter channels in the FH1 that are predicted to be hit by tracks recorded in the central drift chambers. It should be noted the level of discrimination is such that one can identify isolated MIPs in the calorimeter without sophisticated analysis.

For noise and background comparison purposes one can make a separate histogram of those channels that would be hit by a track that is rotated 90 degrees about the beam pipe axis of the calorimeter. This histogram is shown superimposed on the signal plot in Fig 4. Both plots are derived from on-line zero suppressed data. The zero suppressed pulse height is reported, after subtracting a predetermined and down-loaded digital pedestal from the digitized pulse, provided the digitized number, after pedestal subtraction is more than twice the pedestal half width. The zero suppression accounts for the small number of entries in the background/noise plot since, in most cases, the pulse height is less than twice the pedestal half width. The center of the peak at about 75 ADC counts corresponds to 21 MeV deposited in the LAr which is in good agreement with the expected number. The width of the minimum peak has two contributions in addition to electronics noise which is minimal. On the low side of the peak there are tracks that pass only through a fraction of FH1. On the high side one observes the Landau tail which is typical of energy loss processes. Work on these data is still in progress.

7. Conclusion

The DØ central calorimeter has had a successful cosmic ray run. The random noise in the CC is measured to be very low with sigmas between 2 and 11 ADC counts. Coherent noise is not significant even when adding up 5000 channels. Electronic gain changes were less than 0.3 %, and the pedestal changes were less than 0.1 % during the two month cosmic ray run. The LAr purity was very stable over three months and even with nitrogen contamination there was no significant effect on muon signals. The LAr purity monitoring system was sensitive enough to measure nitrogen contamination. The oxygen-equivalent contamination was less than 0.6 ppm. Muon signals are clearly seen.

8. Acknowledgements

I would like to thank my DØ colleagues for their help in preparing this report.

9. References

1. Design report, The DØ Experiment at the Fermilab Antiproton–Proton Collider, November 1984.
2. N.A.Amos,*Proceedings of the First International Conference on High Energy Physics Calorimetry,FNAL, USA*,eds. D. F. Anderson et al.,(World Scientific, Singapore, 1991),p.84.
3. A.L.Spadafora, *Proceedings of the First International Conference on High Energy Physics Calorimetry,FNAL, USA*,eds. D. F. Anderson et al.,(World Scientific, Singapore, 1991),p.77.
4. M.Demarteau,*Proceedings of the First International Conference on High Energy Physics Calorimetry,FNAL, USA*,eds. D. F. Anderson et al.,(World Scientific, Singapore, 1991),p.91.
5. R.D.Schamberger, *Proceedings of the International Conference on Advanced Technology and Particle Physics,* Como, Italy, 1990.
6. P. Franzini,*Proceedings of the First International Conference on High Energy Physics Calorimetry,FNAL, USA*,eds. D. F. Anderson et al.,(World Scientific, Singapore, 1991),p.359.
7. G.C.Blazey, *Proceedings of the First International Conference on High Energy Physics Calorimetry,FNAL, USA*,eds. D. F. Anderson et al.,(World Scientific, Singapore, 1991),p.101.
8. G.C.Blazey and P.Mooney, *Liquid Argon Purity Monitoring in DØ* (to be submitted to NIM).

LIQUID ARGON CALORIMETRY WITH AN ACCORDION GEOMETRY FOR THE LHC

RD3 collaboration
Annecy-BNL-CERN-Milano-Orsay-Saclay-Stockholm

presented by
Giuseppe Costa

Istituto Nazionale di Fisica Nucleare, Sezione di Milano
Milano, Italy

ABSTRACT

Preliminary results on a liquid Argon calorimeter prototype with a novel geometry design, the "accordion", are presented. An energy resolution of $\sim 10\%/\sqrt{E(GeV)}$ and a space resolution of $3.6mm/\sqrt{E(GeV)}$ have been achieved for electrons, using a fast shaping time $t_p(\delta) \simeq 20ns$. The ratio of the muon peak to the total noise is ~ 3.2, this value is in a substancial agreement with the predictions of the data taken with slow shaping time $t_p(\delta) \simeq 100ns$. A large size electromagnetic and hadronic calorimeter prototype with accordion design and projective geometry is forseen to be built and tested in 1992.

1. Introduction

The experiments at the future high luminosity hadron colliders (LHC/SSC) will set strong requirements on electromagnetic and hadronic calorimeters. Speed, high granularity, hermeticity, good space and energy resolutions, radiation hardness should be essential caractheristics of these detectors [1].

Liquid Argon (LAr) calorimetry is, in this context, an interesting technique, that can satisfy most of these requests, moreover is reliable and a number of large LAr devices have been built and used in experiments. Nevertheless the major limitation of a LAr calorimeter is the slow time response; when this detector is operated as ionisation chamber, the signal originated by all the negative ionisation charges is collected in $\sim 400ns$ for a $2mm$ gap.

One way to get a fast response from a LAr calorimeter is to exploit the very fast rise time of the ionisation current, by integrating the current in a time interval $\tau \ll t_d$, using in the electronic chain a charge sensitive preamplifier followed by a bipolar shaper. For high rate operation, the "equal-area" bipolar shaping ensures that the pile-up of several events within the sensitive time of the detector does not,on average, produce a net shift in the measurement of a large signal. The pile-up will,however, contribute as noise to the overall energy resolution of the detector.

It is possible to build a fast shaper with peaking time $t_p(\delta) \sim 20ns$ necessary for the LHC/SSC, but on the same time it needs an intrisecally fast calorimeter cell, in order to have a fast charge tranfer time from the cell to the detector. The charge transfer time is defined from the electrode capacitance C and from the inductance L due to the leads necessary to link together the pads of the tower and to the connections to the preamplifier (PA). The optimum condition for the charge transfer time is achieved in the "critical damp condition" $\tau = 4\sqrt{LC} = 2CR_i$ where R_i is the input resistence of the PA [2].

The novel concept proposed by the RD3 collaboration is the "Accordion" LAr calorimeter [3]. In this new calorimeter design, the converter plates and the readout electrodes are no longer planar but have a longitudinal accordion shape, see fig. (1). In this way we have automatically a connection between successive pads of the same tower. As a consequence, it is easy to connect the preamplifiers directly to the front and the back of each calorimeter tower. With this arrangement the speed is increased, moreover noise, cross talk and dead regions are minimized.

A first prototype of accordion electromagnetic (EM) calorimeter has been built and exposed to electron and muon beams at the CERN SPS in the 1990 from the RD3 collaboration. Very promising results on the analysis of collected data have been already published [4]. The first prototype was equipped with existing shapers with $t_p(\delta) \simeq 100ns$, $t_p \simeq 140ns$ for a triangular signal. In 1991 new prototypes with fast shapers ($t_p(\delta) \simeq 20ns$) have been built and tested at SPS energies.

In this report very preliminary results on the data collected in 1991 are presented. The forthcoming programme of this *R&D* activity concerning the construction and test of an electromagnetic and an hadronic prototype of large size with full projective geometry is also described.

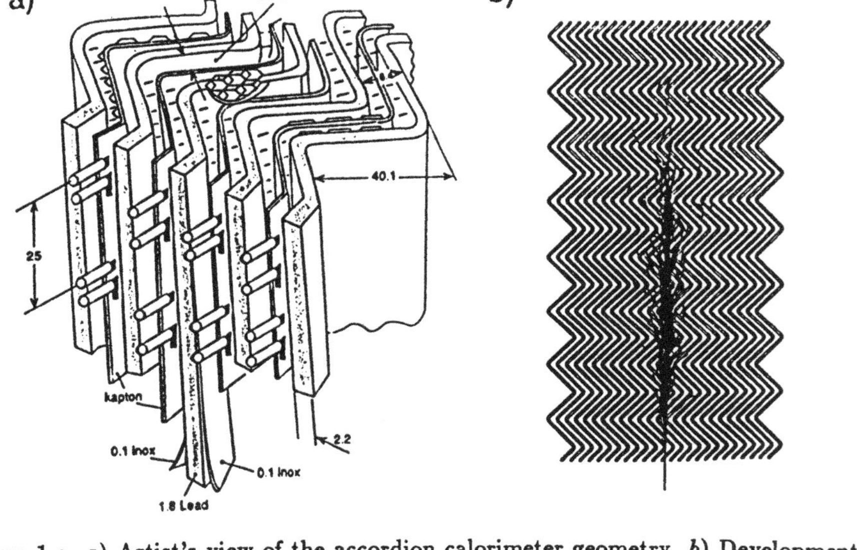

Figure 1 : *a*) Artist's view of the accordion calorimeter geometry. *b*) Development of an electron shower.

2. Calorimeter prototype

An EM calorimeter prototype of $40cm \times 40cm$ transverse section and longitudinal depth $\sim$ 25 radiation lengths (X_0) has been built. The converter plates are $1.8mm$ lead sheets clad in $0.1mm$ of stainless steel. The readout electrodes are multilayer sheets of Kapton-Cu with a total thickness of $\sim 400\mu m$. The two inner conductive layers, connected together, are hetched chemically in longitudinal trips $24mm$ wide and separated by $1mm$ gaps and collect capacitively the fast charge of electrons drifting in the LAr gap. The two external layers have a pad structure of copper and of a resistive coating and are connected to the high voltage to create the electric field ($\sim 10kV/cm$) on the gap. Both converter plates and readout electrodes are bent in an accordeon shape with a pitch of $40.1mm$ and fold angle of $\pm 45°$, see fig. (1). The LAr gap, between converter and electrodes is $1.9mm$; a light honeycomb structure maintains the distance between the electrode and the converter plates .

A calorimeter tower of size ($x \times y$) $2.7cm \times 2.5cm$ is obtained connecting three strips adjacent in x to the same PA. This prototype with "constant angle" fold and "cartesian geometry" consists of two parts: in one half of the calorimeter electrodes with two longitudinal samplings are mounted, the other half uses electrodes with three samplings. In the case of tower with two longitudinal samplings, each cell has $L = 10nH$ and $C = 400pF$ allowing charge transfer time as low as $\tau \simeq 8ns$. On the detector two kinds of cold charge sensitive preamplifiers have been implemented : a PA based on hybrid circuit with input Si-JFET, a PA based on GaAs MESFET. The Equivalent Noise Charge of these PA's, is $\sim 2 \cdot 10^4$ for a $\sim 20ns$ shaping time and power dissipatet is $\sim 70mW$. The electronic chain, out of the cryostat, is completed by a bipolar shaper $t_p(\delta) \simeq 20ns$ ($t_p \simeq 38ns$ for a triangular signal),a sample and hold and a 12 bit ADC read out.

A preliminary "pointing geometry" module similar to the prototype already described, has been built in the 1991. This module uses the same electrode geometry than the "cartesian geometry". Converter and electrodes plates are disposed radially around the axis of the calorimeter. The sampling fraction increases with the depth as the LAr gap becomes bigger and bigger. It is possible to compensate the gap variation by adding low Z dead material whose thickness increases with the depth. This solution gives the advantage to keep constant the electric field along the depth of the calorimeter.

The Monte Carlo simulation shows that the response uniformity is inside $\sim \pm 1.5\%$ wich is slightly worse than the response of the "cartesian geometry" prototype ($\leq \pm 1\%$). The "ponting geometry" prototype consists of two parts; one half of the detector is realized with " compesated gap" and 2 longitudinal sampling electrodes, the other half is built with "opening gap" and 3 longitudinal sampling electrodes.

The contruction of a large size "full projective" EM prototype with "constant gap" geometry is a part of the forthcoming *R&D* programme of the RD3 collaboration. Monte Carlo simulations show that about the same performances are expected from a "constant gap" detector compared to the "cartesian geometry" prototype.

3. Test beam results

3.1. Response to electrons

We present very preliminary results relative to the analysis of data collected in the summer 1991 with the two prototypes described above exposed to electron and muon beams at the CERN SPS. A system of scintillation counters has been used in the trigger to define a beam spot of $\sim 2cm \times 2cm$, three proportional wire chambers (PWC) were used to define the direction of the incident particles. The calorimeter has been installed in the the cryostat of the Helios experiment and mounted on a scanning table.

The energy reconstruction for electrons has been done adding the energies deposited in a 3×3 matrix of contigous read out towers ("nonet"), created around the tower with the maximum energy content. This cell cluster corresponds to a transverse area of $8.1 \times 7.5cm^2$. The lateral leakage, outside this cluster, for an electromagnetic shower is evaluated to be $\sim 7\%$. Due to the peculiar accordion design, the energy response is expected to be dependent on the impact position of the particle in the x direction. In fig.(2) is shown the normalized energy response at $90GeV/c$ as a function of x, the modulation observed is well inside $\sim \pm 1.0\%$ with a $rms \simeq 0.4\%$. Fig.(2) shows also that the Monte Carlo simulation of the accordion calorimeter describes well the observed structure on x. An energy independent correction for the x modulation has been applied at all energies, no correction for possible longitudinal energy leakage outside $25X_0$ is done in this analysis.

The energy distributions have been fitted with a Gaussian shape to evaluate the central value and the width. The response to electrons in a beam spot $\sim 2cm \times 2cm$ is found to be linear to $\sim \pm 1\%$, in the range $5 \leq p \leq 205GeV/c$, see fig.(3a).

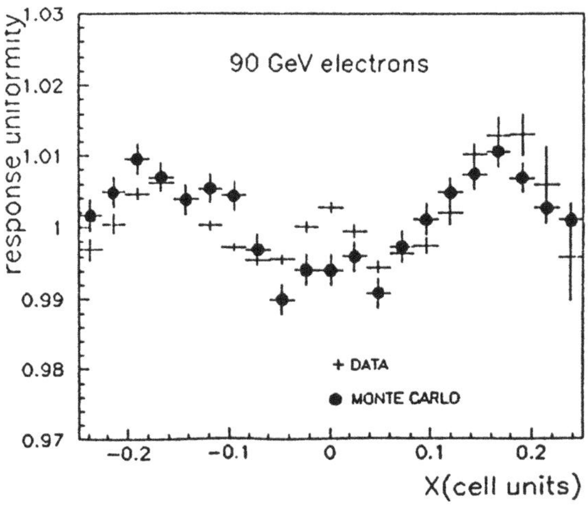

Figure 2 : Relative variation of the calorimeter response over half cell as a function of the electron x position. Full line: experimental points. Black dots: prediction of a Monte Carlo simulation of the accordion.

The values of the energy resolution as a function of the incident particle energy are shown in fig.(3*b*). The behaviour of σ/E can be described with the function :

$$\frac{\sigma}{E} = \frac{k_1}{\sqrt{E}} \oplus k_2 \oplus \frac{k_3}{E}.$$

where the three terms are added in quadrature. The first term is the contribution due to the sampling, the second term reflects the presence of systematic effects and the third one represents the contribution from the electronic noise. The result from the fit to the data is (E measured in GeV):

$$\frac{\sigma}{E} = \frac{(10.1 \pm 0.24)}{\sqrt{E}}\% \oplus (0.2 \pm 0.08)\% \oplus \frac{(0.299 \pm 0.01)}{E}.$$

If the noise contribution is subtracted from the energy resolution, the comparison of the data taken with fast shapers with the results of 1990 data taken with slow shapers, $t_p(\delta) \simeq 100ns$, shows that the fast electronics does not deteriorate the sampling and the constant term in the resolution. The electronic noise, that was negligible with the slow shaping time, $\simeq 4MeV/cell$, increases as expected with the fast shaping time and is measured to be $\simeq 300MeV$ in the "nonet".

A fit to the preliminary results from the analysis of data collected with the "pointing geometry" prototype in the calorimeter region with "opening gap", shows that the energy resolution can be described as:

$$\frac{\sigma}{E} = \frac{(9.3 \pm 0.7)}{\sqrt{E}}\% \oplus (0.7 \pm 0.1)\% \oplus \frac{(0.530 \pm 0.04)}{E}.$$

The constant term, in the "pointing geometry ", is a bit higher than that evaluated for the "cartesian" prototype. The noise, $\sim 530MeV$ for the "nonet" in the "pointing geometry" is ~ 1.8 times the noise measured for the "cartesian " prototype, this is due to the different sampling structure of electrodes and to the different electronic chain adopted; a cold single transistor is used in the section of the "pointing" prototype considered for this analysis.

3.2. Space and angular resolution

The impact point of an electron on the calorimeter can be reconstructed by using the barycentre of the energy cluster. The accuracy of this measure can be determined from the distribution of the difference between the impact point determined from the beam chambers and the point directely reconstructed from the calorimeter. Since in the accordion calorimeter the transverse dimension of the fold is equal to the width of the tower along x, a particle incident on the calorimeter always shares the shower energy among at least two neighboring channels, thus improving the accuracy in the recostruction of the barycentre.

Along the y axis, the accordion calorimeter behaves like a conventional calorimeter, where the position resolution is worse when the particle hits the centre of the cell. The dependence of the space resolution in x, after the PWC resolution has been unfolded, can be described as:

$$\sigma_x = \frac{(3.6 \pm 0.2)\ mm}{\sqrt{E}} + (0.01 \pm 0.03)\ mm$$

i.e. $\sigma_x \simeq 360\mu m$ at $100GeV/c$.

Using the information from each longitudinal sampling, the angular resolution for the electron shower has been measured. Data have been collected with the calorimeter rotated with respect to the position normal to the beam direction. The relative displacement $\Delta x = (x_{front} - x_{back})$ of the shower barycentre in the front and the back sampling is found to be proportional to the rotation angle; from the width of the Δx distribution it is possible to deduce σ_θ. For electron momenta $60 \leq p \leq 120GeV/c$ has been found that σ_θ ranges between 12 $mrad$ and 6 $mrad$.

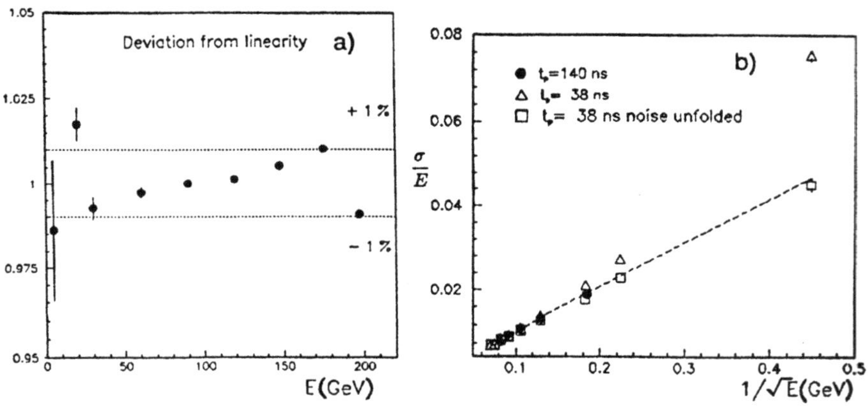

Figure 3 a) : Linearity of the response to electrons for different beam energies.

b) : Energy resolution at different electron energies. Open triangles: data with fast shaping time ($t_p = 38ns$). Open squares: data with fast shaping time and noise unfolded. Black dots: data with slow shaping time ($t_p = 140ns$).

3.3. Response to muons

Data with incident muons have been taken in the test beam to study the calorimeter response to minimum ionizing particles. Since the signal produced by a muon in the calorimeter is always shared at least by two towers adjacent in x, the energy for muons with normal incidence and impact point in a fiducial region of the hit cell is recosntructed adding the signals from cells in the 2×1 tower cluster in the $x \times y$ directions. In fig.(4) is shown the energy spectrum for muons identified in an electron beam of $205GeV/c$ incident in the calorimeter region read out by GaAs preamplifiers. The most probable value of energy loss (1 mip) is $446MeV$, measured with the energy scale determined for electrons. The total electronic noise measured in the 2×1 tower cluster (4 cells) is $140MeV$, i.e. $70MeV/cell$ with a contribution of $\sim 15\%$ due to the fast shaper ($t_p(\delta) \simeq 20ns$). The signal to noise ratio (S/N) is 3.2, to be compared with the value S/N=3.7 expected from the data measured with the slow shaper ($t_p(\delta) \simeq 100ns$) and scaled with the dependence $t_p(\delta)^{3/2}$.

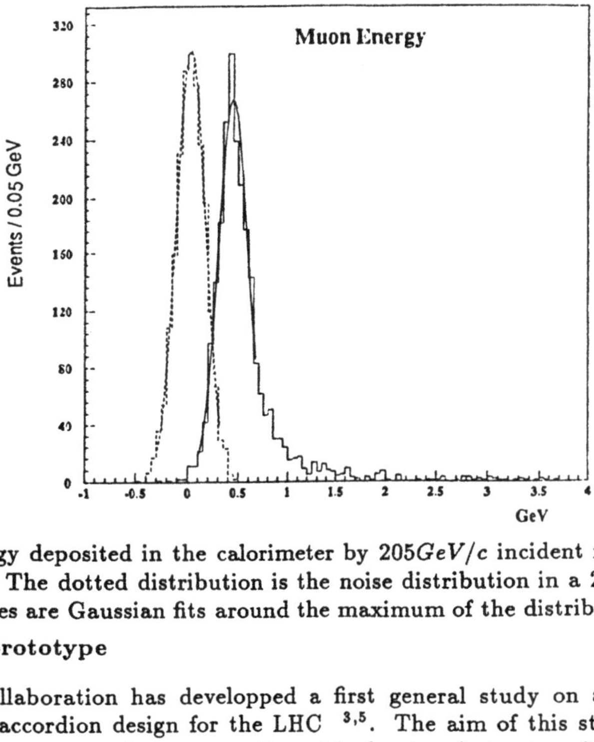

Figure 4 : Energy deposited in the calorimeter by $205GeV/c$ incident muons, full line distribution. The dotted distribution is the noise distribution in a 2×1 tower cluster. The curves are Gaussian fits around the maximum of the distributions.

4. Future prototype

The RD3 collaboration has developped a first general study on a full LAr calorimeter with accordion design for the LHC [3,5]. The aim of this study is the design of a large acceptance electromagnetic and hadronic calorimeter with pointing towers both in azimuth (ϕ) and rapidity (η), with no cracks and full azimuthal coverage. The barrel calorimeter, in particular , has an acceptance $| \Delta\eta | \leq 1.7$ and is divided along the beam direction in two sections of $4m$ length each.

For the hadron calorimeter 8 interaction lengths (λ), iron has been chosen for the converter plates, instead of lead, because of a better rigidity of the material,a lower weight,a minor effect for the multiple scattering of muons detected out of the calorimeter. The granularity $\Delta\eta \times \Delta\phi \simeq 0.049 \times 0.046$ has been fixed to match the electromagnetic calorimeter granularity. The hadron calorimeter is divided in 32 azimuthal sectors with an inner(outer) radius of 2m (3.8m). In the radial direction a sector is segmented in 4 samplings. A sector, in the azimuthal direction, is composed by 4 slices and each slice consists of ground plates and electrodes defining a stucture with 6 or 7 LAr gaps with "electrostatic trasformer" readout [3]. The electromagnetic calorimeter design is based on the evolution of the techniques developped for the small prototypes. The barrel EM calorimeter, inner radius $1.35m$, is divided along the beam direction in two sections; each section $4m$ long, is divided into 40 modules, each module covering 9° in azimuth. Each module is composed of 24 lead converter plates and 24 Kapton-Cu read out electrodes with accordion design. Converter plates and electrodes are disposed radially around the axis of the barrel detector. The calorimeter granularity is $\Delta\eta \times \Delta\phi \sim 0.0176 \times 0.0197$,see fig(5).

An electrode design with the accordion folds inscribed in a cone pointing to the vertex can be adopted to get the best possible performances and to minimize the modulation of the calorimeter response with position. In this case, a "constant LAr gap" is obtained by varying the fold angle with depth. From a detector with pointing geometry and "constant gap" design, about the same performances are expected than for the "cartesian" geometry prototype. In order to test the validity of this calorimeter design, the RD3 collaboration is building a large size prototype. This prototype is a sector, in rapidity and azimuth, of the barrel calorimeter described above. To fully contain a hadronic shower, the hadronic section (7λ) covers 22.5° in azimuth and 0.5 unit of rapidity. The electromagnetic section is a "constant gap" module 2 m long and covering 27° in azimuth. This prototype should be ready in the second half of 1992.

3 samplings

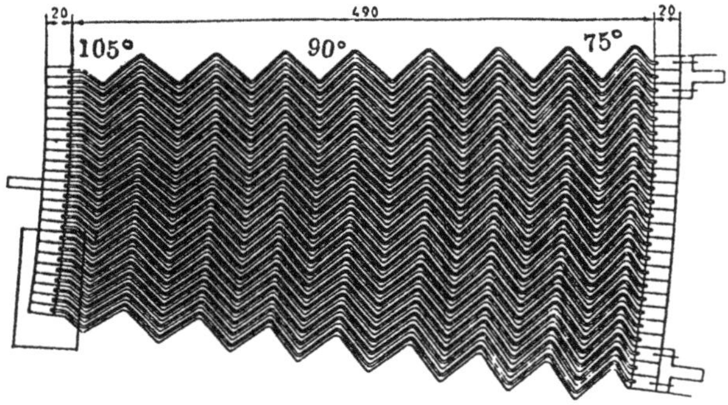

Figure 5 : Converter and readout electrodes in an electromagnetic module with "constant gap" design.

5. References

1. Large Hadron Collider Workshop, Aachen 4-9 October 1990, CERN 90-10, ECFA 90-133.
2. V. Radeka and S. Rescia, Nucl. Instr. and Meth. A265 (1988) 228.
3. B. Aubert et al. (RD3 Collab.), CERN/DRDC/90-31,DRDC-P5 (1990); CERN/DRCD/91-21,DRDC/P5-Add.1(1991).
4. B. Aubert et al. (RD3 Collab.), CERN-PPE/91-73.In pubblication on Nucl. Instr. and Meth.
5. B. Mansoulie, contribution to this conference.

MONTE CARLO SIMULATION OF CALORIMETERS BASED ON LIQUID IONIZATION CHAMBERS

A. GIVERNAUD

CERN, Geneva, Switzerland

Abstract

The collected charge in ionization chambers using a liquid as sensitive medium is calculated by using the GEANT Monte Carlo program, combined with a column recombination theory. This model was used to simulate the upgraded UA1 calorimeter (Uranium/TMP) and was extended to liquid argon ionization chambers.

Introduction

In sampling electromagnetic detectors using liquid noble gases in ionization chambers the collected charge has been shown to be proportional to the deposited enegy at the 1% level [1]. This is also true in warm liquid calorimeters for low ionizing particles. In hadron showers, high ionizing particles are present; high ionizing protons can also be produced in sensitive media containing hydrogen atoms, by neuton interactions; this last phenomena is stronger if uranium is used as absorber. The charge quenching effects are rather small in pure liquid argon, but they are non negligible if methane is added to liquid argon[2,3]. In warm liquids, they are strong, and the Birks constant kB is found to be about one order of magnitude bigger than in scintillators[4]. Moreover kB is found to be a function of the applied high voltage and of the angle between the particle and the electric field[5]; deviations from a simple Birks law have also been reported[6]. Finally, the traping of the drifting electrons by the electronegative impurities are also lowering the collected charge. In these conditions, it seemed to us of big interest to reproduce correctly the measured charge, in a detailed Monte Carlo simulation.

1. Simulation of the collected charge

1.1 The GEANT Monte Carlo

The energy deposited in the sensitive medium is calculated by the GEANT 3.14 program in a full shower simulation where the tracking cuts were set at their lowest values (10 keV).

1.2 From the energy deposited to the collected charge

An ionizing particle crossing the ionization chamber will generate in the sensitive medium N ion pairs escaping primary recombination according to the formula:

$$N = 10^{-2} G(F) \frac{dE}{dx} \tag{1}$$

where G(F), the free electron-ion pair yield, is expressed as the number of pairs produced by 100 eV of absorbed energy. G(F) is a function of the electric field strength and of the temperature of the liquid. The escape probability (as a function of electric field and temperature) was calculated by Onsager[7] for an isolated ion pair, i.e. in the hypothesis where the distance at which most of the electron-ion pairs thermalize is small compared with the average distance between ionization events. The field-dependent escape probability according to Onsager has been calculated for several hydrocarbons[8] . At moderate electric field values (smaller than 20kV/cm), this escape probability f(F), is a linear function of the field F: $f(F) = 1. + \alpha F$ where α depends only of the temperature and of the dielectric constant. The free electron-ion pair yield becomes : $G(F) = G_0(1 + \alpha.F)$ where G_0 is the value of G at 0 V/cm. In the 2244 tetramethylpentane (TMP, C_9H_{20}) G_0=0.74 and α=0.057 cm/kV. This theory is approximately verified for minimum ionizing particles (in TMP, the dE/dx value for such a particle is about 1.6 MeV/cm). As soon as dE/dx increases, the Onsager theory gives an overestimated prediction, even at low fields[9]. A decrease of free ion yield as dE/dx increases, was observed in irradiation with low energy X-rays (dE/dx up to 50 MeV/cm) of several hydrocarbons[10]. This inverse dependence on dE/dx is similar to the one observed in scintillators, where the saturation effect is empirically discribed by the kB parameter of the Birks formula[11]:

$$\frac{dL}{dx} = S. \frac{\frac{dE}{dx}}{1 + kB.\frac{dE}{dx}} \tag{2}$$

where dL is the differential light output and S a proportional (read-out specific) constant between L and E. It is worth noting that this simple parametrisation is not valid for all scintillators[12]. A combination of Birks formula with the Onsager theory has been used : the kB parameter determined by a fit to the data is found to be about one order of magnitude bigger for warm liquid than for usual scintillator[4,5], but the main effect is that kB appears to be a function of the electric field and of the angle between the particle and the electric field vector, as also observed in liquid argon - methane mixtures[2,3]. To descibe these effects in a less empirical way, we are led to consider a track model. Such a theory was developped by Jaffe[14] and later modified by Kramers[15,16] for α paticles.

1.3 The column recombination model

The Langevin assumption of column ionization is that all the ions formed by the passing α particle are collected in a cylinder around the path of the particle. If z is the coordinate along the column axis, the number n_0 of ions per cm^3 is supposed to have a Gaussian distribution: $n_0 = N_0/\pi b^2 \exp(-\rho^2/b^2)$ where N_0 is the initial density and $\rho = \sqrt{(x^2 + y^2}$

); b is a measure of the cylinder radius. This initial distribution n_0 is modified by the recombination (lowering the ion density), by the diffusion of the ions (broadening the cylinder) and by the electric field that separates the ions of different sign from each other. The equation of motion can be written in the symbolic form:

$$\frac{\partial n_{\pm}}{\partial t} = [f] + [d] + [r] \tag{3}$$

The field term, [f] is a function of the electric field itself, F and of sinφ,(where φ is the angle between the column axis and the electric field direction) and also of the ionization concentration gradient. The diffusion term [d] depends on the temperature T, and the recombination term, [r] on $n_+ n_-$, the product of the positive and negative ions densities. To solve the equation of motion, Jaffe, essentially interested by ionization in gases, where the recombination is small, neglected the recombination term. Kramers pointed out that this procedure will no longer be justified in liquids, where the initial ion densities are much larger than in gases. It is then the diffusion term which becomes small compared to [f] (if F is large) and [r]. This assumption is more relevant in liquified gases, at low temperatures. In view of these condiderations, Kramers looked for a solution of the equation of motion without the diffusion term and corrected afterwards for its influence.

We have included the Kramers model in the GEANT Monte Carlo; the dE/dx and the angle φ are given by GEANT at each tracking step in the sensitive medium (TMP). They are the input to the Kramers model which in turn gives the charge value. To otain the collected charge, we have also to take into account the loss of charge due to impurities in the liquid.

2. Warm liquid calorimetry

The warm liquid calorimeters, are using as active medium, a liquid ionization chamber, filled with a room-temperature, non polar hydrocarbon. This technique avoids cryogenic systems, but requires ultra pure liquids. In the UA1 upgraded calorimeter, the TMP obtained after the cleaning procedure[16,17] was extremely pure (lifetimes of 1 ms were measured), the loss of charge appears then to be negligible. The simulation of the uranium/TMP calorimeter is a rather challenging calculation : uranium plays the role of a neutron multiplyer and TMP containing 20 hydrogen atoms (instead of 8 for usual plastic scintillators) gives rise to a high number of (n,p) scatterings.

2.1 α particles

We are comparing the model to the data of Munoz et al.[18] (α particles in warm liquids). These measurements were done for several hydrocarbons in a detector cell with stainless steel electrodes separated by a gap of liquid of 0.07 cm. The values of the electric field in this gap are ramped from 2.3 to 48 kV/cm. The americium α source was deposited on the

center of one electrode. The 5.64 MeV α particles are isotropically emmitted into the gap. A good agreement is found (Fig. 1a) between the measurements and the model for b=31Å, above 7 kV/cm. At the lower values of the electric field, the model founds a smaller charge. This is certainly due to the fact that the diffusion term is not negligilble at low field. A good agreement is also observed for the resolution : at 8.8 kV/cm for instance, the experimental pulse height spectrum has a full width at half maximum (FWHM) of 640 electrons, as the model (corrected for the instrumental resolution) gives a FWHM of 620 electrons.
The value of b, determined from above is our unique parameter and will be kept in this model for all other applications involving TMP as a sensitive medium.

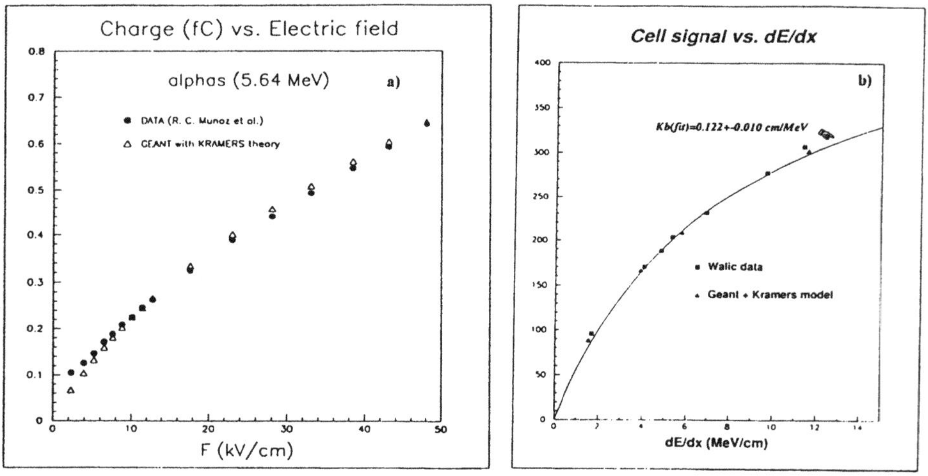

Figure 1. a) α in TMP. b) Signal vs. dE/dx (WALIC data)

2.2 WALIC results

The WALIC collaboration[5] using an ionization chamber similar to thoses of the UA1 U/TMP calorimeter (2 gaps of 1.25mm), has measured the charge induced by a low energy beam (from 200 to 600 Mev/c) at the CERN Synchro cyclotron. Fig. 1b shows the signal in TMP as a function of dE/dx up to 12 Mev/cm. A good agreement between the data and the model (GEANT including the Kramers track model) is observed. The WALIC collaboration has described the data with the Onsager theory and a Birks parametrisation. Fig. 2a shows the variation of kB as a function of the angle φ for an electric field of 12kV/cm in TMP; these data are also well described by our Monte Carlo calculations.

2.3 *Test beam results*

a) The test beam set up

The apparatus was already described[19]. Three U/TMP calorimeter modules were resting on a moveable device and are followed by a module of the iron/scintillator UA1 hadron calorimeter, "C-module", of 4.5 interaction lengths, and by a 1.6m long iron block used as a muon filter. These U/TMP modules are the constituants of the 'gondolas", which are the elements of the central barrel calorimeter. Each of them consists of two stacks of 63 boxes containing TMP, separated by uranium plates. The internal thickness of these boxes is divided in two gaps (d=1.25mm) filled with TMP and separated by a central electrode. The modules have 6 samplings in depth (5,9,15,10,12 and 12 layers for samplings 1 to 6 respectively). Samplings 1 to 4 have 2 mm thick U-plates; they are the constituents of the electromagnetic calorimeter, corresponding to 25 radiation lengths. Samplings 5 and 6 are the hadronic section and have 5 mm thick U-plates. The total of the samplings 1 to 6 represents 2.2 interaction lengths. As each TMP box has 8 electrodes, and as each layer is made up of two boxes, a total of 96 cells per module are connected to the readout. The measurements were made at the SPS (X1) and PS (T7) test beams at CERN. At the SPS, data were taken with muons, electrons and pions between 7 and 70 GeV/c. At the PS only electrons and pions data were collected, between 1 and 10 GeV/c.

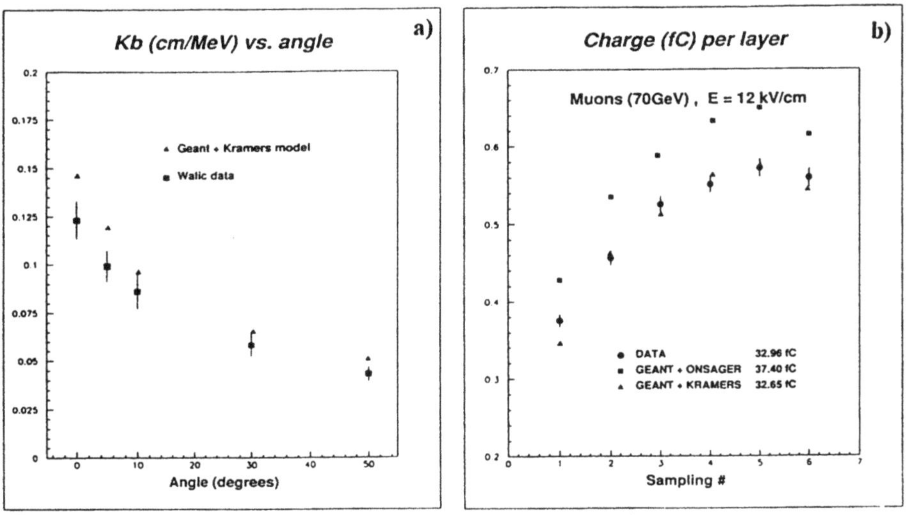

Figure 2. a) kB vs. angle (WALIC data). b) 70 GeV muons in the U/TMP calorimeter

b)Muons

The collected charge per layer in each sampling for 70 GeV muons at an electric field of 12 kV/cm is shown by Fig. 2b. These charge values are normalised to one layer, i.e. we have divided the charge value in each of the 6 samplings by the number of layers per sampling. The collected charge in sampling 1 is the one expected from a minimum ionizing particle (mip). As the μ penetrates deeper into the stack, its behaviour differs from the one of a mip. The same figure shows also a comparison between the measurements and Onsager and Kramers theories. As discussed before, for low ionization densities, the Onsager models agrees well with the data and predicts too large values of the charge as ionization density increases. The opposite behaviour is observed for Kramers model: in sampling 1, a too small amount of charge is predicted, mainly due to the bad description of difuusion, as a very good agreement with the data is observed in sampling 6, where the recombination process is more important.

c) Electrons

A good agreement between the model and the electron data is also observed: the measured charge created by 10 GeV electrons at normal incidence amounts to 317, 360 and 389 femto coulomb (fC) at 4.8, 8.0 and 12 kV/cm respectively as the model gives 318, 361 and 390 fC. The reconstructed energy, E is deduced from the charge Q measured in the 4 samplings of the electromagnetic section: $E(GeV) = \alpha\Sigma_{i=1,4}Q_i(fC)$ The energy independent calibration factor α is 0.0259 GeV/fC for an electric field of 12 kV/cm. The electron resolution as a function of the reconstruted energy for PS and SPS data[19], after substracting the noise, is well described by the formula (quadratic sum) :

$$\left(\frac{\sigma}{E}\right)^2 = \left(\frac{a}{\sqrt{E}}\right)^2 + (b)^2 \qquad (4)$$

With a = 0.124 ; the constant term, b = 0.009, is consistent with the beam resolution. Considering only the energy deposition given by GEANT, we obtain: a = 0.117 and b = 0. and with the collected charge (GEANT combined with the Kramers model): a = 0.122 and b = 0. again, a very good agreement with the data is found.

d) Pions

The relation between the charge deposited in TMP, (or the light output of the C module scintillator) and the incident energy is more difficult to estimate for pions, as their energy deposition can fluctuate over three different calorimeters: the U/TMP electromagnetic section , the U/TMP hadronic section and the C module (5cm thick Fe plates and scintillator). The U/TMP section has been calibrated by electrons, and we keep the same constant α as in the actual calorimeter at the collider, we do not know a priori, the nature of the incident particle. The C module was calibrated in an electron beam; the reconstructed energy is defined as:

$$E(GeV) = \alpha\Sigma_{i=1,4}Q_i(fC) + \beta_1\Sigma_{i=5,6}Q_i(fC) + \beta_2\Sigma_{i=1,2}E_i(GeV) \qquad (5)$$

The calibration factors β_1 and β_2 were determined by minimizing the difference between the reconstructed energy and the incident energy, using a pion beam at the highest momentum (70 GeV/c).

The resolution at 12kV/cm as a function of the reconstructed energy is shown by Fig. 3a. On this figure, we have plotted the SPS data points (10, 30 and 70 GeV/c incident momentum) together with the model predictions at 2, 5, 10, 30, 70 and 100 GeV/c. The noise has been substracted quadratically for the data points and the same method as for electrons was used to fit these data (Eq. 4).

A clear improvement of the resolution as the electric field increases was observed in pion data (and not in electron data)[19]. This effect was already observed on the very forward module (1 cm uranium plates) in a 7 GeV/c beam[20].

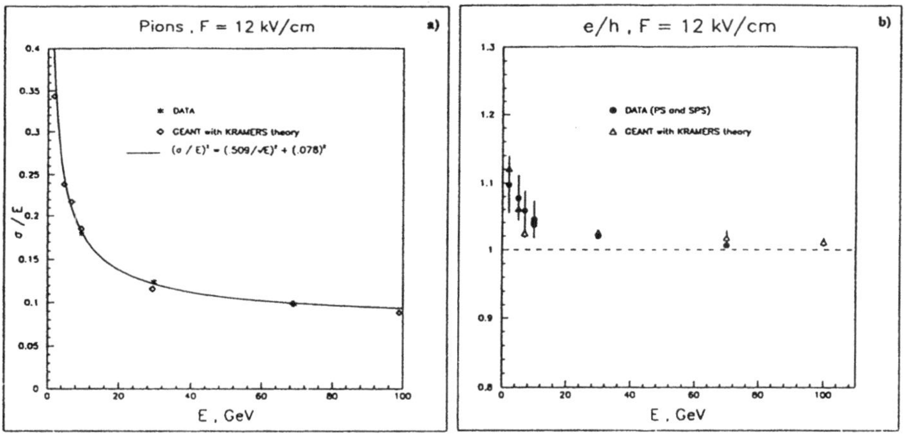

Figure 3. a) Pion energy resolution. b) e/h ratio

The ratio e/h of the reconstructed energy of electrons and pions is plotted on Fig. 3b as a function of the reconstructed energy at 12 kV/cm (PS and SPS data). On this figure, the data are compared with the model predictions at 2, 5, 7, 10, 30, 70, and 100 GeV/c. e/h decreases as the electric field increases[19]; this effect is more evident at low energies, as a bigger part of the hadronic shower is contained in the Uranium/TMP part of the calorimeter. This behaviour is of the same kind as the one observed previously in the pion resolution. The agreement between the model and the data was shown to be better at high field[19].

The track model, based on the Kramers theory, reproduces the charge behaviour with the electric field as well as the dE/dx dependence, provided that the field is not too low (bigger than 7 kV/cm). The improvement of the pion resolution and of the e/h ratio as the electric field increases shows that the charge generated by electrons and pions does not have the same field dependence. This is well explained by the model as a consequence of the field term structure.

3. Large warm liquid calorimeter

The UA1 U/TMP calorimeter with a depth of 2.2 λ is far from containing the hadron showers. As the model gives a good description of the test beam data, it should be able to predict the performance of a larger TMP calorimeter. This simulation work is in progress, and few setups have been studied; they are all 10 λ deep and $1\times1m^2$ wide and segmented in 81 $11\times11cm^2$ towers. The sensitive medium is also TMP, and as before, each layer has two gaps of 1.25 mm. With 5mm uranium (or tungsten) as absorber, 192 layers are needed to achieve 10 λ and the overall depth is about 1.6 m. Above 10 GeV, the e/h ratio decreases from 0.90 to 0.77 as the electric field increases from 8 to 25 kV/cm for the uranium calorimeter and from 1.09 to 0.97 for the tungsten calorimeter. At the same time the resolution as a function of the field fitted between 10 and 100 GeV with Eq. 4 becomes worse for uranium (a=0.42, b=0.015 at 8kV/cm and a=0.45, b=0.045 at 25 kV/cm) while it remains unchanged in the same electric field range for the tungsten calorimeter (a=0.44, and b=0.010).

These results were obtained when considering the whole calorimeter. If only 3x3 towers are selected, as in the test beam studies, the signal is lowered by 15% to 18% while σ/E improves by 8% to 18% in the uranium calorimeter. In the tungsten calorimeter, the signal decreases by 7% to 9% and the resolution is unchanged. Moreover, if the applied time cut is decreased from 2μs (UA1) to 15ns, the pion signal in the uranium calorimeter is lowered by about 10%.

3.1 Time response

Due to the high repetition rate of the future LHC the electron drift has to be fast enough to clear the ionization chamber after each bunch. In pure liquid noble gas, the drift velocity saturates at a field of about 10 kV/cm. In warm liquids no saturation is observed even at 100 kV/cm. This means that by increasing the high voltage and decreasing the gap, a drift time comparable to the bunch crossing time can be achieved. For instance, in the tetramethylsilane (TMS), which is about three times faster than the TMP, a drift time of 23 ns could be obtained with an electric field of 50kV/cm in a 1mm gap[21].

3.2 Radiaton dammage

The radiation doses at LHC will be several orders of magnitude higher than at the present hadron colliders. To study this problem an irradiation test was done in Berlin with gammas of a 21 kCi cobalt source[22]. Two stainless steeel cells, one filled with TMP, the other with TMS were irradiated at two different dose rates (9 and 150 krad/h). The electron lifetime, initially 1ms, was determined after each irradiation by measuring the current from uniform ionization. The results are shown on Fig. 4a. The lifetime decrease is independent of the dose rate, and TMP appears to be more radiation resistant than TMS. These results have to be compared with the calculated doses presented at this workshop[2,3]. In TMS the effective ion mobility remains unchanged, whereas for TMP it decreases by a factor 10 (Fig. 4b) after a 10Mrad irradiation.

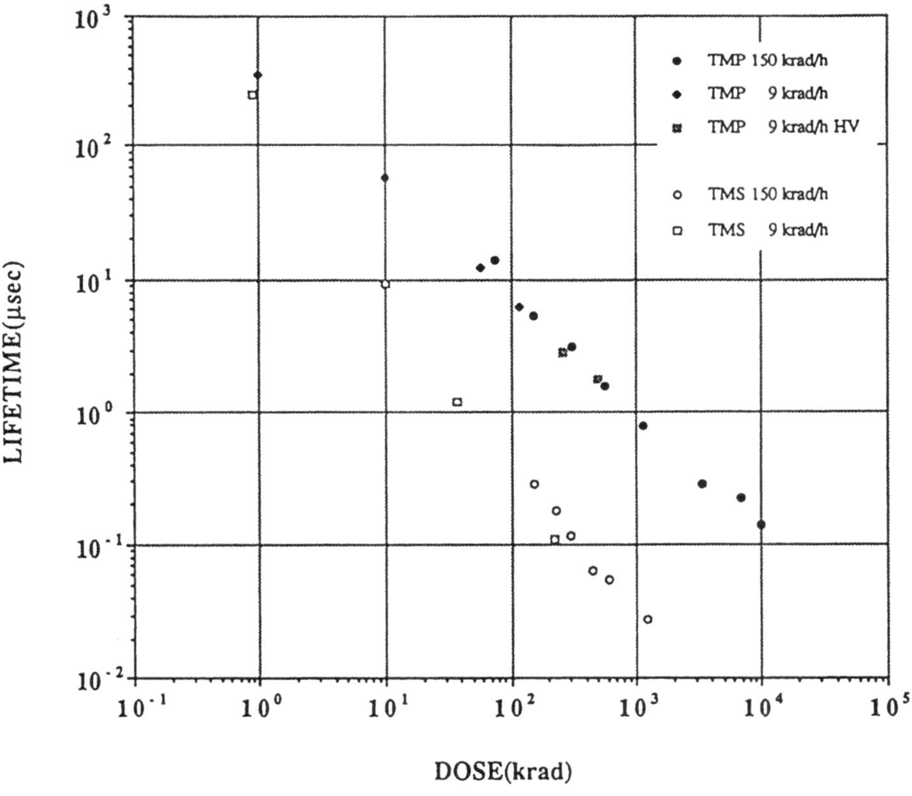

Figure 4. a) Lifetime vs. dose.

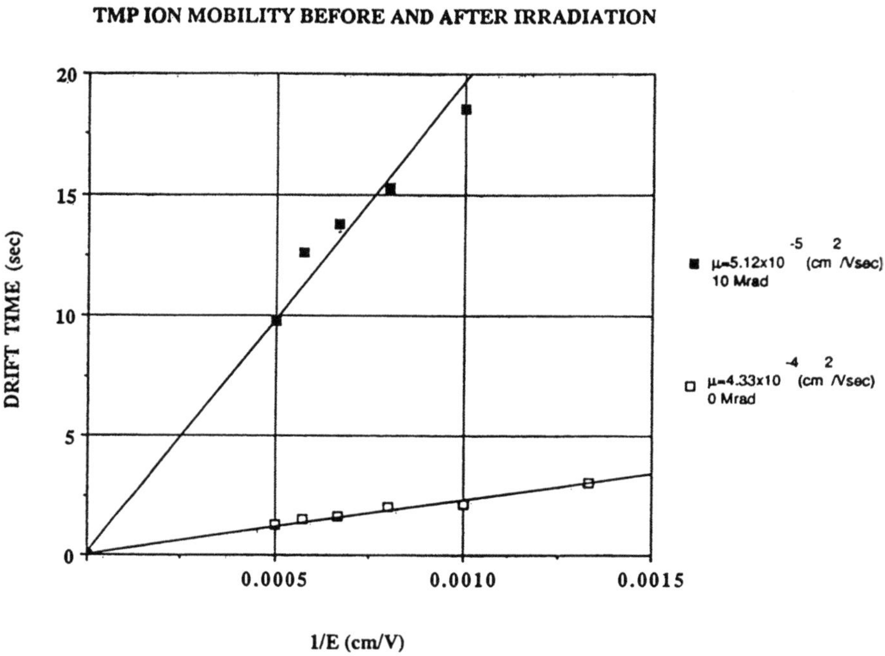

Figure 4. b) Ion mobility in TMP

4. Liquid argon ionization chambers

4.1 Pure liquid argon

In pure liquid argon the charge quenching effects are small. The field-dependant Birks 'constant' is found to be very small[3]. However, the Onsager theory is in strong disagreement with the data[24]. We expect the Kramers approximation to be more relevant in liqid argon (80 °K) than in warm liquids, as the diffusion (depending on the temperature) becomes very small. This model predictions were compared (Fig. 5a) to the results obtained in a gridded ionization chamber with 1cm liquid argon gap[3]. The charge response from α and β particles was measured as a function of the electric field. The data are well described by the model both for α and β with a "column radius" b = 1500Å. At low field, the model gives lower values than the data, as for warm liquids, but this happens at lower field values (<2kV/cm). The high value of b is not surprising, as a most probable ion electron separation of 1540 Å in liquid argon was found[25].

4.2 Liquid argon doped with methane

The addition of methane at a fraction of percent in liquid argon increases the drift velocity, but increases the saturation effects and leads to a bigger e/h ratio in calorimeters using this technique[26,27]. In Fig. 5b are plotted the ratio of charge measured in liquid argon doped with 0.25% methane to the charge measured in pure liquid argon, as a function of the electric field[28]. The Monte Carlo with a column radius b=305Å gives a good description of the data.

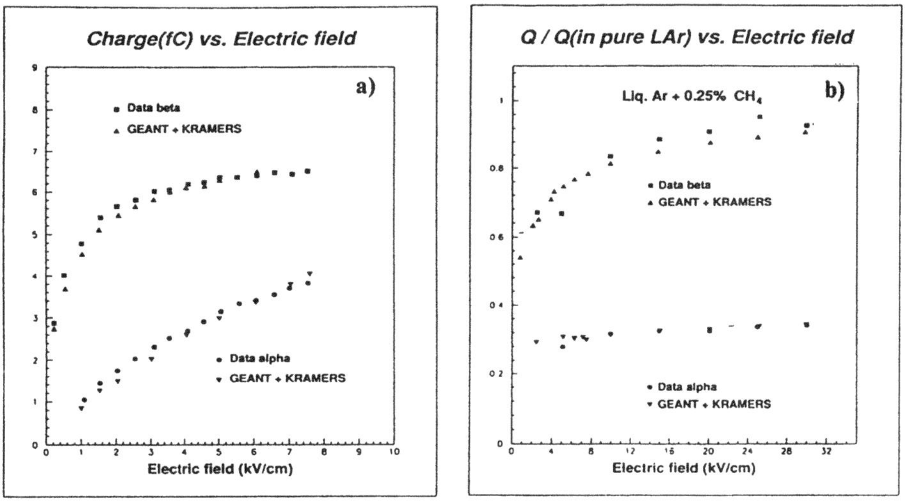

Figure 5. Charge versus field in pure liquid argon (a) and in doped liquid argon (b)

Conclusion

The GEANT Monte Carlo, combined with the column ionization theory, in the Kramers approximation, provides a good and general understanding of the liquid ionization chambers data for several kind of particles and in a wide energy range. The model uses a basic, physical picture and only one parameter, the column "radius", b, is needed. The small value of b in TMP corresponds to a small thermalization radius, increasing the probability of recombination, giving rise to a large charge quenching. A rather modest electron mobility μ at low field is observed, but it increases linearly with electric field and no saturation of the drift velocity is observed. In liquid noble gases, μ is one or two orders of magnitude bigger, but the drift velocity saturates already at 10kV/cm. This is explained by an extended electron state[29]; the thermalization radius is considerably larger than in TMP, (the model needs a large b value) and the charge saturation is small. With an addition of methane at low concentrations in liquid argon, the drift velocity increases with the electric field and does not show saturation (at least below 100kV/cm). However, the thermalization radius of the electrons is reduced and charge saturation effects appear. The b value of 305Å illustrates this behaviour, somewhat "between" TMP and pure liquid argon.

References

1. J. Feltesse , *Proceedings of the Wire chamber conference, Vienna 1989*
2. D. F. Anderson and D.C. Lamb , *NIM A265 (1988) 440*
3. V. Vuillemin et al *CERN -PPE/91-171*
4. S. Ochsenbein , *Contribution to Natoasi Sintra (Portugal) 1987*
5. B. Aubert et al , *NIM A286 (1990) 583*
6. J. Bolz et al , *Karlsruhe Preprint, submitted to NIM*
7. L. Onsager , *Phys. Rev. 54(1938) 554*
8. G. Ambrosini and F. Cavanna , *UA1 TN 90-19 (1990)*
9. R. C. Munoz et al , *Madrid Preprint, CIEMAT-662 submitted to NIM*
10. R.A. Holroyd and T.K. Sham , *J. Chem. Phys. 89, (1985) 2909*
11. J.B. Birks , *Scintillation counters, Mac Graw Hill, New York,1953*
12. R.L. Craun et al. , *NIM 80 (1970) 239*
13. G. Jaffe , *Ann. Physik (4) 42 (1913) 303*
14. H.A. Kramers , *Physica,18 (1952) 665*
15. A.N. Gerritsen , *Physica 14 (1948) 381*
16. S. Ochsenbein et al. , *NIM A273 (1988) 654*
17. F. Diez-Hedo , *Proceedins of the EFCA Study Week, Barcelona 1989,* CERN 89-10, p. 557
18. R.C. Munoz et al. , *J.Chem.Phys. 85 (1986) 1104*
19. A. Givernaud , *Proceedings of the workshop on detector and event simulation,* Amsterdam,1991,p297-311
20. C. Bacci et al. , *NIM A292 (1990) 113*
21. E. Geulig et al. , *CERN/DRDC/90-50,1990*
22. A. Givernaud et al. , *To be submitted to NIM*
23. A. Ferrari , *Proceedings of this workshop*
24. R. T. Scalettar et al. , *Phys. Rev. A 25 (4) (1982) 2419*
25. S. S. S. Huang et al. , *Can. J. Chem. 55 1838 (1977) 2419*
26. P. Cennini et al. , *CERN/DRDC/90-34 (1990)*
27. M. Abolins et al. , *CERN/FERMILAB-Pub-89/38-E*
28. D. Rahm , *Proceedings of the Conference on liquid noble gas detectors ,* Stockholm August 19-25 1991
29. K. Yoshino et al. , *Phys. Rev. A 14 (1) (1976) 438*

IMPACT OF ELECTROMAGNETIC CALORIMETER THICKNESS ON RESOLUTION

CHRISTOPHER HEARTY*
Lawrence Berkeley Laboratory, MS 50A-2160
One Cyclotron Road
Berkeley, CA 94720, USA
bitnet: Hearty@slacvm

Abstract

EGS has been used to study[1] the effect of leakage on EM resolution as a function of EM calorimeter thickness and hadronic calorimeter configuration. The EM calorimeter consists of 4 mm lead plates alternating with active sampling (4 mm of liquid argon). Five cases that are relevant for SDC are considered for HAC1: it constructed from either 8 mm, 16 mm, 20 mm, or 32 mm lead plate (and is noiseless and calibrated) or it is not used at all. Electrons of energies 10 GeV to 400 GeV are used. The impact of leakage is quantified by the quadratic increase in resolution (σ_E/E)—essentially, the increase in the constant term. For this to be less than 0.3% (0.5%), the total EM calorimeter thickness, including preceding dead material, must be $28X_0$ ($26X_0$) if HAC1 is not used or is very coarse, or $22X_0$ ($20X_0$) if a HAC1 of 8 mm lead plate is used. Other cases fall between these two extremes.

References

1. C. Hearty, SDC–91–00084, Sept. 1991.

*This work was supported by a fellowship from the Texas National Research Laboratory Commission and by the Department of Energy under Contract Number DE-AC03-76SF00098.

SIMULATION OF SPACE CHARGE EFFECTS ON IONIZATION CHAMBER SIGNALS

K. LUEBELSMEYER, D. PANDOULAS, M. SCHOENTAG, S. ZITZEN
I.Physikalisches Institut, RWTH Aachen
Sommerfeldstrasse 28, D 51 Aachen, Germany

Abstract
A method of detailed simulation of ionization chamber signals is described, including space charge effects due to ions. The method is applied to a study of signal stability for operation at very high rates.

1. Introduction

Ionization chamber calorimetry is one of the proposed solutions for LHC/SSC detectors. The high rates anticipated at these machines, in combination with the slow drift of ions, may result in space charge accumulations strong enough to alter the electric field prevailing in the medium and thus affect the expected signals. A simulation of such effects is presented in this note.

2. Simulation Method and Results

We obtain the electric field $E(r,t)$ at any time t and point r in the ionization gap by using the method of finite elements to solve Poisson's equation for the electric potential $\phi(r,t)$. The charge density $\rho(r,t)$ in the gap is due to the electrons and ions. Allowing the latter to drift in the electric field for a short interval Δt leads to a new density $\rho(r,t+\Delta t)$ and a new solution $E(r,t+\Delta t)$. Electron-ion recombination during drift is taken into account[1]. Applying Gauss's law to the electric field allows computing the charge $Q(t)$ on the electrodes. Repeating these steps leads to the signal current $I(t)$ by calculating the change in $Q(t)$.

We constructed a "worst case scenario" by considering a 50 GeV electron hitting the same spot of a calorimeter once every 16 ns. The calorimeter consisted of 25 layers of 4mm lead plates interspersed with ionization chambers. The latter were formed by two 2.5mm TMS gaps, separated by a 1mm thick steel electrode (divided laterally in 40.5x40.5 mm^2 pads) and enclosed in 2.5mm thick steel containers.

Electron showers were simulated using the GEANT package and the method above was used to determine the current drawn by an electrode placed at shower maximum. A field of 50 kV/cm was applied to the gap, the electrode potential being held constant. For each shower product traversing the gap the number of electron-ion pairs (assumed uniformly distributed along its path) was computed from the deposited energy. The ion yield[2], as well as the electron drift velocity[3], was parametrized as a function of the electric field. The ion mobility[4] was taken to be 5×10^{-4} cm^2 V^{-1} s^{-1}.

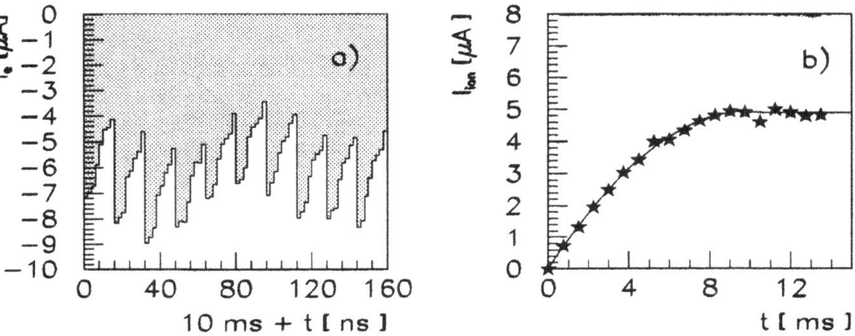

Fig. 1. Electrode current due to a) electrons and b) ions.

Fig. 1a(b) shows the electron (ion) contribution to the electrode current. The ionic component rises to a plateau value of $\sim 5\mu A$ after 8-10 ms, while the electron current shows the typical rise to a maximum value I_{max} every 16ns, i.e. with the arrival of each new shower. For fast electronics, I_{max} can be taken as a measure of the signal. Fig. 2 shows this signal (summed over a 3x3 pad array) to be quite constant for a given shower, unaffected by the presence of the large ionic current.

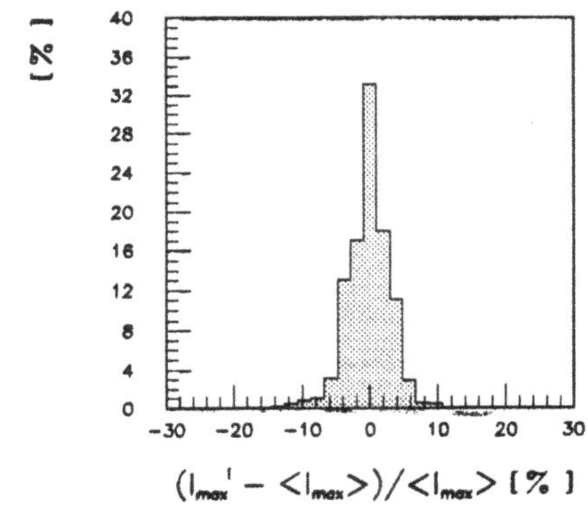

Fig. 2. Deviation probability of the peak electron current from its time average.

3. Conclusions

We developed a method to simulate in detail ionization chamber signals. An application shows that pile-up effects will not affect the stability of TMS signals, even at the highest LHC/SSC luminosities. A similar study for liquid Argon is in progress.

4. References

1. H.S.Massey, E.H.S.Burhop, H.B.Gilbody, in *Electronic and Ionic Impact Phenomena*, **Vol.IV**, Oxford (1974).
2. S.Geer, R.A.Holroyd, F.Ptohos, *NIM* **A301** (1991) 61.
3. U.Sowada, Thesis, Freie Univ. Berlin, Fachbereich Physik (1976).
4. A.O.Allen, M.P.De Haas, A. Hummel, *J. Chem. Phys.* **64** (1976) 2587.

ELECTRON LIFETIME MEASUREMENTS IN TMS

E. GEULIG, W. KARPINSKI, TH. LEHMANN, K. LÜBELSMEYER, R. SIEDLING, W. WALLRAFF
1. Physikalisches Institut, RWTH Aachen,
Sommerfeldstraße 28, D 51 Aachen, Germany

ABSTRACT

The lifetime of electrons in TMS has been measured using both laser driven photoemission generated electrons and. those from the ionization generated by a Bi^{207} source.

1. Introduction

The control of electron lifetimes in warm liquids is of paramount importance for the longterm operation of a warm liquid calorimeter[1,2] at the future hadron colliders. A considerable number of highly reliable, small volume measurement cells are ob- viously required for any large size calorimeter set-up. In this paper first measurements with 2 prototype cells (Laser Cell, Bi^{207} Cell) are presented.

2. Laser Cell

The lifetime measurement in the laser cell is done by measuring the ratio of the charges emerging from (Q_{out}) and entering (Q_{in}) into a 6 mm wide driftspace (fig. 1). The driftspace is defined by two grids of > 98% transparency mounted between the electrodes (separation 8 mm) of a small (Ø 12 mm) ionization chamber. This charge ratio is averaged for 400 pulses from a laser (266 nm) onto the stainless steel cathode.

The increasing charge ratio reflects the field dependent rise of the drift velocity[3]. A common fit to all charge ratios gives an electron lifetime of 19.5 ± 2.1 μs. A data point is taken in ≈ 100s, so a frequent control of the lifetime is easily possible.

For observed charge ratios ≥ .98 the lifetime measurement error is dominated by the uncertainty in the transparency of the grids. The intensity of the laserlight, the field in the driftspace and the electronic noise set a lower limit of the observable charge ratio. Thus using the present set-up the laser cell can measure lifetimes in the range of 0.25 μs to 25 μs.

3. Bi^{207} Cell

A 4 kBq Bi^{207} source (Ø 5 mm) has been mounted in the center of the cathode of a small stainles steel ionization chamber (Ø 10 mm) with a gap of 9.7 mm. The pulse height distribution of the signal charge of individual decays is observed. At the high energy end of this distribution a clear peak due to the 976 keV internal conversion electrons is seen. Electron lifetimes less than the drifttime across the gap lead to a loss in pulseheight. Limitations of this lifetime measuring technique have been discussed by E. Aprile[4] et al., M.L. Lopes[5] et al. and M. Gettert[6].

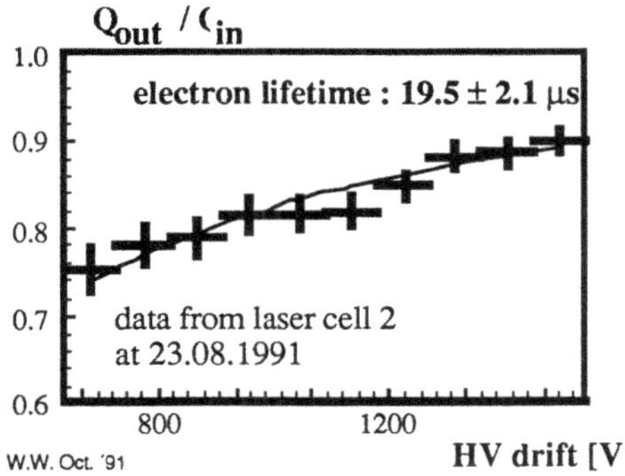

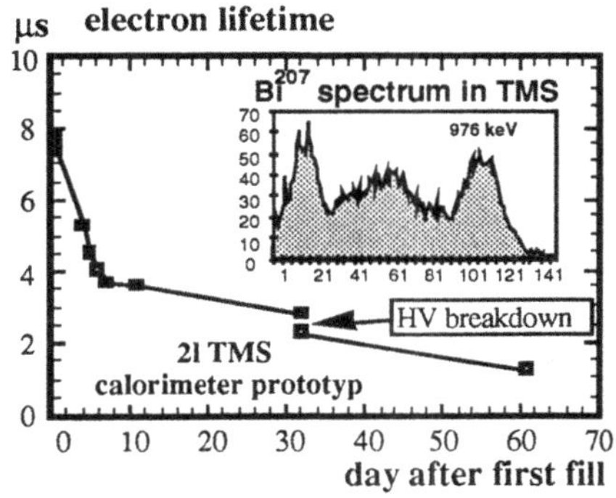

fig. 1 Field dependence of the charge transfer ratio for lasergenerated electrons.

fig. 2. Electron lifetimes measured with the Bi^{207} method.

The ionization chamber and the feedthroughs are mounted on a flange terminating a short tube section with ≈ 100 cm^3 volume. In a test run this lifetime measuring head has been connected to a small TMS calorimeter (vol. 2000 cm^3) for use in a cosmic ray study. Fig.2 shows the evolution of the electron lifetime in this set-up over a period of 9 weeks. The calorimeter under study had been filled with TMS for the very first time. Obviously there are 2 TMS pollution processes: a fast decrease in electron lifetime is observed during the first week, followed by slow decrease in lifetime of about 1% per day.

4. Conclusion

We have used two types of measuring techniques for the determination of electron lifetimes in TMS. The Bi^{207} method is well suited for longterm monitoring whereas the laser method yields precise results over a wide range of lifetimes quickly.

5. Acknowledgement

We are indebted to Dr. D. Schinzel , CERN and Dr. J. Engler, KfK for their technical support as well as their expert advice.

6. References

1. E. Radermacher, in *Large Hadron Collider Workshop, Aachen 4-9 October 1990*, ed. G. Jarlskog and D. Rein (CERN 90-10, vol. III)p. 341 .
2. L* Collaboration, *Letter of Intent to the SSC,* November 1990.
3. N.E. Cipollini and A.O. Allen, J. Chem. Phys. **67** (1977) 131
4. E. Aprile et al. NIM **A261** (1987) 519
5. M.L. Lopes et al. NIM **A271** (1988) 464
6. M. Gettert, Diplomarbeit KfK, Karlsruhe, Oktober 1988

COMPENSATION STUDIES OF A LEAD-TMP CALORIMETER

WALIC COLLABORATION
presented by

Bruno MANSOULIE
DAPNIA/SPP, CEN-SACLAY, FRANCE

Foreword
Most of the material presented in this talk has just been published at the time of the conference: pre-print LAPP-EXP 91-06, submitted to NIM. Due to space restrictions, it was agreed to refer to this paper rather than reproduce its contents in this proceedings. The main conclusions are summarized below.

Results

We have successfully built and operated a prototype calorimeter with TMP as a sensing medium (expt. E795 at FNAL). The modularity, achieved by hanging detector and absorber planes to a frame, has enabled us to take data in 5 different configurations during the 1990 run:

- an electromagnetic configuration with 28 samplings every 1.2 radiation lengths,
- 4 hadronic configurations with different hydrogen contents and total lengths.

In all the configurations, the stability of the electronics and the reproducibility of the mechanics were checked with test-pulse measurements as well as muons.

All the data taken were compared to a detailed Monte-Carlo simulation based on the Geant/Geisha package, and including the saturation properties of the signal in the liquid for highly ionizing particles, according to our previous measurement.

In the electro-magnetic configuration, we found a good linearity, and a resolution of :
$(\sigma/E)^2 = (18.0\%)^2/E + (1.1\%)^2$

With the 4 hadronic configurations, the following conclusions were reached for the e/π ratio:

- e/π varies between 1.1 and 1.4 depending on the beam energy and configuration,
- e/π is not sensitive to the hydrogen content: at the electric field we used ,6.7 kV/cm, saturation effects are dominating,
- e/π is sensitive to saturation of the signal for high ionization density and hence to electric field; it is predicted to reach unity or less as the electric field is increased to 20 kV/cm or more,
- the behaviour of e/π for various configurations is well reproduced by the simulation.

During the 1991 run, one measurement with lead absorbers has been redone with twice as many TMP planes for cross-check, with a result consistent with the 1990 data; the same measurements are now being done with iron absorbers.

CALORIMETER TESTS WITH LIQUID IONIZATION CHAMBERS

J.Engler, M.Gettert, H.Keim, J.Knapp, H.H.Mielke
Kernforschungszentrum and Universität Karlsruhe,
P.O.Box 3640, D-7500 Karlsruhe, Germany

and

A.Arefiev, Yu.Kamyshkov, A.Malinin, A.Savin, K.Shmakov,
E.Tarkovsky, I.Vorobiev
Institute for Theoretical and Experimental Physics
B. Cheremushkinskaya, 117259 Moscow, USSR

Abstract

Configurations of sampling calorimeters with iron, lead and uranium as absorbers have been investigated using liquid ionization chambers as active elements. As liquid tetramethylsilane has been used. Results of beam tests with electrons, pions and muons in the energy range of 2 to 6 GeV are presented and compared with Monte Carlo simlations.

1. Introduction

Sampling calorimeters have become important instruments for the detection of very high energies in the TeV region [1]. As active elements ionization chambers are the best choice for these devices because of several reasons among which are the following: They deliver a signal well proportional to the deposited energy and signal processing is possible over a large dynamic range. This is important particularly in dense shower cores. Also the signal can be calibrated absolutely by injection of charge. Feedback amplification ensures a good stability over long periods of time.

The usual liquid LAr becomes impractical for very large calorimeters. The above considerations have led to the idea to use tetramethylsilane (TMS) as chamber medium, which is liquid at room temperature.In the following we report on a series of measurements which have been performed to study electromagnetic and hadronic showers in iron, lead and uranium absorbers using TMS chambers as active layers. These chambers have been developed to be used in a hadron calorimeter for the cosmic ray project "Kascade" [2] under construction at Karslruhe. The distinctive feature of the chambers is their easy construction suitable for a cheap mass production, which is estimated to be less expensive than a corresponding scintillator version.

2. The experimental setup

2.1 Ionizations chambers

A photograph of a chamber is given in fig. 1. The body is formed by two equally shaped plates, the cover plate containing the filling nozzle and the HV feed-through. The photograph shows the arrangement of the central anodes in six separate strips. The only materials in contact with TMS are stainless steel, ceramics and PTFE spacers.

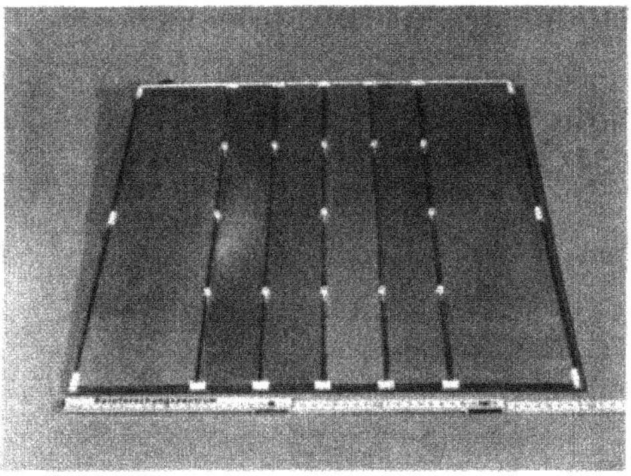

Figure 1: Photograph of the interior of an ionization chamber. The dimensions are 50x50x1 cm^3.

All chamber parts were cleaned in an alkaline ultrasonic bath and rinsed with ultrapure water of at least 18 MΩ cm volume resistivity. After being dried in a 250 °C argon atmosphere, the chambers were assembled in a clean bench and closed immediately by welding.

The chambers were filled with a very clean liquid of a purification system. The impurity of the liquid was equivalent to less than 4 ppb of oxygen. This means that the mean lifetime of free electrons in such a pure TMS is more than 50 µs. The chambers were flushed repeatedly with clean TMS before they were closed. Most of the chambers were pinched off with a clamp, a few, however, were closed by a valve with a Viton gasket. The filling was done at Karlsruhe. The chambers were then shipped to the ITEP institute at Moscow for the beam tests.

A voltage of typically 3 kV was applied to the strip electrodes giving an electric field of about 8 kV/cm. The signals were collected by charge sensitive preamplifiers developed at Karlsruhe using a SK371 as input FET. The first shaping amplifier immediately follows the input circuit on the preamlifier card. This solution was chosen to facilitate signal transmission to the main amplifier by a high signal level. An RC network formed the pulse to a bipolar shape with a rise time of 1 µs. A typical rms noise at zero input capacitance was 300 e^-. For a central strip channel with a capacitive load of 380 pF it typically amounted to 1200 e^-. For a complete calorimeter configuration with 37 chambers a total of 222 channels have been read out.

An important aspect of using ionizations chambers in calorimetry is their easy and convenient electronic calibration by charge injection. It was performed by coupling a step voltage to the preamlifier by a test capacitance of 2.2 pF. We estimate the error in the electronic calibration to be less than 3 %. The chambers were tested before and after the measurements with a 100 MeV triton beam at Karlsruhe. From these measurements, the lifetimes of free electrons could be determined independently and monitored over 2 years. The lifetimes measured were found to range typically between 1 and 4 µs. The corresponding corrections at 3 kV nominal voltage ranged between 4.5 and 17 %.

The electronics gain was checked regularly in particular for each newly stacked calorimeter configuration. During the running periods, the ADC pedestals were continuously monitored by random triggers not associated with particles in the beam line. Hence, for each particular data set the proper pedestal values were available and have been used in the analysis.

2.2 Test beam and calorimeter configurations

The test beam used was a secondary particle beam of the 10 GeV proton accelerator at ITEP, containing pions, electrons and muons up to 6 GeV. The data for the three different particle types were taken simultaneously and selected off-line according to the trigger pattern.

In the course of the measurements different calorimeter configurations were tested. They are listed in table 1. At the beginning, only a limited number of chambers was available. During these measurements, transition curves were scanned for U and Pb absorbers by introducing one chamber into the individual gaps successively. These setups are named Uran 0 and Lead 0. In the second stage, 11 chambers were available to form the configurations Uran 1 and Lead 1 with a coarse sampling and a small total depth. The final tests were performed with a set of 37 chambers and U, Pb and Fe absorbers. In these configurations, fine sampling in the front part of the electromagnetic section and of coarse sampling in the rear part were chosen. In spite of the fact that such a structure will not lead to an ideal compensation, we took this choice to obtain a reasonable electromagnetic resolution. The last configuration was arranged to study the transition [3] or migration effect [4] in a mixed calorimeter of alternating Fe and Pb absorbers. The chambers were positioned such that the odd chambers were faced by lead and the even ones by iron plates. The signal response of the different subsets of chambers could be studied.

Type	electromagnetic (mm)	hadronic (mm)	total lenght (λo)	chambers
Uran 0	92 x 5		4.6	1
Lead 0	104 x 6		3.7	1
Uran 1	11 x 25		2.8	11
Lead 1	11 x 42		2.9	11
Uran 2	10 x 5	27 x 15	5.2	37
Lead 2	10 x 6	27 x 18	4.6	37
Iron 1	16 x 12	4x22 + 17x25	4.9	37
Iron 2	10 x 24	10x22 + 17x25	5.9	37
Lead/Iron	30 x 6 / 4	7 x 18 / 12	3.8	37

Table 1: Calorimeter configurations.

3. Data

In fig.2 an example of transition curves is shown for the three particle types at p=5 GeV/c when using the Iron 2 calorimeter. The transition curves of electrons and pions were fitted with the function

$$f(t) = c\cdot t^{\alpha}\cdot e^{-\beta t} \tag{1}$$

where t is the depth in radiation length X_0 and absorption length λ_0, respectively. In this way the mean energy depostion could be corrected for rear leakage and missing channels in the detector.

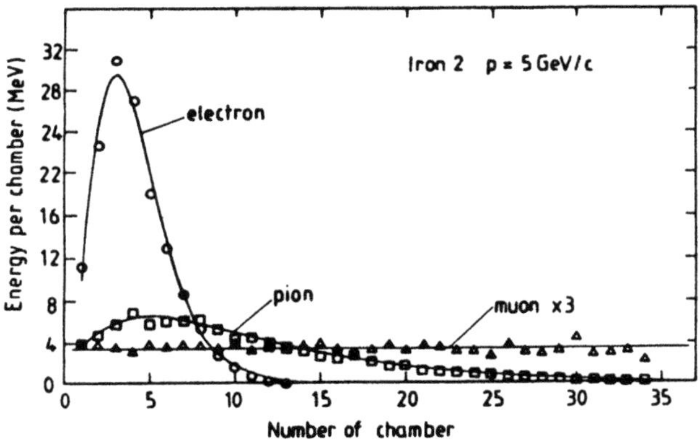

Figure 2: Transition curves of e,π and μ of p = 5GeV/c in the Iron 2 calorimeter. The solid lines represent fits to eq. 1.

3.1 Monte Carlo simulation

To compare our measurements with theoretical expectations, the Monte Carlo code Geant 3.14 was used [5]. For particular questions concerning electromagnetic showers, the data were compared with EGS4 calculations [6] as well. Delta ray production was simulated, and the secondaries were tracked down to 10 keV kinetic energy. The low cuts turned out to be necessary to correctly account for low energy photons and electrons leaving the absorber and producing a signal in the active medium. Hadronic particles were simulated by the Gheisha code [7] as it is implemented in Geant. The chamber geometry with its electrodes and the absorber slabs were taken into account in all details. Also, the non-uniform energy deposit across an ionization chamber gap was implemented in the code.

3.2 Muons

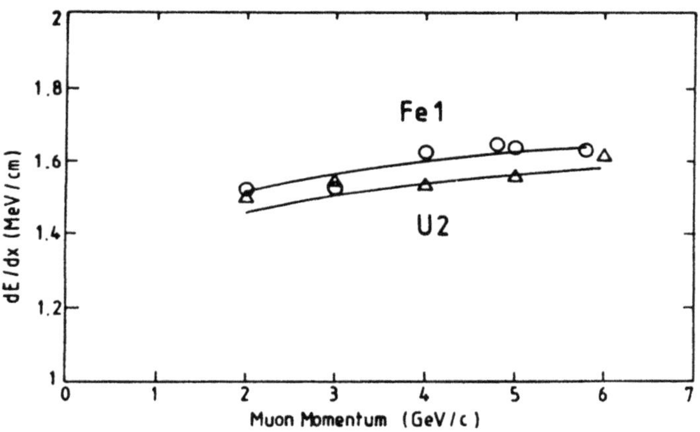

Figure 3: Muon signal in Iron 1 and Uran 2 calorimeters.

How well the chambers have been calibrated is demonstrated in fig. 3, which shows the muon signal of two calorimeter configurations. The solid lines give the theoretical expectations obtained with Geant. The measured values agree with the expected energy deposition within 2%. It can be noticed, that the relativistic rise and even the 3% difference of the muon signal in the two absorber materials are well reproduced by the measurements. We therefore can conclude that the signals are calibrated correctly and that the calorimeter response can be compared directly with the energy deposition predicted by the Monte Carlo simulations.

3.3 Electromagnetic showers

3.3.1 Electron signal

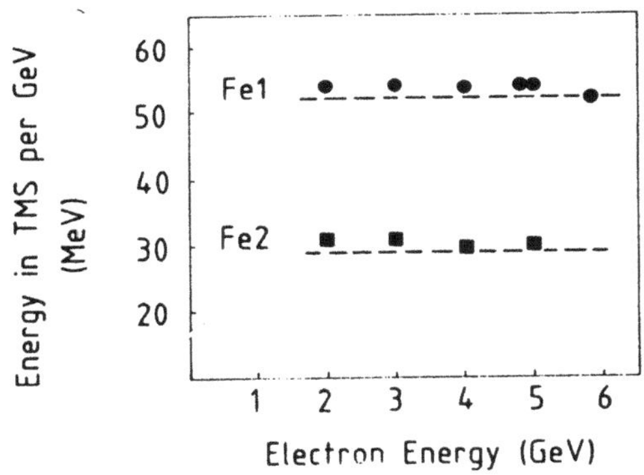

Figure 4:
Electron signal in the two iron calorimeters. The dashed lines represent EGS calculations.

In the analysis, the signals of each layer were weighted with the thickness of the absorber in front of it. The thickness was taken in radiation lengths for electromagnetic showers and in absorption lengths for hadronic showers. The energy measured for electromagnetic showers per 1 GeV primary energy is given in fig. 4, which shows as an example the result of the iron setups. We observe a linearity within 2% in the energy range of 2-6 GeV for all calorimeters. The dashed lines in fig. 4 represent calculations with the EGS code.The mean values agree well with the data within 2 % as demonstrated in the figure. It should be noted that the effect of non-uniform energy depostion in the active medium induced a signal rise of 2 % in the ionization chambers. This effect is caused by low-energy electrons, which deposit their energy in the chamber preferably at the entrance side near the absorber slabs. Electrons are collected on the central anode. Hence, the ionization charge of the low energy electrons influences a higher pulseheight than the charge of particles traversing the chamber gap.

Using the Geant program, we obtain a slightly reduced signal by about 4% compared to EGS4 calculations with the same low cutoff energy for all three absorber materials.

3.3.2 Signal ratio of e / mip

	e/mip	
calorimeters	measurements	M.C.
Iron 1	0.95 ± 0.03	0.92 ± 0.02
Iron 2	0.95 ± 0.03	0.92 ± 0.02
Pb/Fe	0.73 ± 0.05	0.67 ± 0.02
Lead 2	0.63 ± 0.02	0.62 ± 0.02
Uran 2	0.61 ± 0.04	0.60 ± 0.02

Table 2: Signal ratios of e/mip.

For all calorimeter configurations the e/mip ratio was determined. A mip is the signal of a minimal ionizing particle in TMS. In table 2 the results of the measurements and simulations are compared. We observe the distinct decrease of the electron signal with the atomic number Z of the absorber material. The decrease of e/mip with rising Z is caused mainly by the photo-effect of low-energy photons [1,3]. A major part of these photons drops its energy in the passive

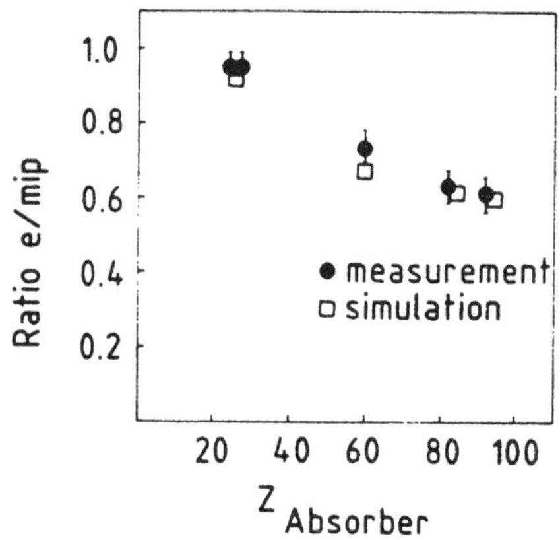

Figure 5: The e/mip ratio versus Z of the absorber material.

absorber. Consequently, less energy is seen in the active layers. It, therefore, has been called migration effect, because the low-energy photons migrate into the absorber material [4]. Simulations and measurements agree well within the errors.

In figure 5, the e/mip ratio is shown versus Z. The mean Z value of the Pb/Fe calorimeter has been determined as a weighted mean of the two materials, the weights being calculated according to the critical energy. A significant decrease from 95% for the two iron calorimeters to 62% for the uranium calorimeter can be noticed. A similar dependence of Z was calculated by Wigmans [1] for calorimeters with scintillators as active elements.

3.3.3 Transition effect

We have investigated, to what extent the surface boundary adjacent to the active layer influences the signal. For this purpose, we have arranged a mixed Pb/Fe calorimeter. The absorber plates were stacked alternately adjacent to odd and even chambers. All odd chambers were embedded in lead and the even chambers were surrounded by iron absorbers. The signal sums in all channels adjacent to lead are compared with those channels surrounded by iron. The signal was found to be higher for shower particles passing from higher Z absorber plates into the ionization chambers. The results are given in table 4 for electron and pion showers together with the Geant simulations. Measurements and simulations agree rather well and show that the transition effect is larger in electromagnetic than in hadronic showers.

	measurements (%)	M.C. (%)
Electrons	4.0 ± 0.7	5.7 ± 2.2
Pions	3.4 ± 0.5	

Table 4: Difference of summed signal of odd (lead embedded) and even (iron embedded) chambers for 5 GeV incident energy.

To make the effect visible, we fitted a shower curve to the mean longitudinal energy deposition according to eq. 1 and subtracted in each layer the value of the fitted curve from the measured data. A characteristic pattern of alternating positive and negative differences appeared as shown in fig.6 for electromagnetic showers.Hadronic showers exhibit a similar behavior. In order to check that no bias influences the analysis, the data of the Iron 2 calorimeter have been treated the same way. Indeed, no effect is observed, the corresponding signal difference amounts to $(-0.2 \pm 1.0)\%$. It should be mentioned that the chamber walls are 1 mm thick and can be penetrated by electrons of a few MeV only. Hence, the observed effect is caused presumably by electrons with an energy in the order of the critical energy, which is 7.2 MeV in lead and 20.5 MeV in iron.

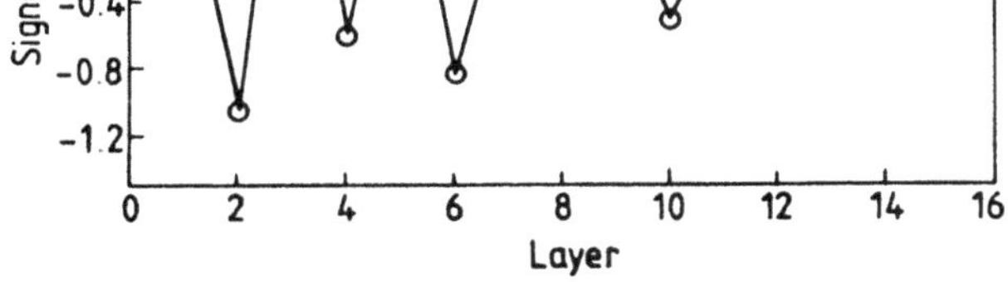

Figure 6: Difference of measured signal and fit of the transition curve for the Pb/Fe calorimeter.

3.4 Hadronic showers

As hadrons negative pions were used throughout the measurements. Our principal aim was to study calorimeter configurations for the three absorber materials for which compensation could be achieved and to determine the corresponding energy resolution.

For future experiments, e.g. in collider or cosmic ray physics [2], it is necessary to rely on calculations up to the 10 TeV energy regime where no test beams are accessible. Therefore, we wanted to check the detector simulation at lower energies as detailed as possible with a view to keep the error of extrapolation small.Of particular interest is the signal response of the TMS ionization chambers to strongly ionizing particles abundantly present in a hadronic cascade.

3.4.1 Signal Yield

The energy seen in TMS per GeV deposited energy by pions is shown in fig. 7 as an example for three calorimeters together with simulations using the Geant code. A reasonable good agreement can be noticed.

However, two remarks have to be made. Firstly, the signal for pions sensitively depends on the signal saturation for non-relativistic hadronic particles, e.g. recoil protons from neutron scattering. For scintillators, this signal damping is usually parametrized by Birks' law[8].

$$S = S_0/(1 + k_B \cdot dE/dx) \qquad (2)$$

Ionization chambers exhibit a different and more complicated dependence[9]. The signal changes with the angle between the particle track and the electric field. In addition, the mean ionization loss dE/dx has to appear with an exponent less than unity to describe the data if Birks' parametrization is used. We used a parametrization following the data given in ref. 9 and extrapolated it smoothly to the full solid angle and to all values of dE/dx needed in the M.C. With this law of signal saturation implemented in the Geant code, the simulations were performed.

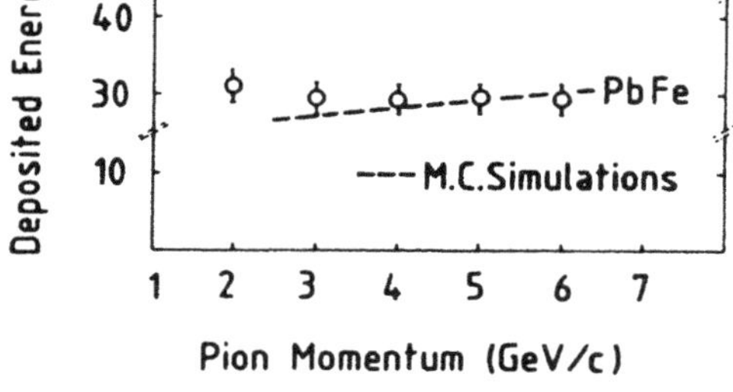

Figure 7: Energy seen in TMS per GeV deposited energy by pions.

As obvious from fig. 7 the signal damping as incorporated reproduces the calorimeter signals satisfactorily, at least for a particle energy of 5 and 6 GeV. We have looked whether the simple ansatz of Birks' law could be used and found that $k_B = 0.2$ gcm-2/MeV reproduces the measured yields equally well.

The second remark concerns signal linearity. The data of fig. 7 show linearity between 2 and 6 GeV within a few per cent. Geant, however, predicts smaller yields of about 10 % at lower energies around 2 and 3 GeV, a feature which is not present in the data. On the contrary, a small signal rise would be expected due to the fact that at lower energies the chance for a pion augments to lose its energy predominantly by ionization without undergoing nuclear interactions. The mean rise of deposited energy in our measurements from 6 to 2 GeV is $\Delta E/E = (3.4 \pm 1.0)\%$. This is in good agreement the signal increase measured by the ZEUS group [10] for π^-, which amounts to $\Delta E/E = 2.5\%$ in the same energy interval. For π^+ and p the authors give a larger increase, namely 7.5 %. This is plausible in view of the lower inelastic cross-section for π^+ and protons compared to π^- in the corresponding energy region.

3.4.2 Density Variation

In a first series of measurements, the transition curves for the Uran 0 and the Lead 0 setups were recorded to study the influence of the mean calorimeter density on compensation. This density could be varied by changing the thickness of the air gap between the absorbers. A smaller mean calorimeter density allows more secondary particles, especially in a hadron shower, to leave the calorimeter or a corresponding fiducial volume in a large calorimeter. For neutron detection the hydrogen content of the active medium is of vital importance. Therefore, during some of the measurements, 4 mm thick plexiglas sheets were introduced into the empty gaps to simulate the missing TMS for the neutrons in the shower. Their thickness was chosen to be equivalent to the H and C content in a TMS layer. The result for uranium is shown in fig. 8. Here the signals of e^- and π^- are plotted for three values of a mean calorimeter

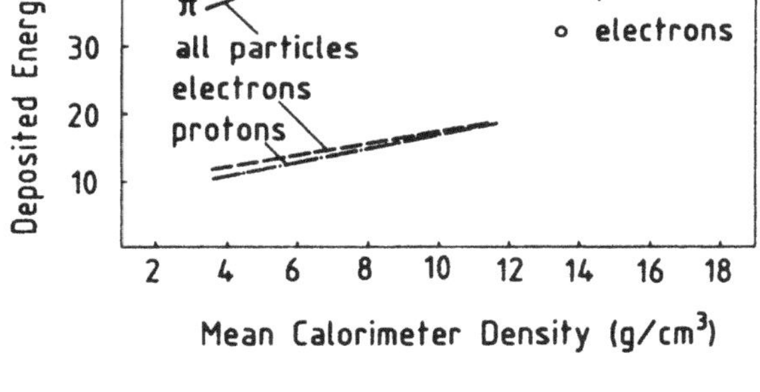

Figure 8: Electron and pion signals versus mean density in an uranium calorimeter. The horizontal solid line gives the simulated response for electrons, the inclined one represents the pion response. The dashed and dot-dashed lines indicate the signal contribution by electrons and protons in a pion shower.

density. One notices that the pion signal varies between 45 MeV and 65 MeV of deposited energy, whereas the electron signal remains practically constant. In these measurements, no compensating sheets of plexiglas were used. The straight line of the Geant simulation is obtained for a configuration with TMS in all gaps. Consequently, a higher signal than predicted by the Monte Carlo method is observed, but the slope is well the same. The simulations, however, coincide very well with the value measured at a density of 10.6 g/cm^3, for which plexiglas had been introduced into the empty gaps (black point).

From these measurements it can be concluded how strongly the pion signal depends on the packing density. Of course it also depends on the thickness ratio of the passive to the active material. The U plates correspond to 9.48 g/cm^2 and the TMS to 0.49 g/cm^2. From fig. 8 it is obvious that compensation would be obtained for this configuration at a mean structure density of 12g/cm^3.

Using the M.C. method it was investigated to what extent the different types of particles present in a cascade are responsible for the signal loss in case of a less densly packed calorimeter. Only the signals induced by secondary electrons and protons show a dependence on density and both equally contribute to the obeserved effect. This is shown in fig. 8 by the dashed and dot-dashed lines. We infer from it that the energy leaking out of the calorimeter or the fiducial volume is escaping in an equal amount as photons and neutrons. The calculations show that the energies of the leaking particles are in the MeV range. This leads to a deficit in low-energy electrons and recoil protons as observed. In case of Pb compensation is achieved at a densitiy of 11 g/cm^3 only. This is an unphysical value in the sense that no space is left for active material to be introduced. In fact, lead compensates only at a much coarser sampling as will be shown below. It is interesting to note that the M.C. simulation corroborates what one would expect, namely, that in the case of Pb photon losses contribute more to the leaking energy than neutron losses.

3.4.3 Sampling thickness

For uranium and lead, we have stacked three distinct calorimeter configurations each with a different sampling step. Thus, the dependence of the e/π ratio on the sampling step can be studied. The results are shown in fig. 9 together with the corresponding simulations. The experimental e/π ratios are plotted versus the thickness of the absorber plates used.

A certain complication arises for the two calorimeters with different sampling steps in the electromagnetic and the hadronic section. The question of which is the equivalent sampling step in this regard has been treated as follows: Monte Carlo simulations were used to simulate the energy deposition ΔE for electrons and pions for a variable sampling thickness d. The product $\Delta E \cdot d$ was plotted versus d, and a linear behavior was observed in the region of $0.3\text{cm} < d < 3\text{cm}$.

The slopes for the electron and pion showers show how strongly they contribute to the e/π ratio. Consequently, we used the slopes as proper weights to determine the mean effective sampling thickness.

In fig. 9 the e/π ratios are plotted accordingly at the corresponding abscissa. The two curves represent Geant simulations with TMS as active layers of 7.5 and 2.5 mm. The TMS thickness has practically no influence on e/π and the measurements are reasonably well reproduced. Thus, we are able to deduce the proper sampling step at which compensation will be achieved. For uranium

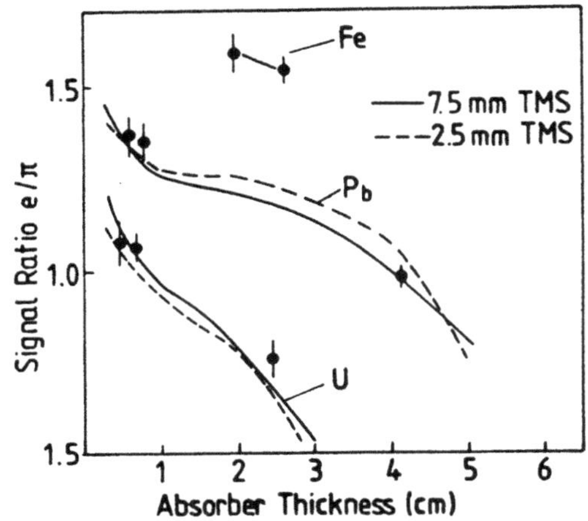

Figure 9: e/π ratio versus different absorber thickness. The curves represent Geant calculations.

this is the case at 8 mm, with the uncertainty being about 1mm. For lead one extrapolates a very coarse sampling of 40 mm, which compensates, but, of course, gives a bad energy resolution. At very high energies, of several hundred GeV, however, even such a coarse sampling could give reasonable resolutions.

Before concluding this paragraph, a remark concerning the two iron calorimeters should be added. Certainly, it is only of academic interest to ask where compensation in an iron calorimeter could be achieved. Nevertheless, the two e/π ratios are plotted in fig. 9, and if one dares to extrapolate a sampling thickness would be obtained which is 12 to 15 cm in good agreement with the value quoted by Wigmans [1]. In the TeV region such a compensating calorimeter would exhibit an energy resolution of a few per cent. In fact, 12 cm iron slabs are used as absorbers in a calorimeter under construction at Karlsruhe for cosmic ray hadrons up to 10 TeV.

3.4.4 Energy resolution

We have determined energy resolutions for the calorimeters with 37 chambers by fitting a Gaussian to the signal distributions. The pedestal fluctuations due to the amplifier noise were subtracted quadratically for each particular experimental data set. As an example the resolutions obtained for the calorimeter Uran 2 are presented in fig. 10 as a function of the momentum p of the incoming particles. Plotted is the standard deviation multiplied by $\sqrt{E}$. It can be noticed that the fluctuations scale reasonably with $\sqrt{E}$, and that they follow the prediction of the M.C. simulation.

Due to the fact that we have read out each chamber individually, resolutions for coarser samplings can be calculated easily using the signal of each second or third chamber only. In this way sampling fluctuations can be separated from the intrinsic fluctuations as was done by Fabjan et al. [11] and discussed in detail by Drews et al [12]. Doubling the sampling step was carried out by summing up the signals of the odd-number and the even-number chambers

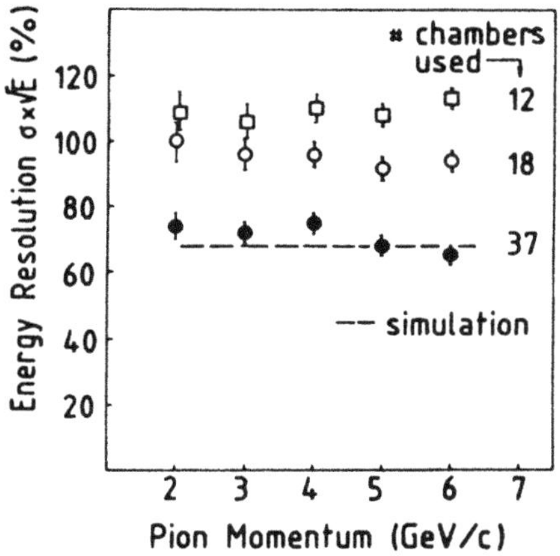

Figure 10: Energy resolution for Uran 2 calorimter versus particle energy. The open symbols give the resolutions when only the signals in each 2nd or 3rd chamber are used. The dashed line represents Geant calculations.

separately and filling it into a common histogram from which the standard deviation was determined. Trebling the sampling step was done in a similar way by histogramming the three signal sums of each third chamber. The resolutions obtained for the Uran 2 calorimeter are shown by the open marks in fig. 10.

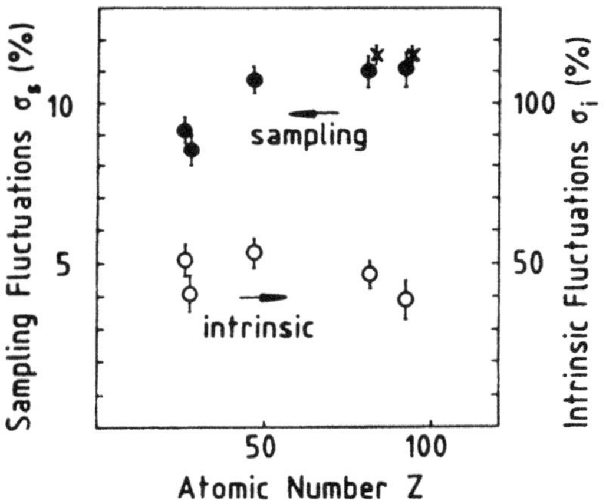

Figure 11: Intrinsic and sampling fluctuation versus atomic number. The dots show this measurment, the crosses give results obtained by Andresen et al. [10].

One notices that no significant deviation from √E scaling can be observed between 2 and 6 GeV. From these values, both the sampling and intrinsic fluctuations σ_s and σ_i are calculated using the ansatz proposed by Fabjan [13]:

$$\sigma_t^2 = \sigma_s^2 + \sigma_i^2 \cdot \Delta E, \tag{3}$$

where ΔE in MeV is the energy deposited by a minimum ionizing particle in a sampling step, i.e. in the active and passive part. The values of σ_s and σ_i obtained this way are shown in fig. 11 versus the mean atomic number Z of the absorber material. The sampling fluctuations for the two iron calorimeters amount to 9 %.

This agrees well with the value derived by Fabian [13] from a compilation of various calorimeter data. For Pb and U, however, higher sampling fluctuations of 11.0% and 11.1% are obtained, and also the mixed Fe/Pb calorimeter gives a large sampling fluctuation of 10.7%. This findings are in accordance with the results of the ZEUS calorimeter group [10], who has published the sampling fluctuations in Pb and U to be $\sigma_S = 11.5\%$ (crosses in fig.11). We can therefore conclude, that the sampling fluctuations rise slightly with the atomic number Z of the absorbing material. The intrinsic fluctuations amount to 50% with an indication of lower values for lead and uranium. Our value of $\sigma_i = 39\%$ for U is higher than the 20% published in ref. 12,13, most probably due to the fact that it has been obtained in a non compensating configuration.

References

1. R. Wigmans, *Nucl. Instr. and Meth.* **A259** (1987) 389.
2. P.Doll et al., The Karlsruhe Cosmic Ray Project Kascade, *KfK-Report* **4686**, (June 1990).
3. W.Flauger, *Nucl. Instr. and Method.* **A 241** (1985) 72.
4. H.Brückmann et al., *DESY-Report* **87-064**, Hamburg, July 1987.
5. R.Brun et al., *CERN-Report* **DD/EE/84-1** (1987).
6. W.R.Nelson et al., *SLAC-Report*-**265** (1985).
7. H.Fesefeldt, RWTH Aachen, *PITHA* **85/02**.
8. J.B.Birks, Theory and Practice of Scintillation Counting, *Pergamon Press* (1964).
9. J. Engler et al., Signal damping of strongly ionizing particles in liquid ionization chambers, to be published in *Nucl. Instr. and Meth.*
10. A. Andresen et al., The ZEUS calorimeter group, *Nucl. Instr. and Meth.* **A290** (1990) 95.
11. C.W. Fabjan et al., *Nucl. Instr. and Meth.* **A141** (1977) 61.
12. G. Drews et al., *Nucl. Instr. and Meth.* **A 290** (1990) 335.
13. C.W. Fabjan, *CERN-EP*/**85-54** (1985).

LARGE SCALE APPLICATIONS OF SILICON DETECTORS IN THE COMPACT MUON SOLENOID (CMS) FOR LHC

E. Radermacher
CERN, Geneva, Switzerland

1. INTRODUCTION

Any detector for the LHC collider should have the capability of triggering on and measuring with sufficient precision muons, electrons, photons, high energy jets and missing energy at a luminosity as high as 2.10^{34} cm^{-2}s^{-1}. For CMS obviously the muon detection has first been optimized choosing a 15 m long superconducting solenoid with 2.9 m inner coil radius and a magnetic field of 4 Tesla in the central gap. Such a magnet gives a muon resolution ranging from 1% at 50 GeV/c to 6% at 3 TeV/c in the central rapidity region using the inner tracking. For the pattern recognition in the inner tracker one aims at $\Delta p_t / p_t \approx 0.1\, p_t$ [TeV] for $|\eta| \leq 2.5$ using four layers of Si microstrip detectors at 25 to 30 cm from the interaction point. Another high priority for CMS is the H -> $\gamma\gamma$ detection implying an electromagnetic energy resolution of $\sigma(E)/E \approx 3\%/\sqrt{E} + 0.5\%$ using a crystal calorimeter. For the hadronic calorimeter an energy resolution of $\leq 80\% / \sqrt{E}$ is sufficient since priority is given to the missing energy resolution, i. e. to the hermeticity or density and compactness of the detector. A perspective view of CMS is given in Fig. 1.

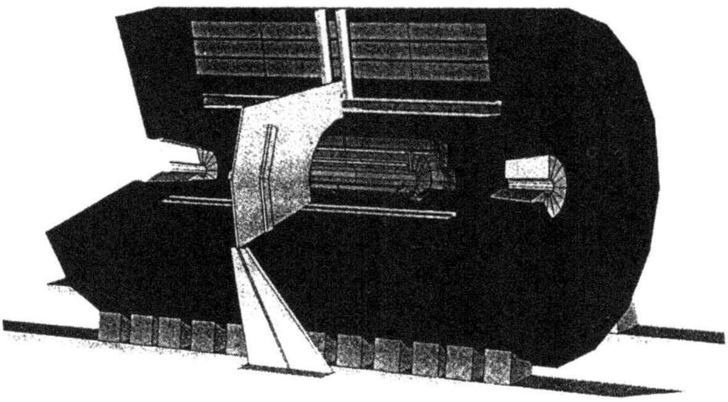

Fig. 1 : Perspective view of the CMS detector.

2. USE OF SILICON IN FUTURE DETECTORS

There are several good reasons for which calorimeters based on Silicon as detecting medium are of considerable interest for an LHC detector, namely the insensitivity to the magnetic field, the high speed of charge collection (10 ns), the stability of the response (1%), the high signal/noise ratio for muons, the long reliability, the ease of calibration, the possibility of fine transverse and longitudinal segmentation and the possibility to obtain maximum hermeticity. In the case of Si microstrip detectors the high coordinate accuracy has to be mentionned. Furthermore, it is actually believed that Silicon is radiation hard up to 10^{13} n / cm^2. Needless to say that the biggest disadvantage of Silicon remains its cost. In the past several Silicon microstrip detectors[1] and calorimeter prototypes[2] have been constructed and studied.

3. USE OF SILICON IN CMS

In CMS Silicon detectors are planned to be used at different places. As already mentionned a inner Si microstrip tracker with a total area of 11 m^2 should be installed around the interaction region. In the crystal electromagnetic calorimeter a Silicon position detector should be incorporated after $\approx 5\, X_0$ having pads of 31 x 2 mm^2 size and a total surface of $\approx$ 140 m^2. Such a detector provides, together with the energy deposition in the crystals an angular resolution of ≈ 7 mrad. In addition let us mention that ≈ 170 m^2 of Silicon photodiodes are needed for the read out of the crystals. Nevertheless the main large scale application of Silicon will be in the hadronic calorimeter. In fact in order to keep CMS dense and compact (the Si layer can be ≈ 3.5 mm) by optimizing the iron of the magnet and the return yoke an application of Silicon is highly wellcome. The available space for the hadronic calorimeter is 1 m starting at a radius of 1.9 m. Choosing for instance a sampling of [4 x (3 mm Pb / 9 mm Fe] per Si read out plane the hadronic resolution will be ≈ 80% / $\sqrt{E}$. The number of interaction length would be ≈ 5.4 in the hadronic part, which, combined with the electromagnetic calorimeter, results in a total of ≈ 6.5 λ_{int}. With 20 layers of detecting medium the total Silicon surface reaches ≈ 3000 m^2 for $|\eta| \leq 2.5$. Due to production yield reasons the pad size will be 2 x 2 cm^2, a surface granularity which is much too fine, therefore allowing a grouping of several pads resulting in a reduction of the read out channels.

4. RESULTS FROM IRRADIATION TESTS ON SILICON

Several workshops have been organized in the past on "Radiation Damage and Large Scale Applications of Silicon Detectors ". A recent overview on radiation damage can be found in ref. 3. The general agreement is that the radiation damage does not seem to be a fundamental limitation for large scale application of Silicon at future high energy colliders. In fact the depletion voltage is still below 300 V for neutron fluences of 10^{14} n / cm^2. The reverse current under neutron radiation shows a linear increase of the order of 10 $\mu A/cm^2$ up to a neutron flux of some 10^{13} n / cm^2 at 20 ^0C for diodes of 300 μm thickness. At fluences of $\geq 2.10^{13}$ n / cm^2 the n-type bulk inverts to apparent p-type bulk depending on the initial doping concentration. Both effects seem to have little influence on the operation of Silicon pads as particle detectors. In fact the time development of the signal current pulse is rather not affected up to fluences of 5.10^{13} n / cm^2 and the collection deficit of charges is ≈ 2.5% per 10^{13} n / cm^2 up to 5.10^{13} n / cm^2. In CMS one expects in the hadron calorimeter neutron fluences < 10^{14} n / cm^2 for $|\eta| \leq 4$.

5. FUTURE PRODUCTION OF SI - WAFERS

A 4" Silicon ingot delivers 12 pads of 2 x 2 cm^2 from a 400 μm slice. This gives 2500 cm^2/kg. A calorimeter with ≈ 3000 m^2 detector plane requires 12 tons of Silicon. Although the price of Silicon approaches today 1 Sfrs/cm^2 the pad production itself is the most expensive part, namely 80 % of the total price. The Joint Institute for Nuclear Research in Dubna together with the Russian industry proposes now a pad production price 8 - 10 times lower than the actual price[4]. This offer may allow now the construction of a large Silicon hadronic calorimeter for the future colliders. In order to prove this offer a proposal for the construction of a prototype with 10 m^2 detector plane will be submitted to the DRDC board at CERN.

6. CONCLUSION

From the present knowledge of Silicon calorimetry one can conclude that this kind of detector is very adequate for the LHC since a compact detector insensitive to the magnetic field and practically not affected by radiation can be achieved. However the price of the silicon pads has to reach an acceptable level, otherwise financial reasons may not yet permit such a construction for the next generation of accelerators.

7. REFERENCES

[1] See for example D. H. Saxon, *Proc. Large Hadron Collider Workshop, Aachen, 1990*, eds. G. Jarlskog and D. Rein, **CERN 90-10**, vol. 1 (1990) 339 and references quoted therein.

[2] E. Borchi et al., *Proc. Int. Conf. Advanced Technology and Particle Physics, Como, 1990*, in *Nucl. Phys. B (Proc. Suppl)* **23A** (1991) 62.
A.L.S. Angelis et al., **CERN - PPE / 91-70**, 19 April 1991.

[3] G. Hall, *Proc. Large Hadron Collider Workshop, Aachen, 1990*, eds. G. Jarlskog and D. Rein, **CERN 90-10**, vol. 3 (1990) 693.
R. Wunstorf et al., *Proc. Large Hadron Collider Workshop, Aachen, 1990*, eds. G. Jarlskog and D. Rein, **CERN 90-10**, vol. 3 (1990) 706.
M. Glaser, F. Lemeilleur and E. Occelli, **CERN / ECP 91-16**, 16 September 1991.

[4] I. Golutvin, *Workshop on "Radiation Damage and Large Scale Applications of Silicon Detectors*, Firenze, July 23-24, 1991, unpublished.

THE ICARUS LIQUID ARGON TPC: A NEW DETECTOR FOR ν_τ SEARCH

P.BENETTI, E.CALLIGARICH, R.DOLFINI, A.GIGLI BERZOLARI, F.MAURI, C.MONTANARI, A.PIAZZOLI, A.RAPPOLDI, G.L.RASELLI, D.SCANNICCHIO
Dipartimento di Fisica e INFN, Università di Pavia, via Bassi 6, Pavia, Italy

A.BETTINI, F.CASAGRANDE, P.CASOLI, S.CENTRO, B.DAINESE, C.DE VECCHI, G.MURATORI, A.PEPATO, F.PIETROPAOLO, P.ROSSI, S.SUZUKI, S.VENTURA
Dipartimento di Fisica e INFN, Università di Padova, via Marzolo 8, Padova, Italy

F. CAVANNA, G.PIANO MORTARI
Dipartimento di Fisica e INFN, Università dell'Aquila, via Vetoio, Coppito (AQ), Italy

P.CENNINI, S.CITTOLIN, C.RUBBIA
CERN, CH-1211, Geneva 23, Switzerland

M.CHENG, D.CLINE, M.ZHOU
Department of Physics, UCLA, Los Angeles, CA 90024, USA

F.GASPARINI
Dipartimento di Fisica e INFN, Università di Udine, via Larga 36, Udine, Italy

L.FORTSON, P.PICCHI, H.WANG
Laboratori Nazionali di Frascati dell'INFN, via E.Fermi 40, Frascati (Roma), Italy

L.PERIALE
Istituto di Cosmo Geofisica del CNR di Torino, cso. Fiume 4, Torino, Italy

ABSTRACT

We describe the characteristics of a novel detector technology which we are successfully testing at CERN: the LAr three-dimensional time projection chamber. We report on the detector performance using data from cosmic rays and a 6 MeV gamma ray source. We give a few examples within the wide range of experimental issues where such a technology can be applied with an emphasis on ν_τ detection.

1. Introduction

A liquid argon time projection chamber (LAr-TPC) working as an electronic bubble-chamber with the ability to provide 3-D imaging of any ionizing event together with a good calorimetric response while continually sensitive and self-triggering was first proposed by C. Rubbia in 1977 [1]. In the following years, experiments were undertaken to verify the feasibility of such a detector and to solve the main technological problems; namely: the purification of liquid argon to the level of 0.1 ppb to allow for long drift

---

Presented by F. Pietropaolo

distances by the ionization electrons; the realization of a nondestructive read-out (3-D imaging) made of several wire planes with a few mm pitch; the development of very low noise preamplifiers since no multiplication occurs in LAr. All these problems had been successfully resolved on small scale tests [2,3,4] by 1989 when the ICARUS collaboration presented a proposal [5] for the construction of a large scale prototype. At present, a complete, 3 ton detector has been working at CERN under stable conditions for several months and its characteristics are described in a previous paper [6].

2. Detector response

The experience with the 3 ton prototype, equipped with complex mechanical and electrical apparata immersed in the liquid and with hundreds of feed-throughs, has shown that the ultrapure liquid argon technique is fully reliable even for large volumes since, after several months of continual operation, no degradation of the very high electron lifetime has been observed.

This detector provides electronic bubble-chamber quality images with millimeter sized "bubbles". But, unlike a bubble-chamber, it can be built with a large, continually sensitive volume and due to an electronic readout, it is self-triggering. The spatial resolution is in the range of 100 μm while an energy resolution of ≈3 % at a few MeV has been indirectly estimated. Ionization and range measurements are used to provide particle identification and the high granularity enables a measurement of particle direction. All these properties make the LAr-TPC a superb homogeneous detector for contained events and for vertex identification. In fact, the detector is essentially bias free and it can detect a very broad class of events: i.e. proton decay and solar neutrinos at Gran Sasso, CP violating interference at the ϕ-factory, ν_τ oscillations in a ν_μ beam or direct detection of the ν_τ at LHC and, like in a bubble-chamber, all kinds of unexpected phenomena.

Cosmic ray muons and 6 MeV monochromatic gamma source events comprise the data collected and are used to study the detector response. An event that illustrates very well the specific characteristics of the detector is presented in fig. 1: a 210 MeV cosmic muon stopping with the subsequent decay to an electron. It shows that the LAr-TPC works as an electromagnetic calorimeter with high granularity (2·2·2 mm^3 cell) and low electronic noise (equivalent to 25 KeV). For example, this detector allows to measure the dE/dx along the muon track (fig. 2) with the increase of ionization near the decay as well as the exact point of the decay and the track of the electron. Also due to the high granularity, the muon lifetime can be determined by measuring the gap along the drift direction between the muon stopping point and the electron starting point.

3. Applications

We believe that the LAr-TPC detector technology is now available for physics both in underground laboratories and at accelerators/colliders. Among the experimental issues mentioned above, we would like to stress the possibility of using such a detector for a ν_τ search in an oscillation experiment at a neutrino beam or for direct search at LHC.

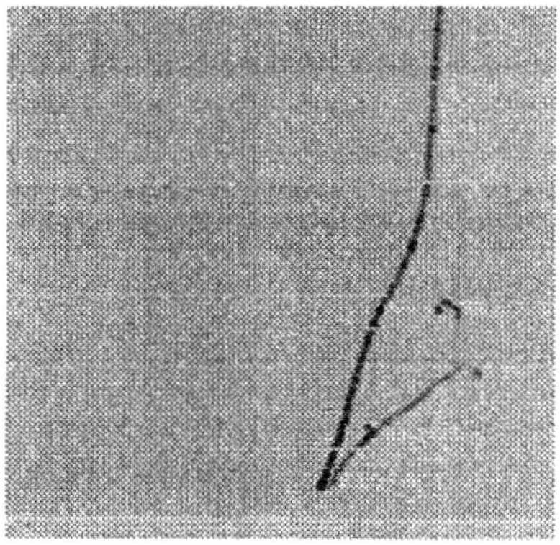

Fig. 1 Muon stopping and successive electron decay as seen by the collection wire plane in a window of 40•40 cm^2. Increasing gray intensity is proportional to the energy deposited on each wire. The total energy deposited along the muon track is 210 MeV, the electron energy is 21 MeV.

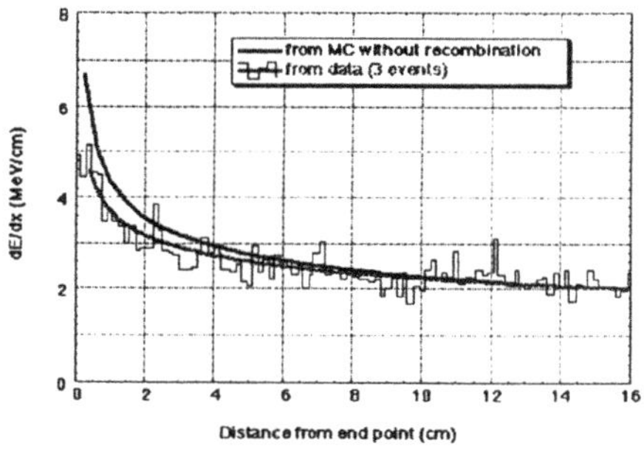

Fig. 2 The dE/dx vs track path of the muon together with the expected behaviour from MC; a saturation effect is visible.

Up to now, the conceptual design for tau neutrino detection aims at detecting the tau decay modes with single charged particles (mainly muons). The identification of a tau decay vertex requires a visible transverse impact parameter (kink) of one particle track with respect to the event vertex and/or a careful analysis of the event kinematics. The transverse decay length is almost invariant with respect to the tau neutrino energy ($\approx$ 100 μm), hence the vertex detector should have a space resolution of about 20 μm. With the various techniques proposed up to now to obtain the required resolution it is possible to assemble only small volumes.

With this type of detector, it should be possible to detect ν_τ charged current interactions by identifying the tau lepton through its decay into different channels. Considering the CERN wide band neutrino beam energy spectrum, the mean tau decay length is about 1 mm leading to events with an average multiplicity of five charged particles. When the decay path is required to be longer than 1 cm, the fraction of tau events selected would be reduced to 2-3 %. However, this reduction in rate is offset by the gain in ease of selection as the events are very clean (low residual multiplicity) because the tau lepton receives most of the neutrino energy.

The tau decay mode into 3 charged pions (BR ≈ 13%) is the easiest to identify using a liquid argon TPC with the following characteristics: large volume (≈ 100 tons); 2 mm pitch of the read-out electrodes (wires or strips); three-dimensional read-out; high resolution in the drift direction (≈ 60 μm, useful for the track separation); energy resolution for the charge deposited over 2 mm by a MIP about 3 % (mainly due to electronic noise and delta-ray production). The basic idea consists of exploiting the last characteristic for the recognition of the tau track inside the event by means of the sudden increase in the total number of charged particles at a finite distance from the main vertex.

In fig. 3 we show a typical event generated by MonteCarlo as it is seen by the LAr TPC. In fig. 4 we show the jump in the ionization along the event development where we have excluded the large angle low energy tracks. This jump corresponds to the ionization due to two minimum ionizing particles.

Fig.3 Simulation of a ν_τ CC interaction followed by τ decaying in $3\pi^\pm$.

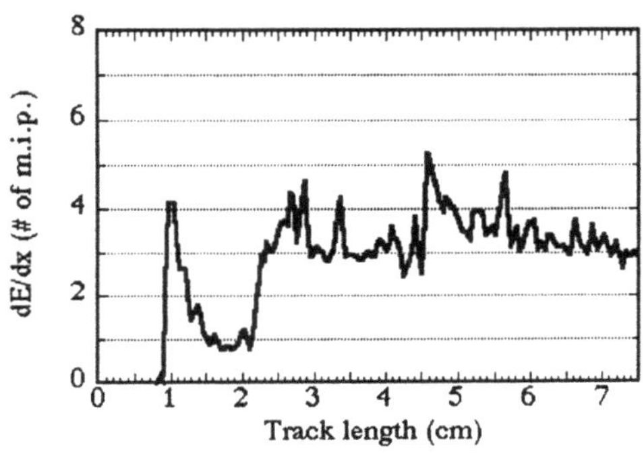

Fig.4 Detection technique consists in identifying the secondary vertex by the sudden increase in the dE/dx (+2 minimum ionizing particles) exiting the main vertex.

With the aim to measure values of $\sin^2(2\theta_{\mu\tau})$ as low as $4 \cdot 10^{-4}$, the tau recognition efficiency mentioned above implies that the experiment should collect about 10^7 ν_μ events corresponding to one year of data taking. This number can be reduced by at least a factor of 10 without reduction in the tau efficiency by exploiting the fact that muons are easily identified (the detection inefficiency is $< 10^{-3}$ as the muons can be tracked up to the point where they escape the detector) as well as using a trigger based on the energy deposition inside the detector. The former requirement eliminates most of the CC interactions while the latter cuts out the low energy NC events.

The most dangerous background for this specific channel is a secondary K^0_s decaying into two charged pions at a distance of 1 to 3 cm from the main vertex superimposed on a charged track. By means of kinematic cuts and exploiting the accurate spatial resolution along the drift coordinate to separate tracks which are close together, we are confident that a rejection factor better than 10^{-6} is easily reachable. Charmed particles are a less dangerous background because the total contribution from charm events which can simulate the signal process is between 10^{-6}-10^{-7}. This is the result of combining the following factors: c-$\bar{c}$ production in NC events is $1.5 \cdot 10^{-3}$, the branching ratio of the charm particle into MIPs which can simulate the jump in ionization level is $\approx 10^{-1}$, and finally the probability of decaying after 1 cm from the interaction vertex is $<10^{-2}$. Background from single charm production is negligible due to the very high probability of detecting the decay muon.

The direct search for ν_τ with the LAr-TPC at LHC is a more well defined experiment than the one just described for neutrino oscillation because the average tau decay length integrated over the neutrino energy spectrum is about 1 cm and because the tau neutrino abundance directly produced at the LHC interaction point is about 3% of the total neutrino flux It follows that the tau detection efficiency in the 3 charged pions channel can be as high as 7 % and the background rejection is easier to handle.

4. References

1. C. Rubbia, *CERN-EP Internal Report* 77-8 (1977).
2. E. Buckley et al., *Nucl. Instr. and Meth.* **A275** (1989) 364.
3. E. Bonetti et al., *Nucl. Instr. and Meth.* **A286** (1990) 135.
4. A. Bettini et al., *Nucl. Instr. and Meth.* **A305** (1991) 177.
5. ICARUS collaboration, *ICARUS I: An Optimized Real Time Detector of Solar Neutrinos* LNF-89/005(R) (1989).
6. A. Bettini et al., *The ICARUS Liquid Argon TPC: a complete imaging device for particle physics* LNGS-91/16 (1991).

HOMOGENEOUS CALORIMETERS AT LHC/SSC

Summary talk

P. LECOQ
CERN / PPE, Geneva, Switzerland

ABSTRACT
Very good resolution for electromagnetic calorimetry can only be obtained with homogeneous calorimeters. The importance of having such a calorimeter at LHC/SSC, together with the conditions which will guarantee the ultimate precision of the measurements are discussed here. The two possible approaches noble liquids and inorganic scintillators, are described and the latest results are presented.

1. Introduction

The International High Energy Physics community is on its way to make a major investment of the order of 15 billion dollars for 2 accelerators and 4 large experiments to explore a new domain of physics. One expected outcome of this effort is the discovery of the standart model Higgs (SM) or the minimum supersymetric standart model Higgses (MSSM). Outside of the Standart Model and its simple extensions, the physics discovery potential should be maintained at a very high level for the detection of possible new phenomena in a completely new energy range, at the 1 TeV scale.

It is well recognized that homogeneous calorimeters have very good performances in terms of energy and position resolution.The need for having at least one detector of this kind at LHC and SSC is stressed in the second section and ref[1]. Section 3 describes the conditions to be met to guarantee the best possible resolution, and particularly the smallest constant term[1,2]. Sections 4 and 5 describe the two possible approaches for homogeneous calorimetry, namely the noble liquids (liquid Xenon and Krypton)[3,4] and the inorganic scintillators, crystals and glasses[5,6,7].

2. Need for high resolution calorimetry at LHC/SSC

Several studies have already shown that a Standart Model Higgs having a mass in the range 80 to 130 GeV/c^2 can only be detected through its $\gamma\gamma$ decay mode[8,9,2]. The direct production cross-section, multiplied by the branching ratio into 2γ, yield event rates of the order of several thousands in one LHC year. But the backgrounds are enormous, dominated by a combination of π° inside jets, and the so called "irreducible background", with the direct production of 2 photons via the $q\bar{q}$, gg and higher order diagrams. The width of the Higgs in this mass range being extremely small ($\approx$ 10 MeV), the signal to background ratio (S/B) will be entirely dominated by the calorimeter energy resolution.

For the associated production

$$pp \rightarrow WH^\circ \rightarrow \gamma\gamma + l\nu \qquad \text{and} \qquad pp \rightarrow t\bar{t}H^\circ \rightarrow \gamma\gamma + l\nu$$

where the 2 photons can be tagged by a high energy electron or muon, the production rate could reach a few tens of events per year at both LHC and SSC. The background, dominated by the process

$$pp \rightarrow t\bar{t}\gamma\gamma$$

although much smaller than for the direct production, can be of the same order as the signal[10] for a mass resolution of 3%.

As pointed out by Kunszt and Zwirner, there is also a possibility to detect the MSSM neutral Higgses (h° and H°), at least in a limited region of the parameter space via their $\gamma\gamma$ decay modes[11]. The rates are at best equal to those of the SM Higgs, but have a good chance to be smaller, probably only accessible to a detector where the performance has been pushed to the ultimate limit.

The discovery potential of a new particle is expressed in terms of significance ($S/\sqrt{B}$). A very interesting work has been made in plotting the significance for the SM $H^\circ \rightarrow \gamma\gamma$ at 100 GeV/c^2 as a function of the calorimeter performance[12]. The energy resolution of an electromagnetic calorimeter is commonly parametrized in the following way:

$$\sigma(E)/E = a_0/E \oplus a/\sqrt{E} \oplus b$$

where a_0 is the noise term, generally negligible above a few GeV
a is the statistical term (sampling or fluctuations of all sorts)
b is the constant term

The significance is plotted in Fig 1 as a function of a and b for a calorimeter at 2 meter distance from the vertex. The following remarks can be made:

- with such a high background, a significance of less than 5 gives no hope to see the signal.
- realistic conditions correspond to the case where the pile-up in one bunch crossing at LHC has been added, and where some kind of vertex reconstruction with a precision of about 1 cm can be achieved. The latter condition, which is mandatory for calorimeters at relatively small radius, can be somehow relaxed, if one can go at distances of 3 meters or above.
- it is obviously very important to keep the constant term as low as possible if the energy dependant term "a" is as small as 2%. In that case one looses almost 2 degrees in significance by going from b = 0.5% to b = 1%
- if the energy dependant term "a" is of the order of 10% , as can be expected from the best sampling calorimeters, there is no much need to have a very small constant term, as the significance is obviously dominated by the"a" term.
- there is a gain in significance of 1.3 for "a" going from 10% to 2% if the "b" term is 1%. This gain in significance is as large as 2.7, if the "b" term is 0.5%.

It is therefore very important to have both the energy dependant term "a" and the constant term "b" as low as possible and to achieve an energy resolution of typically:

$$\sigma(E)/E = 2\%/\sqrt{E} \oplus 0.5\%$$

3. Conditions to obtain the best possible performances

An homogeneous calorimeter is not limited by sampling fluctuations and an energy dependant term of the resolution "a" as small as 2% has been currently achieved on several large size calorimeters. It is much more difficult to achieve a constant term "b" of ≈ 0.5%. On large systems like the L3 BGO calorimeter, one recognizes usually 3 contributions to the constant term[13]:

$$b^2 = b_L^2 \oplus b_F^2 \oplus b_C^2$$

The b_L term represents the fluctuations due to the energy leakage. The front leakage due to backscattered particles has a negligible contribution above a few GeV. The rear leakage can be easily controlled by having sufficient material for a full longitudinal energy containment. For the photon energies up to 100 GeV, at least 25 radiation lengths are necessary to maintain the leakage term within reasonable limits. The side leakage is more difficult to control. It is determined by the number of cells one has to sum-up to reconstruct the energy. On the other hand this number is limited by the electronic noise and chiefly by the pile-up at LHC if the shower spreads too much laterally. Low noise electronics and high density material will of course limit this contribution. A particular attention must be paid to all gaps, walls, and dead material in front, which may have a important contribution to the leakage term. Detectors requiring cryogenics must be very carefully designed to minimize the effect of the cryostat. All leakage contributions can be well reproduced by MonteCarlo and a b_L term of $\approx$ 0.3% can be achieved with a crystal calorimeter.

The b_F term is associated to non uniformities. They can result from non homogeneities in the active material, like variation of doping concentration in non intrinsic scintillators. Temperature gradient can be the dominant factor of non uniformity when there is a strong dependence of the light yield with the temperature (BGO for instance has a large temperature coefficient of - 1.5% / °C). Crystals like Cerium Fluoride (CeF_3), with a temperature coefficient as small as 1‰ / °C at room temperature will be insensitive to this effect. Other sources of non uniformity are associated to the light collection . Pointing geometry implies cells of pyramidal shape. A strong non uniformity is induced by the focussing effect of these cells, particularly if the refraction index of the medium is high; CeF_3, with an index of 1.6 is here again much better than BGO (n = 2.15). Finally much attention should be paid to avoid large absorption of the light in the medium. This is perhaps the most important problem for noble liquids in the scintillation mode, because of the emission spectrum in the UV and the long pathlength due to the relatively small density of these liquids. For dense crystals with an emission spectrum in the visible, a b_F term as small as 0.25% should be possible.

The last contribution b_C corresponds to the intercalibration errors. Frequent calibrations with an efficient monitoring system are necessary to maintain this contribution below 0.3%. The L3 BGO has proven that a well designed fiber monitoring system can maintain the intercalibration within 0.3% for long periods of time. The high luminosity pp machines will also provide excellent means for a continuous calibration of the detector. At LHC a rate of 10 Hz is expected for electron pairs from Z° decays in the rapidity acceptance of ± 2, which should allow a weekly calibration with about 100 electrons per channel. If there is a good inner tracker, individual e^+ or e^- can be used (E/p matching) at a rate which could be as high as 100 per channel and per day. Finally one should be able to take advantage of the copious number of minimum ionizing particles for an almost continuous intercalibration control of all the channels.

Taking all these contributions together it seems that obtaining a constant term b = 0.5%, although difficult, is not out of reach for a well designed homogeneous calorimeter.

The pile-up problem will impose severe additional constraints, particularly for LHC. Not only the detector should be fast enough to allow the readout in one bunch crossing only, but the shower should be fully contained in an area as small as $\Delta\eta \times \Delta\phi = 0.06 \times 0.06$. In order to allow some centre of gravity position measurement, the granularity must then be $\approx \Delta\eta \times \Delta\phi = 0.02 \times 0.02$. This implies a very dense material for having a small Moliere radius, and a large internal radius of the calorimeter, at least 2 meters in a low magnetic field. In the case of a very strong field, the pile-up from charged tracks will be much reduced at the calorimeter level and a radius of 1 to 1.5 meter is possible.

In order to resolve the vertex ambiguity due to the pile-up and to allow a good photon angle measurement for a precise mass reconstruction of the Higgs, the calorimeter should be segmented in depth. A 2 segment crystal calorimeter (5 and 20 radiation lengths) can give a precision of about 7 mrads, if a compact position detector is placed between the 2 segments. Moreover, this will allow a π^o identification up to about 50 GeV/c, which is essential at LHC, where isolation cuts will be very inefficient.

4. Noble liquids

There is an increasing interest for using noble liquids in calorimetry both in the ionization and scintillation mode. The properties of the three most interesting one are listed in the table 1.

		LAr	LKr	LXe
Density	g / cm^3	1.39	2.45	3.06
Radiation length	cm	14.3	4.76	2.77
Moliere radius (PDG)	cm	7.3	4.7	4.1
Photons / MeV	$\times 10^3$	-	19	26
Light in fast component	%	8	1	77
Decay constant fast	nsec	6.5	2	2
slow	nsec	1100	85	30
λ peak	nm	130	150	175
Refractive index @ 170 nm		-	1.41	≈1.6
dE/dx <mip>	MeV/cm	2.11	3.45	3.89
Drift vel. (10KV/cm)	cm/µsec	0.5	0.5	0.3
Fano factor		0.11	0.06	0.05
Dielectric constant		1.51	1.66	1.95
Working temperature	°K	94	124	160

Table 1: Properties of noble liquids

Only liquid Krypton (LKr) and liquid Xenon (LXe) are dense enough to be considered for homogeneous calorimetry.

In the ionization mode, very interesting results have been obtained on LKr prototypes by the Novosibirsk group[14] and very recently by the NA48 collaboration[15]. The charge is collected on copper strips deposited on thin kapton foils. The Novosibirsk group has obtained an energy resolution of 1.7% for positrons of 1.2 GeV. The preliminary results from the NA48 collaboration are slightly worse than expected for the energy resolution ($4.8\%/\sqrt{E}$ instead of $3\%/\sqrt{E}$ expected), probably because of high electronics noise, too low drift field and bad intercalibration of the cells. On the other hand they have obtained very good position resolution (σ = 0.84 mm at 50 GeV) and excellent timing resolution (0.7 nsec at 120 GeV).

The main disadvantage of the ionization mode is that the charge collection time is an intrinsically slow process. But these liquids are also very efficient scintillators, unfortunately in the vacuum UV. LXe has the shortest radiation length, the shortest scintillation decay time, the longest wavelength and the highest working temperature, but its cost and availability are certainly a major limitation for large volumes. A small quantity of LXe in LKr (cheaper and 10 times more abundant than LXe) will act as a wavelength shifter and most important will shorten the relatively long decay time of LKr to acceptable values. A detailed analysis of the decay time shows that there are actually 3 components, the sharing of which depends on the ionization density of the particle. For instance the slow component (45 nsec) of LXe disappears for α particle excitation. Another interesting point is that the Birk coefficient is very low for these liquids, which means that the linearity of the light emission as a function of the deposited energy density is extremely good. In fact there is even a slight enhancement of the light yield at very high densities of deposited energy. This effect, combined with a clever treatment of the different time constants could open an attractive way to tune the e/π ratio.

Another approach is to combine the light and charge readout. This scheme should give the best energy resolution, as the two measurements are essentially independent and the 2 signals are anti-correlated[16]. A set-up is being worked out by T.Ypsilantis (see for instance ref[8]), where the photons induce a fast signal on a CsI+TMAE photocathode and produce photo-electrons which are injected in the liquid. The charges from the shower itself and the photocathode drift across the gap to a wire plane where they are digitized to give an excellent localization of the shower in two dimensions.

The main advantages of noble liquids are their radiation resistance, their modest purity requirements, at least when used in the scintillation mode, and their fast scintillation with very large light yield. They offer also a large flexibility for lateral and longitudinal segmentation. On the other hand, the need for a cryostat will lead to some material in front of the calorimeter which will deteriorate the resolution. Moreover, the density of these liquids is rather marginal (even for LXe) and yields a large Moliere radius which may severely increase the pile-up problems at LHC, unless the detector is installed at very large distance from the vertex, at least 3 meters, as proposed by W. Wallraff[4]. Another problem is the attenuation length for vacuum UV light which will introduce strong non uniformities in long cells (25 X_0 = 70 cm for LXe and 120 cm for LKr). This effect could be reduced with a multi longitudinal segmentation, as proposed by M. Chen[8]. Recent tests have been made with a 60 cm long long LXe cell, readout at both ends by silicon drift photodiodes[3]. The walls have been mirrored with a double layer of 70 nm Al and 30 nm magnesium fluoride. The reflectivity of the side walls was about 85% for the wavelength of LXe (175 nm) with a slight dependence on the angle of incidence. A beam of 5 GeV electrons was hitting the cell, either longitudinally or transversally at different longitudinal positions. In both cases the ratio of the signals seen by the photodiodes at both ends was recorded and compared with Monte Carlo expectations. As can be seen on Fig 2 the results are compatible with an attenuation length of 40 cm. This number is still too low to guarantee a small constant term with a standart readout scheme, and solutions must be found to maintain the uniformity at an acceptable level.

5. Inorganic scintillators (crystals and glasses)

In many respects, crystal calorimetry can be considered as a mature and rather conservative technique. Several large scale crystal calorimeters have been built in the past (Crystal Ball, Chloe, Cusb, L3 BGO, Crystal Barrel, etc...) and a large experience is being progressively build-up in matters such as mechanical structures, light collection, monitoring and calibration of such detectors. A large R&D effort has been launched by the "Crystal Clear" collaboration[17] in order to develop inorganic scintillators for LHC and to study economic production methods.

The most promising candidates are listed in table 2, with some of their characteristics, compared to the BGO and Barium Fluoride (BaF_2).More information and a detailed bibliography can also be found in ref[5,18].

Crystal	Density	Radiation length cm	Moliere radius cm	Decay time nsec	Peak emission nm	Radiation hardness Gy
BGO	7.13	1.11	2.33	300	480	10
BaF2	4.89	2.06	3.39	0.6 620	210 310	$> 10^4$
CeF3	6.16	1.68	2.63	5 20	300 340	$> 10^4$
YAP:Ce	5.35	2.83	2.82	35	390	$> 10^5$
GSO:0.5%Ce	6.71	1.39	2.42	55	435	$> 10^6$
GSO:2.5%Ce	6.71	1.39	2.42	31	450	$> 10^6$
ThF4	6.32	1.18	2.71	10 , 30	315,330,450	?
BaLiF3	5.24	2.13	3.13	2	435	?
LiYbF4	6.09	1.56	2.70	≤ 25	≈ 450	$> 10^5$
BaYb2F8	6.99	1.29	2.37	-	-	10^3 to 10^4
PbF2	7.77	.93	2.21	-	-	?

Table 2: Some promising crystals studied by the "Crystal Clear" collaboration

The criteria for a good scintillator to be used in the severe environment of LHC are the following:

- High density to reduce the total volume and the lateral shower size (pile-up)
- Fast scintillation with a few hundred pe^- / Mev in the visible to allow readout with photodiodes
- Radiation hardness up to 1 Mrad per year at least
- Good prospects for a low price.

One solution is to start from crystals already approaching the desired performances and to develop cheap production methods. Another approach is to work on potentially cheap materials in order to meet the specifications. The "Crystal Clear" collaboration is working on these two lines.

5.1. Crystals with good performances

Several crystals are already recognized as good candidates for high rate calorimetry. The most promising one is certainly the Cerium Fluoride (CeF_3). It has a high density (6.16 g / cm^3) due to the high concentration of Ce^{3+} ions in the lattice, which yields a small radiation length (1.68 cm) and a very small Moliere radius (2.6 cm). The scintillation mechanism is simple and well understood, involving the fast, intense and temperature independent transition 5d $\rightarrow$ 4f from the Ce^{3+} ion. Two broad emission bands centred at 300 nm and 340 nm produce a light yield which corresponds to about 500 pe^- / MeV and can be easily readout by silicon photodiodes. The global decay time constant is 20 nsec, with a fast component of less than 10 nsec at the shortest wavelengths.

Much progress has been made recently for the radiation hardness of CeF_3, which is now very good for strong irradiations of up to 1000 Gy in 4 hours . With some treatment of the raw material before the crystal growth, the attenuation length after such an irradiation is larger than 2 meters. The radiation hardness depends very much on the quality of the raw material. In inorganic scintillators, the scintillation efficiency is generally not affected by the damage, but impurities, even in very small quantities, can create colour centres which absorb a fraction of the emitted light. A large effort is on the way to identify the specific impurities which are responsible for the damage in CeF_3 and to find economical ways to remove them from the raw material. Another very promising approach from the economical point of view is being developed by the group from the Lebedev Institute in Moscow[6]. They use a method which has already proven to be successful for the BGO[19]: when electric charges are trapped by some crystal defects or impurities and create colour centres, a specific doping can allow a charge transfer from the traps to some other states where a non radiative relaxation is possible. Fig 3 shows the spectacular improvement obtained on a rather low quality CeF_3 crystal when doping with 0.67% of barium. The group from Bombay has also tested the neutron damage on CeF_3 with the Apsara reactor from the Bhabha Atomic Research Centre. A 1 MeV neutron fluence of 10^{12} n/cm^2 in 30 minutes did not create any visible damage on a 1 cm^3 CeF_3 crystal. A weak induced luminescence was observed for a few hours after the irradiation.

A similar test on Cerium doped Gadolinium Silicate (GSO:Ce), showed a much stronger activation (expected from the Gadolinium) which also disappeared in about one day. This crystal is very attractive because of its very high density (6.71 g / cm^3), very high light yield (more than twice the light yield of BGO) and exceptional radiation hardness up to 1 Grad[20]. Unfortunately the high melting temperature (1900 °C) combined with a very low pulling rate ($\approx$ 1mm / hour) to allow the Cerium to enter into the lattice, give little chance to cut the production price to an affordable level for a large detector. This crystal will certainly be a very good candidate for positron emission tomographs.

Other interesting candidates are Cerium doped Yttrium Aluminate (YAP:Ce) which is a very promising laser material, Thorium fluoride (ThF_4) and mixed compounds based on BaF_2 combined with lead, yttrium or yterbium. The "Crystal Clear" collaboration is presently studying BaY_2F_8 and $BaYb_2F_8$ as a possible alternative to BaF_2. The latter has a very high density (6.99 g / cm^3) but we have recently found that a high temperature phase transition could make the production of large crystals difficult. On the other hand, BaY_2F_8 has a modest density of 5.08 g / cm^3 but can be easily produced and is a well known laser material. When doped with 1.39% Thulium, it produces as much light as BGO, with 94% in two fast components of respectively 1.33 nsec and 7.6 nsec, and only 6% in a slow component of 100 nsec decay time[6]. This is of course a major improvement as compared with BaF_2 where 25% only of the light is going to a very fast component (0.6 nsec) and 75% in a very slow component (620nsec).

5.2. Is a cheap solution possible?

The factors building-up the price of crystals are being systematically analysed for each of the candidates. In some cases, the raw material of the desired purity can represent a substantial fraction of the cost. For this reason, all crystals based on the very rare Lutetium are discarded, although a lot of them are very fast cross-luminescent materials ($KLuF_4$,K_2LuF_5). Cerium is much more favourable, as it is the most common rare earth, with a natural abundance of 46.1 g / ton, comparable to that of nitrogen and tin : 46.3 and 40 g / ton respectively. Only the purification of Cerium against the other rare earths may be difficult and expensive, but our first studies show that this is not a critical parameter. One should also avoid crystals with too high melting temperature and low pulling rate, for which the cost of the energy, rather incompressible, will be prohibitive. As mentioned above, that maybe the case for GSO. For dense materials the investment costs can be reduced to minimum, as furnaces of limited size only are needed, as well as all their associated equipment (heating components, crucibles). A very good point for fluoride crystals, is that they can be grown in most of the cases in graphite crucibles, much cheaper of course than platinum or rubidium crucibles commonly used for oxydes.

The "Crystal Clear" collaboration has undertaken a systematic work to understand the optical properties of Lead Fluoride (PbF_2). The abundance of the raw material and the low melting point (822 °C) should make this crystal one of the cheapest. This crystal is considered at the moment as a good Cerenkov material, as no luminescence has been detected so far. A large variety of scintillating ions has been introduced in the PbF_2 lattice in order to understand the quenching mechanism for the scintillation of the Pb^{2+} ion. Similar studies are under way with Calcium fluoride (CaF_2), having the same crystalline structure as Lead fluoride, to compare the properties of the doping ions in these two crystals. Some other lead based structures are also investigated like, Pb_2ZnF_6,$Pb_3Al_5F_{19}$,$PbCsF_3$.

Another very attractive solution for calorimeters could be to use scintillating glasses, as their production cost may be much reduced as compared to crystals. By some substitutions of heavy metals in fluorozirconate and fluoroaluminate glasses, densities of respectively 5.72 g / cm^3 and 5.92 g / cm^3 have been recently obtained and prospects to go up to 6.5 to 7 g / cm^3 are encouraging. Batches of 10 to 20 kg are feasible with the standart techniques in use for fluoride glasses. With some development, batches of about 100 kg could be obtained. That may have a direct consequence in lowering the production cost. In addition the glasses offer the unique advantage over crystals, that moulding, hot pressing and extrusion techniques can be used to shape the blocks. This is of course very important if one considers relatively cheap materials,for which the mechanical processing will represent a significant, if not the dominant, fraction of the total cost.

6. Conclusions

In the present state of the analysis, an homogeneous electromagnetic calorimeter stands the best chance to not miss a SM Higgs boson in the intermediate mass range at LHC and SSC. Moreover, the very high performances of noble liquid or crystal calorimeters place them in the best position to explore the intermediate region for possible MSSM neutral Higgs boson and the complete new energy domain opened by LHC and SSC in the TeV scale, with the best discovery potential. It would be a pity, if not a mistake, if none of the 4 large experiments foreseen at LHC and SSC is considering having such a calorimeter in its set-up.

7. Acknowledgements

The author is very grateful to all the participants of the session on homogeneous calorimetry at the Capri conference, as well as to all the other colleagues working in this field, for very fruitful discussions.

8. References

1. M. Schneegans, Proceedings of the Second International Conference on Calorimetry in High Energy Physics, Capri, October 1991, *World Scientific*
2. R.Y. Zhu, *CALT-68-1707*
3. T. Virdee, Proceedings of the Second International Conference on Calorimetry in High Energy Physics, Capri, October 1991, *World Scientific*
4. W. Wallraff, Proceedings of the Second International Conference on Calorimetry in High Energy Physics, Capri, October 1991, *World Scientific*
5. P. Lecoq et al, Proceedings of the Second International Conference on Calorimetry in High Energy Physics, Capri, October 1991, *World Scientific*
6. A.A. Asseev et al, Proceedings of the Second International Conference on Calorimetry in High Energy Physics, Capri, October 1991, *World Scientific*
7. S. Banerjee et al, Proceedings of the Second International Conference on Calorimetry in High Energy Physics, Capri, October 1991, *World Scientific*
8. G. Jarlskog et al, Proceedings of the Large Hadron Collider Workshop, Aachen, October 1990, *CERN 90-10, ECFA 90-133*
9. M. Schneegans, Proceedings of the XXVIe Rencontres de Moriond, Les Arcs, March 1991, *LAPP-EXP-91-04*
10. A. Ballestrero, E. Maina, *DFTT 29/91, June 1991*
11. Z. Kunszt et al, *ETH-TH/91-17, DTP/91/40, June 1991*
12. C. Seez, Proceedings of the Second International Conference on Calorimetry in High Energy Physics, Capri, October 1991, *World Scientific*
13. F. Ferroni, Proceedings of the Second International Conference on Calorimetry in High Energy Physics, Capri, October 1991, *World Scientific*
14. V.M. Aulchenko et al, Novosibirsk Preprint 89-124
15. D. Schinzel, private communication
16. T. Doke, *Experimental techniques in High Energy Physics,Ed. T. Ferbel,Addison Wesley*
17. R&D Proposal for the study of new fast and radiation hard scintillators for calorimetry at LHC
    Crystal Clear Collaboration , CERN / DRDC P27 / 91-15, project RD-18
18. P. Lecoq, M. Schüssler and M. Schneegans, in *Proceedings of the 5th Pisa meeting on advanced detectors, to be published in NIM CERN / PPE 91-124*
19. P. Lecoq, P.J. Li, B. Rostaing, *NIM A300 (1991) 240-258*
    Z.Y. Wei et al, *NIM A297 (1990) 163*
    Z. W. Yin et al, *NIM A275,(1989) 273*
    G.J. Britvich et al, *IFVE 91-134 Protvino*
20. M. Kobayashi et al, *KEK preprints 90-133 (oct 1990) and 91-1 (apr 91)*

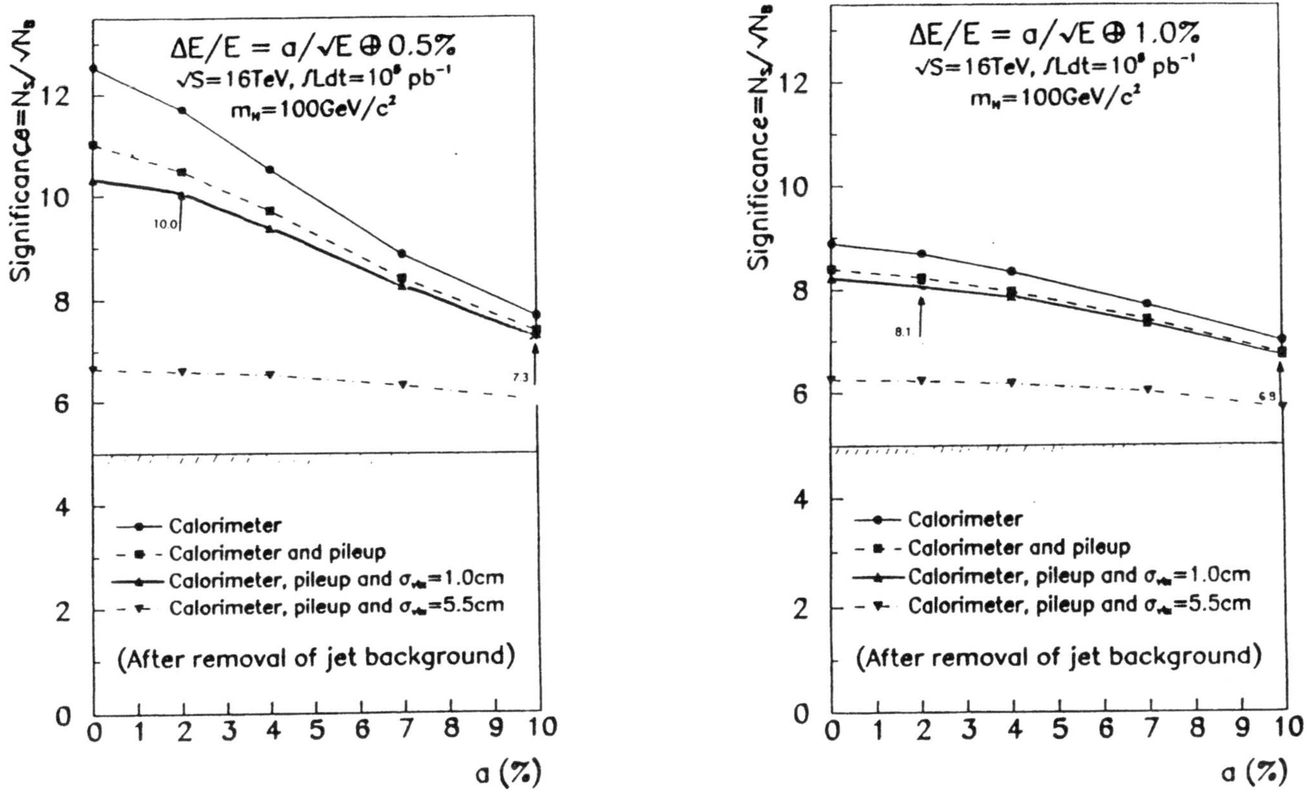

Fig 1: Significance for a 100 GeV Higgs as a function of resolution parameters

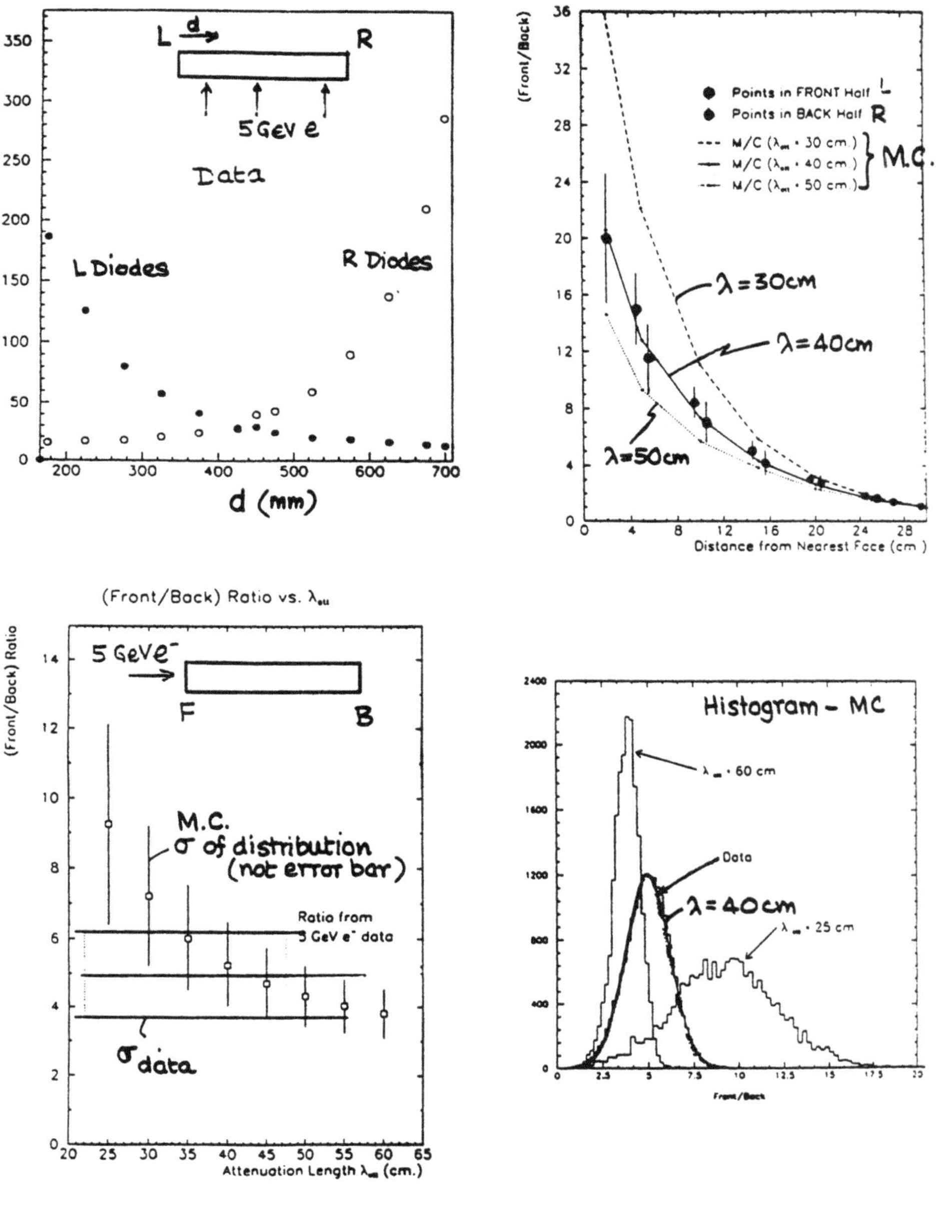

Fig2: Attenuation length in liquid Xenon

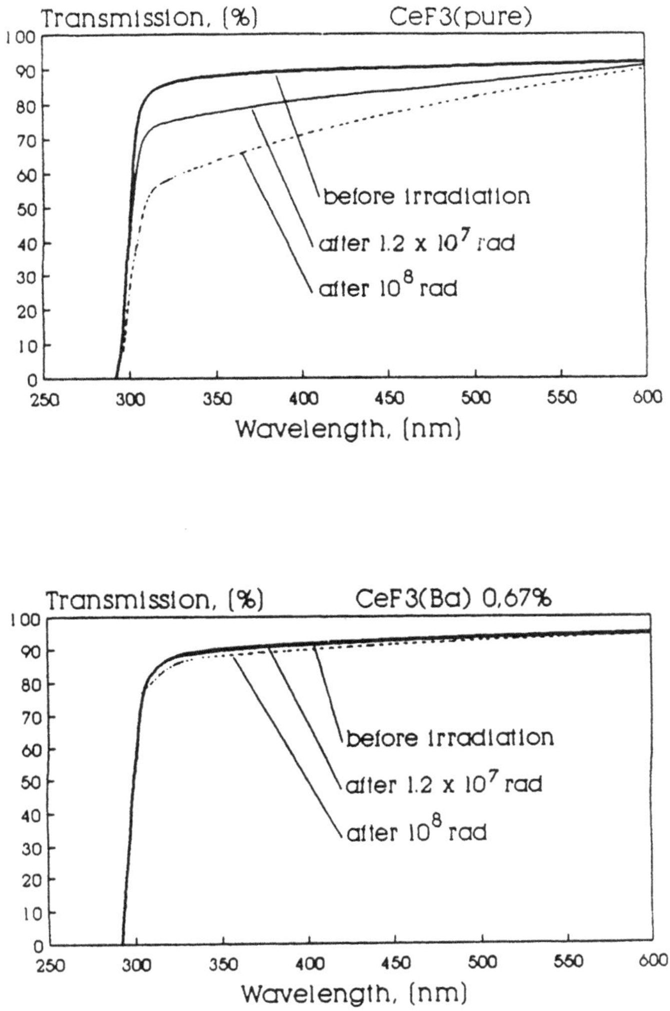

Fig 3: Effect of doping on CeF3 radiation hardness

HOMOGENEOUS CALORIMETRY AT FUTURE pp COLLIDERS

M. SCHNEEGANS
LAPP/Annecy/France

ABSTRACT

The need for high resolution calorimetry at SSC and LHC has been stressed many times, in particular for completing the mass scan of the Higgs boson by detecting its 2γ decay mode in the intermediate mass region. To reach the ultimate performance in mass resolution, good energy and photon angle measurements are necessary in addition to pizero identification for rejecting the QCD background. To reach very high resolution in the energy range of the photons from a Higgs, both the a and b-component of the resolution must be low. Only homogeneous calorimetry allows to have a low a-term. To have a very low b-term is difficult for any calorimeter and can only be obtained in a 'dedicated' detector. The limiting contributions to the b-term are reviewed and some parameters for an homogeneous detector at LHC are discussed.

1. Introduction

In an homogeneous detector, the electron or photon showers are 'fully' contained in the sensitive medium. The energy is then revealed by scintillation or Cerenkov light, both collected in highly transparent materials, or by ionisation in conductive liquids, where charges are collected. Homogeneous detectors are well known to give excellent energy resolution: for example, the INa (Tl) calorimeter of 'Crystal Ball' gave: $0.27\% * E^{-1/4}$ and the BGO detector of L3: $2.0\%/\sqrt{E} \oplus 0.5\%$ in a test beam. To allow compact detectors at LHC/SSC, with at least 25 X^0, high density liquids or solids are needed, such as liquid Xe (X^0=2.8cm), liquid Kr(4.8cm), BaF_2(2.05cm), CeF_3(1.68cm), PbF_2(0.93cm), or scintillating glasses (1-2.5cm). For details on the properties and R&D status of such scintillators, see Ref.1,2.

High e/γ energy resolution is very useful in several domains of physics at pp colliders. The main argument, generally used to stress the high level of performance required, is the detection of the 2γ mode of the Standard Model Higgs boson, for which the natural width is completely dominated by the instrumental resolution (energy and angular) below Higgs masses of ~200 GeV/c^2. Extensive studies, for SSC[3] and LHC[4], have shown that the 2γ signal rates were comfortable for masses ranging from 80 to 160 GeV, but that the background was very high. After kinematical cuts and QCD background rejection, the Signal over Background rates were still of the order of ~1/20, even with a very high performance detector. For more detailed discussions and references, see Ref.5,6.

Let us add two comments:

- We should not forget the processes in which the SM Higgs is produced in association with a W boson or ttbar, both resulting in a eγγ signature. Recent calculations[7;8] confirm the small signal rates but yield background rates rather larger than reported previously[4;9]. These rates, together with the fact that large uncertainties in the estimates persist, push us again to aim for the best detector performance possible.

- Hopes to detect the scalar Higgs bosons h^0 and H^0 of the Minimal Supersymmetry Standart Model (MSSM) in large regions of parameter space, via their γγ decay mode in the intermediate mass region, were recently confirmed[10]. Nevertheless, the rates are at best equal to those of the SM Higgs, but generally lower. This shows that to stand a chance of detecting also a MSSM Higgs, one should aim not just at a 'good' detector, but push its performance to the ultimate limit!

As will be discussed in section 3, only an homogeneous and 'dedicated' calorimeter gives high chances for detecting such objects.

2. Constraints and specifications

Besides the 'normal' constraints expected at a multi-TeV pp collider, work at luminosities of nearly $2*10^{34}$ $cm^{-2}s^{-1}$ as envisaged at LHC, brings additional difficulties[4], due to the pile-up of 20-30 events in each bunch crossing (15ns):

- the energy measurement will often be contaminated by additional particles. This will determine the angular granularity needed, thus also the Moliere radius (cylinder of 90% shower containment) desirable and the minimum distance of the calorimeter to the interaction region.
- an isolation cone around each photon, which seems to be very powerful in rejecting the photon background coming from pizeros in jets, when dealing with one event at the time, becomes much less efficient in the case of many events superposed. An additional tool for rejecting the QCD background is then necessary: the identification of the pizeros by separation of the 2 photons, up to pizero energies of ~50 GeV.
- an ambiguity in the 2γ vertex z-coordinate (σ_z~5.5cm, as usualy assumed). This will cause a degradation of the mass resolution, inacceptable for a high energy resolution detector at moderate distance from the interaction region. A determination of the photon angle in the detector itself to ~7mrad should give an acceptable contribution to the mass width and allow some rejection of the 2γ combinatorial background.

We can now list the requirements on a good homogeneous calorimeter optimized for the Ho->2γ detection at LHC:

a) Excellent energy resolution: Small a-term: ≤3% and b-term: ≤0.5%. These numbers will be justified in the next section.

b) Fine angular granularity: As indicated by Monte-Carlo calculations[11] on the effect of pile-up, a maximum full shower containment cell size for an 'acceptable' contamination of the energy measurement is $\Delta\eta*\Delta\phi$= 0.1*0.1, which corresponds to individual cells of about 0.03*0.03. This implies a small Moliere radius of the sensitive material or/and a large internal radius of the detector.

c) Good photon angle measurement: 5-10 mrad for photons at large theta angles. This can only be obtained by longitudinal segmentation of the towers in at least 2 pieces. A passive preshower section would jeopardize the energy resolution.

d) Good 2-photon separation for π^0 identification: To identify pizeros up to a momentum of ~50 GeV/c, a 5mrad angular separation power for 2 photons is necessary. The tower granularity will in general yield a 10-30mrad separation. A position detector inside the active volume of the calorimeter (after 3-6X^o) should allow to separate photons distant by a few mm.

e) A fast readout: integration over 1 bunch crossing is desirable to minimize the contamination of the energy measurement. This of course implies a short decay time of the scintillator light response.

f) Radiation hard components: All components should be able to stand gamma doses larger than 1Mrad/year in the central region and up to 10 Mrad/year at η=3.

3. Comments on energy resolution

For a Higgs mass of 80-150 GeV, photons down to ~25 GeV should be measured. The results of calculations[12] performed for LHC in the case of a 100 GeV Higgs mass are summarized in Fig.1. The optimum significance is calculated for each point in the appropriate mass window. The large QCD background is supposed to be suppressed to a negligible level by isolation cuts and pizero identification. The effects of pile-up, vertex smearing and detector energy resolution are clearly seen. Following the realistic curve (triangle point up), one can observe that:

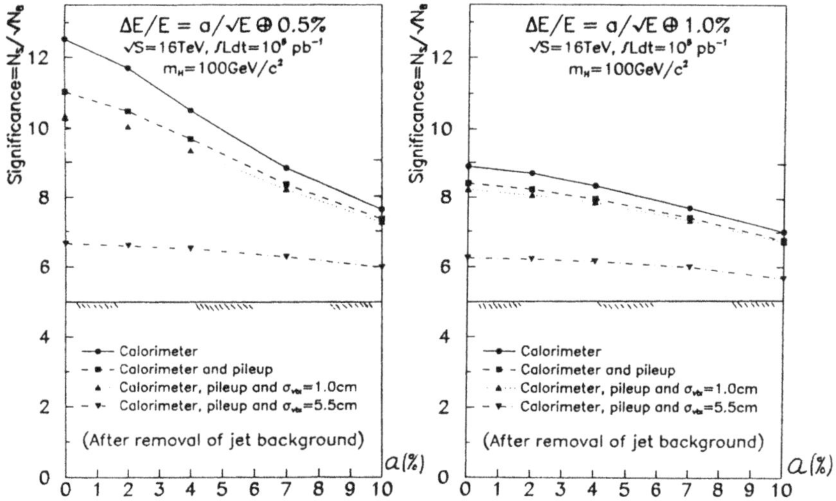

Fig.1: Significance of a 100 GeV mass Higgs decaying into 2γ in various conditions.

- if, for b=0.5%, one lets the a-term go from 2 to 10% (from homogeneous to a good sampling calorimeter), one looses nearly 3 units of significance,

- if, for a=2%, one lets the b-term go from 0.5 to 1%, one looses nearly 2 units of significance.

- for a detector having a=10%, there is little gain (for Ho->2γ) in making a large effort to go down to b=0.5%.

Simulations[12] of a 100 GeV Higgs signal superposed on a realistic background show clearly that a significance of 5 or 6 gives very little chance to recognize a Higgs signal for S/B ~1/20. One can conclude that in order to reach degrees of significance well above 5, it is necessary to have **both a** and **b**-terms **low**.

The **a**-term is usually low for an homogeneous calorimeter; 1-3% is normally obtained if one has enough scintillation light to make photostatistics and electronic noise contributions negligible above a few GeV energy. A **b**-term of 0.5% is very difficult to achieve for any calorimeter! A high priority must be given to the energy resolution at all stages: choice of the design parameters, details of the construction, calibration methods and frequency, fine corrections at the analysis level. Only a 'dedicated' detector, in which no compromise is made with other detectors or constraints such as hadron compensation, material in front, mechanics, space, cost, can hope to reach b~0.5%.

The **b**-term of the energy resolution can be decomposed in the following way:

$$b^2 = b_L{}^2 + b_F{}^2 + b_C{}^2$$

To have b~0.5%, each of the contributions should be <0.3%.

b_L: This term contains all the shower leakages:

- backward: not much to do, but it is a small contribution above 10 GeV,

- side: small if N (in ΣN channels) is large enough. But, we should keep N small because of electronic noise and contamination,

- forward: this term can be made small by taking enough X^0s (for gammas)

b_L also contains contributions of the gaps and walls between channels and material between longitudinal segments (position detector) or in front.

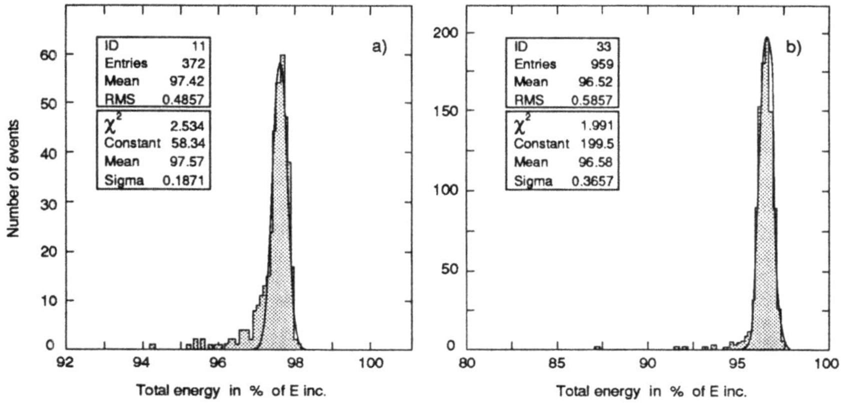

Fig.2: Monte-Carlo results on energy resolution for 50 GeV photons in 25 X^0, $3*3cm^2$ CeF3 crystals, without (a) and with (b) material Attention! Different scales.

It can be seen in Fig.2 that by summing up 25 crystals, a resolution of 0.2% is reached, increasing to 0.38% when realistic walls and some material between 2 segments are introduced. With some effort to minimize the material and/or 1-2 more X^0 in crystal length, a b_L of 0.3% seems accessible.

b_F: This term contains all other fluctuations, such as:

- intrinsic shower fluctuations (negligible),
- scintillation yield fluctuations which can be due to bad stoechiometry of the constituants, to boiling in noble liquids, to doping gradients in doped crystals, or to temperature gradient along crystals (Ex: BGO has a -1.5% light yield dependance per deg.C). The obvious solution to the last point is to look for scintillators with output nearly independant of t at room temperature.
- light collection non-uniformity due to absorption (local or gradient) and focussing by a tapered tower shape.

All these possible non-uniformities must be limited to a few percent over the tower length in order to keep their contribution to the b-term less than 0.3%.

b_C: Frequent, precise (0.3%) calibrations of each channel are necessary to keep b_C contribution at the desired level. At LHC/SSC, physics should help us in providing several fast methods:

- Electron pairs from Z^0 should yield a good way to improve the calibration constants using the Z^0 mass constraint. Rates of 10Hz should yield ~100 electrons per channel in about 1 week at LHC.
- If a very good momentum resolution can be obtained in the tracking system (~1% at 100 GeV), one can hope to collect ~100 electrons per channel in 1 or 2 days at LHC! This possibility has to be investigated with great care; if it is confirmed, it would give us calibration and monitoring of all channels to the desired precision over all the useful energy range.
- Many hadron tracks (MIPs) should be usable in each event at LHC/SSC leaving a few hundred MeV in crystals or noble liquids. One should be able to collect ~200 MIPs per channel in a few hours at LHC. This would yield a very useful monitoring, in particular if one has a longitudinal segmentation.

In addition to physics, one can foresee monitoring systems:

- Xe light distributed by optical fibers has proved in the BGO/L3 calorimeter to reach the 0.3% level,

- an RFQ (radio-frequency quadrupole) system may also give a fast way of monitoring. Such a system will soon be tested in L3.

Detailed simulations of the physics calibration methods have to be performed in the conditions of each LHC project before one can defend a calibration scheme insuring the 0.3% level.

4. Discussion of design criteria

The tasks and performance expected from an homogeneous calorimeter at SSC/LHC being now defined, one can discuss possible design parameters for such a detector:

As was said earlier, a tower depth of 25 X^0 (maybe 26) should be adequate. Lateral dimensions of towers should match the Moliere radius of the scintillator for e/γ position resolution (<80% of the shower in the central crystal). For a large detector radius (3m or more) and high density materials, this will lead to angular granularities of the order of 0.01*0.01 with a very large number of towers.

To keep rates per tower constant over all the rapidity range, one should have constant eta=phi steps. On a cylinder, this leads to quasi-square towers nearly equal in section. A cylinder up to eta=2 being very long for large internal radii, a conical part between eta~1 and 2 is more economical, decreases the tower front face staggering and still makes possible a constant momentum resolution of the tracking system (for an illustration of such a concept, see Fig.3). In the conical part, one can either keep constant rapidity steps (same number of towers in phi), but let the tower section decrease progressively or keep more equal tower sections by decreasing the number of towers in phi.

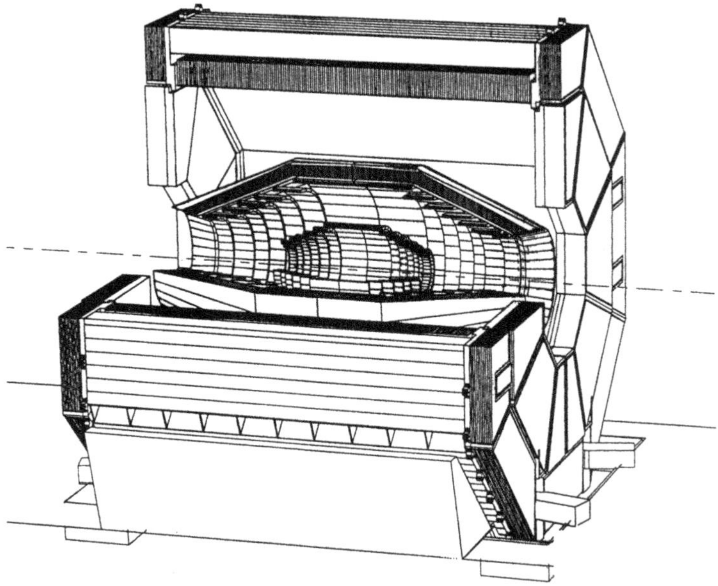

Fig.3: Cylinder + Cone crystal (CeF_3) configuration for a 0.7T solenoidal field and R_i =3m. The smaller shell inside is a tracking layer at 1/2 distance (L_3/LHC, e/γ option).

The choice of the internal radius of the calorimeter is a crucial issue. Simple Monte-Carlo pile-up studies[11] indicate that a shower should be fully contained in a cell: $\Delta\eta*\Delta\phi$~0.06, for having 90% of clean energy measurements. This leads, for CeF_3 (Moliere radius=2.6cm) for instance, to a radius of ~1.3m. But, a solenoidal field will spiralize charged particles below a given p_T, thus decreasing the charged particle contribution to the pile-up. Two approaches can be mentionned:

- high field (4T), Ri=1.4m, p_T cut =0.85 GeV/c
- low field (0.7T), Ri=3m, p_T cut =0.32 GeV/c

The first approach gives better cleaning of the charged particles, but the larger radius of the second solution will improve the shower isolation (for a given Moliere radius).

As was seen on Fig.1, the ambiguity on the z-coordinate of the vertex at high luminosity will cause an unacceptable smearing of the mass peak, at least for radii of 1-2m. Longitudinal segmentation into 2 towers should allow reconstruction of the photon directions to better than 8mrad either using the 2 barycenters or a position detector between segments and the back barycenter. For larger calorimeter radii, such as 3m or more the vertex ambiguity seems to cause a tolerable loss in Higgs signal significance. On the other hand, determination of the vertex by the photon angles will have poor precision.

As was shown some time ago[4;12], identification of a large fraction of the pizeros at LHC requires separation of the gammas by ~5mrad. As this is not possible from the towers alone, a position detector between the 2 segments is needed. At radii of 3m or more, it is possible that the 2-photon separation from towers is good enough taking into account the higher efficiency of isolation cuts at these distances.

The choice of the internal radius is also relevant for the quality of the tracking. Good momentum resolution is important for electron identification by E/p matching and essential for fast calibration of all towers as discussed previously. In the two field/radius cases mentionned above, the $B*L^2$, which matters in the case of sagitta measurement, is approximately equal and a resolution of ~1% at 100GeV can be expected.

5. Conclusion

At least one high resolution and dedicated calorimeter should be present at SSC/LHC, essentially to complete the SM or MSSM Higgs coverage in the intermediate mass region. We have shown that it must be an homogeneous calorimeter, optimized on photon detection by taking great care of all contributions to the b-term of the energy resolution and by providing adequate photon angle determination and 2-photon separation. The design and construction of such a calorimeter looks feasible, but before being more affirmative, some major issues must be clarified, such as calibration by physics, fast electronic readout and trigger.

6. Reference

1. P.Lecoq, M.Schneegans and M.Schüssler, Progress in the development of new scintillators for future HEP experiments, *Proc. of the 5th Pisa Conference on advanced detectors*, Elba, May 91 *and Prep. CERN-PPE/91-124,*
2. This Conference (Homogeneous Calorimetry Session),
3. R.Y.Zhu, *Rep.*CALT-68-1707,
4. C.Seez et al, *Proc. of the ECFA/LHC Workshop,* **Vol II,**Aachen,Oct 90,
5. M.Schneegans, Can one miss a SM Higgs in the intermediate region, *Proc. of the XXVIe Rencontres de Moriond,* Les Arcs,March 91 *and Prep.*LAPP-Exp-91.04,
6. D.Ferrere, M.Lebeau, P.Lecoq, M.Schneegans and M.Vivargent, High resolution crystal calorimetry at LHC, *Proc. of the 5th Pisa Conference on advanced detectors*, Elba, May 91 *and Prep.* LAPP-Exp-91.05,
7. A.Ballestrero and E.Maina, *Rep.*DFTT 29/91,
8. Z.Kunszt et al, *Rep.*ETH-TH/91-17, DTP/91/40,June 91,
9. J.F.Gunion, *Rep.*UCD-91-2,Jan 91,
10. Z.Kunszt and F.Zwirner, recent talks in L3 and CMS,.
11. J.Colas et al, *Proc. of the ECFA/LHC Workshop,* **Vol I,**Aachen,Oct 90,
12. C.Seez, this conference (Physics Session),
13. Simulations in the frame of the 'Crystal Clear Collaboration'.

PROSPECTS FOR LARGE SCINTILLATING NOBLE LIQUID CALORIMETERS

W. WALLRAFF
1. Physikalisches Institut, RWTH Aachen,
Sommerfeldstraße 28, D 51 Aachen, Germany

ABSTRACT

Scintillating noble liquids, especially liquid Xenon (LXe) and liquid Krypton (LKr), provide unique possibilities for the construction of fully active, high resolution electromagnetic calorimeters. We discuss fundamental properties of electromagnetic shower detectors based on scintillating noble liquids. Some variants of a very large calorimeter inside the L3 magnet used for an LHC experiment are presented. This device can fullfill the very stringent requirements for the detection of the two photon decay channel of the Higgs boson produced in pp interactions at the LHC.

1. Introduction

There is general agreement[1] that a dedicated photon detector is highly desirable for closing the detection gap for Higgs production beween the mass range accessible at LEP and that mass range where so called "gold plated" 4 lepton decay events from the decay H -> $Z^{\circ}Z^{\circ}$ can be observed providing the yardstick for the capabilities of general purpose experiments at the future TeV pp colliders. This mass ($80 \leq m_H \leq 140$ GeV) region is especially important for vindicating or rejecting supersymmetric versions of the standard model.

Since the total width of the Higgs for masses in our region of interest is less than 10 MeV the discovery potential of an experiment is tightly coupled to the mass resolution of its photondetector. The discovery of the Higgs may resemble in more than one way to the discovery of the J or Υ in high intensity fixed target experiments. Thus the hunt is based on a rare ($B(H\to\gamma\gamma)=10^{-3}$) but clearly identifiable decay mode that allows for high resolution mass spectroscopy at the limit of the natural width of the object searched for.

All high resolution, 4 π electromagnetic calorimeters (Xtal Ball, L3, CLEO II) so far use scintillation in cristalline materials. But none of the materials (NaJ, BGO, CsJ) on which our large detector experience is based, can be used in the pp collider environment (see review by T. Virdee at this conference). In consequence a wider search for radiators of totally active electromagnetic calorimeters has started. So far scintillating noble liquids (LKr, LXe) constitute the only viable alternative (first proposed in 1987[2]) to inorganic crystals (see P. Lecoq, this conference). The L* collaboration[3] was the first to consider in earnest the application of this new technique.

2. Scintillating Noble Liquids

In table 1. we give an overview of the physical and particle detection properties of the heavy noble gases. Xenon and Krypton are especially interesting due to their short radiation length and acceptable Molière radius. Both have very high scintillation output and short decay times. The use of these scintillators has been pioneered by the Nuclear Physics community mostly before the introduction of Ge or Si solid state detectors[4].

Table 1. Properties of scintillating noble liquids.

		Ar	Kr	Xe	
abundance		1.3	10^{-4}	10^{-5}	Vol% in air
radiation length	X_0	14	4.7	2.8	cm
Moliere radius	R_M	7.4	4.8	4.0	cm
absorption length		84	61	55	cm
dE/dx		2.11	3.2	3.79	MeV/cm
(Birks constant)$^{-1}$		10200	?	30600	MeV/cm
density		1.4	2.4	3.06	g/cm^3
boiling point		-185.9	-153.4	-108.2	°C
gas/liquid volume ratio		785	620	520	at 1013 mbar
scintillation decay time		6.5	2	3	ns (fast component)
fast intensity fraction		8	1	77	%
scintillation decay time		1100	85	22	ns (slow component)
total photon yield		$>10^4$	$>10^4$	$>4\cdot10^4$	MeV^{-1}
emission peak wave length		130	150	170	nm

The L* collaboration has investigated the world resources together with the air liquification industry showing that at least 15m^3 of LXe can possibly be made available[5] if a 5 - 8 year procurement program is acceptable. But much larger detectors have to be made from Krypton or Argon.

The short wavelength photons from noble liquid scintillation are most conveniently caught with solid state photodetectors placed inside the radiator itself. Several variants of this approach have been investigated by the L* Xenon study group[6]. The light collection efficiency and the reflection properties of wall coatings have been recently investigated for a Xe cell at CERN (see T. Virdee, this conference).

It has been experimentally shown that the excitation leading to the slow part of the Argon scintillation can be drained away to faster scintillations if small amounts of Xenon are added[7]. One can expect that a similar effect will be observed with LKr. For any quantitative analysis of pile up effects it is mandatory that a measurement of the scintillation of doped Krypton is performed.

Fig. 1 shows the energy deposit (calculated with GEANT) of 10 (50) GeV electrons in a 25 X_0 LKr radiator. The fluctuations of the energy deposit are less than

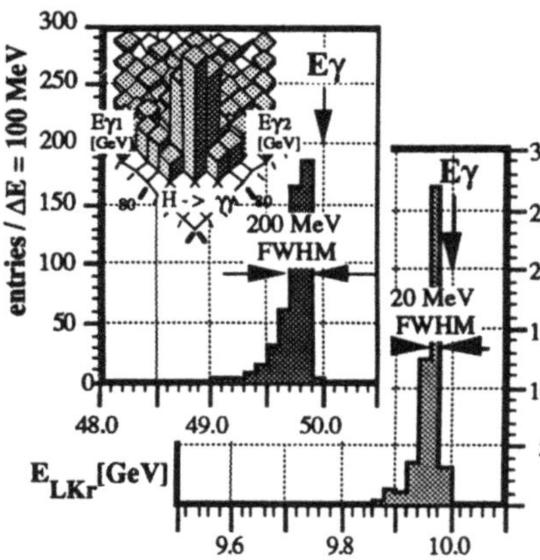

fig. 1. energy spectra for a 25 X_0 depth LKr-calorimeter of 10 x 10 R_M^2 crossection.

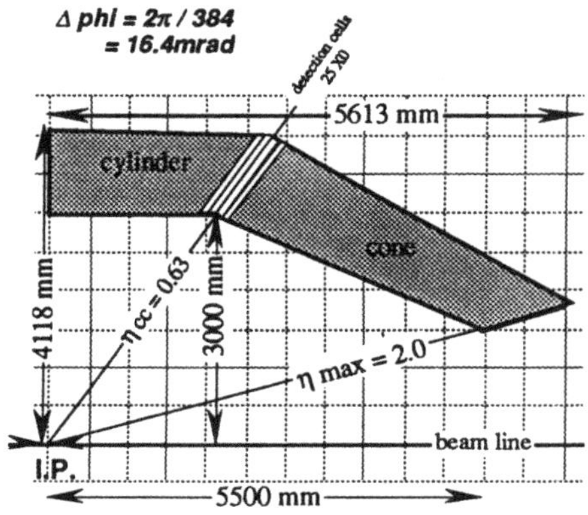

fig. 2. Large LKr calorimeter (r,z) cut

0.2% for the most important energy range (see insert showing the correlation of gamma energies for m_H =100 GeV produced at the LHC). There is in mean a 3.8% rear leakage at 50 GeV. It is possible to correct for this leakage by measuring the energydeposit in 3 successive cells (5,10,10 X_0). It is crucial to have a homogeneous light collection efficiency in order to preserve this high resolution. The direct signal of the showerparticles inside the solid state photodiodes (η_{phot} > 0.5) is no problem[8] due to the very high scintillation efficiency of the noble liquids (> 10^9 photoelectrons / 50 GeV e.m. shower).

3. Large Scintillating Noble Liquid Calorimeters

A very large high resolution Krypton calorimeter can be conceived as a 3 m inner radius cylinder covering $|\eta| < (0.6)$ 1.0 with conical frontends for the region (0.6) $1.0 \leq |\eta| \leq 2.0$ (2.5). A large radius calorimeter has a number of distinctive advantages:

1. The occupancy per cell decreases quadratically with the radius (assuming constant cell cross section (for example: R_M x R_M)).
2. The invariant mass resolution is less affected by mismeasured photon angles due to the extended luminous region (σ_z = 5.7 cm at LHC).
3. π^0 identification will be possible up to high energies .

The detector layout shown in Fig. 2 covers the region $|\eta| \leq 2.0$. For a depth of 25 X_0 the total volume is 200 m^3. The sensitive area is 133 m^2. The towersize matches approximately the Moliereradius of LKr resulting in 85000 detection cells of size $\Delta\eta$ x $\Delta\phi = (0.016)^2$. It is understood that the Kr is doped with up to 5% of Xe in order to shorten the scintillation decaytime to values suitable for running at 60 MHz repetition rate.

Maintaining constant cross section (hence approximately constant volume of ≈3 l per cell) leads to a progressively decreasing number of cells per circumference (from 384 at $|\eta| \leq .63$ to 208 at the front tip of the cone $|\eta| = 2$). In consequence the cell hit rate shoots up from 320 kHz to 1.2 MHz at LHC design luminosity of $1.65 * 10^{34}$ $cm^{-2}s^{-1}$ (fig. 3 a). The constant occupancy approach requires 37% more cells on the otherwise unchanged detector surface (fig. 3 b)

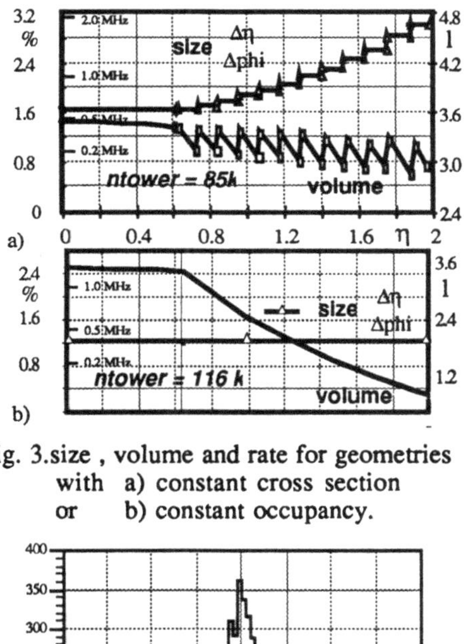

fig. 3.size , volume and rate for geometries with a) constant cross section or b) constant occupancy.

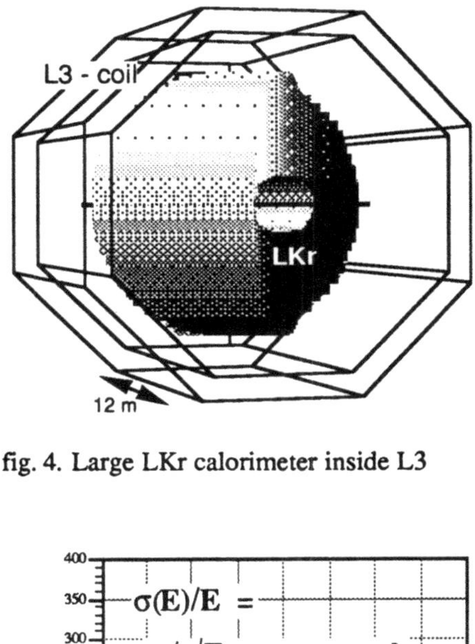

fig. 4. Large LKr calorimeter inside L3

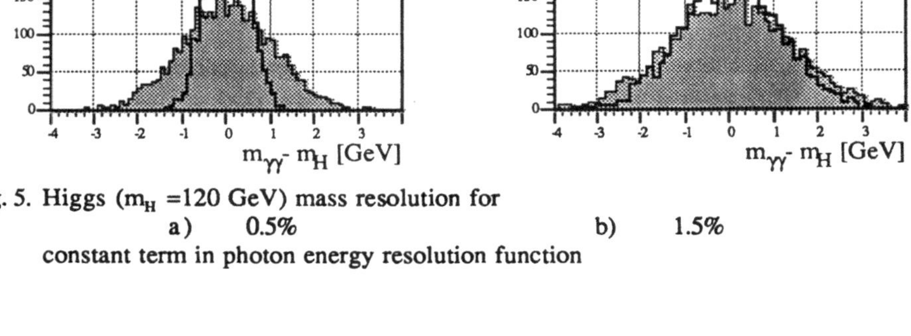

fig. 5. Higgs (m_H =120 GeV) mass resolution for a) 0.5% b) 1.5% constant term in photon energy resolution function

This LKr detector fits well inside the L3 magnet (fig.4). Adding a simple tracking system allows for sign determination of electrons from the high mass Higgs decays H -> Z°Z° -> 4e . The tracker enjoys the same benefit as the calorimeter. At radius $2.5 \leq r \leq 3$ m the occupancy of straw tubes of up to 1m length is less than 10^{-3}.

As for any large cell number system intercalibration is of utmost importance in maintaining the resolution. Noble liquid system can be calibrated at low energy (<10 MeV) absolutely[9] by using alfa sources (Po^{210}). High resolution has to be paired with an very low constant term in the resolution function (see fig.5). So far only totally active calorimeters have achieved constant terms in the vicinity of 1%.

4. Conclusion

Scintillating noble liquid calorimeters can in principle fulfill the requirements for an excellent photon detector at the LHC. The fundamental properties of this type of scintillator are well understood. However a great deal of work is needed to provide efficient and homogeneous light collection in the near UV. It is clear that a series of prototypes with increasing size have to built and thoroughly investigated, before a $100m^3$ detector construction project can be started.

5. References

1. G. Kalmus, summary talk of the *Large Hadron Collider Workshop, Aachen 4-9 October 1990* , ed. G. Jarlskog and D. Rein (CERN 90-10, vol. I) p.581
2. W. Wallraff, in the *LAA Project , Cern 25 June 1987,* p. 295 ;
   M. Chen et al. in *Experiments, Detectors and Experimental Areas for the Super Collider, Berkeley July 1987,* ed. R. Donaldson and M.G.D. Gilchriese, (World Scientific Publishing Co., Singapore 1988), p.574 .
3. M.Chen, W. Wallraff et al. NIM. **A267** (1988) 43 ;
   L* Collaboration, *Expression of Interest to the SSC,* May1990, p. 63 ;
   M. Chen et al. in *Detector Research and Development for the SSC, Fort Worth , October 15 - 18,1990,* ed. T. Dombeck, V. Kelly, G. P. Yost, (World Scientific Publishing Co., Singapore 1991), p.401 ;
   L* Collaboration, *Letter of Intent to the SSC,* November 1990, p.34
4. A. Sayres and C.S. Wu, Rev. Sci. Instr. **28** (1977) 758 ;
   T. Doke, Portugal Phys. **12, 1-2** (1981) 9 , NIM **196** (1982) 87 ;
   M. Mutterer, NIM **196** (1982) 73
5. L* Collaboration, *Letter of Intent to the SSC,* November 1990, p.36
6. T. Doke et al. MIT-LNS Tech. Rep. no. 175, May1990
7. S. Kubota et al. NIM **196** (1982) 101
8. W. Wallraff, in *Conference on Very High Resolution Electromagnetic Calorimetry, Stockholm 9 - 11 November 1989*
9. L* Collaboration, *Letter of Intent to the SSC,* November 1990, p.35

NEW RESULTS FROM THE "CRYSTAL CLEAR" COLLABORATION

P. LECOQ, M. SCHÜSSLER
CERN / PPE, Geneva, Switzerland

D. FERRERE, M. SCHNEEGANS
LAPP, Annecy, France

ABSTRACT
In the frame of the CERN project RD-18, the "Crystal Clear" collaboration is investigating new dense, radiation hard and fast inorganic scintillators to be used as the components of very good resolution electromagnetic calorimeters at the new high luminosity accelerators. The most promising research lines are described here, new results are given as well as some comments about mass scale production problems and cost considerations.

1. Introduction

The challenging physics of the future proton-proton machines, LHC and SSC, will be very demanding on the detector performances. There is a particular interest in having a high performance electromagnetic calorimeter which could for instance give the best chance to detect a Higgs boson in the intermediate mass range ($M_Z \leq M_H \leq 2M_Z$).

Calorimetry based on inorganic scintillators can be considered as a mature technique for such a high quality detector, although the scintillators used in previous experiments do not fulfill the requirements imposed by the future colliders. At LHC and SSC, the scintillation must be fast and the optical properties must remain unchanged under strong radiation. The pile-up problem is eased by using a dense crystal, allowing a fine granularity.

The "Crystal Clear"collaboration is now developing and testing a large number of different crystals and glasses in this prospect[1,2]. New results are given here for the most promising candidates. The problem of mass scale production and cost is also addressed.

2. Selection of the best scintillators

The criteria for the choice of the scintillator are based on the density, the scintillation properties and the radiation hardness [2]. The cost is also an important issue and will be discussed in paragraph 5.

2.1. Density

The compactness of the calorimeter is essential in order to reduce the detector volume and cost. Moreover the lateral containment of the showers has to be kept as small as possible (small Moliere radius) in order to minimize the contamination of the energy measurement due to pile-up. A small Moliere radius will also improve the precision of the reconstructed impact point, and help for the identification and rejection of the numerous π^0 produced at LHC/SSC. Finally a compact material will reduce the lateral spread of the shower in a high magnetic field.

2.2. Scintillation properties

The pile-up will impose severe constraints to the readout, particularly at LHC. It is therefore essential to collect as much signal as possible within one bunch crossing (15 nsec), to keep a good signal to noise ratio in the electronics chain, and to reduce the fluctuations due to the timing jitter. Decay times of the order of the bunch crossing time or even less are necessary. Only crystals with optically allowed transitions (like the transition 5d $\rightarrow$ 4f for Ce^{3+}) can give rise to a fast scintillation. In some cases, more energetic transitions, involving the core electronic states of the cations, can produce a very fast ($\approx$ 1 nsec) UV emission, like in barium fluoride[2,3].

As the electromagnetic calorimeter is likely to work in a magnetic field, a readout by photodiodes is desired. This implies a sufficient light yield (a few 100 pe^- / MeV of deposited energy), and an emission wavelength above 250 to 300nm, where the quantum efficiency of the phodiodes is still high. A light emission in the visible spectrum will also ease the problems of light collection.

The energy resolution of the calorimeter will strongly depend on all possible sources of non uniformity. The light collection in a pointing geometry will introduce a non uniformity due to the focussing effect, which depends on the refractive index of the crystal. Fluoride crystals and glasses, with refractive index around 1.5, will limit this effect to a much smaller value (and therefore make it much easier to correct) than for the BGO (index 2.15). In addition the scintillation yield should be as independant as possible from the temperature. Large temperature coefficients increase the complexity of the detector and of the software corrections, and temperature gradients between the front and back face of the crystals introduce non uniformities affecting the constant term of the resolution.

2.3. Radiation resistance

It is now well established that the most significant damage in inorganic scintillators results from the formation of colour centres in the bulk of the material, which absorb part of the scintillation light on its path to the photodetector[4]. A short radiation length will therefore reduce the total attenuation for a given damage. In addition, as most of the colour centres absorb mostly in the UV, crystals emitting light in the visible will be less severely damaged. Some surface effect may also occur but it is generally very small and saturates quickly with the dose.

The formation of colour centres is directly associated to defects or impurities, even in very small quantities. A large effort should be made to purify the raw materials to the best quality. However, in some cases, a specific doping of the crystal has proved to be an efficient and economical way of significantly increasing the radiation hardness[5].

The "Crystal Clear" collaboration is involved in a vigorous program to investigate a large number of existing scintillating crystals and develop new crystals and glasses in order to fulfil the above criteria. The most significant ones are listed in table 1 and described in ref [2]. New results and comments are given here on Cerium fluoride (CeF_3), Lead fluoride (PbF_2) and fluoride glasses.

3. Cerium fluoride

Since its rediscovery as a good scintillator for high rate calorimetry[6], much progress has been made on this crystal. The initial motivation was based on its high density (6.16 g / cm^3) due to the compactness of the lattice, which yields a small radiation length (1.68 cm) and a very small Moliere radius (2.6 cm). The scintillation mechanism is simple, involving the fast, intense and temperature independant transition 5d → 4f from the Ce^{3+} ion. Two broad emission bands centered at 300 nm and 340 nm produce a light yield which corresponds to about 500 pe^{-} / MeV (45% to 50% of the BGO with a R5094B photomultiplier gated at a 5 μsec).

The situation concerning the decay time constant of the emitted light looks rather complex. A analysis of the pulse shape on a digital oscilloscope shows a global time constant of 20 nsec with no afterglow, and a weak component of about 5 nsec[2]. For a more detailed analysis, we have used the delayed single photon counting method[7] with a monochromator to select a 30 nm wide emission bandwith. In the UV part of the emission spectrum, near the absorption band edge at 290 nm, two decay time constants are clearly visible, of respectively 8 nsec and 18 nsec (Fig 1). At longer wavelength, the fast component progressively disappears and the decay time of the slower component increases. On the low energy side of the emission spectrum, only the slower component remains, with a much slower rise time of about 8 nsec and a decay time constant of 33 nsec. This strongly suggests that the fast decay of the Cerium in normal sites excites other Ce^{3+} ions in different perturbed sites which are responsible for the slower component. It must be noticed that the distribution of the perturbed sites is such that most of the light is emitted between 300 and 320 nm, where the slow time constant is about 20 nsec, which confirms the global measurement with the digital scope.

The growth of large crystals is a major issue for use in High Energy Physics. Recently, pieces of 50 mm diameter and 150 mm in length have been grown, and an attempt to grow a 70 mm diameter and 250 mm long crystal is under way. The quality of the raw material is the dominant factor for the radiation hardness. An 8 cm long crystal, irradiated from the side (1000 Gy ^{60}Co) shows a strong dependance of the damage on the position along the crystal (Fig 2). The first half of the crystal is pure and shows no damage at all. This means that during the growth, the impurity concentration progressively increases up to a level where some of the impurities eventually enter the lattice in some specific sites allowed by the crystal field. That explains the sudden increase of the damage in the middle of the crystal, followed by a saturation when all the sites are occupied. It can be concluded from this measurement that the impurity responsible for the damage is probably not a rare earth. Rare earths enter easily the CeF_3 lattice and are uniformely distributed along the crystal, as we have checked with the Nd^{3+} ion. This is a good point to know, as the purification of Cerium against other rare earths is a very difficult and expensive process. The damage results most probably from the oxydation of another impurity like $Y^{2+} \rightarrow Y^{3+}$, as can be deduced from the strong polarisation of the damage we have observed. Work is in progress to identify this impurity and to develop an economical purification process of the raw material, oriented only to the impurities responsible for the production of color centres under radiation.

3. Lead fluoride

Lead fluoride is a very attractive crystal because of its high density (7.77 g / cm^3). Crystals of 24 cm in length (already available today) would be sufficient to contain 100 GeV γ showers (1 X_0 = 0.93 cm) and the Moliere radius is as small as 2.21 cm. Furnaces to grow crystals of this size are relatively cheap and exist already in large quantities, as well as crucibles and heating elements. Moreover the abundance of the raw material and the low melting point (822 °C) make this crystal one of the cheapest. This crystal is considered at the moment as a good Cerenkov material[8], as no luminescence has been detected so far.

The "Crystal Clear" collaboration has undertaken a systematic work to try to induce some scintillation in this crystal. A large variety of scintillating ions has been introduced in the PbF_2 lattice in order to understand the quenching mechanism for the scintillation of the Pb^{2+} ion. Similar studies are under way with Calcium fluoride (CaF_2), having the same crystalline structure as Lead fluoride, to compare the properties of the doping ions in these two crystals. A very interesting perovskite (CsPbF3) could be the ideal solution to the problem of scintillation of Lead fluoride.

4. Fluoride glasses

Glasses offer an attractive solution for calorimeters, as the production cost can be expected to be lower as for crystals. Fluorozirconate and fluoroaluminate glasses seem to offer the best perspectives in terms of radiation hardness. Unfortunately. the density of these glasses was limited to 4 to 4.5 g / cm^3 until recently. By some substitutions of heavy metals in the glasses, densities of respectively 5.72 g / cm^3 and 5.92 g / cm^3 have been recently obtained for a fluorozirconate and a fluoroaluminate. Prospects to go up to 6.5 to 7 g / cm^3 are encouraging.

Batches of 10 to 20 kg are feasible with the standart techniques in use for fluoride glasses. With some development, batches of about 100 kg could be obtained. That may have a direct consequence in lowering the production cost. In addition the glasses offer the unique advantage over crystals, that moulding, hot pressing and extrusion techniques may be used to shape the blocks. This is of course very important if one considers that for all the relatively cheap materials described in this report, the mechanical processing cost will represent a significant, if not the dominant fraction of the total cost.

5. Conclusions

Among the large variety of dense, radiation hard and fast scintillating crystals and glasses being studied by the "Crystal Clear" collaboration, Cerium fluoride, Lead fluoride and some fluoride glasses deserve a particular interest, mainly because of their attractive technical characteristics as well as their interesting economical potential.

6. Acknowlegements

The authors are particularly gratefull to R. Sparrow from Optovac (USA) for the procurement of many samples of Cerium and Lead fluoride grown in different conditions to satisfy our requirements. We also want to thank G. Mazé from the company "Le Verre Fluoré" (France) for his effort in producing fluoride glasses according to our specifications. Finally we want to thank all the members of the "Crystal Clear" collaboration and the funding agencies supporting our project, without which this work could not have been made.

7. References

1. R&D Proposal for the study of new fast and radiation hard scintillators for calorimetry at LHC
Crystal Clear Collaboration , CERN / DRDC P27 / 91-15, project RD-18
2. P. Lecoq, M. Schüssler and M. Schneegans, in *Proceedings of the 5th Pisa meeting on advanced detectors, to be published in NIM CERN / PPE 91-124*
3. Y.Valbis et al, *JETP Lett. 42 (1985) 172*
S. Kubota et al, *NIM A242 (1986) 291*
4. P. Lecoq, P.J. Li, B. Rostaing, *NIM A300 (1991) 240-258*
P. Lecoq, *Proceedings of the ECFA workshop on LHC in Aachen, CERN 90-10, ECFA90-133 Dec90, Vol 3, P651*
5. Z.Y. Wei et al, *NIM A297 (1990) 163*
Z. W. Yin et al, *NIM A275,(1989) 273*
G.J. Britvich et al, *IFVE 91-134 Protvino*
6. W.W. Moses and S. E. Derenzo, *IEEE Trans. Nucl. Sci. NS-36 (1989) 137*
D.F. Anderson, *IEEE Trans. Nucl. Sci. NS-36 (1989) 137*
NIM A287 (1990) 606
7. L.M. Bollinger and G.E. Thomas, *Rev. Sci. Instr. 32 (1961) 1044*
8. D.F. Anderson et al, *NIM A290 (1990) 385*

Table 1: ***SOME CRYSTALS STUDIED BY "CRYSTAL CLEAR" COLLABORATION***

Crystal	Density	Radiation length cm	Moliere radius cm	Decay time nsec	Peak emission nm	Radiation hardness(Gy)
BGO	7.13	1.11	2.33	300	480	10
BaF2	4.89	2.06	3.39	0.6 620	210 310	$> 10^4$
CeF3	6.16	1.68	2.63	5 20	300 340	$> 10^4$
YAP:Ce	5.35	2.83	2.82	35	390	$> 10^5$
GSO:0.5%Ce	6.71	1.39	2.42	55	435	$> 10^6$
GSO:2.5%Ce	6.71	1.39	2.42	31	450	$> 10^6$
ThF4	6.32	1.18	2.71	10 , 30	315,330,450	?
BaLiF3	5.24	2.13	3.13	2	435	?
LiYbF4	6.09	1.56	2.70	≤ 25	≈ 450	$> 10^5$
BaYb2F8	6.99	1.29	2.37	-	-	10^3 to 10^4
PbF2	7.77	.93	2.21	-	-	?

Also : **Fluoroaluminate and fluorozirconate glasses**

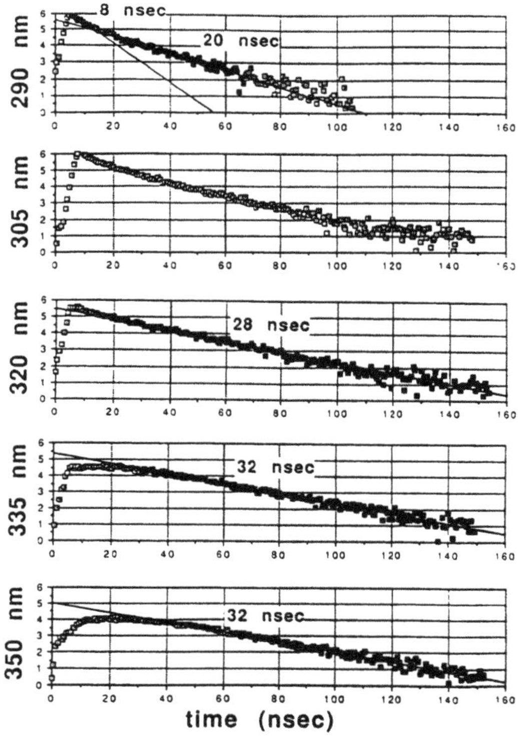

Fig 1.: CeF_3 logarithmic decay curves with 511 keV γ-excitation for different emission wavelengths

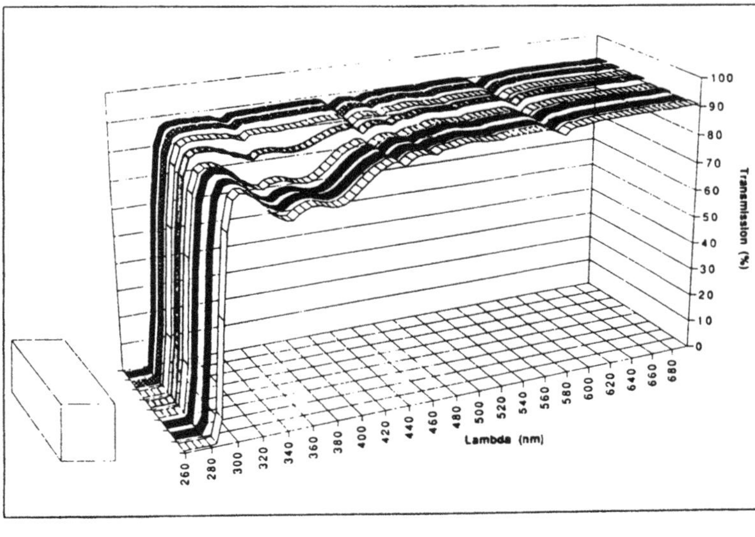

Fig. 2: Transmission of CeF_3 after 1000 Gy Co-60 along an 8 cm long crystal

NEW RADIATION HARD HEAVY CRYSTALS FOR E.M. CALORIMETRY

A. A. ASSEEV, E. G. DEVITSIN, A. A. KOMAR, V. A. KOZLOV,
Y. I. HOVSEPYAN, S. Y. POTASHOV, K. A. SOKOLOVSKY, T. V. UVAROVA

P. N. Lebedev Physical Institute of the Academy of Sciences of the USSR, Moscow, USSR

Abstract

The scintillation and radiation properties of some heavy crystals for e.m. calorimetry are described. The samples were irradiated by a ^{60}Co source and the reduction of transmission of crystals were studied. There were found the dopants which cause the increase of radiation hardness of initial crystal up to 100 Mrad dose. Also the light yield and decay times of the doped crystals were investigated.

1 Introduction

Construction of the electromagnetic calorimeters for experimental setups of future accelerators like LHC, SSC and UNK requires use of new radiation hard fast materials, such as heavy scintillators and Cherenkov radiators, because of high luminosities expected for these accelerators.

2 $BaYb_2F_8$

Along with the properties of heavy scintillators as possible candidates for homogeneous electromagnetic calorimeters we studied the radiation hardness and optical properties of various Cherenkov radiators. Among them a heavy fluoride $BaYb_2F_8$ has drawn our attention. The $BaYb_2F_8$ crystals of 10 mm in diameter and 40 mm thick were grown in the Institute of Crystallography of the Academy of Sciences of the USSR, Moscow. The $BaYb_2F_8$ is the optically transparent material, has good mechanical properties, is easily polished and not hygroscopic. Besides this the crystal possesses a significant property: its rigid structure provides a higher resistivity to ionizing radiation.

Measurements of radiation hardness of pure $BaYb_2F_8$ as well as of $BaYb_2F_8$ crystals doped with Ce, Cd, Tm, Gd, Tb, Pr, Ho, Tm^+, Er up to 2.5% by weight were carried out using 330 rad/s ^{60}Co source. Transmission curves for the samples of 10 mm in diameter and 5 mm thick were measured with the help of a spectrometer before and after irradiation. In present work we demonstrate the results for $BaYb_2F_8$ doped with different activators.

Fig. 1 shows the influence on the transmission of the dose of 7.3×10^7 rad for $BaYb_2F_8$ doped with different elements, mainly rare earths. All the dopants cause the increase of radiation hardness of the main crystal. The most interesting results have been obtained for Pr, Tb, Tm dopants. The measured transmission reduction after irradiation were for Pr, Tb and Tm - 0.5%, 3%, 4.3% respectively.

3 CeF_3

The first results on CeF_3 have been published in ref. [1,2]. In current work the radiation and luminescent properties of CeF_3 doped with BaF_2, CaF_2, SrF_2 up to 1.5% by weight are studied.

Decay times of CeF_3 crystals were investigated using the delayed coincidence technique [3]. The obtained scintillation decay curves were fitted by the exponents in order to get decay constants. Such a doping leads to a decrease in the decay constant from 32.9 ns for the CeF_3(pure) to 27.0 ns for CeF_3(0.12% Sr).

There have also been performed some measurements of radiation hardness of doped CeF_3 crystals by exposing them to 1.2×10^7 rad and 10^8 rad of γ-rays of ^{60}Co at the rate of 330 rad/sec. The light transmission of crystals was measured before and after irradiation (fig. 2).

The transmission reduction of 11% was obtained for CeF_3(pure) in peak emission spectra (340 nm) after the exposed dose of 1.2×10^7 rad. Such a reduction for CeF_3(Ca) and CeF_3(Sr) reached 5% for the same dose. There was no reduction in transmission spectra for CeF_3(Ba) after the dose of 1.2×10^7 rad and a small reduction for 10^8 rad.

4 CsI

The first results on CsI(pure) investigations have been published in ref.[4]. According to this publication the short decay constant was 10 ns and the total light output reached to 5-8% of that of NaI(Tl).

In this work we also have studied the luminescent properties and radiation hardness of CsI crystals doped with CsBr of various concentrations up to 9.6% by weight. The samples used were 20 mm in diameter and 10 mm thick.

There are three short and the long decay components. To the point the short and the long decay components has their peak emission at 305 nm and at 440 nm respectively.

The total light output of the CsI(pure) crystal was 6.5% and of the CsI crystals doped with CsBr of various concentrations reached up to 16% of that of NaI(Tl).

The investigations of CsI(Br) crystal radiation hardness was performed under the same conditions as the above mentioned crystals. The dose exposed was 1.34×10^8 rad (see fig. 3). It is obvious that when the initial CsI crystal is doped of CsBr the radiation hardness increases (especially for 1% and 7% concentrations of CsBr).

5 BaY_2F_8

Recently we have investigated the scintillation properties of the well known lazer crystal BaY_2F_8. This crystal can be easily manufactured by industry. Pure BaY_2F_8 crystal have the scintillation light output only about 0.25% of that of

NaI(Tl) but when the crystal is doped by Tm(1.39% by weight) the light output as mach as 7% of that of NaI(Tl) can be reached.

We have also studied light decay constants of BaY_2F_8 crystal (see Table 1). The long decay constants of BaY_2F_8 has only 6.2% of the total crystal light output. This crystal because of its similar characteristics of density and radiation length to these of BaF_2 may be the good substitution of BaF_2 in e. m. calorimetry in high energy physics.

Table 1

Comparison of the main properties of scintillators

	BaF_2	BaY_2F_8
Density, g/cm^3	4.88	5.08
Radiation length, cm	2.1	2.02
Light yield (NaI=100%)	20	7
Decay constants, ns short long	 0.6 (25%) 620 (75%)	 1.33 (82.6%) 7.6 (11.2%) 100 (6.2%)

6 References

[1]. D.F.Anderson, *IEEE Trans. Nucl. Sci.* **NS-36**(1989) 137

[2]. W.W.Moses and S.E.Derenzo, *IEEE Trans. Nucl. Sci* **NS-36**(1989) 173.

[3]. L.M.Bollinger, G.E.Thomas, *Rev.Sci.Instr.* **32**,1044-1050 1961.

[4]. S.Kubota et all, *NIM* **A268**(1988) 275

7 Figure captions

Fig.1 Transmission curves for the dose of 7.3×10^7 rad for $BaYb_2F_8$(pure) and $BaYb_2F_8$ doped with different mixtures.

Fig.2 Transmission spectra of CeF_3(pure) and CeF_3+ dopants before and after irradiation.

Fig.3 Transmission spectra of CsI crystals doped with CsBr before and after irradiation.

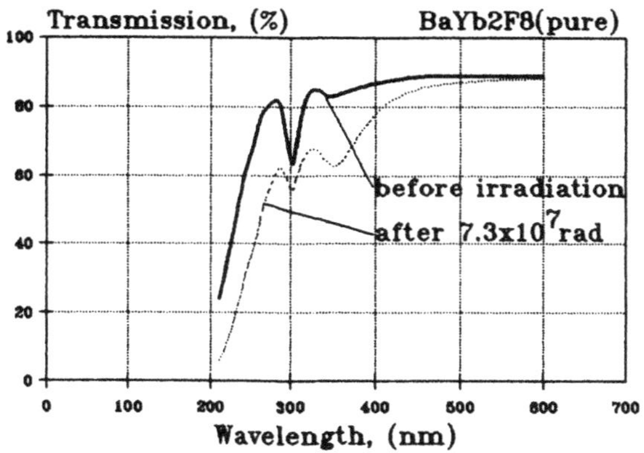

Fig.1a

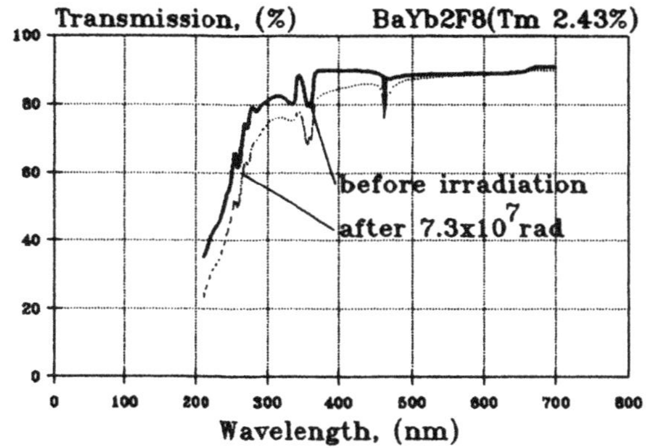

Fig. 1b

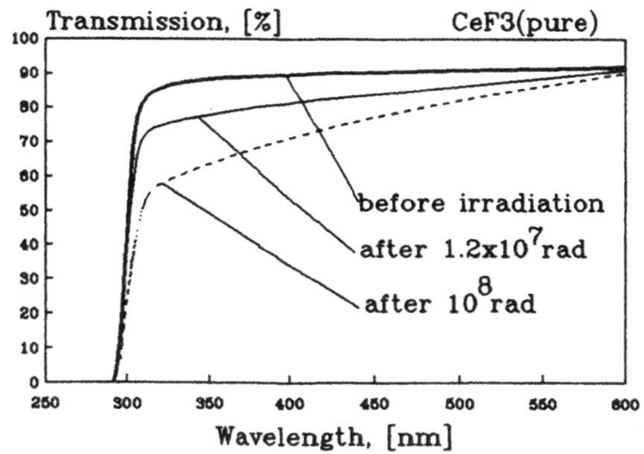

Fig. 2a

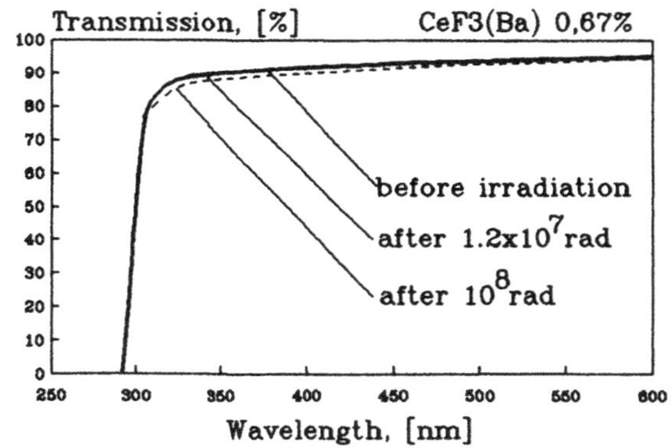

Fig.2b

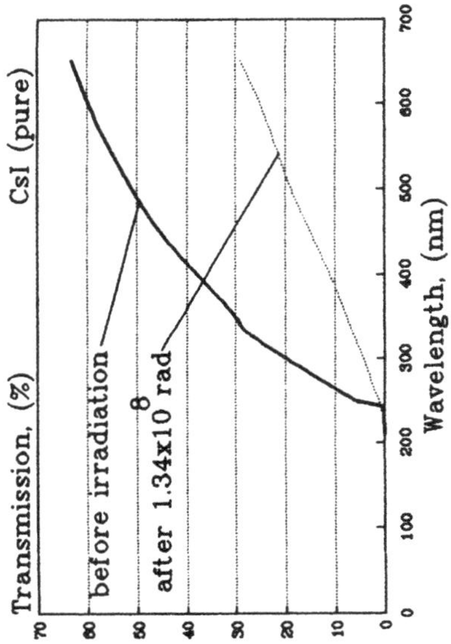

Transmission, (%)
CsI (pure)
before irradiation
after 1.34x10^8 rad
Wavelength, (nm)
70
60
50
40
30
20
10
0
0
100
200
300
400
500
600
700

Fig.3a

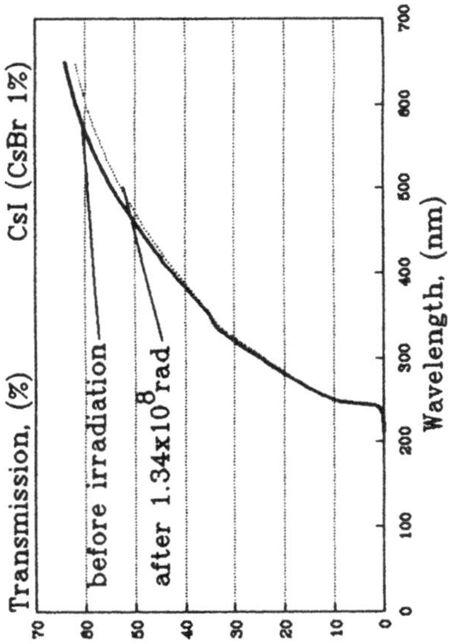

Transmission, (%)
CsI (CsBr 1%)
before irradiation
after 1.34x10^8 rad
Wavelength, (nm)
70
60
50
40
30
20
10
0
0
100
200
300
400
500
600
700

Fig.3b

Radiation Hardness Study of Scintillating Crystals [1]

S.Banerjee, P.K.Malhotra, R.Raghavan and S.C.Tonwar
Tata Institute of Fundamental Research, Bombay
(Presented by A.Gurtu)

Abstract

We have studied the effect of sustained radiation to MeV neutrons and photons in various scintillating crystals. The neutrons are obtained from the research reactor Apsara in the Bhabha Atomic Research Centre, Bombay. The photons are obtained from a gamma chamber equipped with Co^{60} source. The change in the transmission as well as the scintillation properties have been studied.

1. Irradiation Facilities

Apsara is a swimming pool type research reactor in the Bhabha Atomic Research Centre (BARC). It uses a core of uranium rods enriched with U^{235} in a pool filled with natural water. Cadmium rods are used as controllers and moderators. The reactor normally operates at 300 KW. Two rows, close to the core, are used for placing samples. Neutron flux has been measured in one of these positions (E8) [1] and this position has been used for the present work. Flux of neutrons with energy above 1 MeV has been measured to be $6 \times 10^{10} cm^{-2} sec^{-1}$ in E8 with reactor operating at 400 KW. The thermal neutron flux is $\sim$ 25 times higher. The sample to be irradiated in this position was enclosed in 0.5mm thick cadmium cover so that thermal to MeV neutron flux becomes 1:100.

A cadmium column (SNIF) at 47 cm from the core provides fast neutron flux of $2.5 \times 10^{8} cm^{-2} sec^{-1}$ with the reactor operating at 1 MW. The irradiation chamber has $\sim$2 cm thick lead wall. The gamma contamination as well as the slow neutron component is small (7% or less).

A gamma chamber, situated at the BARC, is equipped with several Co^{60} sources giving MeV photon flux. The chamber normally operates at 3.33 KRad / minute.

[1] Work carried out in the framework of the Crystal Clear Collaboration

2. Measuring Facilities

The transmission properties were measured using a Shimadzu UV2100 double beam Spectrophotometer at TIFR. Transmission measurements were done in the range 190 – 900 nm in steps of 0.1–5.0 nm. The instrument has an accuracy of ±0.3% and reproduciblity of ±0.1%. The sample compartment is of size 15cm × 26cm ×11cm with a maximum allowable path length of 100 mm. All measurements were carried out using a slit width of 2 nm at medium speed which gives a small noise level.

Scintillating properties of the crystals were studied by their response to known sources like Cs^{137} (661 KeV gamma line) or Co^{60} (1.17 and 1.33 MeV gamma lines). Crystals were mounted on a RCA 8575 photomultiplier operated between 1.5–1.7 KV and the output from the PM fed to a qVT multichannel analyser with 1024 channels (each channel corresponding to 0.25 pc). The internal trigger mode had a threshold set at 6 mV and a gate width of 1 μs. The MCA was read out to a PC using qVT–CAMAC and GPIB interfaces.

3. Irradiations

We have studied the radiation hardness properties of two BaF_2, one CeF_3 and one Gd_2SiO_5 crystal. The details of the irradiations of these crystals are summarized in table 1. The BaF_2 crystals were irradiated twice, the second time after annealing.

Table 1: Irradiation details of the crystals

Crystal	Size of the crystal	Period of Irradiation (min.)	Neutron Fluence (cm^{-2})	Thermal to fast n ratio	Gamma Irradiation (K rad)
BaF_2 (CERN)	1.5cm×1.5cm ×3.0cm	20.0 440.0	10^{13} 10^{12}	1 : 100 1 : 1000	40 0.3
BaF_2 (BARC)	3.8cm (dia) ×2cm	30.0 440.0	 10^{12}	 1 : 1000	100 0.3
CeF_3	1cm×1cm×1cm	30.0	10^{12}	1 : 1000	0.3
Gd_2SiO_5	1cm×1cm×3cm	30.0	10^{12}	1 : 1000	0.3

4. Results

4.1 Induced radioactivity

The induced radioactivity levels were measured at various time intervals after the irradiations. The details of these measurements are summarized in table 2. Neutron irradiation induces measurable radioactivity in all the crystals. The Gd_2SiO_5 crystal shows a relatively high level of radioactivity owing to the presence of gadolinium. Half lives of the induced radioactivity are quite small for all these crystals.

Table 2: Induced radioactivity in the crystals

Crystal	Exposure	After 0 day (mr/h)	After 1 day (mr/h)	After 7 days (mr/h)	After 30 days (mr/h)
BaF_2	1	20		1.0	0.02
(CERN)	2	40	0.03		
BaF_2	1	0			
(BARC)	2	30	0.03		
CeF_3	1	40	0.02		
Gd_2SiO_5	1	150	0.3		

4.2 Transmission properties

Both the BaF_2 crystals turned violet after the first irradiation. The transmission dropped over the entire spectrum, more significantly in visible and uv regions as shown in figure 1a for the CERN crystal. There is a slow recovery with time.

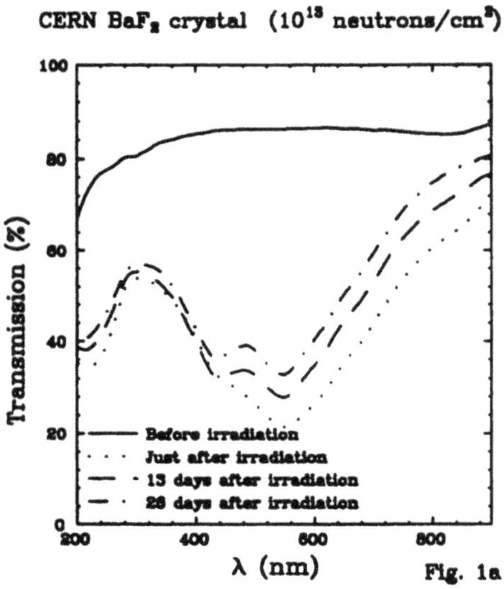

Fig. 1a

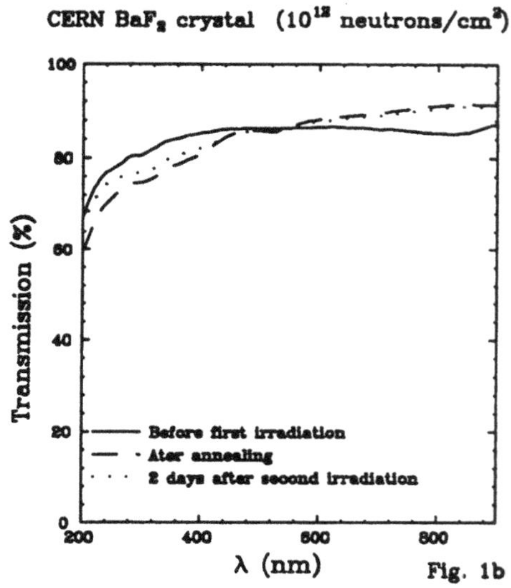

Fig. 1b

The two crystals become transparent after annealing at 300° for 30 minutes. The transmission property in the uv region was not completely regained (figure 1b).

The irradiation to 10^{12} neutrons cm^{-2} (the second irradiation to the BaF_2 crystals and the first irradiation to CeF_3 and Gd_2SiO_5 crystals) neither produced any visible damage nor any deterioration of transmission properties of the crystals (figures 1b, 2). The small increase in the transmission of the CeF_3 crystal after irradiation is partially due to the uncertainty in the measurement induced by the positioning error.

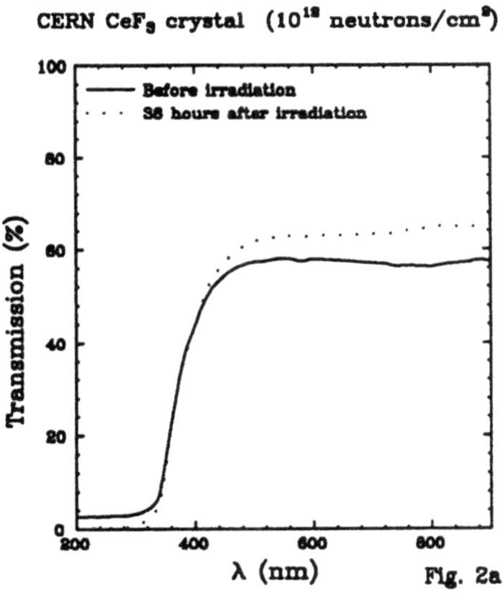

Fig. 2a

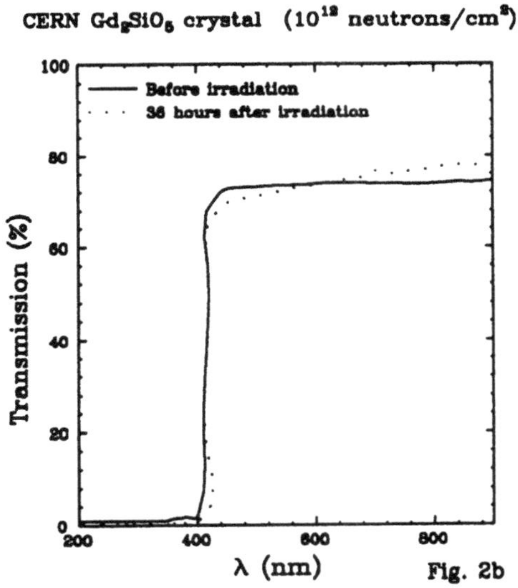

Fig. 2b

4.3 Scintillation properties

Before the irradiations, all four crystals showed clear signals from the Cs^{137} 667 KeV and Co^{60} 1.17 MeV gamma lines. The BaF_2 crystals have typical energy resolutions of 10–20% at 661 KeV (figure 3a). The energy resolution of the Gd_2SiO_5 crystal is found to be the best and it could clearly resolve the two Co^{60} lines at 1.17 and 1.33 MeV (figure 3b).

After the irradiations, the energy spectra were taken with and without a radioactive source. All the crystals showed luminiscence which decreased slowly with time. No appreciable change in the resolution was observed.

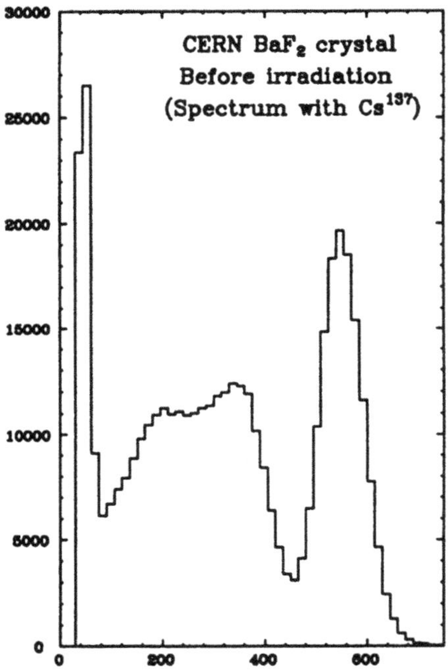

Fig. 3a

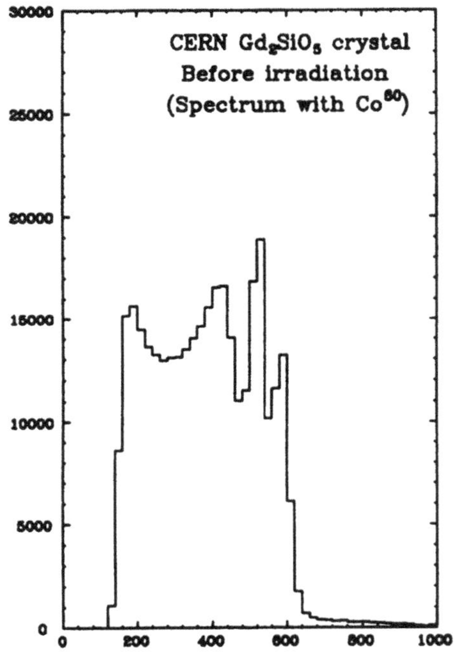

Fig. 3b

References

[1] H.K.Dravid *et al.*, Measurement of Neutron Spectrum at Apsara Core Edge, BARC/I-808 (1984).

OBSERVATION OF THE UV SCINTILLATION LIGHT FROM HIGH ENERGY ELECTRON SHOWERS IN LIQUID XENON †

Tejinder S. VIRDEE
Blackett Laboratory, Imperial College, London SW7 2BZ, U. K.

ABSTRACT

The UV scintillation light in liquid xenon has been observed using silicon drift photodiodes. A high quantum efficiency for UV light has been measured. The diodes and the preamplifiers were immersed in the liquid. The scintillation light yield is found to be (1.8 ±0.5) 10^4 photons per MeV. A light attenuation length of ≈ 40 cm has been measured.

1. Introduction

The use of liquefied noble gases, like xenon (LXe) and krypton (LKr), for high resolution (and also high speed) calorimetry has recently evoked a great deal of interest, especially in connection with future hadron colliders[1]. Noble liquids are believed to be able to withstand large radiation doses though the radiation hardness of a system will probably be limited by the hardness of the read-out elements. High energy e.m. showers can be measured by the collection of the ionization electrons[2] and/or by the detecting the UV scintillation light. The latter offers the advantages of fast signals, insensitivity to magnetic fields and the need for ony modest liquid purity. There is also the possibility (not yet demonstrated) of obtaining e/h closer to 1 by exploiting the different scintillation decay constants for lightly and heavily ionizing particles (2 and ~25 ns respectively), and the absence of saturation of the scintillation light yield up to very high LET radiation.

The relevant properties of some noble liquids are listed in Table 1. Of the noble liquids, LXe has the shortest radiation length, the smallest scintillation decay time constant, the highest emission wavelength and the highest operating temperature.

The main problem of the light detection method is the spatial non-uniformity of light collection due to imperfect reflectivity of cell surfaces and the finite attenuation length of light. The non-uniformity should be minimized as it can lead to a sizeable systematic contribution to the energy resolution.

2. Experimental Details

The experimental setup is shown in Fig.1. The scintillation light generated in the liquid xenon filled cell of dimensions 7.4 x 7.4 x 60 cm^3 was detected by silicon drift photodiodes (SDPD). The various components of the system are described below.

2.1 *Mechanical and Optical Properties of the Light Collecting Cell*

The 3 mm thick Al cell walls were made into mirrors by depositing, under vacuum, a layer of 70 nm of pure aluminium followed by 30 nm of Mg F_2 to prevent oxidation.

Table 1. Properties of noble liquids.

		LAr	LKr	LXe
Density	g/cm³	1.39	2.45	3.06
$<dE/dx>_{mip}$	MeV / cm	2.11	3.45	3.89
Radiation Length	cm	14.3	4.76	2.77
Moliere Radius*	cm	7.3	4.7	4.1
Photons/MeV	k		19	26
% light in fast component		8	1	77
Decay Const. Fast	ns	6.5	2	2
Slow	ns	1100	85	25
λ peak	nm	130	150	175
Refractive Index @ 170nm		-	1.41	1.60
Drift vel (10kV/cm)	cm/μs	0.5	0.5	0.3
Fano Factor		0.11	0.06	0.05
Dielectric Constant		1.51	1.66	1.95
Temperature	°K	94	124	160

* PDG definition

Reflectivity was measured to be ≈ 85% at 175 nm. Four 1 cm² SDPD's were placed at each end of the rectangular tube. The preamplifiers were placed behind the photodiodes.

2.2 *The Si Drift Photodiodes and the Preamplifiers*

Silicon drift photodiodes are photodiodes based on the principle of the solid state drift chamber[3]. The active area (1cm²) was subdivided into five 2mm wide cells with a central anode in each cell to collect the electron signal. The five cells of each diode were connected in

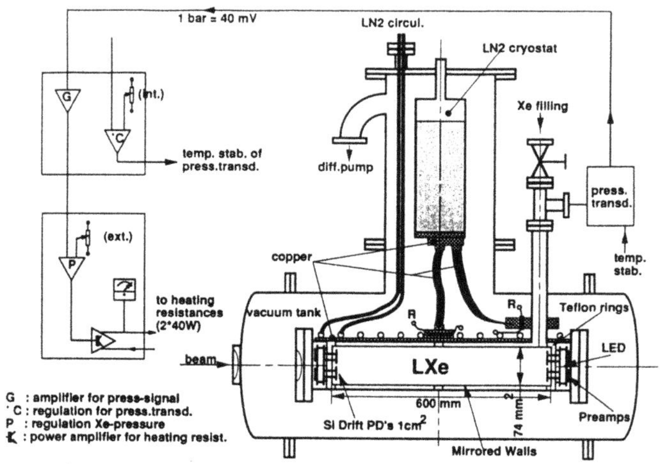

Fig 1. The experimental setup.

parallel. These diodes have a considerably lower capacitance than conventional silicon photodiodes. The drift velocity in silicon increases with decreasing temperature and a collection time of ≈ 50 ns is observed at –91 ^{o}C for 1mm drift.

The external quantum efficiency (Q), defined as the number of electrons detected per incident photon, is given by $Q = T \eta F$. The transmittance, T, as a function of the incident angle for light of wavelength 175 nm is shown in Fig. 2. At 175 nm the internal quantum yield, η, is ~1.80. F is the probability that a minority carrier generated at a certain distance from the surface will cross the junction before recombination. The measurement of the quantum efficiency of a typical DSPD is shown in Fig.3.

The readout electronics consisted of Laben 5254 charge sensitive preamplifiers followed by shaping amplifiers with a peaking time of 0.6 μs. An E.N.C. of ~ 1000 electrons was measured during the experimental run. Neither the diodes nor the electronics were optimized for these tests and full advantage was not taken of the speed of the diodes.

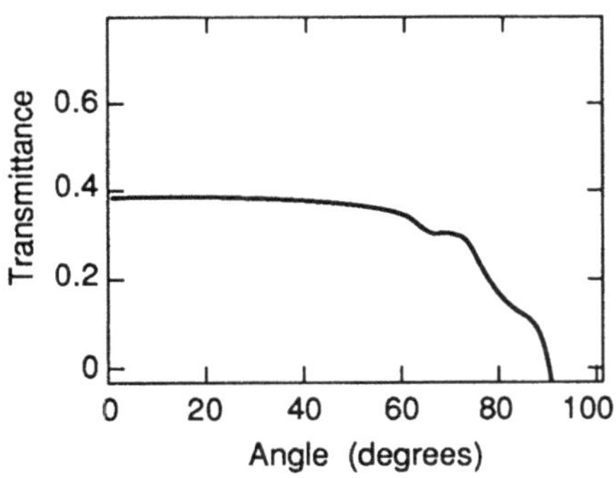

Fig. 2. The transmittance of 175 nm light into SDPD as a function of angle of incidence

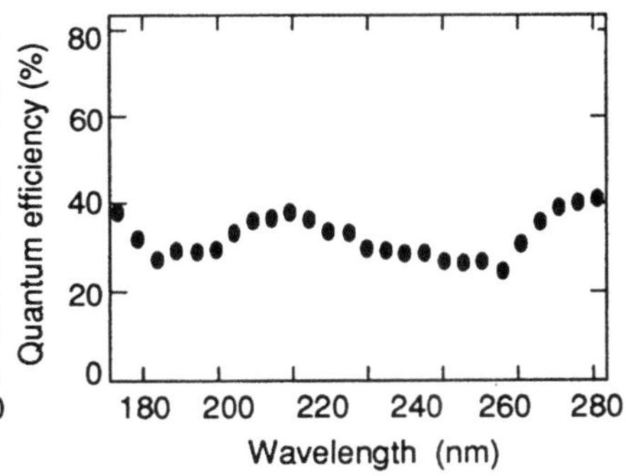

Fig.3. The quantum efficiency of a typical SDPD.

2.3 The refractive index of liquid xenon

Measurements of the refractive index over the wavelength range 350 - 650 nm are reported in reference 4. A modified form of the Lorentz-Lorenz equation gives a very good description of the data and extrapolating down to a wavelength of 175 nm. gives a value of 1.60 for the refractive index (Fig. 4). This value should be treated with caution as it is very sensitive to the functional form of the extrapolation. We are not aware of any measurement of the refractive index of liquid xenon at a wavelength of 175 nm.

2.4 The Cryogenic System and Temperature Control

Xenon gas was purified by passing it through Oxisorb®, silica-gel and molecular sieves and finally a stainless steel dust filter with a pore-size of 0.5 μm. The cooling power to the inner vessel was provided by two copper braids connecting the cryostat to the inner vessel. Copper blocks containing a 40 W resistor are soldered at the ends. Both copper braids delivered a total colling power of 150 W. The temperature of liquid xenon can be deduced from the vapour pressure. The signal from a pressure transducer was amplified and fed into a comparator whose reference could be set externally. The output of the comparator was used to alter the current through the heating resistors R. After some time the pressure, and hence the temperature, of xenon stabilizes to the reference value. A stability of ±10 mbar was achieved. The cooling system proved to be very reliable and stable but insufficient for fast liquifaction. Therefore, an auxilliary liquid nitrogen cooling loop was used during the liquefaction. The nitrogen flow was regulated such that the xenon pressure always stayed above the triple point. After liquefaction the auxilliary sytem was switched off and disconnected.

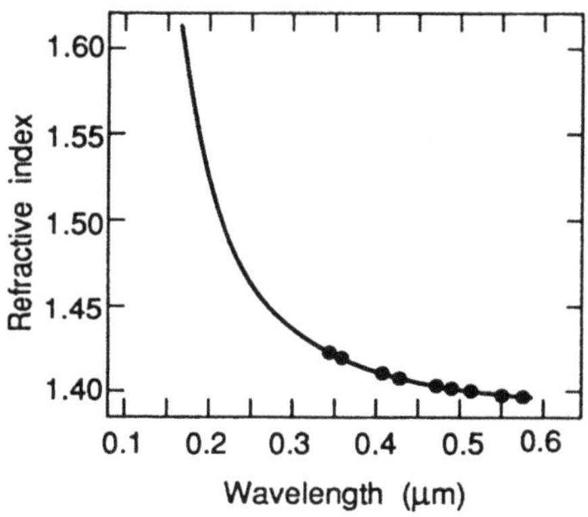

Fig. 4. The refractive index of liquid xenon versus wavelength

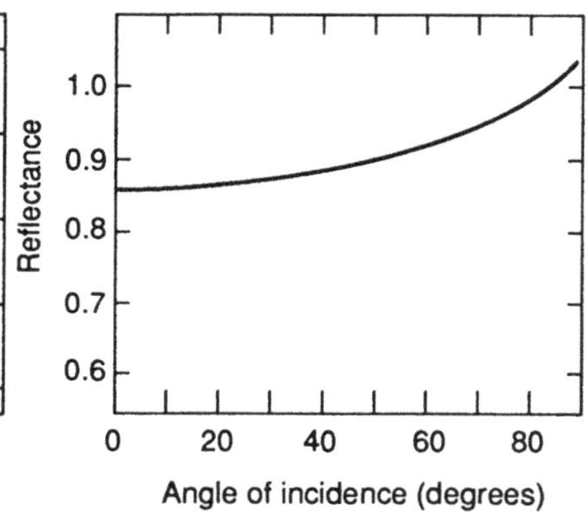

Fig. 5. Side wall reflectance versus angle of incidence for 175 nm light.

2.5 *Light Collection Monte Carlo*

The operation of the test cell was simulated using the Monte Carlo program GEANT. We introduced into the Monte Carlo our estimation of the reflectivity of the side walls and the transmittance of light into the DSPD's as a function of the angle of incidence. Shower development in the experimental setup and the propagation light and its collection by the eight SDPD was simulated.

The light sensitive volume was divided into cubes with sides of 1cm. 100,000 light rays, with an isotropic angular distribution, were generated in the centre of each cube. Using the appropriate values for the reflectivity of side walls (Fig.5), the transmittance into the SDPD, and the chosen attenuation length a decision was made as to whether the photon was considered to be detected.

The energy deposited in the cubes inside the sensitive volume of the cell was recorded for 5000 showers of an energy of 5 GeV. For each shower, the number of photons emitted at the position of the centre of the cubes was taken to be proportional to the energy deposited in each of the cubes. This was then weighted by 8 values of light collection efficiency, one for each diode position, computed for the appropriate source position. The sum of these numbers, over all the 2940 cubes, gave the predicted diode signals for that shower.

3. Results and Analysis

Data were taken in the CERN-PS T7 test beam which was used to deliver 5 GeV electrons.

3.1 Light Yield

The total signal observed, for 10000 events, is shown in Fig. 6. The number of photons, n_γ, emitted per unit energy deposited in LXe can be expressed as $n_\gamma = ne / (E f_E f_l \eta F)$ where E is the incident particle energy in MeV, n_e (= 3.2.10^5) is the observed number of photoelectrons, f_E (= 0.77± 0.05) is the fraction of shower energy deposited in the sensitive volume, f_l (=[5.8 ± 1.0]10^{-3}) is the fraction of the emitted light that is absorbed by SDPD's. T enters into the estimation of f_l and thus a measurement of the external quantum efficiency Q (=0.40±0.1) is sufficient to obtain ηF. The error in the calibration of the pre-amplifiers is estimated to be ~ 20%. We find that n_γ = (1.8 ±0.5).10^4 photons / MeV. This is consistent with previous determinations of the light yield of ~ 2.6 . 10^4 photons / MeV [5].

3.2 Attenuation Length

The apparatus was rotated about the vertical by 90° and electrons of an energy of 5 GeV were fired perpendicularly into the long sides. The apparatus, placed on a mobile platform, was moved horizontally in steps of 5 cm. One thousand events were recorded for each position. The ratio of the signal observed by the diodes (l.h.s./r.h.s or r.h.s./l.h.s.) is plotted in Fig. 7. The large difference in the observed signal close to, and far away, from the diode planes is mainly due to two effects, namely non-uniformity of light collection and attenuation of light. Response of the left and right hand side diodes was estimated in the Monte Carlo for the different entry positions and is compared with data in Fig 7. All the diodes were assumed to have the same quantum efficiency. A satisfactory description of the data is obtained for an attenuation length (λ) of 40 cm.

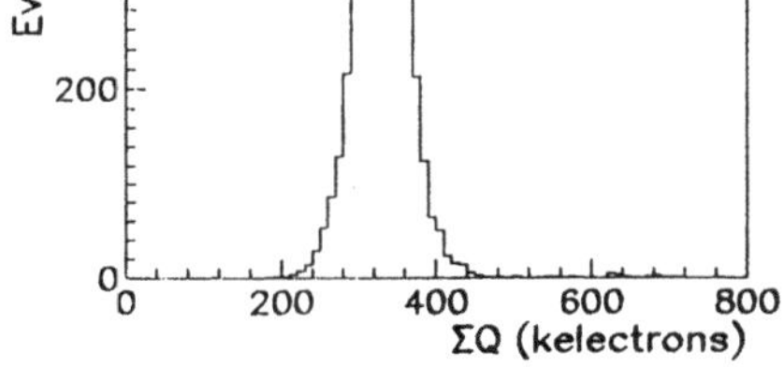

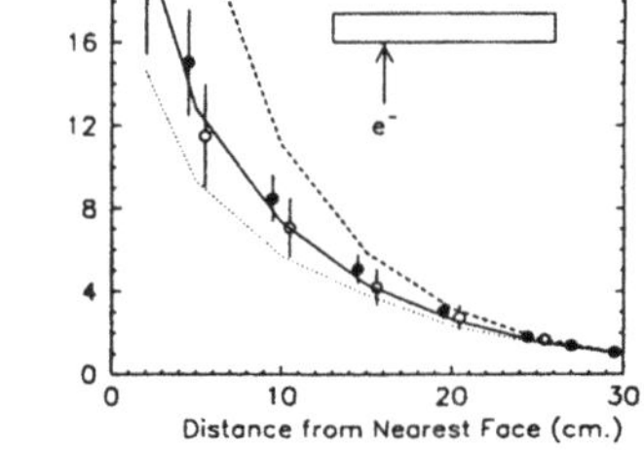

Fig. 6. The observed signal for 5 GeV electrons.

Fig. 7. The ratio of the signal observed by the diodes (l.h.s./r.h.s or r.h.s./l.h.s.)

An independent check of λ can be obtained by comparing the ratio of the observed signals in the front and rear diodes for the run with 5 GeV electrons in which the electrons were incident normal to the plane of the diodes. The predicted value of the ratio is plotted as a function of λ in Fig.8. The vertical bars represent the r.m.s. of the simulated distribution. Again λ ≈ 40 cm is favoured. It is not known if our value of λ is limited by contamination of liquid or is an intrinsic property of the liquid. An attenuation length of ≈ 10 cm is reported in ref. 6.

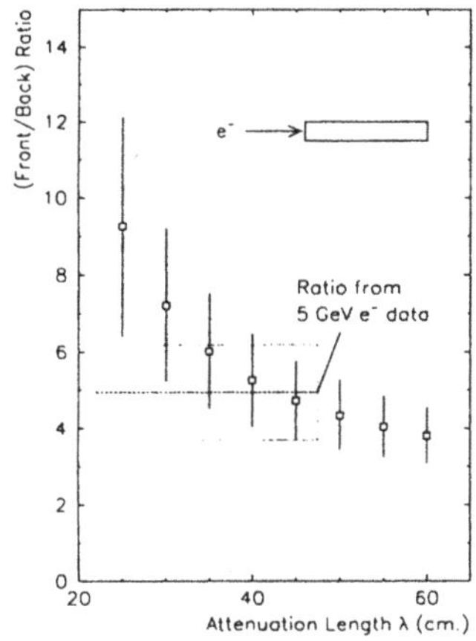

Fig. 8. The ratio of the observed signal in the front and rear SDPD's.

3.3 Energy Resolution

In the runs where the beam is incident along the longitudinal axis of the cell a substantial fraction of the shower energy leaks out. The energy resolution is however dominated by the non-uniformity of light collection. The contributions to σ/E are: 3.0 % due to energy leakage alone, and 8.0 % after inclusion of the non-uniformity of light collection. This agrees reasonably with the experimentally observed resolution of σ/E = 9.5 %. The intrinsic energy resolution in liquid xenon is expected to be much better. The best measurement of the energy resolution using Si photodiodes has been reported to be ~ 0.6 % for an energy deposit of 1.64 GeV from 117 MeV/n ^{14}N ion beam [7].

Acknowledgements

Several useful discussions with M. Chen and T. Ypsilantis are gratefully acknowledged.

† *A. Braem, A. Gonidec, D. Schinzel, W. Seidl* - CERN, Geneva, Switzerland
E. Clayton, G. Davies, G. Hall, R. Payne, S. Roe, C. Seez, J. Striebig, T. S. Virdee - Blackett Laboratory, Imperial College, London SW7 2BZ, UK.
D. Cockerill - Rutherford Appleton Laboratory, Chilton, UK.

References

1. M. Chen et al., Nucl. Instr. and Meth., **A267** (1988) 43.
   J. Seguinot et al., CERN-LAA/89-1 (1989), CERN-LAA/PC/90-24 (1990).
   E. Clayton et al., IC/HEP/90-9, (1990), Blackett Lab., Imperial College, U.K.
2. V.M.Aulchenko et al., Nucl. Instr. and Meth., **A289** (1990) .
   A. Baranov et al., CERN EP/90-03, 1990.
3. B.S. Avset et al., Nucl. Instr. and Meth. **A288** (1990) 131.
4. A.C. Sinnock and B.L. Smith, Phys. Rev. **181** (1969) 1297.
5. E. Aprile et al., Proc. of IEEE 1989 Nuclear Science Symposium, San Fransisco, 1990.
   J. Seguinot et al., CERN/AT-CR/91-08, 1991.
6. T. Ypsilantis, CERN 90-10, Proceedings of ECFA LHC Workshop, Aachen, 1990.
7. M. Chen, private communication.

SUMMARY ON PHYSICS REQUIREMENTS

M. Della Negra

CERN/PPE

ABSTRACT

Progress on the understanding of the physics requirements for calorimetry at LHC/SSC since the 1990 Aachen workshop has been discussed in this workshop. This report is a summary of the main conclusions reached in the parallel session on "physics requirements". Full write-up of the different contributions are also available in these proceedings.

1. Introduction

Calorimeters will play a major role in LHC/SSC detectors: they are needed for the detection of electrons, photons, jets and missing transverse energy, which, together with the muons, are the building blocks for the signature of the wanted physics signals at this new energy scale: Higgs, top, Susy etc... At the minimum the performance of the calorimeters should be matched to the wanted physics signatures as known of today. As we will see this is very demanding for the e/γ calorimeters, less demanding for the hadronic calorimeters. This talk is a summary of the presentations made in the parallel session on "physics requirements". Our main reference was the proceedings of the LHC Aachen workshop [1]. Most of the presentations were in fact continuations of studies started for Aachen. We first reviewed the requirements on e/γ calorimetry: J.P. Repellin [2] and D. Froidevaux [3] discussed the problem of electron detection; C. Seez [4] and A. Gurtu [5] discussed the problem of photon detection in the context of the 2-photon decay of the intermediate mass Higgs. H. Plothow-Besch [6] discussed the effect of dead material in front of an e/γ calorimer. Finally L. Poggioli [7] reviewed the requirements for hadronic calorimetry coming from jet physics and missing transverse energy measurements. Interesting but more technical contributions by C. Charlot [8] and D. Groom [9] are not covered in this report.

2. Electron detection

The problem of electron detection in hadronic colliders is a problem of rejection against QCD jets. What is the needed level of rejection? This depends on the topology of the final state: number of electrons, additional jets, missing transverse energy. In the goldplated 4 electron channel for Higgs search the background is from 4 jets or Z+ 2jets, in both cases the rejection needed to achieve S/B = 10 is $3\ 10^3$ per jet. In the 2 electron channel for inclusive Z production, the background is from 2 jets. For typical cuts $P_T > 20$ GeV, $|\eta| < 2$ the ratio ee/jj is of order 500 pb/ 330 μb = $1.5\ 10^{-6}$. Again the rejection needed to achieve S/B = 10 is $3\ 10^3$ per jet. The most demanding rejection factor comes from the one electron channel (inclusive), for $P_T > 50$ GeV and $|\eta| < 2$ e/j is of order 10^{-4} (Fig. 1) dominated by bottom decays (non-isolated electrons), for isolated electrons from inclusive W production the ratio e/j is 10^{-5}. This means that rejections of 10^6 per jet must be achieved to detect inclusive isolated electrons with less than 10% background contamination. This is a formidable task at 10^{34} luminosity. Rejections of 10^5 have been achieved at low luminosity by UA1 and UA2. Asking for additional jets and missing transverse energy may gain a factor 10 to 100 [10].

J-P. Repellin [2] showed that rejections of 10^3 against jets of 50 GeV ($|\eta| = 0$) can be achieved by asking that the jet looks like a single isolated electromagnetic shower in the e/γ calorimeter ($\Delta\eta \times \Delta\phi$ = 0.02x0.02). For a coarser segmentation of 0.04x0.04 the rejection decreases only by a factor 2. The electromagnetic triggering jet consists always of a leading π^0 sometimes accompanied by soft photons.

In the no magnetic field case (B=0) half of the jets have a track pointing to the cluster therefore faking an electron.

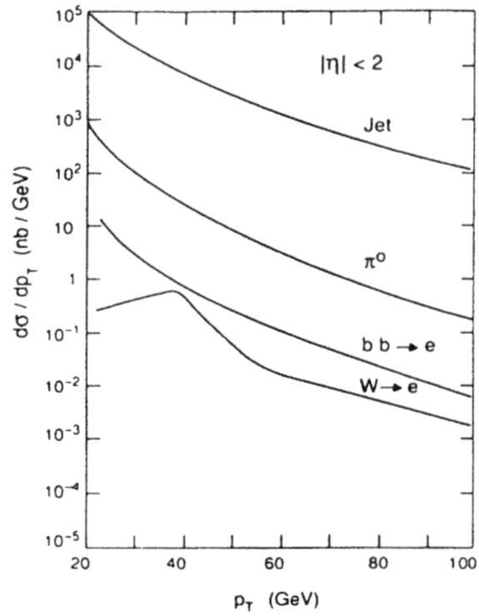

Fig.1 $d\sigma/dP_T$ in $|\eta| <2$ for jets, π^o and electrons from bottom and W decays versus P_T at $\sqrt{s}$ = 16 TeV.

To reject these "π^o jets" by a further factor of 10^3 is a difficult task. Tracking information is needed in addition. D. Froidevaux [3] presented detailed studies on electron identification with and without magnetic field. Without magnetic field (B=0) electrons can be faked from the overlap of a charged hadron with a π^o. In this case preshowers and/or transition radiators are needed providing each similar rejections of 10^2. With magnetic field ($B\neq 0$) overlaps are removed to a negligible level by asking E= P . In all cases the absolute limit (10^{-6}) of jet rejection at low momenta (10 GeV P_T) is given by $\pi^- \to \pi^0$ charge exchange: $P(jet \to e) = P(jet \to \pi^-) x P(\pi^- \to \pi^0) = 10^{-3} x 10^{-3}$.

After removal of these fake electrons one is left with the real electron background from Dalitz and conversions. The Dalitz background is negligible compared to conversions for most LHC/SSC trackers with material thicknesses varying between 5% and 10% of radiation lengths (r.l.). Fig.2 shows the rejection against conversions from π^0's with P_T > 10 GeV that can be achieved as a function of the magnetic field (from 0 to 4 Tesla) requiring successively: i) the converted electron to point to a calorimeter cluster above 10 GeV threshold, ii) E/P< 1.2, iii) outer isolation, i-e only one track in $\Delta\eta x \Delta\phi$ = 0.05x0.05 at R=1m and iv) inner isolation, i-e only one track in $\Delta\eta x \Delta\phi$ = 0.024x0.024 at R=0.3m. In this study it is assumed that the calorimeter starts at R=1m and that conversions occuring after 1% r.l. (i-e inside the sensitive volume of the tracker) are recognized. In these conditions a rejection factor of 40 against converted electrons is needed to achieve $P(jet \to e) = P(jet \to \pi^0) x P(\pi^0 \to e) < 10^{-6}$.

From Fig.2 one concludes that the rejection against conversions is adequate at all values of the magnetic field for P_T > 10 GeV. At B=0 outer isolation is very powerful (preshower technique). For large field values (B > 2 Tesla) the cut E/P < 1.2 combined with isolation provides the necessary rejection. Similar rejections are obtained at higher P_T values with the E/P cut becoming more and more efficient. It should be noted that for $B\neq 0$ the efficiency of the E/P cut is dominated by Bremsstrahlung effects. For a tracking volume of radius 1m an inner detector at R= 0.35m of 5% r.l. implies 92% efficiency for the cut E/P < 1.2 due to Bremstrahlung.

The isolation cuts to reject jets faking electrons, background electrons from bottom decays and conversions play a crucial role in electron identification. At high luminosity pile-up effects forbids the use of standard cuts on ΣE_T in a cone corresponding to the jet opening. The isolation cuts used in

Fig.2 are multiplicity cuts in very small cones (occupancies < 10%) adapted to a high luminosity detector without magnetic field. With magnetic field one would tempt to recognize all high P_T tracks in a larger cone around the electron and apply isolation cuts on those. This is very demanding on the performance of the tracker, but would result in more efficient isolation cuts This indicates that with magnetic field one would aim for a much better (i-e more expensive) inner tracker.

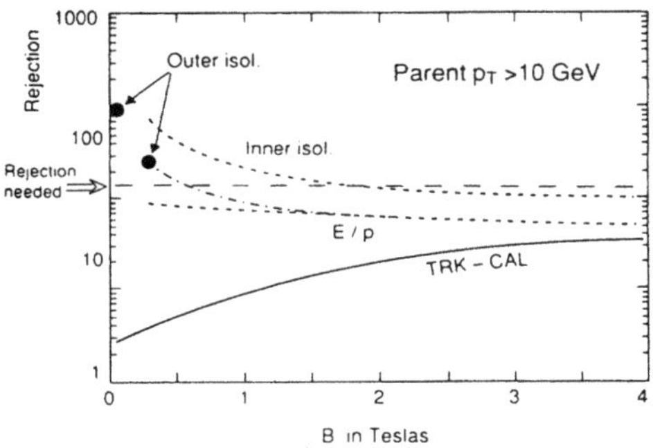

Fig.2 Rejections against converted electrons from π^0 with P_T > 10 GeV as function of the magnetic field.

3. Photon detection

The most stringent requirement on the performance of the e/γ calorimeter comes from the desire to detect the 2 photon decay of the intermediate mass Higgs. In Aachen C. Seez et al.[11] concluded that a $H \rightarrow \gamma\gamma$ signal could be seen at LHC provided: i) the energy of the photon is reconstructed to better than 1%, ii) the direction of the photon is reconstructed with an rms error of 7 mrad and iii) the QCD jet background is reduced well below the $\gamma\gamma$ continuum, requiring a rejection factor of order 10^4 per jet.

At this workshop C. Seez [4] presented new simulation results, including π^0 rejection with a position detector, obtained since Aachen. Fig. 3 shows after all cuts how the $H \rightarrow \gamma\gamma$ signal (M_H = 100 GeV) would look like after one year of run at LHC (10^5 pb^{-1}) with a very precise e/γ calorimeter ($\sigma/E = 2\%/\sqrt{E} \oplus 0.5\%$ and $\sigma(\theta_\gamma)$ = 7mrad, Fig. 3a,b) and with a standard e/γ calorimeter ($\sigma/E = 10\%/\sqrt{E} \oplus 1\%$ and $\sigma(\theta_\gamma)$ = 7mrad, Fig. 3c,d). Clearly a very precise e/γ calorimeter is needed.

In Fig.3 the jet background is assumed to be negligible compared to the 2γ continuum. To achieve the necessary rejection it was argued in Aachen that the LHC detector should allow : i) to reduce the jet to a few leading π^0's by using isolation cuts (see previous discussion), ii) to further reject π^0 jets by recognizing, with the help of a highly segmented position detector after about 4 r.l. inside the e/γ calorimeter, that the triggering electromagnetic cluster is not consistent with a single electromagnetic shower.

Fig. 4 shows a comparison of the jet background and the irreducible 2γ background as a function of the 2γ mass. The isolation cut used to reduce the jet background in this analysis is the following: measure photon energy in a nonet (3x3) cluster $\Delta\eta x \Delta\phi$ = 0.1x0.1, reject if any charged track with P_T > 2 GeV is found in $\Delta\eta x \Delta\phi$ = 0.2x0.2, reject if any additional electromagnetic cluster with E_T > 2 GeV is found in $\Delta\eta x \Delta\phi$ = 0.2x0.2 (but outside $\Delta\eta x \Delta\phi$ = 0.1x0.1). After isolation the two backgrounds are comparable. Realistic simulations of a position detector located at R = 1.5m, looking at the detailed transverse profile of the cluster after 4 r.l. with strips of 2mm, show that the π^0 background (at least two photon showers) can be indeed further reduced by a factor 7, leaving as only background for $H \rightarrow \gamma\gamma$ search 2γ QCD production (Fig. 4).

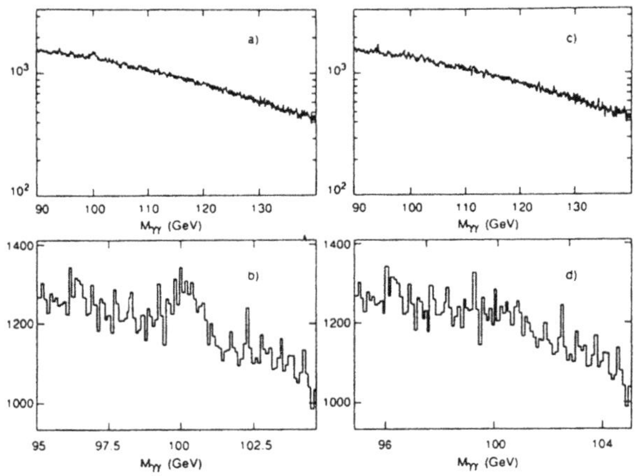

Fig.3 H→γγ signal (M_H = 100 GeV, 10^5 pb^{-1}) superimposed to the 2γ continuum
a) $\sigma/E = 2\%/\sqrt{E} \oplus 0.5\%$ and $\sigma(\theta_\gamma)$ = 7mrad, b) same with expanded scale
c) $\sigma/E = 10\%/\sqrt{E} \oplus 1\%$ and $\sigma(\theta_\gamma)$ = 7mrad, d) same with expanded scale

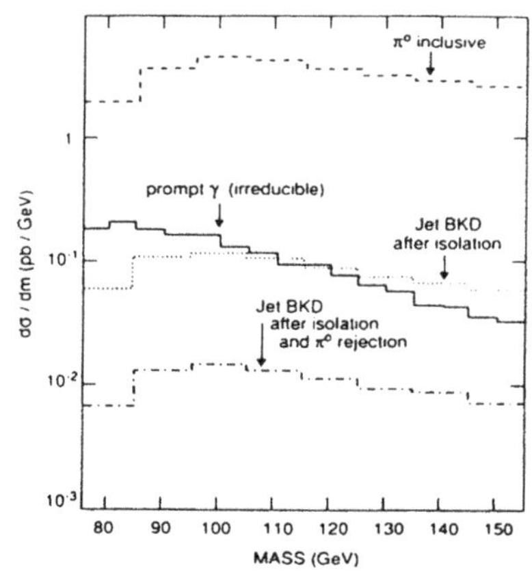

Fig.4 $d\sigma/dm_{\gamma\gamma}$ for 2γ continuum compared to 2 jet background after isolation and single shower selection with a position detector at R = 1.5m.

CeF_3 crystals are good candidates for a very precise calorimeter [11]. A. Gurtu [5] simulated the reponse to 50 GeV photons of a matrix of 7x7 crystals of CeF_3 . Each crystal has a transverse section of 2.5cmx2.5cm and two longitudinal segments: segment 1 has 5 r.l., segment 2 has 20 r.l., between the two segments there is a position detector of 0.05 r.l. giving the position of a 50 GeV photon to ± 150μm. The dead material between crystals is 300μm Carbon. Full Geant simulation gives for a 5x5 crystal cluster a resolution of 0.38%. On concludes that the contributions to the constant term due to side leakage, rear leakage and dead materials are less than 0.38% and are acceptable for H→γγ search. The photon direction is obtained from two points: point 1 is given by the position detector, point 2 is given by a weighted average over the back segments. An rms angular resolution of about 7mrad is obtained [5] on the photon direction, which again is adequate for $H \rightarrow \gamma\gamma$ search.

4. Dead materials in front of the e/γ calorimeter

H. Plothow-Besch [6] reported on full Geant simulations of the energy resolution of different e/γ calorimeter systems located inside or outside magnetic fields. In particular the effect of dead materials, thin coils or cryostat wall for example, in front of the calorimeter was simulated. She compared the responses of three types of e/γ calorimeters with identical cells $\Delta\eta x \Delta\phi = 0.02x0.02$ to 40 GeV electrons at normal incidence : 1) Liquid Argon (3mm Pb/ 6mm LAr), 2) Pb/Scintillator (4mm Pb/1mm Sci and 1mm Pb/1mm Sci) and 3) CeF_3 crystals. The main conclusions of her studies are:

1)The energy resolution of these calorimeters is not affected by magnetic fields. Even the very precise CeF_3 calorimeter is not affected by a strong magnetic field of 4 Tesla provided the energy is integrated over a large enough cluster of 5x5 cells. With 3x3 cells clustering a small effect due to the increase of the side leakage in the direction perpendicular to the magnetic field is observed.

2) For a liquid Argon calorimeter inside a magnetic field of 2 Tesla it is mandatory to reduce the thickness of the cryostat and the distance between the dead material and the first active layer to a minimum. Otherwise a non negligible fraction of the shower fragments will be trapped by the magnetic field in a non instrumented region, resulting in considerable energy resolution degradation. The effect becomes even more important in the barrel region as one goes away from rapidity 0.

5. Jets and missing transverse energy measurements

What are the constraints put on hadronic calorimetry from jet physics and missing E_T measurements? L. Poggioli [7] reviewed a number facts coming from Aachen studies [13] as well as from SSC studies also reported at this workshop by D. Green [14].

It is possible to correct the jet cross-section for resolution and non-linearity effects, provided these effects are perfectly known. To a certain extent the response to jets can be inferred from test beam measurements with single particles. However uncertainties on the fragmentation and maximum energy available with existing test beam facilities make it impossible to guarantee the knowledge of the jet response up to 4 TeV, which is about the statistical limit for jet measurements in one LHC-year at 10^{34} . This is an argument in favour of a linear (i-e compensating) rather than precise calorimeter. On the other hand jet response can be calibrated in situ by looking at the P_T balance of two jet events in which one of the jet is a π^0 jet or a single photon.

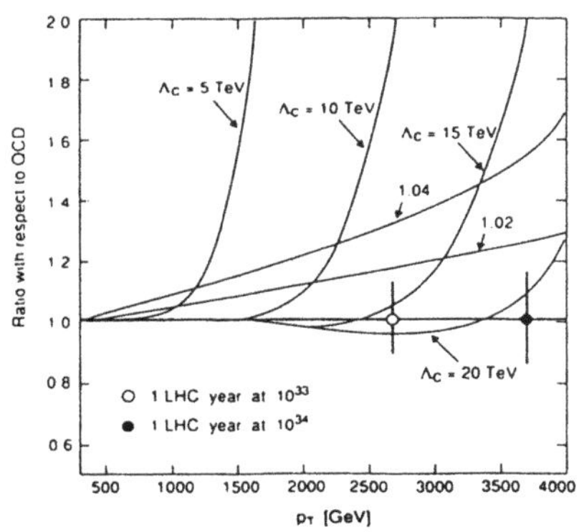

Fig.5 Ratio of measured jet cross-section with respect to QCD as a function of jet E_T.

Fig. 5 shows that the effect of a 4% systematic difference on the jet response between 500 GeV and 4 TeV is equivalent to a deviation from the QCD cross-section by a factor 1.7 at 4 TeV. In terms of quark compositeness this systematic error on the jet energy scale would reduce the sensitivity at LHC on Λ_c from 18 TeV to 15 TeV. The SDC collaboration [15] has shown that with a non compensating calorimeter of intrinsic e/h = 1.3 and for a fixed cone aperture of $\Delta R = 0.7$ the jet response could vary by 2% between 1 TeV and 5 TeV due to uncertainties in the fragmentation on the hadron momentum spectrum and the neutral/charged ratio. In their case a non compensating calorimeter with e/h = 1.3 reduces their sensitivity on Λ_c from 25-30 TeV to 20-25 TeV.

For dijet mass measurements the general conclusion from LHC and SSC studies [1,7,9] is that the intrinsic resolution of the hadronic calorimeter is not critical: Intrinsic effects due to the jet algorithm, confusion with the underlying event, magnetic field and at 10^{34} luminosity pile-up fluctuations dominate the mass resolution independently of the calorimeter precision.

For missing transverse energy measurements the hadronic resolution is not critical either. L. Poggioli [7] discussed three examples:

Example 1: $t\bar{t} \rightarrow e\nu$ + jets; at 10^{33} luminosity a cut $E_T^{miss} > 50$ GeV brings a rejection factor of about 50 against QCD jets, provided jets are detected up to pseudorapidities of $\eta = 4$. This factor is rather independent of the resolution but depends strongly on the η coverage and on pileup effects at higher luminosities.

Example 2: $H \rightarrow ZZ \rightarrow ll\nu\nu$; this channel extends the mass range for Higgs search up to 800 GeV with a typical jacobian peak in E_T^{miss} at about 300 GeV. The irreducible background is from the ZZ continuum. The instrumental background is from Z + jets in which a jet escape detection due to a crack or to the forward hole. The shape and magnitude of this instrumental background is entirely dominated by the η coverage. If jets can be detected up to $\eta = 4$, the the Z+jet background becomes negligible for $E_T^{miss} > 200$ GeV irrespective of the resolution and pileup effects at high luminosity.

Example 3: SUSY gluinos $\rightarrow$ jets + E_T^{miss}; The situation is very much like in example 2: Fig. 6 shows before analysis cuts the E_T^{miss} spectrum from gluinos of 1 TeV mass (the signal) compared to the irreducible background of genuine missing transverse energy predicted from standard model (SM) processes ($W \rightarrow e\nu$, $Z \rightarrow \nu\nu$ etc...) and to the instrumental background due to mismeasured 3 jet events in a calorimeter system covering $\eta < 4.5$ [16]. The instrumental background has been evaluated for two different calorimeter resolutions and includes crack effects. The slope of the E_T^{miss} spectrum is very different for the two backgrounds. One concludes that in the signal region ($E_T^{miss} > 400$ GeV) the 3 jet background is negligible compared to the SM background and that resolution effects are unimportant.

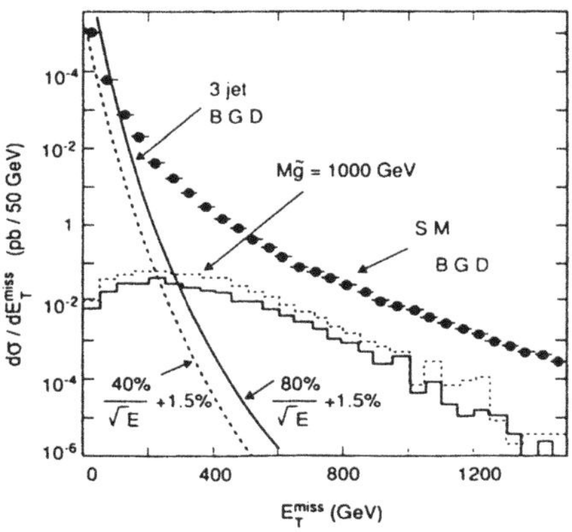

Fig.6 E_T^{miss} distributions from 1 TeV gluinos (solid and dashed histograms are for $\tan\beta = 2$ and 10), Standard Model background (solid points) and 3 jet backgrounds for two calorimeter resolutions (solid and dashed curves are for $80\%/\sqrt{E} + 1.5\%$ and $40\%/\sqrt{E} + 1.5\%$).

REFERENCES

[1] Aachen LHC workshop, CERN 90-10

[2 J-P Repellin, "Electromagnetic jets selected in a simulated calorimeter", contribution to this workshop.

[3] D. Froidevaux, "Electron identification at LHC/SSC with or without magnetic field", contribution to this workshop.

[4] C. Seez, "Calorimeter requirements for detection of the 2-photon decays of an intermediate mass Higgs at LHC", contribution to this workshop.

[5] A. Gurtu, "Energy resolution and gamma direction using crystals", contribution to this workshop.

[6] H. Plothow-Besch, "Effect of material in front and of magnetic field on the energy resolution of an electromagnetic calorimeter", contribution to this workshop.

[7] L. Poggioli, "Physics requirements on hadronic calorimetry at LHC/SSC", contribution to this workshop.

[8] C. Charlot, "A fast calorimeter simulation using Geant", contribution to this workshop.

[9] D. Groom, "Four component approximation to calorimeter resolution", contribution to this workshop.

[10] Aachen LHC workshop, CERN 90-10, vol. II p.365

[11] Aachen LHC workshop, CERN 90-10, vol. II p.474

[12] P.Lecoq, "Summary talk on homogeneous calorimeters", contribution to this workshop.

[13] Aachen LHC workshop, CERN 90-10, vol. II p.547

[14] D.R. Green, "Physics requirements for SSC/LHC calorimeters", invited talk to this workshop.

[15] SDC Letter of Intent, 30 November 1990

[16] Aachen LHC workshop, CERN 90-10, vol. II p.621

ENERGY RESOLUTION STUDIES OF DIFFERENT TYPES OF ELECTROMAGNETIC CALORIMETERS AS A FUNCTION OF RADIATION LENGTHS AND MAGNETIC FIELD EFFECTS

H. PLOTHOW-BESCH
CERN - PPE, CH - 1211 Genève 23, Switzerland

ABSTRACT

The energy resolution of different types of electromagnetic calorimeters has been studied as a function of the amount of radiation length in front of the calorimeters and in the presence and the absence of a magnetic field. As a result of this study the amount of dead material in front of a high precision calorimeter should be kept at minimum, in particular in the presence of a magnetic field. In view of the best energy resolution, given the possibility to recover the energy resolution by adding active layer(s) in front of the electromagnetic calorimeter, the preferred solution is a calorimeter outside a magnetic field.

1. Introduction

The physics requirements for an electromagnetic calorimeter at the LHC or SSC are

- good energy resolution,
- linearity,
- precise shower position (for precise impact measurements of electrons and photons),
- good angular resolution,
- good electron-hadron separation (to reduce the $\pi^{\pm}$, π^{0} background faking electrons),
- fast timing,
- radiation hardness.

Any dead material such as i.e. a magnetic coil or a cryostat in front of an electromagnetic (EM) calorimeter results in a deterioration of the energy resolution, σ/E. This is especially true for a high resolution EM calorimeter, such as a lead/liquid-argon or a lead/scintillating-fiber calorimeter with resolutions $\sigma/E \leq 10\%/\sqrt{E}$, where E is the energy of the showering particle. The deterioration becomes worse in the forward region due to the increasing amount of dead material from the coil or cryostat walls with polar angle.

In the following we describe a study which concentrates on the energy resolution of different types of EM calorimeters in the presence and the absence of a magnetic field and on the energy resolution as a function of dead material (radiation length X_0) in front of the EM calorimeter again in the presence or absence of a magnetic field.

2. The effect of a magnetic field on the energy resolution

Four different calorimeter types have been studied:

- a lead/liquid-argon calorimeter (Pb-LAr),
- a compensating lead/scintillating-fiber calorimeter with a 4:1 ratio of lead to scintillator,
- a non-compensating lead/scintillating-fiber calorimeter (Pb-Sci) with a 1:1 ratio of lead

to scintillator, and
- a CeF_3 crystal calorimeter.

The first three calorimeters are sampling calorimeters while the last one is a total absorbing calorimeter. In case of the sampling calorimeters (Pb-LAr and Pb-Sci) a cylindrical plate geometry has been simulated. In case of the Pb-LAr calorimeter the configuration of this geometry consisted of 3mm lead plates with 6mm LAr in between, in case of the 4:1 Pb-Sci calorimeter of 4mm lead and 1mm scintillator plates, in case of the 1:1 Pb-Sci calorimeter of 1mm lead and 1mm scintillator plates and in case of the CeF_3 crystal calorimeter of a pointing tower structure. For all calorimeters similar cell sizes for one cell of $\Delta\phi \cdot \Delta\eta = 0.02 \cdot 0.02$ have been used corresponding to about 3x3 cm^2 at a radius of 1.5 m. Two scenarios for each calorimeter type have been studied:

A - a superconducting coil with additional dead material (optional) in front of the EM calorimeter. In this configuration the EM calorimeter is *outside* a magnetic field.

B - a superconducting coil after the EM calorimeter with additional dead material at a distance of 10cm in front of the EM calorimeter. This additional material has been used as a possibility to simulate effects due to the increasing amount of material with polar angle. In this configuration the EM calorimeter is *inside* a magnetic field.

For all configurations a full GEANT[1] shower simulation of electrons of 40 GeV energy at 90° incident angle has been performed. In particular all materials such as lead, liquid-argon, plastic scintillators, CeF_3 crystals, steel and/or aluminium of the cryostats of a Pb-LAr calorimeter or a magnet coil and the aluminium of the coil conductor and its support have been taken into account in the simulation of the energy loss. The event structure has been generated in the SLUG[2] environment, and for parts of the different geometries the DICE program as developed by Ref. 3 has been used. No electronics noise, no cracks from support structures, no impact point depen-dence and no pile-up effects have been included in the study. Note that we are looking for relative differences on the energy resolution, therefore these effects should be less important.

The intrinsic energy resolution, σ/E, of the four calorimeter types for 40 GeV electrons at 90° incident angle is shown in Fig. 1a in the absence of a magnetic field and in Fig. 1b in the presence of a magnetic field of 2 Tesla (4 Tesla in case of the CeF_3 crystal calorimeter). The resolution σ/E for summing over 3x3, 5x5 or 7x7 cells is indicated by different labels. As can be seen from the Figures, there is little change in the energy resolution at 90° incident angle even in the presence of a high magnetic field of 4 Tesla, except in the configuration of 3x3 cells for the CeF_3 crystal calorimeter. Fig. 1c shows the experimental measurements obtained in testbeams for the different calorimeter types[4,5,6]. Since there are no measurements for the CeF_3 crystal calorimeter available yet, results for BGO crystals[7] are given instead (the CeF_3 crystals are expected to give similar resolutions). In general there is good agreement between measured and simulated data.

3. The effect of dead material in front of an electromagnetic calorimeter on the energy resolution

The energy resolution as a function of a variable amount of dead material in front of an EM calorimeter has been studied for the Pb-LAr and the Pb-Sci (1:1) calorimeters. This study becomes very relevant if the additional radiation length (X_o) of a cryostat in case of the Pb-LAr calorimeter and/or of a superconducting coil in front of the EM calorimeters is taken into account in the energy resolution, and it becomes particular relevant for the energy resolution at small polar angles ($\theta \leq 40^o$ or pseudo-rapidity $|\eta| > 1$). It should be noted that in this study the simulation of a preshower detector at $\theta = 90^o$ mainly serves as a simulation for the increase of amount of materials at small polar angles.

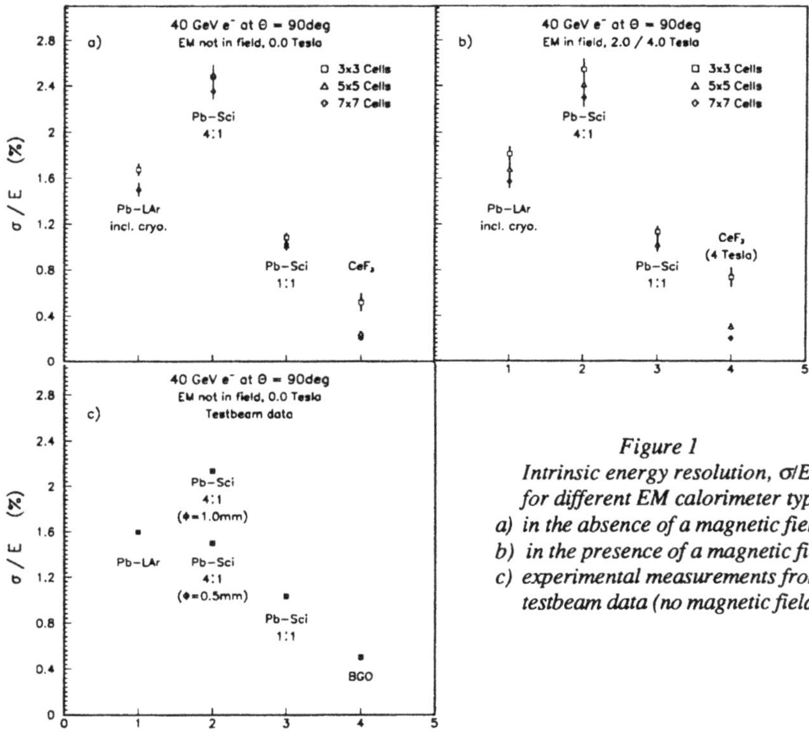

Figure 1
Intrinsic energy resolution, σ/E, for different EM calorimeter types
a) in the absence of a magnetic field
b) in the presence of a magnetic field
c) experimental measurements from testbeam data (no magnetic field)

The results of the study are shown in Figs. 2a and 2b for the Pb-LAr calorimeter in the absence and the presence of a magnetic field, respectively. For a cone size of 3x3 cells which corresponds to a cluster size of 10.5x10.5 cm^2, the energy resolution in the absence of a magnetic field is $\sigma/E = 10.6\%/\sqrt{E}$ at $\theta = 90^o$ if only the material from the cryostat is taken into account. It deteriorates to $\sigma/E = 16.1\%/\sqrt{E}$ if Pb material corresponding to 1.7 X_o is added. As indicated on Fig. 2a by the arrows, this behaviour could be interpreted as the energy resolution for a Pb-LAr calorimeter at $\eta = 0$ where only its cryostat and a superconducting coil in front of the calorimeter has been taken into account. The point at $X_o = 2.6$ on Fig. 2a would then correspond to the same configuration of a Pb-LAr calorimeter with its cryostat and a coil in front at $\eta = 1.1$. In this configuration the visible energy deposited in the argon drops by 14% at $\eta = 1.3$ with respect to $\eta = 0$. The energy resolution gets much worse in the presence of a magnetic field as can be seen in Fig. 2b. This is due to several effects. The main effect comes from the opening of the electromagnetic shower in the magnetic field due to increasing shower fluctuations if dead material is added. This effect can be corrected for if more cells are added together, i.e. 7x7 cells, as indicated on Fig. 2b by different labels (pile-up effects as expected at LHC luminosities have been ignored). Another effect comes from the necessary distance of dead and active material in case of a Pb-LAr calorimeter. Because of the space needed for the heat shielding and for the readout-electronics which is foreseen to be put in front and the back of the active medium, low momentum showering particles (i.e. $p \lesssim 100$ MeV in a 2 Tesla magnetic field) may escape the energy measurement in the calorimeter due to curling. In the presence of a magnetic field of 2 Tesla and for a cone size of 3x3 cells the energy resolution of the Pb-LAr calorimeter including only its cryostat is $\sigma/E = 12.0\%/\sqrt{E}$ at $\eta = 0$ and deteriorates to $\sigma/E = 26.0\%/\sqrt{E}$ at $\eta = 1.5$.

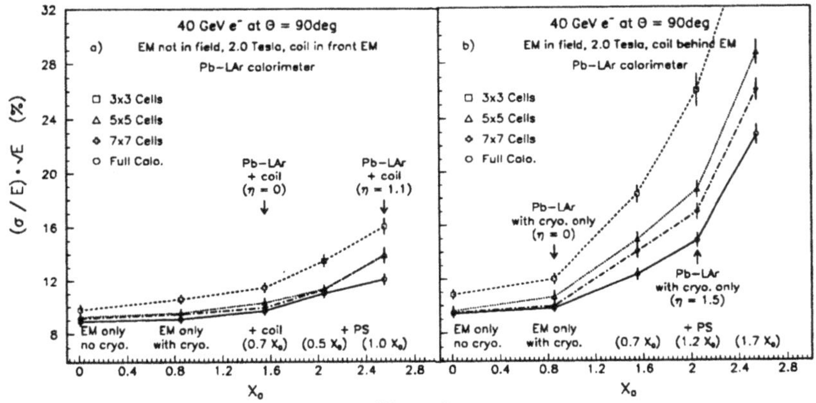

Figure 2

Energy resolution for a Lead-Liquid Argon calorimeter as a function of radiation length X_o

a) EM outside the magnetic field b) EM inside the magnetic field

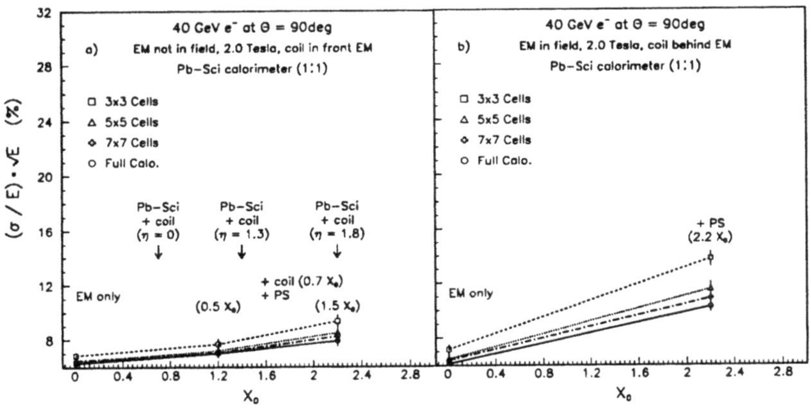

Figure 3

Energy resolution for a Lead-Scintillator calorimeter as a function of radiation length X_o

a) EM outside the magnetic field b) EM inside the magnetic field

Similar results are shown in Figs. 3a and 3b for a Pb-Sci (1:1) calorimeter in the absence and the presence of a magnetic field, respectively. Because of the dense calorimeter (1mm lead and 1mm scintillator plates) the apparent deterioration due to additional material in front of the EM part in the absence and the presence of a magnetic field is smaller compared to the Pb-LAr calorimeter in a very similar geometrical configuration.

It is worse mentioning that the deterioration of the energy resolution is less important for photons compared to electrons.

It has been shown that the deterioration of the energy resolution due to dead material in front of a high resolution EM calorimeter may be corrected for, at least to a large fraction, by adding active layer(s)[8] in front. How well this procedure works in the presence of a high magnetic field needs to be demonstrated.

4. Conclusions

The energy resolution for different types of electromagnetic calorimeters in the presence and the absence of a magnetic field and the energy resolution as a function of radiation length X_0 in front of the electromagnetic calorimeter also in the presence or absence of a magnetic field has been studied. A strong dependence of the energy resolution on the amount of radiation length X_0 in front of an electromagnetic calorimeter has been found in the presence of a high magnetic field which can be interpreted as a deterioration of the energy resolution as a function of η. The effect can be reduced if the cone size chosen for the energy reconstruction is enlarged. Bigger cones show smaller effects because the fluctuations due to the opening of the shower in the magnetic field is somewhat compensated by the bigger cone. At LHC luminosities the enlargement of the cone has to be balanced against pile-up effects. It also has been found that for a fixed cone size the deterioration of the energy resolution gets worse if there is a large gap between the dead material and the active part of the calorimeter. For dense calorimeters the deterioration is less.

As a general conclusion, the total amount of material in front of an electromagnetic calorimeter should be carefully surveyed and kept as low as possible. The preferred solution would be a very dense calorimeter with active layer(s) in front of the electromagnetic part and a precise position device close to shower maximum after 3-5 X_0.

References

1. R. Brun, F. Bryant, M. Maire, A.C. McPherson, P. Zanarini, *GEANT3*, DD/EE/84-1 (1987)
2. R.S. DeWolf, *Simulations for LHC: SLUG, GENZ and related packages*, Proc. of the 3rd Workshop on Detector and Event Simulation in High Energy Physics, Amsterdam (1991).
3. A. Dell'Acqua and M. Nessi, *the DICE program*, private comunication.
4. B. Aubert et al., *Performance of a Liquid Argon Electromagnetic Calorimeter with an 'Accordion' Geometry*, CERN-PPE/91-73, accepted for publication in Nucl. Instr. Meth. A.
5. D. Acosta et al., *Electron, Pion and Multiparticle Detection with a Lead/Scintillating Fiber Calorimeter*, CERN-PPE/91-85, accepted for publication in Nucl. Instr. Meth. A.
6. D.W. Herzog et al., *A High-Resolution Lead/Scintillating Fiber Electromagnetic Calorimeter*, Nucl. Instr. Meth. **A 294** (1990) 446.
7. J.A. Bakken et al., Nucl. Instr. Meth. **A 254** (1987) 535;
   F. Ferroni, *Design, Realisation, Performances of the L3 BGO EM calorimeter*, these proceedings.
8. G.S. Abrams et al., *Performance of the Lead/Liquid Argon Shower Counter System of the Mark II Detector at Spear*, IEEE Nucl. Sc. NS-27, 59 (1980);
   H. Hirayama, Nucl. Instr. Meth. **A 302** (1991) 427.

Energy Resolution and Photon Direction using Crystal Calorimetry [1]

A. Gurtu, CERN, Geneva and TIFR, Bombay
K. Sudhakar, Aachen I and TIFR, Bombay

Abstract

We have carried out simulation studies to estimate the energy resolution for photons incident on a crystal matrix of BaF_2,CeF_3 and PbF_2 crystals. Each crystal is longitudinally segmented into two parts of 5 and 20 radiation lengths with a position detector for charged particles in between. It is shown that sampling and constant terms $\sim$ 2.0 % and 0.35 % respectively can be achieved. For 50 GeV photons the CeF_3 crystal matrix gives an energy resolution of 0.38% and the use of the position detector allows one to determine the photon angle to within 7 mrad which is somewhat better than the necessary condition for the detection of a Higgs $\longrightarrow \gamma\gamma$ of 80 - 130 GeV mass.

1. Introduction

Crystal calorimetry is one of the serious candidates for electromagnetic energy measurement for the next generation of high energy physics experiments being planned at the LHC and SSC. The disadvantage of an overall non-compensating calorimeter is offset by the extremely precise measurement of the electromagnetic component. This is important for the possible detection of a Higgs $\longrightarrow \gamma\gamma$ decay in the Higgs mass range 80 – 150 GeV. In this simulation study we examine the issue of energy resolution and the precision with which one can determine the γ direction using a realistic crystal matrix.

2. Geometry

The GEANT 3.14 package has been used. The geometry setup includes

1. Tapered crystals pointing back to an "interaction vertex" (convergence distance)
2. Carbon layer between crystals

[1]Work carried out in the framework of the Crystal Clear Collaboration

3. Longitudinal segmentation of crystals with a position detector in between

The various inputs needed are

1. Front face (size) of crystal
2. Convergence distance of the crystal matrix
3. Carbon layer thickness (T_{carb})
4. Lengths of crystal and position detector segments ($L_{seg1}, L_{seg2}, L_{pd}$)

Two simulation studies have been carried out, one on energy resolution and the second on the photon angle determination. Details of the inputs used in these two studies are given in table 1. In both studies the first and second crystal segments were 5 and 20 radiation lengths and the convergence distance of the crystal matrix was taken as 100 cm. Since the position detector would likely use silicon mounted on aluminum the material for this has been taken to be aluminum. The matrix consists of a 7 × 7 array.

Table 1: Geometry Inputs

Study	Front face (cm^2)	L_{pd} (cm)	T_{carb} (μm)	Crystals used	E_γ (GeV)
Resolution	3 × 3	0.90 Al	500	CeF_3,BaF_2,PbF_2	10, 50, 200
γ angle	2.5 ×2.5	0.45 Al	300	CeF_3	50

3. Energy Resolution Study

For this study the number of events generated at 10,50 and 200 GeV are 1000, 500 and 200 respectively. The energy resolution is taken to be the spread in the energy deposited in 5 × 5 crystal array centered on the highest energy crystal. Since the conversion distance of a photon after its entry into the crystal can vary significantly, rear leakage of energy can correspondingly vary from event to event. This leads to a degradation of the energy resolution for photons in comparison to electrons. Having longitudinal segmentation of the crystal can alleviate this problem. Requiring that there is a minimum amount of energy deposit in the first crystal segment ensures that the photon was converted well within it. We select events with $\frac{E_1}{E_2} \geq F_{cut}$ where E_1, E_2 are the energy deposits in the first and second crystal segments and F_{cut} is a photon energy dependent cut–off ranging between 0.05 at 10 GeV to 0.01 at 50 and 200 GeV. This leads to a loss of ~ 10 % events and improves the energy resolution particularly at low energy. The values of energy resolution thus obtained are given in table 2. The values of A_{fit}, B_{fit} are also given where

$$\frac{\sigma(E)}{E} = \frac{A_{fit}}{\sqrt{E}} \oplus B_{fit}$$

In this fit the sampling and constant terms have been added in quadrature. As one sees from the table, inspite of the degradation of the energy resolution due to the presence of dead material (500 μm carbon layer between crystals and a 0.1 radiation length position detector between crystal segments) the constant term is less than 0.4 % in all the crystals and the sampling term is $\sim$ 2 %.

Table 2: Energy Resolution ($\frac{\sigma(E)}{E}$ in %)

Crystal	E_γ 10 GeV	E_γ 50 GeV	E_γ 200 GeV	A_{fit}	B_{fit}
BaF_2	0.69	0.55	0.34	2.1±0.4	0.34±0.06
CeF_3	0.67	0.49	0.36	1.9±0.3	0.35±0.04
PbF_2	0.68	0.47	0.33	2.0±0.3	0.32±0.04

4. Determination of photon angle

Details of the basic crystal geometry have been given in table 1 under the "γ angle" heading. The incident photon has been smeared uniformly over the front face of the central crystal. One thousand 50 GeV photons are simulated in a CeF_3 crystal matrix. This is the photon energy relevant in the study of Higgs $\longrightarrow \gamma\gamma$ decay in the Higgs mass range 80 – 150 GeV.

The position detector has been assumed to have an accuracy of 150 μm in X and Y (the Z direction being along the crystal axis). Charged particles with kinetic energy $\geq$ 1 MeV entering the position detector are used to determine the shower centre (the tracking cut-off values for GEANT have been set at 100 keV). The photon direction is determined by the lever arm method. The position detector provides the X,Y and Z coordinates at one space point. The second crystal segment (20 radiation lengths) of the 5 $\times$ 5 crystal array centred on the highest energy crystal provides the second space point. The X and Y are determined by the barycentric method. The Z coordinate is the energy weighted mean averaged over all the events.

The distribution of energy deposited in both crystal segments in a 5 $\times$ 5 array is shown in fig 1 along with a Gaussian fit. The energy resolution is 0.38 %. Note that this value is for the geometry corresponding to the γ angle study as given in table 1 (compared to the geometry used for the energy resolution study here one has less dead material in the crystal matrix).

The distribution of $|\Delta\theta|$, which is the difference of the true and reconstructed space angle is shown in fig 2. The rms value from the Gaussian fit is 7 mrad. Fig 3 shows the intercept on the absolute Z–axis (along the direction of the central crystal) when the photon direction is projected backward in the R–Z plane. It is a Gaussian with an rms of ~ 0.8 cm.

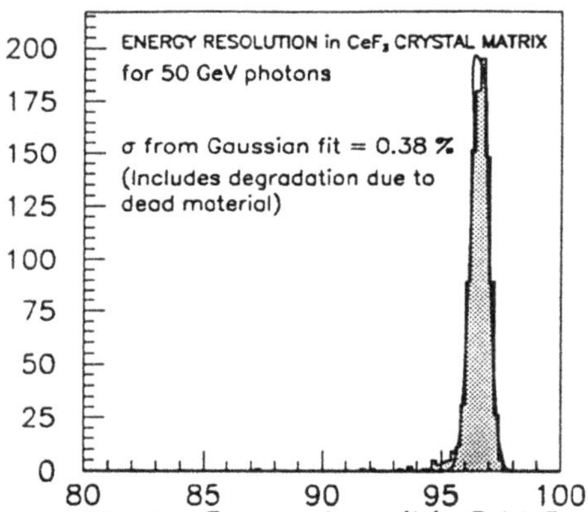

Fig. 1 Energy deposit in 5 X 5 Crystal Array (in %)

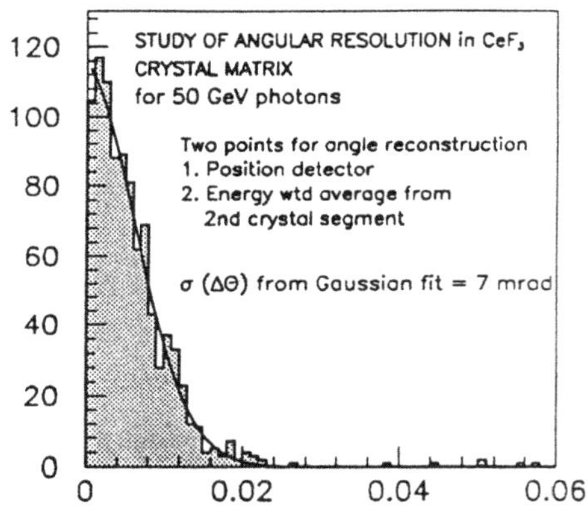

Fig. 2 $|\Theta_{rec} - \Theta_{true}|$ (rad)

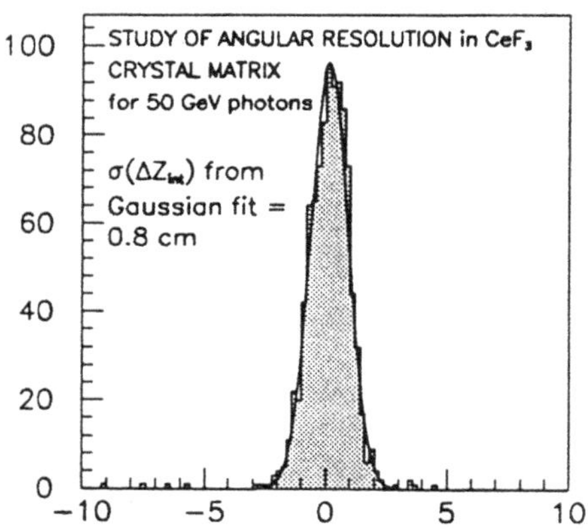

Fig. 3 Intercept on Z axis of reconstructed γ (cm)

5. Summary and Conclusion

Simulation studies of energy resolution and photon angle determination in a crystal calorimeter have been carried out in the GEANT framework. Owing to its high resolution it is a serious candidate for measuring the electromagnetic energy component in LHC/SSC experiments. As pointed out by C. Seez[1], in order to detect the Higgs $\longrightarrow \gamma\gamma$ mode in the 80 – 150 GeV mass range, the constant term in the photon energy resolution must be better than ~ 0.5 % and the reconstruction of the photon angle must have an rms error better than ~ 7 mrad.

In this study it is shown that all the leading crystal candidates (BaF_2,CeF_3,PbF_2)[2] are capable of providing the requisite energy resolution. This despite the fact that the crystal matrix has been assumed to have a rather large amount of dead material in it in the form of carbon layer around each crystal and a position detector between the two longitudinal crystal segments.

For reconstructing the photon angle a CeF_3matrix with 5 and 20 radiation length longitudinal crystal segments and a position detector having an accuracy of 150 μm in each of the transverse directions has been considered. It is shown that an rms of 7 mrad is a realistic proposition. This corresponds to a determination of the Z coordinate of the interaction vertex to ~ 0.8 cm.

To conclude, a segmented crystal calorimeter with a position detector between segments is capable of detecting the Higgs $\longrightarrow \gamma\gamma$ decay mode in an LHC/SSC experiment.

References

1. C.Seez, Requirements for H$\longrightarrow 2\gamma$ detection : Presentation at this conference

2. P.Lecoq, Report from parallel sessions : Presentation at this conference

SIMULATION AND STUDY OF ELECTROMAGNETIC JETS

J.C. Chollet J.-P. Repellin
Laboratoire de l'Accélérateur Linéaire
Orsay, France

ABSTRACT

A study of the particle composition of simulated jets satisfying an electromagnetic selection based on calorimetry is presented.

1. Introduction

The identification of electrons and photons in a apparatus of the type considered for LHC experimentation relies upon the selection of localized electromagnetic energy deposition in a calorimeter[1]. As a result of this selection based on calorimetry, jets are still the main contribution to background well above the genuine electron and photon signal. The purpose of this study is to investigate to what extent this background is dominated by high p_T isolated π^0 , and what is the charged tracks multiplicity.

2. Simulation and selection of electromagnetic jets

Two samples of Pythia jets have been simulated with transverse momentum threshold of 20 GeV/c (3900 jets) and 50 GeV/c (7400 jets) and around pseudorapidity zero. The shower developpment of the particles has then been simulated with Geant in a simplified cylindric calorimeter of radius 1 metre made of homogenous lead with density of one half. The calorimeter is segmented on cells of transverse dimensions $\Delta\Phi = 0.021$, $\Delta\eta = 0.02$ and in two depths, an electromagnetic compartment, 25 radiation lengths deep, followed by a hadronic compartment of 8 interaction lengths. No magnetic field was assumed.

These simulations have been used to study the electron and photon selection in three different calorimeter granularity configurations, fine, medium and coarse as summarized in Table 1. A more detailed description can be found in Ref. 2.

The selection of electromagnetic objects is done in 3 levels. First a trigger requires a minimum electromagnetic energy (say 18 GeV/c and 44 GeV/c for 20 GeV/c and 50 GeV/c jets respectively) in a $\Delta\Phi\ \Delta\eta$ area as indicated in the third column of Table1. A second level requires a maximum energy leakage in either the hadronic compartment or the electromagnetic cells sourrounding the trigger cells, over an area as given in the fourth column of Table 1. The last level consists in the requirement of both previous conditions.

Table 1

segmentation	em	had	trigger	isolation	20 GeV/c	30 GeV/c
fine	.021 ×.02	.063 ×.06	.063 ×.06	.19 ×.18	4 jets	9 jets
medium	.042 ×.04	.084 ×.08	.084 ×.08	.17 ×.16	8 "	16 "
coarse	.063 ×.06	.13 ×.12	.13 ×.12	.25 ×.24	22 "	25 "

The number of jets which pass all the cuts are given in the last 2 columns of table 1.

An example of the acceptance of 50 GeV/c jets is shown in fig. 1. The upper curve results from the trigger selection, the second and third result from electromagnetic isolation and hadronic leakage cut respectively, the fourth gives the full rejection. The dotted line gives the overall rejection when rising the trigger cut from 44 to 50 GeV/c. The cuts applied on the hadron leakage and electromagnetic isolation are in GeV/c for the three segmentations:

3.5 / 7.0   3.0 / 5.0   5.0 / 5.0

They are such that the efficiency for 20 GeV/c electron or photon selection are 95% in step 2 and 90% in step 3, and for pile-up conditions corresponding to a luminosity of 2 x 10^{34} in a liquid argon calorimeter with a peaking time of 50 ns

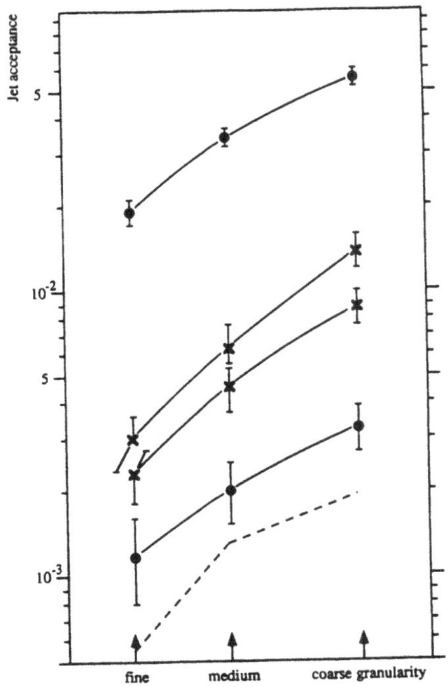

Figure 1 Calorimeter acceptance of 50 GeV/c jets for 3 granularities and various selections

Besides the limited statistics, there are uncertainties on the jet simulation and the choice of cuts, but a rejection against jets close to a factor 10^3 seems acheavable with a fine granularity calorimeter. Similar rejections are achieved against 20 GeV/c jets.

3. Photon component in electromagnetic jets

The jets which pass the electromagnetic selection are indeed mainly composed of photons from π^0, η and K^0 decays. This is seen in fig. 2 which shows, for the 50 GeV/c sample, the ratio of the sum of the energy of all the photons in a cone $\sqrt{\Delta\Phi^2 + \Delta\eta^2}$ (of size 40, 80 and 120 mrad for the three granularities respectively) to the electromagnetic trigger energy, as a function of the latter.

Of the electromagnetic jets in the various samples, 95% have a leading photon and for those jets, the distributions of the number of extra photons with E_T larger than 0.5 GeV in the cones defined above are shown in fig. 3 as a function of the energy of the next to leading photon for the 3 segmentations. As expected, for all the 50 GeV/c jets within the present statistics, the leading photon is always accompanied in the cone by the second photon of the π^0 parent (note that in the 50 GeV/c jet sample, 2 jets out of 25 selected with the coarse granularity contain an η particle).

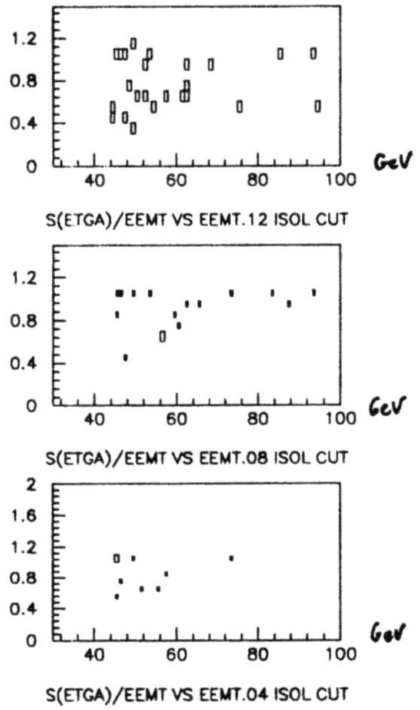

Figure 2 Proportion of photon energy in the trigger cell energy

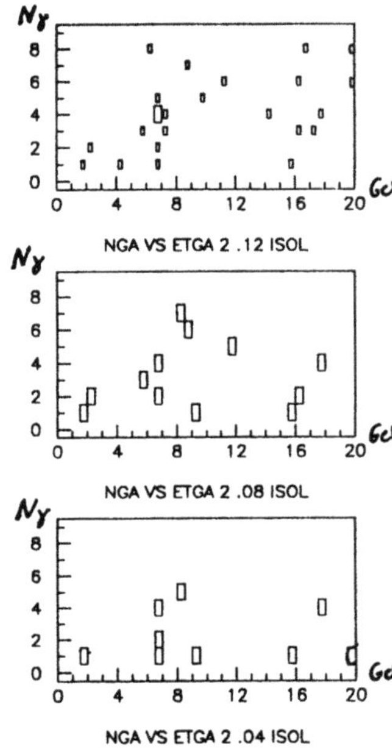

Figure 3 Number of photon accompaning the leading photon

The proportion of jets with three or more photons does not depend strongly on the level of selection, as seen in fig. 4. Naturally it depends on the granularity. For 50 GeV/c jets, fig. 4b shows that it varies from about 60% for the 40 mrad cone and fine granularity selection (lower curve) to about 85% for the 120 mrad cone and coarse granularity selection (upper curve). The proportion is somewhat lower for the 20 GeV/c jet sample as expected and observed in fig. 4a.

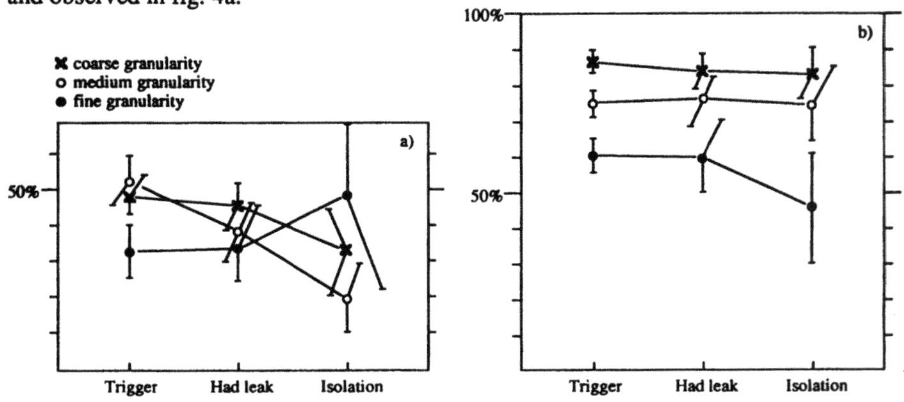

Figure 4 Proportion of jets with more than 2 photons, a) 20 and b) 50 GeV/c jets

For electromagnetic jets containing at least a third photon beside the two photons of the leading π^0, one can calculate the angle it forms with the highest p_T photon of the π^0. The distribution of this angle is shown as a function of the p_T of the third photon for jets selected with the coarse granularity in fig. 5a for the 20 GeV/c sample (8 jets with ≥3 photons out of 22) and in fig. 5b for the 50 GeV/c sample (20 jets with ≥3 photons out of 25). In the figure, the full boxes correspond to the photon of those jets which also pass the fine granularity selection and are whithin the 40 mrad cone.

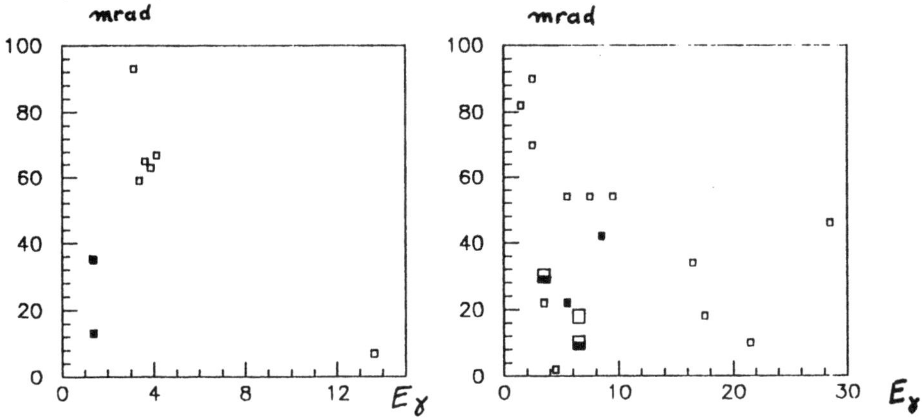

Figure 5 Angular separation and energy of the photon accompaning the leading π^0

In summary, isolated high p_T π^0 are the main contribution to electromagnetic jets selected with a calorimeter as expected, but multi π^0 also contribute substantially, especially for high p_T jets and for calorimeters with a coarse granularity. A preshower or shower max detector added to an electromagnetic calorimeter to distinguish isolated photons from π^0 at large p_T should also be adequate to reject multi π^0 triggers. Larger statistics would be required to confirm the present indication that such devices could compensate the loss of rejection of a coarse calorimeter granularity.

4. Charged particles accompanying electromagnetic jets

The jets satisfying elctromagnetic requirements are biased towards a low multiplicity and low energy charged particle content.The charged multiplicities observed with no magnetic field in the 50 GeV/c jet sample are as follows after the full electromagnetic selection:

coarse granularity and 120 mrad cone
    out of 25 jets, 3 have no charged track and 8 have one
medium granularity and 80 mrad cone
    out of 16 jets 3 have no charged track and 9 have one
fine granularity and 40 mrad cone
    out of 9 jets 4 have no charged track and 4 have one

This is summarized in fig. 6. The upper curve gives as fig.1 the jet acceptance of the electromagnetic selection, the medium curve shows what would be interpreted as an electron candidate following the requirement of one and only one charged track and the lower curve shows what would be interpreted as a photon resulting from a no track requirement.

It is interesting to mention that among the 5 jets having one or more track in the 40 mrad cone associated with the fine granularity, 4 have a track with a p_T between 10 and 30 GeV/c. For 20 GeV/c jets the even more limited statistics indicates that the p_T of the tracks is significantly lower (between 2 and 6 GeV/c).

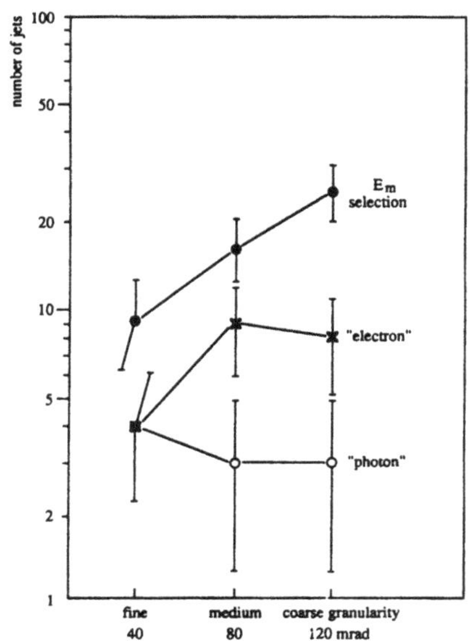

Figure 6 Fraction of jets with zero or one track (50 GeV/c sample)

Electron signature.

In the case of no magnetic field, asking for a track facing the trigger cells rejects at most only half of the jets. Pile-up at high luminosity can only worsen the rejection. A better matching of the track with the shower position as measured in the calorimeter is necessary.

In the case of a magnetic field the matching of the track momentum with the electromagnetic energy provides the necessary rejection. With the approximation that the jet selection used in this study is not affected significantly by the field, all the selected jets would not survive an E/P cut as loose as 1.5.

Photon signature

The detection of charged particles can be used to improve the photon selection . From these results, in a configuration with no magnetic field, a total rejection of 2×10^3 can be achieved by requiring no track facing the trigger cells. But such a cut, applied ignoring the particle momentum, must be used with care in the presence of pile-up. In such a case, a cut limited to tracks of momentum larger than some threshold, could still be performant, even in the presence of a magnetic field. For example a cut at 10 GeV/c would still improve by a factor 2 the calorimeter selection.

References

1. J Colas etal . Calorimetry at the LHC . Large Hadron Collider Workshop, Aachen 1990, vol 1 p 370
2. J.P Repellin. Electron detection in LHC calorimeters . Large Hadron Collider Workshop, Aachen 1990, vol 3 p 402

PHYSICS REQUIREMENTS ON HADRONIC CALORIMETRY AT LHC/SSC

Luc Poggioli, CERN, Geneva, Switzerland

Abstract

We investigate the three major tasks to be fulfilled by a hadron calorimeter at LHC/SSC, i.e. inclusive jets cross section measurement, di-jet mass measurement and missing E_t measurement.

For these three tasks we first address the physics channels involved and then we extract the requirements for the hadron calorimeter. We focus on jet energy resolution, coverage in η, hermiticity and granularity.

1 Introduction

Hadron calorimetry plays a crucial role in all the sectors of the Standard Model physics and beyond. The hadron calorimeter should provide jets and missing E_t measurement and the capability to trigger on these signatures.

For both physics and calorimeter design, the major features are the intrinsic response, coverage in η, hermiticity and granularity. Intrinsic response gathers features as jet energy resolution, linearity of the response and gaussian shape of the measured energy. One should remind that these three criteria are "de facto" optimized for a compensating calorimeter.

There are other aspects which are of prime importance for a hadron calorimeter at LHC, i.e. radiation hardness, calibration and trigger issues and for integration to a full detector, i.e neutron fluence, muon detection and return flux (choice of the absorber) and price of course.

One should first mention two "a priori" parameters which can deteriorate the jet measurement, i.e. the pile-up and the eventual magnetic field. The pile-up contribution is important for jets because of the size of the showers. Its contribution is constant in E_t and goes like $\sqrt{\mathcal{L}}$. As a consequence it can be neglected for high p_t jets and at $\mathcal{L} = 10^{33} cm^{-2} s^{-1}$, whereas at 2×10^{34} for a standard calorimeter and a jet of $E_t = 100\ GeV$ it contributes as much as a scaling term $60\%/\sqrt{E}$ in the resolution.

Concerning the magnetic field few studies are going on. The opening of the jet cone and low momentum fragments missing the calorimeter due to the field will disturb the jet measurement. This effect is probably dominated by the pile-up one at high luminosity.

2 Inclusive jets cross section

This measurement allows test of QCD at high p_t or Q^2. Any deviation to the inclusive jets cross section w.r.t. QCD prediction (i.e. enhancement of the cross-section at large p_t) is a signature for compositeness.

The rapidity coverage is not crucial here. Working at rapidities $|\ \eta\ | \leq 1$ yields reasonable statistics and since the $d\sigma/dp_t$ vs p_t is less steep for small rapidities one is thus less sensitive to systematics.

Since one is looking at very high p_t jets (Fig. 1) the pile-up has a negligible contribution to the jet energy resolution and one can use large cones for reconstructing the jets ($\Delta R \simeq 0.7$).

More critical could be the energy resolution (mostly the constant term since the scaling term in $1/\sqrt{E}$ is negligible at high p_t) and the non-gaussian response of the calorimeter. These 2 effects populate the high and low p_t region around the initial value, but due to the very steep dependance of the cross-section with p_t they populate more the high p_t region thus faking compositeness.

It turns out [1] that this effect is negligible for a standard calorimeter.

More crucial is the possible non-linearity of the calorimeter. In this case the *full* distribution is shifted towards the high p_t region. As an example a non-linearity of 4 % between 500 and 4 TeV can fake compositeness with $\Lambda_c = 15\ TeV$ (Fig. 1).

It seems extremely difficult to rely on Monte-Carlo to determine the non-linearity of the calorimeter. Test beam data are also limited in energy. But for example a 4 TeV jet will be made of particles with maximum energy around a few hundred of GeV thus allowing the use of test beam data.

Moreover it can be seen (Fig. 2) that the overall uncertainty is dominated by structure functions uncertainty [2] which will be measured more accurately at HERA.

As a conclusion the inclusive jets cross-section measurement does not put stringent requirements on the hadronic calorimeter.

3 Di-jet mass measurement

Di-jet mass measurement allows one to observe decays of resonances into 2 jets thus enhancing the statistics (branching ratios higher than for leptons). One has of course in that case to face a much higher background (jet jet continuum).

3.1 Low p_t jets

In this subsection we deal with processes like intermediate mass Higgs $H \rightarrow b\bar{b}$, $Z(\rightarrow l^+l^-)\ Z$ or $W(\rightarrow jj)$ in the continuum, $t \rightarrow Wb$ and SUSY $H^+ \rightarrow c\bar{s}$.

A rapidity coverage up to $|\eta| = 3$ is sufficient.

The energy resolution plays a role in the last 2 cases [3] which can be studied only at low luminosities (Fig. 3). In the first 2 cases, because of b-fragmentation effects or Z/W overlaps, the resolution is less critical [4].

On these low p_t jets however the pile-up effect takes over and implies to reduce ΔR down to 0.4-0.3 and dominates over jet energy resolution contribution [4] (Fig. 4).

3.2 High p_t jets

Here we deal with heavy Higgs production $H \rightarrow W(\rightarrow l\nu)W(\rightarrow jj)$ or $H \rightarrow Z(\rightarrow l^+l^-)Z(\rightarrow jj)$, and also extra gauge boson production Z' (2 TeV) $\rightarrow jj$. The $|\eta|$ coverage is not crucial because the final jets are centrally produced. If one wants to tag the jets radiated by the W which is compulsory to reject the QCD background one needs a coverage down to $|\eta| = 4$ [5].

Because of the large boost of the produced resonances the final jets are close in space and thus merge [5]. A good granularity can be required but more likely the granularity in the electromagnetic part is sufficient.
This jet merging deteriorates the jet-jet mass resolution to a point where a good jet resolution doesn't help [4].
If the pile-up contributes here less because of the higher p_t it deteriorates the mass resolution as a crude calorimeter (i.e. $\sigma/E = 100\%\sqrt{E} \oplus 4\%$) does [4].
The $Z' \rightarrow jj$ study is still preliminary but one can expect a significant effect of the hadronic energy resolution (pile-up negligible). Anyway the Z' mass can be reconstructed with more accuracy in the leptonic decay mode.
As a conclusion, for di-jet measurement no strong requirement is made for the calorimeter but a very good η coverage is needed for quark tagging.

4 Missing transverse energy measurement

$\not{E}_t$ is a clear signature for various processes as top semi- leptonic decay ($t\,\bar{t} \rightarrow e\nu +$ jets), heavy Higgs decay $H \rightarrow Z(\rightarrow l^+l^-)Z(\rightarrow \nu\bar{\nu})$, and SUSY particles production (squarks, gluinos).
$\not{E}_t$ can be faked by essentially 4 factors, i.e. $\mid \eta \mid$ coverage, cracks, pile-up and jet energy resolution (including non-gaussian tails for non-compensating calorimeters). These 4 factors have never been studied altogether up to now, which make the conclusions difficult to draw.
In the first case ($t\bar{t}$ production) the dominant effect is the η coverage [3]. It can be seen on Fig. 5 that one needs to have a rapidity coverage down to $\mid \eta \mid = 4$ to keep the contribution of the $\not{E}_t$ created by mismeasured 2 jets events below the $t\bar{t}$ signal. In Fig. 5 is also displayed the pile-up contribution to the fake $\not{E}_t$. The dependance on the calorimeter resolution is found to be quite small.
In the second case (neutrinos production from Z decay) the most important parameter is the η coverage, which has to be at least $\mid \eta \mid \leq 5$ to be able to observe the Higgs signal above the Z + 2 jets background. The pile-up contribution has been evaluated (Fig. 6) [6]. According to this study, one can expect the resolution and cracks contributions to be less important. It has nevertheless to be investigated.
Finally for SUSY studies the rapidity coverage is crucial ($\mid \eta \mid \leq 4.5$). Cracks modelling à la CDF and resolution effects have been investigated [7]. It turns out that the cracks are important as well as the resolution of the calorimeter (Fig. 7). One should mention that the pile-up effect has not been included in this study. Moreover the slope of the $\not{E}_t$ shown in Fig. 7 and coming from mismeasured 3 jets events is much steeper that the one coming from the Standard Model background allowing rather easy rejection of the multi-jets events background.
As a conclusion of this section, the η coverage and the absence of cracks are the most important criteria to be fulfilled by the calorimeter. Studying cracks contribution is quite difficult because an accurate design of the full calorimeter is needed.

5 Conclusion

Hadron calorimetry plays an important role in the study of the Standard Model and beyond.

Up to now, no strong requirements on the intrinsic performances (spoiled by pile-up, fragmentation, magnetic field) are observed.

A very good rapidity coverage and a good hermiticity are likely to be the most important requirements.

6 Acknowledgements

I would like to thank F. and M. Nessi, D. Froidevaux and L. Fayard for many helpful discussions.

References

[1] F. and M. Nessi, EAGLE studies (1991).

[2] SDC Letter of Intent (1991).

[3] L. Fayard et al., Aachen Workshop, Vol. 2 (1990).

[4] L. Poggioli, Aachen Workshop, Vol. 2 (1990).

[5] M. Seymour, Aachen Workshop, Vol. 2 (1990).

[6] I. Zuckerman, EAGLE studies (1991).

[7] P. Sphicas et al., Aachen Workshop, Vol. 2 (1990).

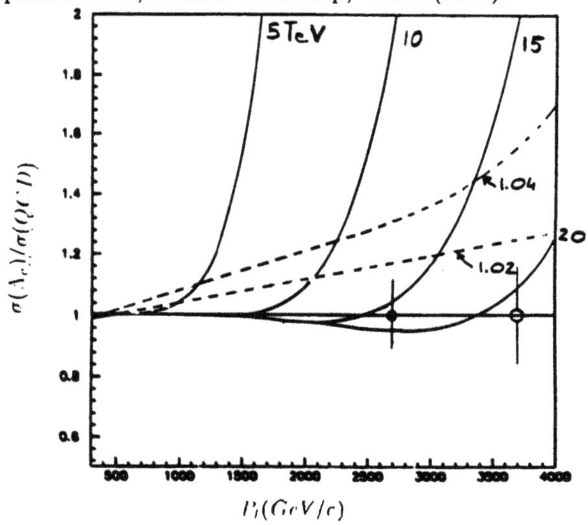

Figure 1: Deviation from QCD for different Λ_c values. The dashed lines show the effect on non-linearities $E(4TeV)/E(500GeV) = x$. Are also shown the statistical sensitivity for $10^4 pb^{-1}$ (black circle) and $10^5 pb^{-1}$ (open circle)

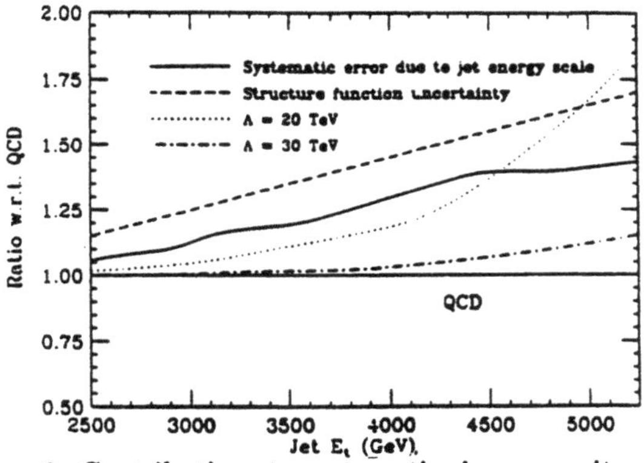

Figure 2: Contributions to systematics in compositeness studies

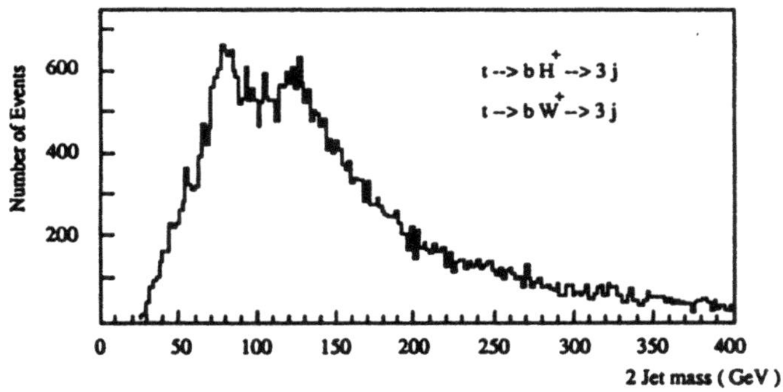

Figure 3: Mass distribution of 2 out of 3 jets from $t \to bH^+$, $H^+ \to c\bar{s}$ and $t \to bW^+$ for an ideal calorimeter. The 2 peaks merge for $\sigma/E = 100\%/E^{1/2} \oplus 4\%$

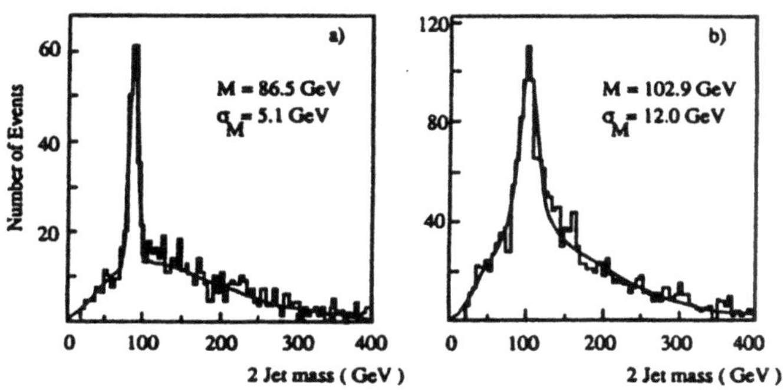

Figure 4: Pile-up effect on 2 jet mass resolution for low p_t $Z \to jj$ for an ideal calo. a) without pile-up b) with 40 MB events. The calorimeter resolution is not included

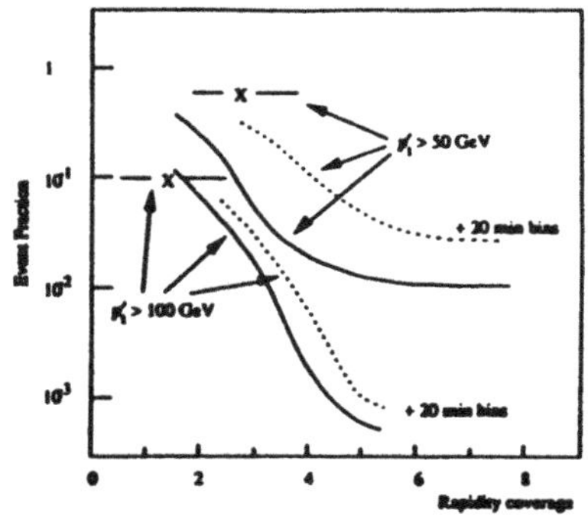

Figure 5: Acceptance for $t\bar{t} \rightarrow e\nu + jets$ and the QCD background rejection factor on $\not{E}_t$ vs η coverage. Pile-up contribution is included

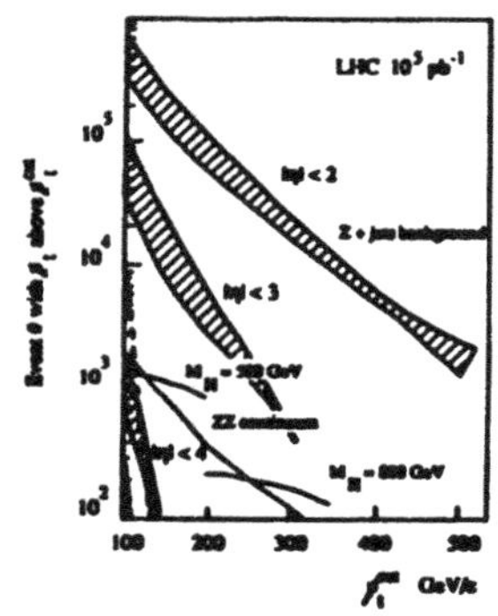

Figure 6: $\not{E}_t$ for $H \rightarrow ZZ \rightarrow ee\nu\nu$ signal and QCD background from ZZ continuum vs η coverage. Pile-up contribution from 60 MB added is shown for η=4 (dashed) and η=5 (dotted)

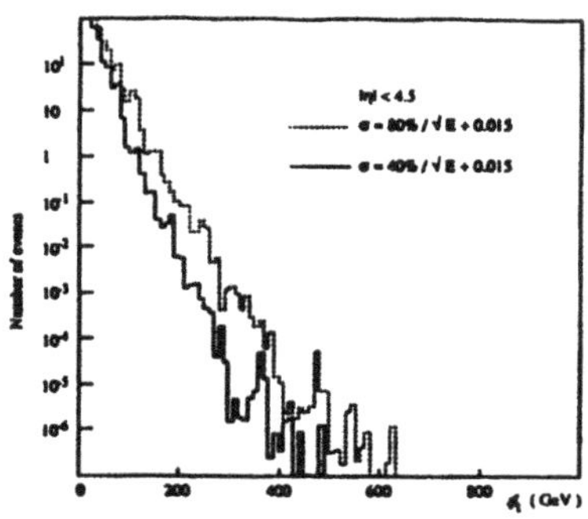

Figure 7: $\not{E}_t$ for 3-jets events as function of the calorimeter resolution (cracks included). Pile-up is not included

CALORIMETER REQUIREMENTS FOR DETECTION OF AN INTERMEDIATE MASS NEUTRAL HIGGS BOSON AT LHC VIA ITS TWO PHOTON DECAY MODE

C. J. SEEZ
Blackett Laboratory, Imperial College of Science and Technology, London SW7, UK.

ABSTRACT

The calorimeter requirements for detection of $H^0 \rightarrow \gamma\gamma$ at LHC are summarised. The significance of the expected signal is illustrated graphically. A Monte Carlo study of π^0 rejection with a silicon strip detector placed at $5X_0$ in a CeF_3 crystal calorimeter is described.

1. Summary of detector requirements

The calorimeter requirements for detection of an intermediate mass neutral Higgs boson at LHC via its two photon decay mode are implicit in the two graphs of figure 1. These two graphs show the significance of the signal expected, for m_H=100 GeV/c^2, after one year's running at LHC (10^5 pb^{-1} at $\sqrt{s}$=16TeV), as a function of e.m. calorimeter resolution[1].

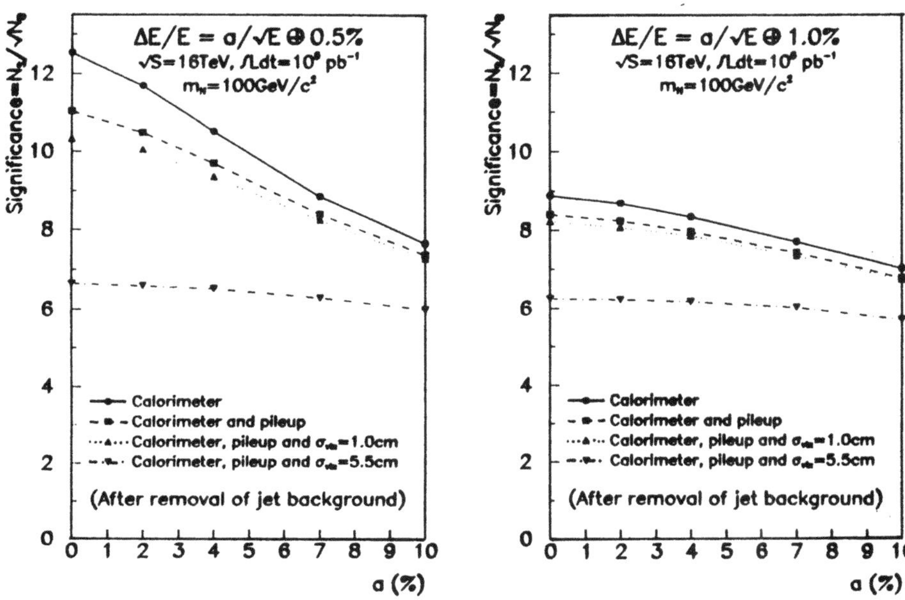

Figure 1: $H^0 \rightarrow \gamma\gamma$ signal significance as a function of calorimeter performance

The cross-sections of the signal and of the irreducible background used are summarised in table 1. All cross-sections are calculated using the HMRSB structure functions[2], and with m_{top}=150 GeV/c^2. The cross-section for Higgs production used here includes the contribution from the W-fusion process (qq→qqH) which was not included in the previous analysis; the values of $N_S/\sqrt{N_B}$ shown in fig 1 are thus about 20% higher than those previously published. Note that for a higher mass H^0 (up to $m_H \approx 160$ GeV/c^2) the significance of the signal expected is larger since the product of cross-section and branching ratio for $H^0 \rightarrow \gamma\gamma$ increases to a peak at about 135 GeV/c^2 before decreasing, while the cross-sections of the background processes decrease monotonically.

The bremsstrahlung contribution to the background which was previously computed rather inelegantly, has now been recomputed for $m_{\gamma\gamma}$=100 GeV/c^2. It is reassuring that the cross-section found was less than 15% higher than the value given here. PYTHIA 5.5 (with JETSET 7.3) was used for this calculation since it has been shown recently that the parton showering in JETSET gives a rather good description of bremsstrahlung at these energies[3].

For further details of the analysis the reader is referred to the previously published work.

Signal: $H^0 \rightarrow \gamma\gamma$

m_H (GeV/c^2)	σ.Br (fb)
80	45.0
100	55.3
150	33.7

Background: $d\sigma/dm_{\gamma\gamma}$ (fb/GeV) of di-photons (after cuts)

$m_{\gamma\gamma}$ (GeV/c^2)	$qq \rightarrow \gamma\gamma$	$gg \rightarrow \gamma\gamma$	Brems	Total
80	31.7	84.7	87.5	204.
100	30.8	60.8	61.4	153.
150	12.0	16.1	17.0	45.

kinematic cuts: $|\eta| \leq 2$, $p_T^1 > 40$, $p_T^2 > 25$GeV/c, $p_T^1/(p_T^1+p_T^2) < 0.7$

Table 1: The signal and background cross-sections

The calorimeter requirements implied by the graphs shown in figure 1 can be summarised as follows:

- a very high resolution e.m. calorimeter is required, and, in particular, the constant term of the resolution function needs to be controlled to ~ 0.5 %,

- it is necessary to reconstruct the photon showers by integrating over only a small region of $\Delta\phi \times \Delta\eta \times \Delta t$, since pileup causes:
  - smearing of the reconstructed mass
  - rejection of signal events when an isolation cut is used,
- powerful π^0 rejection is needed. After applying an isolation cut the remaining rejection can be acheived by a shower position detector at 5 X_0 as is shown below,
- it is necessary to locate the longitudinal vertex position with more precision than is given by a knowledge of the average interaction point; for this purpose it is necessary to reconstruct the photon direction with an RMS error of no more than about 7 mrad,
- the coverage of calorimeter should be up to $|\eta| \sim 2$; in fact the signal significance changes only slowly with acceptance, increasing or decreasing by about 10% for an increase or decrease of the pseudorapidity coverage by 0.5.

2. Visual representation of the expected signals

A visual representation of the signals gives an interesting indication of the meaning of the value of $N_S/\sqrt{N_B}$. First the irreducible background, as a function of $m_{\gamma\gamma}$, after all cuts, was parameterized by fitting a polynomial. Using this parameterization a histogram was made of the number of background events expected for 10^5 pb^{-1} in 100 MeV/c^2 bins with appropriate random Poisson fluctuations.

Higgs events (m_H=100 GeV/c^2) generated with the ISAJET Monte Carlo program for the same integrated luminosity were added to the histogram after simulation of the analysis cuts and detector response for a calorimeter having a resolution function $\Delta E/E = 2\%/\sqrt{(E/GeV)} \oplus 0.5\%$ ($\oplus$ here denotes quadratic summation), pileup for an integration over $\Delta\phi \times \Delta\eta \times \Delta t = 0.1 \times 0.1 \times 1$ crossing (at an instantaneous luminosity of 10^{34}cm^{-2}s^{-1}, $\sigma_{inelastic}$= 60mb), and for a calorimeter giving the photon direction with an RMS error of 7mrad. The resulting mass spectrum is shown in figure 2(a). A fit was made to this mass spectrum (excluding the signal region), and then subtracted - the result is shown in fig 2(b). A window with the optimum width for such a calorimeter (Δm=1.5 GeV/c^2) was scanned over the background subtracted histogram in steps of 100 MeV/c^2, and the value of $N_S/\sqrt{N_B}$ plotted. The result is shown in figure 2(c). One regains the expected peak significance of 10.1. The same exercise was performed for a calorimeter having a resolution function given by $\Delta E/E = 10\%/\sqrt{(E/GeV)} \oplus 0.5\%$, but with its other characteristics as before. The results are shown in figure 3.

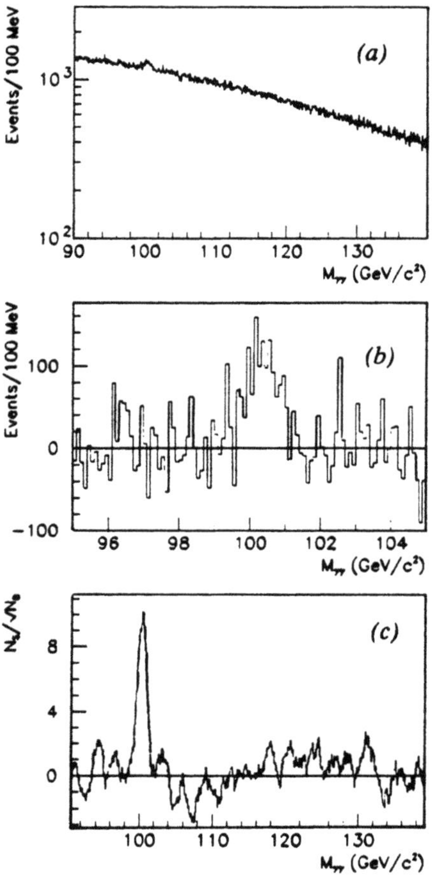

Figure 2: $N_S/\sqrt{N_B}=10.1$

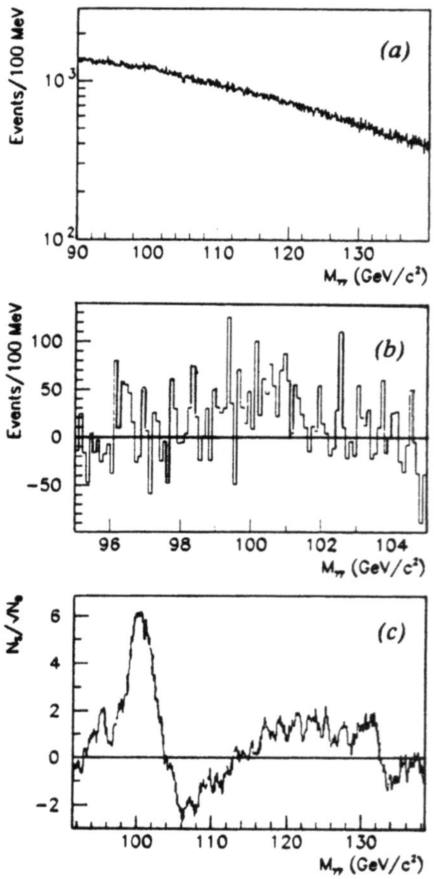

Figure 3: $N_S/\sqrt{N_B}=6.1$

3. π^0 rejection with a shower position detector

Isolation cuts can reduce the jet background to the level of the irreducible background. Since the uncertainty on this background is large (~ 5) it is necessary to aim at reducing the level by a further significant factor in order to be confident that the signal will not be lost. In the previously published analysis it was assumed that a shower position detector could provide a significant rejection against isolated π^0s. This assumption has now been tested in a Monte Carlo study in which photon showers were generated by GEANT 3.14. A calorimeter constructed from cerium fluoride crystals (X_0=1.68cm, ρ=6.16 g/cm^2) was assumed and a 4 Tesla magnetic field was applied transverse to the incoming photon direction. The field significantly spreads the shower in one of its transverse projections.

The energy deposited in two 300 μm thick planes of silicon placed at a depth of $5X_0$ was recorded for showers initiated by incident photons at 10 energies between 5 and 100 GeV. An incident neutral pion could then be simulated by taking two shower records of appropriate energy, spacing them by an appropriate distance, and binning them with a granularity corresponding to the chosen silicon strip width. The strip width chosen was 2mm in the direction parallel to the magnetic field, and 4mm in the orthogonal direction (choosing narrower strips was found to give no improvement in final results). Random noise with an RMS width corresponding to 2500 electrons (i.e. about 10% of a m.i.p.) was added to these neutral pion signals (increasing this noise by a factor of 4 did not seriously degrade the final results).

The energy profiles seen in the silicon strips were compared with a parametrisation of the signal expected from a single photon to produce a quantity χ^2 defined by:

$$\chi^2 \equiv \frac{\Sigma\,(E_i - f_i)^2}{n\ \ \sigma_i^2}$$

where E_i is the energy found in strip i, f_i is the energy expected from the parametrisation of the energy profile using the functional form $f(x) = A\exp(-|x-\langle x\rangle|/s)$, and σ_i is the error on E_i. The sum is run over n (taken to be 21) strips. The values of the function parameters A and <x> were calculated for each event from the total pulse hieght and the weighted mean position in the strips. The error, σ_i, was defined so that the χ^2 function behaves in a similar way for all photon energies. Two values of the function parameter s, one for each projection, were fixed by minimizing the average value of χ^2 over all event samples.

Figure 4 shows the fraction of neutral pions rejected, by a cut on χ^2 that rejects 10% of single photons, as a function of the separation of the two photons in the two projections. It can be seen that a separation in the direction parallel to the magnetic field is more easily detected than a separation in the other projection where the shower profiles are spread by the field. The photons used here have energies of 15 and 10 GeV.

When the separation between the two photons is given a random orientation and the cut

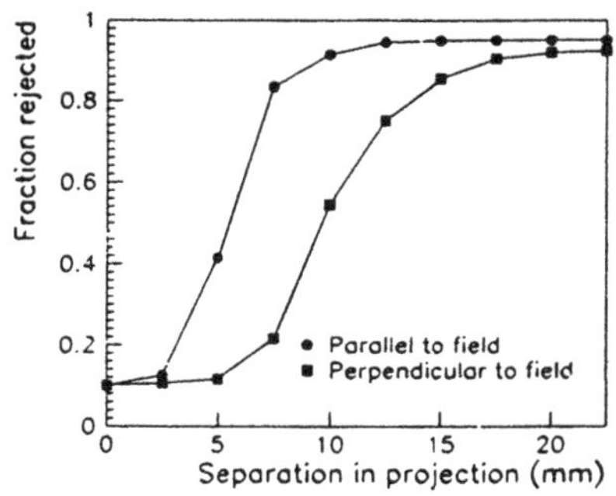

Figure 4: Rejection of shower pairs versus their separation in projection

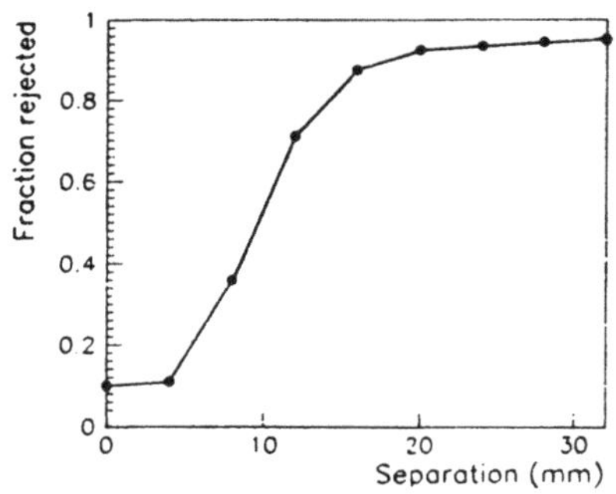

Figure 5: Rejection of shower pairs versus their separation (both projections used)

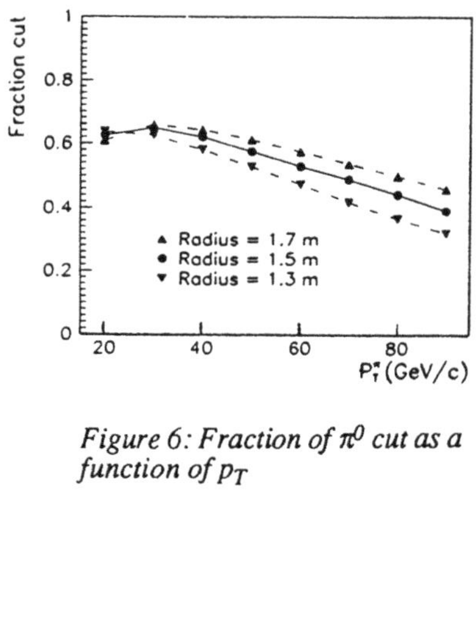

Figure 6: Fraction of π^0 cut as a function of p_T

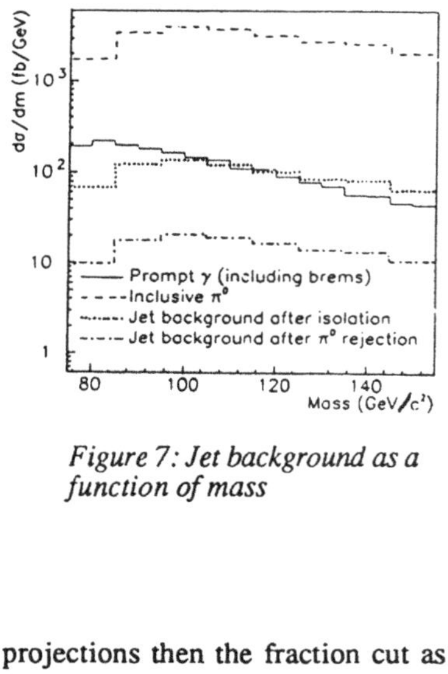

Figure 7: Jet background as a function of mass

is made on the average χ^2 calculated in the two projections then the fraction cut as a function of separation is as shown in figure 5.

In order to calculate the rejection of neutral pions as a function of transverse momentum 50000 π^0s were generated at each of a series of p_T values. The pseudorapidity distribution was approximated by a flat probability function with $|\eta|<2$. The χ^2 cut was set at a value such that approximately 6% of single photons, at all energies between 20 and 100 GeV, failed the cut. The energy profiles in the position detector were tested against the χ^2 cut if the photon separation was less than 3cm in both projections. If the energy of either photon was less than 5GeV it was assumed to be undetected and hence the π^0 was not rejected. Figure 6 shows the rejection power as a function of p_T calculated by the monte-carlo, for three different radii. Applying these rejection factors to events containing two isolated π^0s one can calculate the cross-section of the background remaining, after using such a position detector, as a function of $m_{\gamma\gamma}$. The result is shown in figure 7 for the case where the shower position detector is at a radius of 1.5 metres. It can be seen that the position detector provides a significant rejection against the isolated π^0 background, reducing it to a level well below that of the irreducible background.

It should be noted that although the χ^2 technique used here is simple, providing a rapid answer for this study, it probably does not achieve the maximum rejection power available.

4. Higgs production in association with a charged lepton

The neutral Higgs boson can be produced in association with a $t\bar{t}$ pair or a W boson which gives an isolated muon or electron in the final state. Calorimeter requirements for the two photon decay channel may be modified when the Higgs is produced in association with an isolated muon or electron. Calculations of the cross-sections of both this signal and its dominant backgrounds have recently been published[4]. At present it appears that a somewhat more significant signal might be expected in this channel than in the simple unassociated channel. However more work needs to be done before a fully realistic comparison between the two channels can be made.

5. Acknowledgements

I would like to thank E. F. Clayton who made the CeF_3 calorimeter representation in GEANT, and generated the showers libraries used here. I also thank T. S. Virdee for numerous discussions about this work.

6. References

1. C. Seez, T. Virdee, *et al*, Proceedings of the Large Hadron Collider Workshop, Aachen, 4-9 October 1990, Ed: G. Jarlskog, D. Rein, **II** (1990) 474
2. P. Harriman, A. Martin, R. Roberts and W. Stirling, Phys. Rev. **D42** (1990) 798
3. D. Decamp *et al*, ALEPH collaboration, Phys. Lett. **B264** (1991) 476
4. Z. Kunszt, Z. Trócsányi and W. J. Stirling, ETH-TH/ 91-17, DTP/91/40 (1991)
   A. Ballestrero and E. Maina, DFTT 29/91 (1991)

ELECTRON IDENTIFICATION AT LHC WITH AND WITHOUT MAGNETIC FIELD

D. Froidevaux[1] and A. Poppleton[2]
[1) RD6 Collaboration, 2) RD2 Collaboration]

CERN, Geneva, Switzerland

1. Introduction

Even at luminosities as high as those foreseen at the LHC machine, electrons will be relatively easy to measure with electromagnetic calorimetry. The electron energy measurements will be carried out over small areas, typically of the order of 0.1 x 0.1 in $\Delta\eta$ x $\Delta\phi$, where η is the pseudorapidity and ϕ the azimuth, and therefore distorsions to these energy measurements due to pile–up are expected to be small.

Electrons will however be extremely difficult to identify, and electron identification studies for LHC have to concentrate mainly on background studies and detector capabilities to reject these backgrounds. As we shall see below, calorimetry is one of the key elements in electron identification at these machines, but is not sufficiently powerful in itself to achieve unambiguous electron identification.

In the following, after a reminder of the basic electron identification requirements at LHC, we shall concentrate mainly on a comparative study of the electron identification abilities of detectors with or without central magnetic field.

2. Basic electron identification requirements

2.1 Acceptance

As has been shown in previous studies [1], electron identification is required over a large solid angle, i.e. down to small angles with respect to the proton beams ($|\eta| \leq 3$), and over a large range of transverse momenta, p_T, typically $10 \leq p_T \leq 2000$ GeV. This is illustrated below with three examples:

a) Let us consider the case of a Higgs boson, H, of mass $m_H = 120$ GeV, decaying into four leptons (electrons or muons). The acceptance of a detector with lepton coverage over $|\eta| \leq 3$, is ~ 4% for $p_T > 20$ GeV and increases to ~ 32% for $p_T > 10$ GeV [2]. Since the expected event rates are very low, this particular channel will benefit greatly from lepton detection down to $p_T \approx 10$ GeV or even lower. Similar conclusions were reached in studies of gluino cascade decays, producing Z bosons in the final state with very low event rates [3].

b) In the case of top quark physics, which can be studied at luminosities as low as 10^{32} to 10^{33} cm^{-2}s^{-1}, one needs to identify if possible single electrons of $p_T \geq 40$ GeV at central rapidities [4].

c) For the search for additional vector bosons, Z', decaying into electron pairs, one should achieve electron identification and measurements for p_T as large as ≈ 2000 GeV [5].

2.2 Rejection against QCD jets

Here we take two examples from previous studies to illustrate the rejections needed against QCD jets in order to bring fake electron backgrounds safely below the signals expected at LHC.

a) Taking into account all kinematic cuts and selection efficiencies, the expected signal for $H \rightarrow Z^*Z^* \rightarrow eeee$, with $m_H \approx 130$ GeV, is 0.2 fb. Two types of events containing QCD jets are potential backgrounds to this channel:
   - 4–jet events with a cross–section of ~ 1 mb = 10^{12} fb, within the same cuts and including pile–up at a luminosity of 10^{34} cm^{-2}s^{-1}.
   - $Z \rightarrow e^+e^-$ events accompanied by jets, with a cross–section of the order of 500 pb.

   To reject such events well below the expected level of signal events, a rejection of 3000 per jet is sufficient, but is needed even for events containing two real electrons.

b) Single electrons from top decay or Higgs decay ($H \rightarrow WW \rightarrow e\nu jj$) are expected at levels much below that of electrons from b–quark and W decays. One can therefore raise the question of jet rejection in the case of a search for inclusive electrons, as illustrated in Fig. 1 [6], which shows the expected event rates per unit rapidity at $L = 10^{34}$ cm^{-2}s^{-1} at LHC as a function of p_T, for jets, π^o, γ and electrons from b–quark and W decays. We first note that π^o/jet $\geq 10^{-3}$ for $p_T \approx 40$ GeV, which shows that one cannot expect rejections better than 10^3 from calorimetry alone. We also note that e/jet $\approx 10^{-5}$ at $p_T \approx 40$ GeV, a ratio much less favourable than that found at the CERN $p\bar{p}$ collider, where e/jet $\approx 10^{-3}$ at $p_T \approx 40$ GeV.

We conclude from Fig. 1 that, in order to bring jet backgrounds safely below the level of single electrons, one needs rejections at LHC of:

a) ~ 10^6 against QCD jets
b) $\geq 10^3$ against π^o/η mesons
c) ≥ 10 against π^o/η Dalitz decays.

2.3 *Trigger rates*

The expected trigger rates are reasonable for electron pairs, assuming that the level 1 trigger is provided by the calorimeter alone. As an example, for $p_T > 20$ GeV and $|\eta| < 2$, at $L = 10^{34}$ cm^{-2}s^{-1}, we expect a level 1 trigger rate of ~ 1000 H_z for a calorimeter rejection of 100 per jet, which drops to ~ 10 Hz for the maximum attainable calorimeter rejection of 1000 [7]. This latter rate, if at all achievable, is only a factor 2 to 3 larger than the expected trigger rate from $Z \rightarrow e^+e^-$ in the same geometrical acceptance.

On the other hand, a single electron trigger will be difficult at LHC, since the level 1 trigger rate expected would be ~ 5000 Hz for a calorimeter rejection of 1000 and for $p_T > 40$ GeV and $|\eta| < 2$. Additional rejection from tracking detectors will be certainly needed at higher trigger levels.

3. Backgrounds from fake electrons

Such backgrounds in hadron colliders arise from two sources:

- single charged hadrons (e.g. through charge exchange);
- charged hadron overlap with one or several photons from π^o/η decay.

3.1 *Detector with central magnetic field*

In this case, as demonstrated by UA1 and CDF at lower energies, hadronic backgrounds are expected to be small, i.e. at a level of ~ 10^{-6} of the jet cross–section, well below the expected level for electrons from b–quark and W decay. Although this can be estimated from the ratio $\pi^{\pm}$/jet $\approx 10^{-3}$ and the probability that a charged pion produces an electromagnetic shower in the calorimeter, some redundancy in electron identification may be needed at LHC in order to control these backgrounds at high luminosity.

3.2 *Detector without central magnetic field*

Past experience in UA2 has shown that hadronic backgrounds are very dangerous in this case and have to be dealt with by the inner tracking system. As an example, a combination of preshower technique [6] and transition radiation [8], which could provide independent rejections of ~ 100 each against these backgrounds, would bring them down to a level well below that of prompt electrons.

4. Backgrounds from electron pairs

We shall now concentrate on the potentially most dangerous background to a prompt electron signal, namely electron pairs from conversions of photons from π^0/η decay or from π^0/η Dalitz decays. In this context, we have performed a simple comparative study of 'generic' detectors with and without central magnetic field, in order to evaluate their respective expected performances against these electron pair backgrounds. In particular, we have assumed that the parent π^0/η spectrum is the expected single π^0/η spectrum, and have ignored more complicated topologies containing multi–π^0 final states [7].

4.1 *Electron pair production*

We first generated a parent (π^0 or η) with a p_T–spectrum $d\sigma/dp_T = p_T^{-5.5}$, as suggested by Fig. 1, for $p_T > p_T^0$ and $|\eta| < 2$. The parent p_T threshold, p_T^0, was usually chosen to be 10 GeV (the most difficult case). This parent was then allowed to decay following the recipe of Table 1. The ratio η/π^0 was chosen to be 0.55, and external conversion production was normalized to the average number of photons per produced π^0, including $\eta \to \gamma\gamma$, $\eta \to 3\pi^0 \to 6\gamma$ and a 6% contribution from direct photons.

The branching fraction for external conversions is computed assuming 1% X_0 at $|\eta| = 0$, a minimum corresponding to the vacuum pipe, some air and the support of the first tracking layer.

Channel	Parent production rate	Branching fraction	Total weight	Fraction of all e^+e^- pairs
External conversions for 1% X_0 over $\|\eta\| < 2$	3.49 (average number of photons per produced π^0)	$1.56\ 10^{-2}$	$5.43\ 10^{-2}$	72.0%
$\pi^0 \to \gamma e^+e^-$	1	$1.20\ 10^{-2}$	$1.20\ 10^{-2}$	15.9%
$\eta \to 3\pi^0 \to 5\gamma + e^+e^-$	0.55	$1.15\ 10^{-2}$	$0.63\ 10^{-2}$	8.4%
$\eta \to \gamma e^+e^-$	0.55	$0.50\ 10^{-2}$	$0.28\ 10^{-2}$	3.7%

Table 1: production of e^+e^- pairs

From Table 1, it appears clearly that external conversions will be the dominant source of production of electron pairs, even for as little as 1% X_0 of material in the detector.

4.2 *Detector simulation*

After generating e^+e^- pairs with the above procedure, we performed a simple detector simulation, based on two detector models corresponding to possible configurations at LHC, as described below:

a) An electromagnetic calorimeter with inner radius, R = 130 cm, and a granularity $\Delta\eta \times \Delta\phi = 0.02 \times 0.02$, selected candidate electron clusters, under the conditions that the generated electron or positron, properly tracked through the inner detector (see below), pointed to the cluster and that the transverse energy E_T, collected over $\Delta\eta \times \Delta\phi = 0.06 \times 0.06$, was greater than p_T^0, the p_T–threshold for the parent particle generation. This first selection was performed for both detector models used, and is labelled TRK–CAL in the plots described below.

b) In the case of a solenoidal central magnetic field, a momentum measurement was simulated in the inner tracking region, using two measurements of 100 μ accuracy at radii of 30 cm and 100 cm. With the knowledge of the transverse vertex position, the resulting transverse momentum measurement accuracy was $\sigma_{p_T}/p_T = 0.003 \cdot p_T/B$, where p_T is in GeV and B in Tesla. As shown below, the results do not depend significantly on the value assumed for the momentum measurement accuracy. A second selection, labelled E/p below, was then performed requiring the measured cluster energy E to match the measured electron or positron momentum p within $E/p < 1.2$.

c) In this step a charged particle isolation cut was applied at R = 100 cm, labelled Outer Isolation below, requiring only one charged particle to be present in a cone of area $\Delta\eta \times \Delta\phi = 0.05 \times 0.05$, which matches the area over which the energy is collected in the electromagnetic calorimeter and corresponds to roughly 10% occupancy at a luminosity of $2\,10^{34}\ \mathrm{cm}^{-2}\mathrm{s}^{-1}$. This cut will reject close by electron pairs and is therefore most efficient for B = 0 as will be shown below. Some care was taken to include multiple scattering, since it turns out to be the limiting factor in the efficiency of this cut for rejecting conversions at B = 0.

d) In a final step, for completeness, an additional charged particle isolation cut was applied at R = 30 cm for $B \neq 0$ over a somewhat smaller solid angle, $\Delta\eta \times \Delta\phi = 0.024 \times 0.024$, to account for the higher expected occupancy in a central magnetic field. This cut was labelled Inner Isolation in the figures presented below.

4.3 *Rejection of conversions*

Figure 2 shows the expected rejection of electron pairs from photon conversions, for a parent p_T threshold of 10 GeV, as a function of the strength of the central magnetic field. The rejection of 40 needed to bring such a background down to the desired level is also indicated, assuming 1% X_0 of material in front of the first sensitive tracking layer.

The rejection obtained from the TRK–CAL cut rises rapidly from 2 at B = 0 (the total parent energy is often collected in $\Delta\eta \times \Delta\phi = 0.06 \times 0.06$) to 15 at B = 4T, where the calorimeter basically selects mainly isolated electrons which therefore very often do not pass the threshold. This is confirmed by the curve which includes in addition the E/p cut for $B \neq 0$. This cut essentially results in an overall rejection independent of B, basically due to the fact that, after this cut, most of the remaining events correspond to isolated high–p_T electrons (asymmetric conversions) which are either selected by the E/p cut (low field) or the calorimeter (high field). Fig. 3 shows the E/p distribution expected for B = 0.5 T, 1.0 T and 2.0 T. A stricter cut than the one used ($E/p < 1.2$) is seen to be rather ineffective, the more so the higher the field, where one expects better momentum accuracy. In addition, bremsstrahlung effects which were not explicitly included in this study, are expected to result in significant losses of real electron signals for E/p cuts stricter than the one applied.

Figure 2 shows that the Outer Isolation cut provides a powerful rejection against conversions at B = 0, but very little for $B \neq 0$. The Inner Isolation cut applied for $B \neq 0$ rejects to some extent conversions produced in the support of the inner tracking layer.

Figure 4 shows, for B = 2T, how these rejection factors vary as a function of the choice of the parent p_T–threshold. As expected from kinematics, the calorimeter rejection is much worse at higher p_T, but E/p recovers most of this loss, and the charged particle isolation also becomes more efficient at rejecting asymmetric conversions at high p_T.

4.4 *Rejection against Dalitz decays*

Fig. 5 shows the expected rejection against Dalitz decays, for a parent p_T larger than 10 GeV, as a function of B. Also shown is the rejection of 20 needed against this background to bring it down to the desired level. The features of the various curves are much the same as for the case of external conversions (Fig. 2), except for the charged particle isolation cuts, which are much less powerful because of Dalitz kinematics.

4.5 *Additional rejection*

The results shown in Figs. 2 and 5 for conversions and Dalitz decays show that the desired rejection can be achieved, but they are based on many optimistic assumptions:

- The electron efficiency has not been thoroughly investigated (bremsstrahlung, tracking detectors...);
- the chosen isolation cones are quite large;
- a minimum amount of material (1% X_0) was assumed in front of the first tracking layer.

For these reasons, we have investigated whether the quoted rejections can be improved upon by searching for the second electron and/or nearby photon(s).

Figure 6 shows the energy spectrum of the second electron after the first three selection cuts described in Section 4.2 have been applied, for B = 0 and B = 2T and for a parent p_T threshold of 10 GeV. For B = 0, the average energy of this second electron is expected to be 0.5 GeV, precluding any hope to identify it in the LHC environment. For B = 2T, the higher expected average energy of 4.0 GeV is probably sufficient to allow some additional rejection to be achieved, but any quantitative estimate implies detector specific simulations beyond the scope of this contribution.

Figure 7 shows the angular distance at the vertex between the electron candidate chosen and the largest energy photon in the event for B = 0 and B = 2T. As expected, the presence of the field forces this photon away from the electron because of the E/p cut, but also forces it to much lower energies as shown in Fig. 8.

In summary, additional rejections can be obtained by searching for a second electron of 4 GeV average energy for B = 2 T, and by searching for a photon of average energy 20 GeV at an average distance of 5 cm from the electron for B = 0.

5. Conclusions

The most dangerous backgrounds to electron signals at LHC arise from electron pairs produced by external conversions and Dalitz decays. An adequate rejection against the latter can be achieved in a straightforward way by 'generic' detectors independently of the presence and strength of a central magnetic field. In the case of external photon conversions, we can conclude that:

a) such backgrounds are much easier to reject in the absence of a central magnetic field, particularly at low p_T (10 to 20 GeV);

b) if the detector material is $\leq 1\%$ X_0 in front of the first sensitive tracking layer, an adequate rejection can also be achieved in the presence of a central magnetic field, independently of its value;

c) as a corollary, any tracking detector with more than 1% X_0 of material should provide a rejection against conversions inside its own material, given the difficult experimental environment at LHC/SSC.

The choice of a central magnetic field can certainly be justified on the basis of broader arguments, such as an obviously richer physics potential, at least at moderate luminosities. The practicality of such a choice hinges however on the crucial point of the feasability of inner tracking in the very difficult environment of these hadronic machines, a topic which is beyond the scope of a calorimeter conference.

References

1. T. Akesson and E. Fernandez, Identification of Electrons at LHC, *Proceedings of the Large Hadron Collider Workshop, Aachen 4–9 October 1990*, CERN 90–10, Vol. I, p. 420.

2. D. Froidevaux, Experimental Review of the Search for the Higgs Boson, *Proceedings of the large Hadron Collider Workshop, Aachen 4–9 October 1990*, CERN 90–10, Vol. III, p. 444.

3. C. Albajar et al., Experimental Aspects of Gluino and Squark Searches, *Proceedings of the Large Hadron Collider Workshop, Aachen 4–9 October 1990*, CERN 90–10, Vol. III, p. 621.

4. L. Fayard and G. Unal, Prospects for Top Search and Top Mass Measurement, *Proceedings of the Large Hadron Collider Workshop, Aachen 4–9 October 1990*, CERN 90–10, Vol. III, p. 360.

6. P. Camarri et al., Z' into Leptons, Proceedings of the Large Hadron Collider Workshop, Aachen 4–9 October 1990,, CERN 90–10, Vol. III, p. 704.

6. A proposal to study a Tracking/Preshower Detector for the LHC, CERN/DRDC/ 90–27 (02.09.90).

7. J.–C. Chollet and J.–P. Repellin, these Proceedings.

8. Integrated high rate transition radiation detector and tracking chamber for the LHC, CERN/DRDC/90–38 (29.08.90).

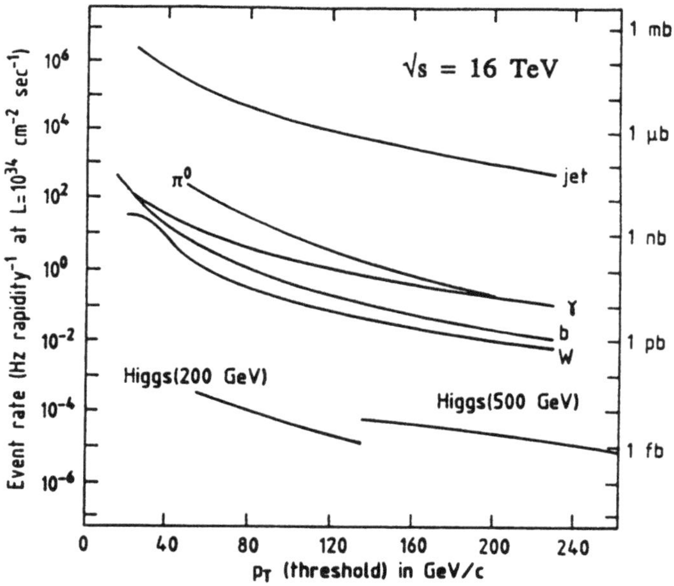

√s = 16 TeV
Event rate (Hz rapidity^-1 at L=10^34 cm^-2 sec^-1)
10^6
10^4
10^2
10^0
10^-2
10^-4
10^-6
1 mb
1 μb
1 nb
1 pb
1 fb
jet
π^0
γ
b
W
Higgs(200 GeV)
Higgs(500 GeV)
0
40
80
120
160
200
240
p_T (threshold) in GeV/c

Fig. 1

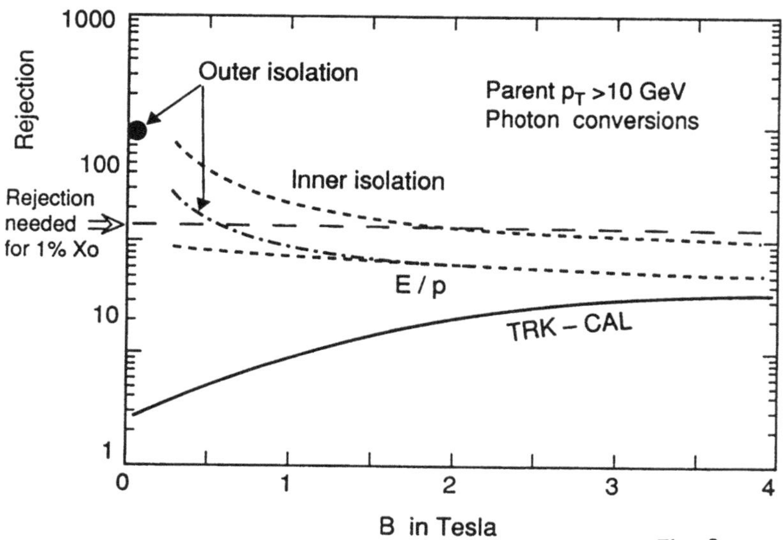

1000
100
10
1
Rejection
Outer isolation
Parent p_T >10 GeV
Photon conversions
Inner isolation
Rejection
needed
for 1% Xo
E / p
TRK – CAL
0
1
2
3
4
B in Tesla

Fig. 2

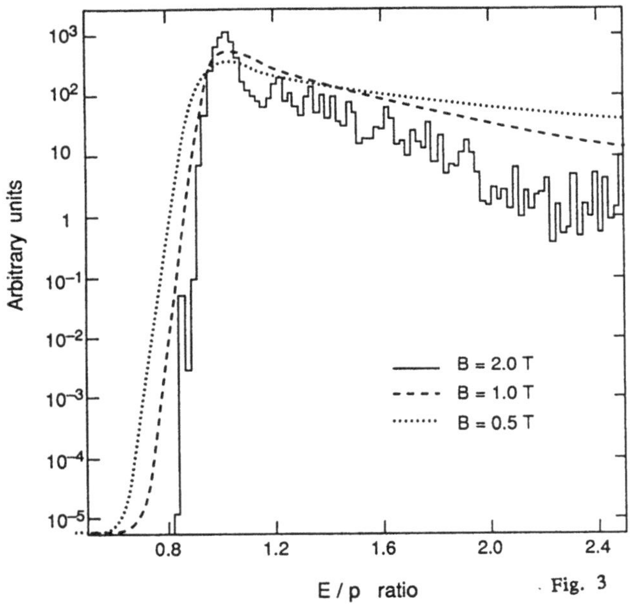
10³
10²
10
1
10⁻¹
10⁻²
10⁻³
10⁻⁴
10⁻⁵
Arbitrary units
B = 2.0 T
B = 1.0 T
B = 0.5 T
0.8
1.2
1.6
2.0
2.4
E / p ratio

Fig. 3

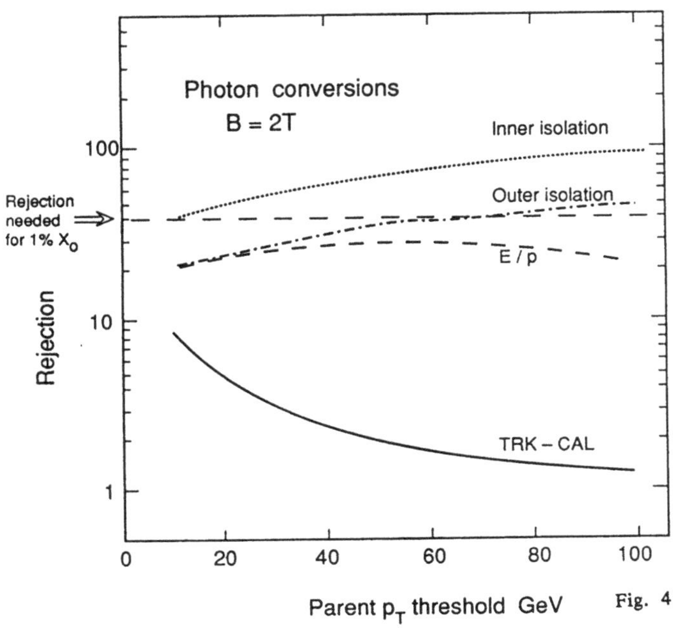
Photon conversions
B = 2T
Inner isolation
Outer isolation
E / p
TRK – CAL
Rejection needed for 1% X₀
Rejection
100
10
1
0
20
40
60
80
100
Parent p_T threshold GeV

Fig. 4

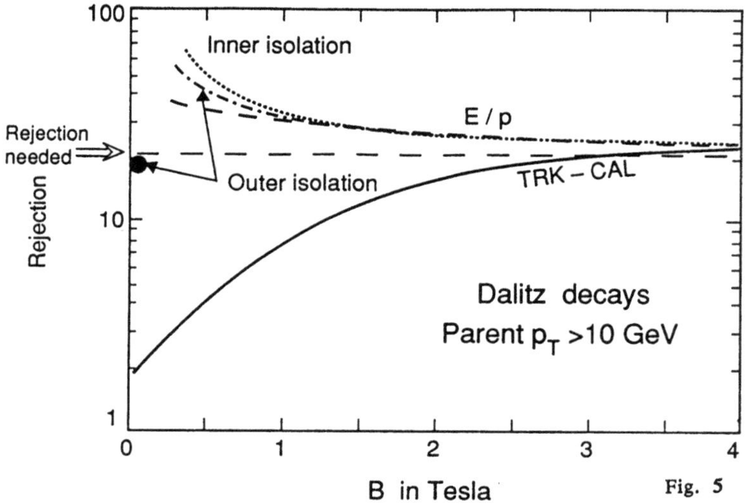

100
10
1
0
1
2
3
4
Rejection
Rejection needed
Inner isolation
E / p
Outer isolation
TRK – CAL
Dalitz decays
Parent p_T >10 GeV
B in Tesla

Fig. 5

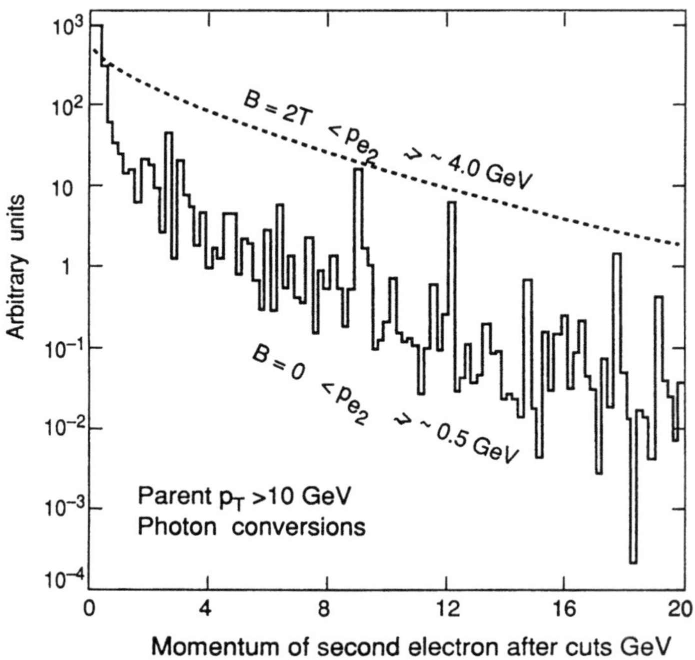

10^3
10^2
10
1
10^-1
10^-2
10^-3
10^-4
0
4
8
12
16
20
Arbitrary units
B = 2T <pe2> ~ 4.0 GeV
B = 0 <pe2> ~ 0.5 GeV
Parent p_T >10 GeV
Photon conversions
Momentum of second electron after cuts GeV

Fig. 6

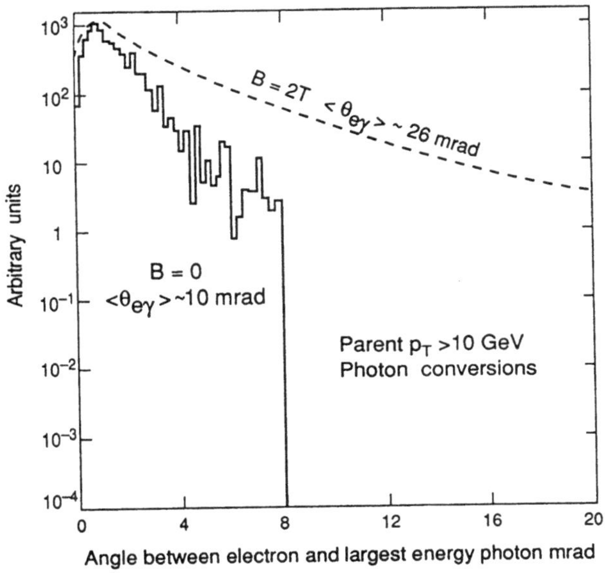

Arbitrary units
10^3
10^2
10
1
10^{-1}
10^{-2}
10^{-3}
10^{-4}
0
4
8
12
16
20
Angle between electron and largest energy photon mrad
B = 2T < θeγ > ~ 26 mrad
B = 0
<θeγ>~10 mrad
Parent pT >10 GeV
Photon conversions

Fig. 7

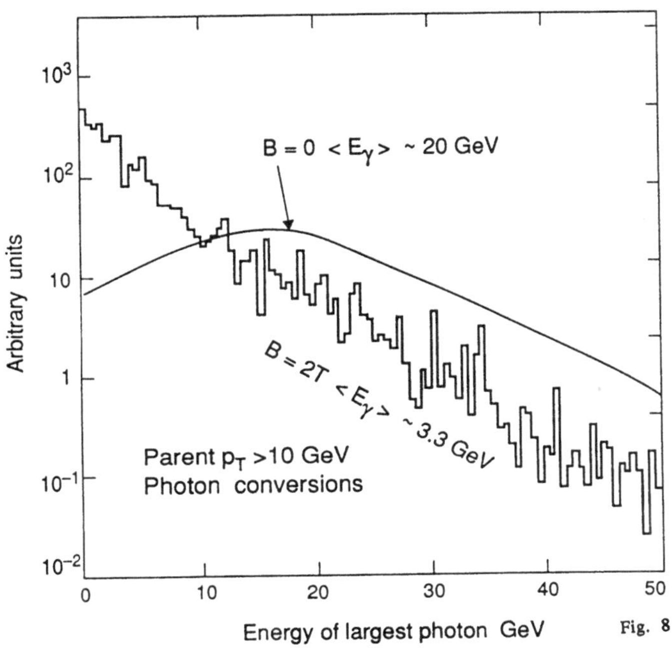

Arbitrary units
10^3
10^2
10
1
10^{-1}
10^{-2}
0
10
20
30
40
50
Energy of largest photon GeV
B = 0 < Eγ > ~ 20 GeV
B = 2T < Eγ > ~ 3.3 GeV
Parent pT >10 GeV
Photon conversions

Fig. 8

FOUR-COMPONENT APPROXIMATION TO CALORIMETER RESOLUTION*

D. E. Groom
Lawrence Berkeley Laboratory 50-308, Berkeley CA 94720, USA

ABSTRACT

A hadron with energy E cascades in a uniform calorimeter. With some probability, energy E_e enters the electromagnetic sector, mostly through π^0 production. Other distributions describe the conversion of electromagnetic energy to visible signal, hadronic energy to visible signal, and the conversion of their sum to an output signal. We describe the combined frequency function for all four processes. It turns out that they are uncorrelated, so that variances add in quadrature. As expected, the variance contribution due to the initial π^0 production is proportional to the degree of noncompensation in the Gaussian approximation. However, (a) there is a slow ($\ln E$) decrease in the "constant term", and (b) the familiar $1/\sqrt{E}$ for the sampling term is not quite correct.

1. Introduction

Conventional wisdom says that the main contributions to the resolution of a hadron calorimeter come from a sampling term $C_1/\sqrt{E}$ and a constant term C_2. These are either combined linearly or in quadrature, at the discretion of the writer. The consequences of this choice are far from academic, since results are usually quoted without respect to how the fits were made. Typically, the "constant term" is more than 60% higher when the fit is made assuming linear addition. One reason for this study was to look for correlation which might justify linear addition. None was identified.

In previous papers I have considered the behavior of the means of various distributions, arguing that the mean response of a simple calorimeter is described by only three quantities, and that only one of these is energy dependent.[1,2] The picture should also yield a description of resolution, and this is done in the present study. The function describing the distribution of the electromagnetic fraction in a cascade is also explored in more detail than before.

As in other papers, I restrict the discussion to the response of a uniform calorimeter to an incident high-energy hadron with energy E. By "uniform" I mean that the structure of the calorimeter (usually a sandwich structure) is the same from front to back, and that the dimensions are sufficient to contain the cascade.

2. Formulation of the problem

The logic flow in a hadronic cascade is shown in Fig. 2. It is simplified in two respects: escaped energy (carried by albedo, neutrinos, and muons) is neglected, and complicated processes are lumped together. For example, in the low-energy hadronic sector, the effects of nuclear binding losses, saturation, and gate width are combined in relating the hadronic energy E_h to the visible hadronic energy $E_h^{\rm vis}$.

Energy is transferred to the electromagnetic sector (the EGS world) in high-energy collisions, mostly through π^0 production. In a given cascade the energy transfer is E_e, but the mean value is $F_{\pi^0}E$.* With some efficiency (ϵ_e, on the average) this energy appears as a visible signal $E_e^{\rm vis}$. Similarly, $E_h = E - E_e$ enters the hadronic sector, and $E_h^{\rm vis}$ contributes to the visible signal. E_h contains all of the hadronic

* This work was supported by the U. S. Department of Energy under Contract No. DE-AC03-76SF00098.

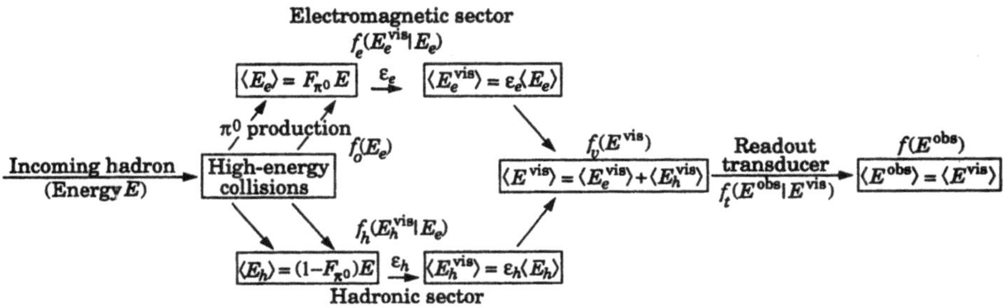

FIG. 2. Logic flow in a hadron-induced cascade in a calorimeter. Transfer of energy to the electromagnetic sector occurs in many high-energy collisions. ϵ_h (ϵ_e) is the efficiency with which energy in the hadronic (electromagnetic) sector is converted to visible energy. Included are all effects of saturation, gate width, *etc.* The visible energy is converted to the energy (signal) finally observed by a photomultiplier or other readout transducer, which introduces further fluctuations.

energy below the π^0 threshold as well as ionization loss by above-threshold hadrons. It also contains the energy of photons from nuclear processes, since this energy scales with the incident energy in the same way as do other hadronic contributions.

The four distributions under consideration are as follows:

1. The distribution of E_e, given by the frequency function (fr.f.) $f_0(E_e)$. (It is convenient to leave the incident energy E implicit.)
2. The distribution of visible electromagnetic energy is given by the conditional fr.f. $f_e(E_e^{vis}|E_e)$.
3. Similarly, the distribution of visible hadronic energy is given by $f_h(E_h^{vis}|E_e)$. Since $E_h = E - E_e$. It is sufficient to express the condition as on E_e.
4. Finally, the distribution of observed signal, or energy, is obtained by folding this with whatever photoelectron statistics or electronic noise distribution characterizes the readout transducer, as described by the conditional fr.f. $f_t(E^{obs}|E^{vis})$.

Separation of the final distribution is somewhat artificial, but it has the function of separating the "in principle" resolution from the "real" resolution as degraded by the output transducer.

3. The frequency function $f_0(E_e)$

The function $f_0(E_e)$ has been generated with the aid of the high-energy hadronic cascade simulation program FLUKA90.[3] Typical distributions, at 20 GeV and 200 GeV, are shown in Fig. 3. As expected, the mean moves to the right with increasing energy, as more of the hadronic energy is transferred to the electromagnetic sector as the number of hadronic generations increases.

In Fig. 4 are shown the energy dependences of the mean divided by incident energy (F_{π^0}) and the square root of the variance divided by energy ($\sigma_{nc} \equiv \sigma_0/E$). The curves drawn through the means are of the form $[1-(E/E_0)^{1-m}]$.[1] The solid curve is a least-squares fit (excluding the 10 GeV point) for which $m = 0.866$ and $E_0 = 0.764$ GeV. Nominal values from the Tuscaloosa paper ($m = 0.85$ and $E = 1$

* We reserve f_{π^0} for the energy transfer to the electromagnetic sector in a single collision.

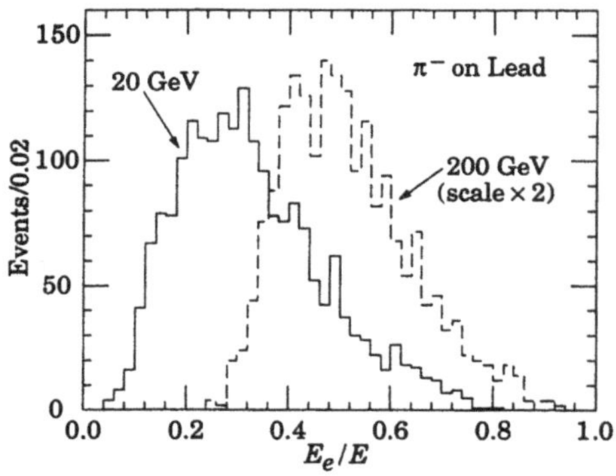

FIG. 3. Distributions of E_e for 20 GeV and 200 GeV negative pions incident on lead, as simulated with FLUKA90.

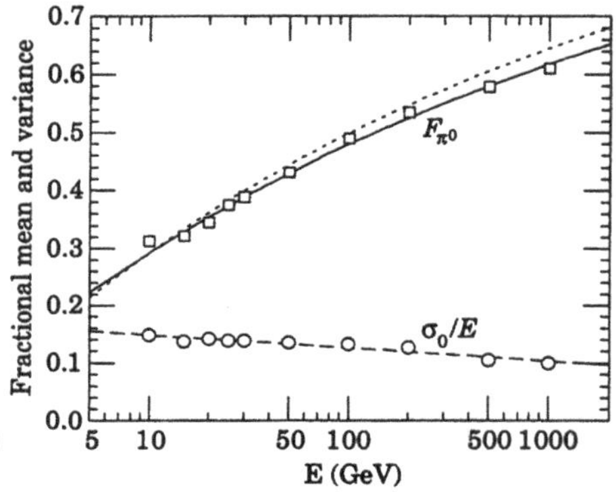

FIG. 4. Energy behavior of the mean $F_{\pi^0}E$ and width σ_0, scaled by the energy E, as simulated using FLUKA90 for negative pions on lead. Fits are discussed in the text.

GeV) are used for the dotted curve.

The width σ_0/E is remarkably constant. This is the noncompensation contribution to the famous "constant term."[4] Dan Green has observed that most of the variance is simply the result of fluctuations in the number of π^0's produced in the first collision. Since the multiplicity increases logarithmically with energy, σ_0/E should decrease slightly as the energy increases. The dashed line is a least-squares fit, $0.171 - 0.010 \ln E$. This function has the value 0.14 at 20 GeV and 0.12 at 200 GeV.

The "coefficient of skewness" μ_3/σ_0^3 (where μ_3 is the third moment about the mean) is constant to within the Monte Carlo statistics and has a value near 0.6. The mean and width have been studied previously.[1,5] The present method is more sensitive since the distributions of E_e are studied directly.

4. Characteristic functions

The analysis is considerably simplified by the use of characteristic functions (c.f.'s), where the c.f. $\phi(u)$ associated with the fr.f. $f(x)$ is essentially its Fourier transform, or the expectation value of $\exp(iux)$:[6] $\phi(u) = \boldsymbol{E}(e^{iux}) = \int e^{iux} f(x)dx$. Several of its useful properties are exploited here.

If $f_1(x)$ and $f_2(y)$ have c.f.'s $\phi_1(u)$ and $\phi_2(u)$, then the c.f. of the weighted sum $ax + by$ is $\phi_1(au)\phi_2(bu)$.

Let the (partial) c.f. corresponding to the conditional fr.f. $f_2(x|z)$ be $\phi_2(u|z)$, and the fr.f. of z be $f_1(z)$. The c.f. after integration over the intermediate value is given by

$$\phi(u) = \int f_1(z)\phi_2(u|z)dz \quad . \tag{1}$$

Suppose we can write ϕ_2 in the form $\phi_2(u|z) = A(u)\exp(ig(u)z)$. Then

$$\phi(u) = A(u)\phi_1(g(u)) \quad . \tag{2}$$

The c.f. of a Gaussian fr.f. with mean m and variance σ^2 is $\phi(u) = \exp(imu -$

$\frac{1}{2}\sigma^2u^2)$. The simplest skewed distribution is

$$\phi(u) = \exp(imu - \tfrac{1}{2}\sigma^2u^2 - \tfrac{i}{3!}\mu_3u^3) \ , \tag{3}$$

where μ_3 is the third moment about the mean.

5. Results

The fr.f. of the total visible energy, $f_v(E^{\mathrm{vis}})$, can be calculated from f_0 and the two conditional fr.f.'s. Only the total visible energy ($E^{\mathrm{vis}} = E_e^{\mathrm{vis}} + E_h^{\mathrm{vis}}$) can be observed. The problem is to find the fr.f. $f_v(E^{\mathrm{vis}})$ for an incident hadron with energy E. We first combine f_e and f_h:

$$f^{\mathrm{vis}}(E^{\mathrm{vis}}|E_e) = \int f_e(E_e^{\mathrm{vis}}|E_e)f_h(E^{\mathrm{vis}} - E_e^{\mathrm{vis}}|E_e)dE_e^{\mathrm{vis}} \tag{4}$$

This fr.f. is then summed over intermediate values of E_e:

$$f_v(E^{\mathrm{vis}}) = \int f_0(E_e)f^{\mathrm{vis}}(E^{\mathrm{vis}}|E_e)dE_e \ . \tag{5}$$

We use the same labels to identify the c.f.'s corresponding to the fr.f.'s ($\phi^{\mathrm{vis}}(u|E_e)$ corresponds to $f^{\mathrm{vis}}(E^{\mathrm{vis}}|E_e)$, *etc.*). Then, via the properties given in Section 4, we have

$$\phi^{\mathrm{vis}}(u|E_e) = \phi_e(u|E_e)\phi_h(u|E_e) \ , \tag{6}$$

and

$$\phi_v(u) = \int f_0(E_e)\phi^{\mathrm{vis}}(u|E_e)dE_e \ . \tag{7}$$

The electromagnetic response is expected to be fairly Gaussian. The hadronic response is less well known, but because the variance is dominated by a small number of collisions with large nuclear binding energy losses, it is likely to deviate substantially from a Gaussian fr.f. However, the fr.f.'s are not yet available, so I make a Gaussian approximation here as well.

For an ensemble of events with electromagnetic energy E_e, we would not expect the variance of the electromagnetic signal to be constant, but instead to be proportional to E_e. The variance of the the hadronic fraction should be proportional to $(E - E_e)$. Accordingly, we consider the c.f.'s

$$\phi_e(u|E_e) = e^{iu\epsilon_e E_e - \frac{1}{2}\sigma_{e0}^2\epsilon_e E_e u^2} \quad \text{and} \quad \phi_h(u|E_e) = e^{iu\epsilon_h(E-E_e) - \frac{1}{2}\sigma_{h0}^2\epsilon_h(E-E_e)u^2} \ , \tag{8}$$

so that

$$\phi^{\mathrm{vis}}(u|E_e) = e^{iu\epsilon_h E - \frac{1}{2}\sigma_{h0}^2\epsilon_h E u^2} \times e^{iE_e(u\Delta\epsilon - \frac{i}{2}u^2\Delta\sigma^2)} \ , \tag{9}$$

where $\Delta\epsilon \equiv \epsilon_e - \epsilon_h$, and $\Delta\sigma^2 \equiv \sigma_{h0}^2\epsilon_h - \sigma_{e0}^2\epsilon_e$. With this definition, $\Delta\sigma^2$ is usually positive. Noting that Eq. (9) has the form $A(u)\exp(ig(u)z)$ discussed above, we see that

$$\phi_v(u) = e^{iu\epsilon_h E - \frac{1}{2}\sigma_{h0}^2\epsilon_h E u^2}\phi_0(u\Delta\epsilon - \tfrac{i}{2}u^2\Delta\sigma^2) \ , \tag{10}$$

where ϕ_0 is the c.f. corresponding to f_0.

The fr.f. of E_e was discussed in Section 3. Based on that analysis (and with reference to Eq. (3)), it is approximated as $\phi_0(u) \approx \exp(iF_{\pi^0}Eu - \frac{1}{2}\sigma_0^2u^2 - \frac{i}{3!}\mu_{30}u^3)$, where F_{π^0} and σ_0 are given in Section 3, and $\mu_{30} \approx 0.6\sigma_0^3$. Then Eq. (10) becomes

$$\phi_v(u) = \exp\Big(iu\epsilon_e E\{\epsilon_h/\epsilon_e + F_{\pi^0}\Delta\epsilon/\epsilon_e\}$$

$$- \tfrac{1}{2}u^2\{\epsilon_e E[(1-F_{\pi^0})(\sigma^2_{h0}\epsilon_h/\epsilon_e - \sigma^2_{e0}) + \sigma^2_{e0}] + (1-\epsilon_h/\epsilon_e)^2\sigma_0^2\epsilon_e^2\}$$
$$- \tfrac{i}{3!}u^3\{\mu_{30}(\Delta\epsilon)^3 - 3\sigma_0^2\Delta\sigma^2\Delta\epsilon\} + \ldots\Big) \; . \quad (11)$$

(Since only the first three moments of f_0 were evaluated, terms of order u^4 and higher are dropped.)

Two housekeeping chores are necessary before interpreting these results: In the first place, it was pointed out in Section 3 that σ_0 is proportional to energy except for a slow logarithmic decrease: $\sigma_0 \approx \sigma_{nc}E$. We therefore make this substitution. Secondly, with the notation used above, the mean visible energy for an incident electron (or for a hadron if $\epsilon_e = \epsilon_h$) is $\epsilon_e E$. In a real situation, one calibrates the device so that the measured mean is E. Accordingly, we replace $\epsilon_e E$ by E.

With these substitutions and the aid of Eq. (3), the mean, variance, and skewness of the fr.f. given by Eq. (11) can be written down. After some rearrangement, the mean can be written as

$$\langle E^{\text{vis}} \rangle = E[1 - (1 - F_{\pi^0})(1 - \epsilon_h/\epsilon_e)] \; . \quad (12)$$

This is a convenient form, since the second term, the product of the hadronic fraction and the degree of noncompensation, expresses the energy-dependent correction for noncompensation. The quantity in square brackets is the "linearity curve" (the ratio of response to energy) for hadronic response, and, alternatively, the reciprocal of the usual "e/π" ratio.

Similarly, the variance is the coefficient of $\frac{1}{2}u^2$ in Eq. (11), or the quantity in curly brackets. With the substitutions given above, the fractional variance becomes

$$\left(\frac{\sigma}{E}\right)^2 = \frac{(1 - F_{\pi^0})(\sigma^2_{h0}\epsilon_h/\epsilon_e - \sigma^2_{e0}) + \sigma^2_{e0}}{E} + (1 - \epsilon_h/\epsilon_e)^2\sigma^2_{nc} \; . \quad (13)$$

This is *almost* the usual form for the resolution, with three important differences:

1. There is no cross term, so the contributions add in quadrature.
2. The "constant term" is explicitly proportional to the degree of noncompensation. We have previously concluded that the contribution (in quadrature) to the fractional resolution σ/E was $(1 - \epsilon_h/\epsilon_e) \times 14\%$. In this analysis σ_{nc} decreases from 14% at 20 GeV to 12% at 200 GeV.
3. If F_{π^0} were not a function of energy, the numerator of the first term would be just the square of the usual coefficient of $1/\sqrt{E}$. However, F_{π^0} *is* energy dependent. This means that the familiar $1/\sqrt{E}$ dependence never did exist!

To understand this statement, it is important to remember the relative size of the different variances. For purposes of evaluating SDC detector requirements, the design goal assumed in the LoI was $\sigma_{had}/\sigma_{EM} = 0.7/0.25 = 2.8$. We identify this ratio with $(\sigma_{h0}\sqrt{\epsilon_h/\epsilon_e})/\sigma_{e0}$. The hadronic variance (scaled by ϵ_h/ϵ_e) is thus about 8 times larger than the electromagnetic variance, and to a good approximation σ^2_{e0} can be dropped. Eq. (13) then becomes

$$\left(\frac{\sigma}{E}\right)^2 \approx \frac{(1-F_{\pi^0})\sigma^2_{h0}\epsilon_h/\epsilon_e}{E} + (1-\epsilon_h/\epsilon_e)^2\sigma^2_{nc} \approx \frac{(E/E_0)^{m-1}\sigma^2_{h0}\epsilon_h/\epsilon_e}{E} + (1-\epsilon_h/\epsilon_e)^2\sigma^2_{nc}$$

$$\approx \left(\frac{\sigma'}{E^{0.57}}\right)^2 + (1 - \epsilon_h/\epsilon_e)^2\sigma_{nc}^2 \quad . \tag{14}$$

The physical reason for the small deviation from $1/\sqrt{E}$ behavior is simple. As the incident energy increases, more and more of the cascade energy moves from the hadronic to the electromagnetic sector. The resolution improves for this reason alone, and so improves more rapidly with energy than would be expected just from the sampling statistics.

6. Inclusion of transducer contributions

For illustrative purposes we also take the fr.f. $f_t(E^{\rm obs}|E^{\rm vis})$ as Gaussian, with mean $E^{\rm vis}$ (by definition) and variance $\sigma_t^2 E^{\rm vis}$. Following the same arguments as above, the c.f. of the final fr.f., $\phi(u)$, is obtained by evaluating $\phi_v(u)$ as given by Eq. (11) at $u + \frac{i}{2}\sigma_t^2 u^2$. The mean remains the same (by definition) and the variance increases by $\sigma_t^2 \langle E^{\rm vis}\rangle$. The third moment about the mean is increased by $3\sigma_t^2\sigma^2$, where σ^2 is the variance of f_v as given by Eq. (13). The change in skewness comes about because the variance is proportional to the input signal; the same "feedup" of moments has occurred in earlier parts of this discussion.

The inclusion of f_t is mathematically hollow, but it is of importance for designers: The burden on the experimenter is to make σ_t^2 smaller than σ^2 (for example, by efficient light collection and good photocathode sensitivity, in the case of photomultiplier readout), but there is no point in making σ_t^2 very *much* smaller than σ^2.

7. Summary

It follows from the mathematical structure of the problem as presented here that the sampling and noncompensation contributions to the resolution are not correlated. Nor have I been able to identify any mechanism (usually via a multivariate distribution) which would provide correlation. On the other hand, it has been found that the sampling term decreases a little faster than $1/\sqrt{E}$, and that the noncompensation contribution decreases slowly with increasing energy.

It will be of great interest to understand the fr.f. $f_h(E_h^{\rm vis}|E_e)$, as this will permit a reasonably complete quantitative solution to the problem.

8. References

1. D. E. Groom, in *Proc. of the Workshop on Calorimetry for the Superconducting Super Collider,* Tuscaloosa, Alabama, 13–17 March 1989, ed. R. Donaldson and M. G. D. Gilchriese, World Scientific (June 1990), 59–75.
2. D. E. Groom, "Why Waste All Those MIPS on Compensation Studies?," *Proc. of the 1990 DPF Summer Study on High Energy Physics Research Directions for the Decade,* Snowmass CO, June 25–July 13, 1990, ed. by E. L. Berger and R. Craven (to be published) (1991).
3. A. Ferrari, A. King, and G. R. Stevenson, oral tradition (1991).
4. R. Wigmans, Nucl. Instrum. Meth. **A265**, 275 (1988).
5. D. E. Groom, "Contributions of Albedo and Noncompensation to Calorimeter Resolution," *Proc. of the 1990 DPF Summer Study on High Energy Physics Research Directions for the Decade,* Snowmass CO, June 25–July 13, 1990, ed. by E. L. Berger and R. Craven (to be published) (1991).
6. H. Cramér, *Mathematical Methods of Statistics,* Princeton Univ. Press, New Jersey (1958).

A FAST CALORIMETER SIMULATION USING GEANT

C. Charlot
LPNHE, Ecole Polytechnique and CNRS-IN2P3
Palaiseau, France

ABSTRACT

We present here a fast Monte Carlo algorithm based on GEANT for calorimetry simulation. It is based on a parametrization of electronic showers longitudinal and transversal profiles, including fluctuations. This has been applied to the case of a lead - scintillating fibers calorimeter.

1 Introduction

The design and optimization of detectors for future LHC/SSC experiments requires the simulation of detector response by means of the Monte Carlo method. In the case of calorimeter, existing simulation tools as GEANT[1] can provide a reliable description, by following all the particles produced during the shower development down to a given threshold. However, such a procedure implies a prohibitive CPU time consumption at LHC/SSC energies, since this time is proportionnal to the shower energy.

On the other hand, there are studies that do not need a complete simulation of all the shower properties, but rather some global description of the energy deposition with its fluctuations. Such a description can be obtained by means of a fast Monte Carlo using a parametrization of the showers, deduced from full shower simulations or, if possible, from existing data. The shower fluctuations are introduced by smearing the parameters, and by fluctuating the number of hits that simulate the energy deposition. A number of parametrization programs have already been proposed[2].

We present here an improved version of the ALEPH parametrization, which has been adapted for the simulation of lead - scintillating fibers calorimeters.

2 Procedure

As the calculation of tracking boundaries leads already to an intrinsic high CPU time consumption in case of sampling calorimeter, the first operation consists in replacing the detailed, but repetitive, sampling structure by a single homogeneous equivalent medium (which will be called from hereafter the mean medium), whose properties are an average of the detailed medium properties.

Since some differences exist between the global properties of showers induced by electrons and by photons, only the electron and positron energy deposition are parametrized here. This means that in any particle induced shower, the standard GEANT Monte Carlo is used to follow the showering particles until an electron or a positron is produced. Once one of the latter particle is uncountered, the control is given to the parametrization algorithms, to produce a set of hits corresponding to the energy deposition that the particle would have generated.

The parameters which are needed for the fast Monte Carlo have been extracted from reference data samples. We have used a set of samples of 1000 showers generated with GEANT (version 3.14) in the mean medium. These samples where obtained by initializing electromagnetic showers with electrons of 0.1, 0.3, 1, 10, 50 and 100 GeV entering the calorimeter under normal incidence. The electron and photon threshold where set to 1 MeV for incident energies greatest than 1 GeV, and lowered to 100 KeV for lower energies. In addition,we have used a set of 3 samples of electronic showers of 1, 10, and 100 GeV generated using the detailed sampling structure of the lead - scintillating fibers calorimeter, with a cut-off of 10 KeV, in order to compare and fit the mean medium description. Finally, available data have been used to refine the tuning of the parameters.

These procedure has been applied to the case of the SPACAL lead - scintillating fibers calorimeter[3].

3 Simulation of the longitudinal profile

The longitudinal profile is described according to a gamma function:

$$f(z) = E_{inc}\frac{\beta^{\alpha}}{\Gamma(\alpha)}z^{\alpha-1}exp(-\beta z) \tag{1}$$

with z in units of radiation length (X_0).

Such a distribution can be studied by means of the first and second order moments m and σ. Instead of m and σ, it is more convenient to use the variables $\frac{1}{m}$ and $\frac{\sigma^2}{m^2}$ whose distributions appears to be uncorrelated and gaussian-like.

The event by event longitudinal profile is then generated by generating a value for these two quantities. In order to provide an interpolation for any energy, the energy dependence of the mean values and sigma of their distributions are fitted with logarithmic functions.

A comparison of the longitudinal profile distribution obtain with the fast Monte Carlo and with a full simulation in detailed medium for 10 GeV is presented on Fig. 1.

Figure 1: *Mean longitudinal profiles obtained using the parametrisation (full line) and using the full simulation in detailed medium (dashed line) for 10 GeV electrons*

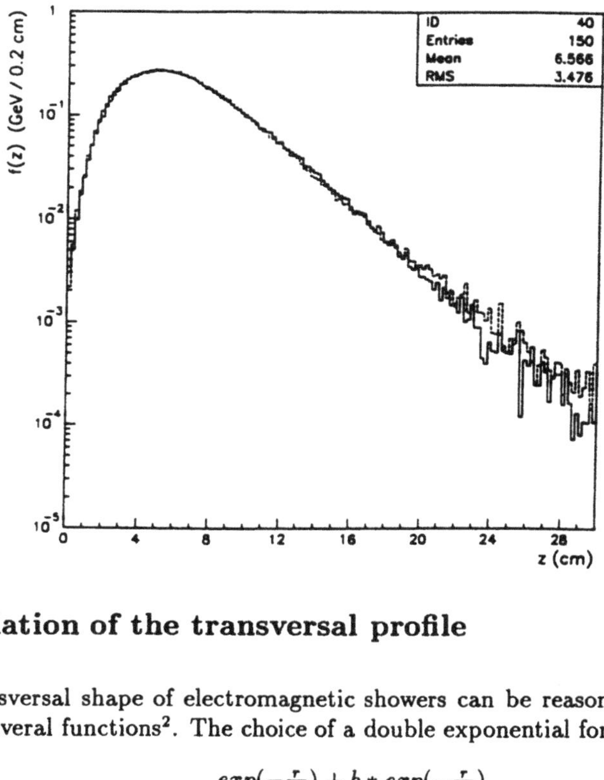

4 Simulation of the transversal profile

The transversal shape of electromagnetic showers can be reasonnably well described by several functions[2]. The choice of a double exponential form

$$g(r) = \frac{exp(-\frac{r}{\mu_1}) + b * exp(-\frac{r}{\mu_2})}{\mu_1 + b\mu_2} \tag{2}$$

with r in units of X_0, allows a simple description of the transversal profile as a function of the depth. The three parameters μ_1, μ_2, and b can be parametrized as functions of energy and depth using the reference data samples by the following empirical formulae :

$$\mu_1(z, E) = 0.025 * exp(\frac{z}{p_1(E)})^{\frac{1}{2}} \tag{3}$$

and

$$b(z, E) = 0.0003 * exp(p_2(E) * z) \tag{4}$$

The second slope parameter μ_2 can be kept constant at all energies. The parameters $p_1(E)$ and $p_2(E)$ are fitted by logarithmic functions of the energy.

A comparison between the transversal profile obtained with the presented parametrisation and from the data is presented on Fig. 2 for 150 GeV electrons.

Figure 2: *Transversal profile distribution obtained using the parametrisation and compared to 150 GeV electron data*

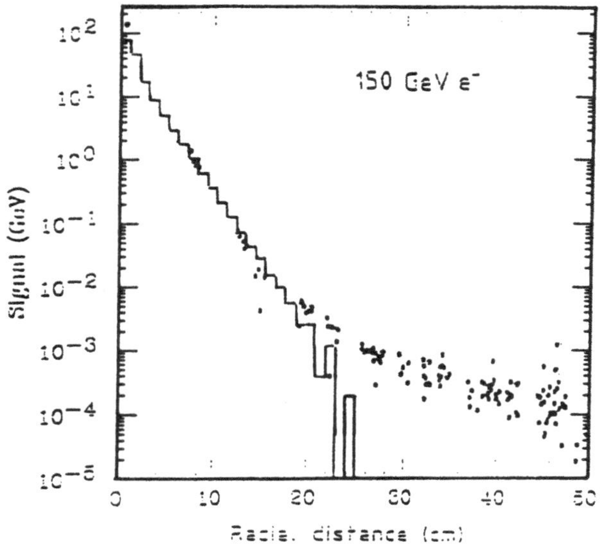

5 Fluctuations of the energy deposition

Global fluctuations of the profiles can be generated by smearing the profile parameters around their mean values, as described in section 3.

Sampling fluctuations of the energy deposition are introduced in the fast Monte Carlo by fluctuating the position of the energy hits. An energy E is deposited by generating a Poisson distributed number N of hits of energy ϵ according to the profile distributions. Assuming the energy resolution to be simulated is given by :

$$\frac{\sigma_E}{E} = \frac{\alpha}{\sqrt{E}} \tag{5}$$

and substituting E by $N\epsilon$ with $\sigma_N = \sqrt{N}$, we get the hits energy :

$$\epsilon = \alpha^2 \tag{6}$$

In the case of the SPACAL lead - scintillating fibers calorimeter, this leads to a hit energy of 17 MeV, or 59 hits per GeV.

A constant term to be added quadratically can be introduced in the Monte Carlo by directly fluctuating the incident energy of the particle around the nominal value.

6 Normalisation of the hadron response

Although the hadronic shower development remains under GEANT control, the hadronic response has to be normalize according to the $\frac{e}{h}$ ratio of the calorimeter to be simulated.

The hadron normalization is achieved in two steps. Firstly, we correct the mip response with respect to electron. In the case of the SPACAL calorimeter, an $\frac{e}{mip}$ value of 0.71 is determined using muon data. We further assume that the compensation is achieved by low energy neutrons, and hence normalize the neutron response in order to reproduce the experimentally measured $\frac{e}{h}$ value.

Figure 3: *Fit of energy resolution for hadrons on calorimeter data. The line is a fit from SPACAL data corresponding to* $\frac{\sigma}{E} = \frac{.306}{\sqrt{E}} + .010$

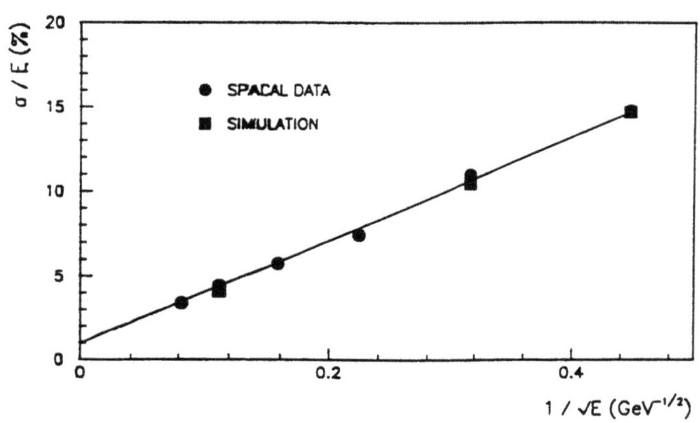

Finally, one introduces sampling fluctuations in order to match the desired hadronic energy resolution. Here, the fluctuations to be introduced are small, the energy resolution being dominated by the fluctuations of the hadronic processes involved in the cascade. As in the electromagnetic case, a constant term, if not resulting from non compensation, can be introduced. The Fig. 3 shows how the experimental hadron resolution can be reproduced in the fast Monte Carlo.

7 Performances of the fast Monte Carlo

As shown on Fig. 1-2 the fast Monte Carlo is able to reproduce well enough the electronic shower development. For this it needs external inputs from full simulation or from existing data.

Such a fast Monte Carlo cannot however describe the real shower development in very inhomogeneous regions of the calorimeter like boundaries between modules or crack between calorimeter elements. The optimization of such procedures would consist in using only the full shower simulation when approaching such regions, and switch on the parametrisation elsewhere.

Concerning the CPU time consumption, the fast Monte Carlo needs about 9 msec per GeV for an electromagnetic shower on CCIN2P3-IBM 3090. As a comparison, the time needed to generate the reference samples with the mean medium is about 120 msec per GeV with the mentionned thresholds, and the time needed to generate full showers in the detailed medium is about 6 sec per GeV on the same IBM. As for full shower simulation this time is proportionnal to the shower energy.

References

1. R. Brun et al., GEANT-CERN Program Library W5013.
   For a general review of the existing Monte Carlo for calorimeter simulation, see R. Brun in these proceedings.

2. A. De Angelis et al., Nucl. Instr. and Meth. A271 (1988) 455 .
   Aleph handbook 1989, the ALEPH collaboration, ALEPH 89-77, CERN note 89-03, april 1989.
   Particle Data Group, G.P. Yost et al., Review of particle properties, Phys. Lett. B 239 (1990).
   G. Grindhammer et al., Nucl. Instr. and Meth. A290 (1990) 469.
   J. Del Poso et al., DESY preprint, DESY 91-005, Feb. 1991.

3. D. Acosta et al., Nucl. Instr. and Meth. A294 (1990) 193.
   D. Acosta et al., Nucl. Instr. and Meth. A308 (1991) 481.

SUMMARY OF THE RADIATION DAMAGE SESSION

A. Ferrari
Istituto Nazionale di Fisica Nucleare INFN, Sez. di Milano

1. Introduction

The radiation environment of the future colliders will result in major constraints on the technological choiches for the detectors. A careful and extensive R&D programme must be carried out to select the proper solutions in order to assure that detectors will survive in such an environment for the their whole life time.

Examples of such studies have been reported in a few of the contributed talks to this session. Other talks dealt with the predictions of the radiation environment.

2. Radiation damage to scintillator tiles and scintillating fibres

The four talks concerning with fibres and scintillators are summarized below: only few points are briefly discussed since the reader can look at the full papers.
"Radiation Radiation Tolerant Plastic Scintillators / Result of a Low Dose Irradiation of Selected Plastic Scintillating Fibres", presented by K. Johnson: the speaker presented also results obtained by C. Zorn at CEBAF who could not come to the Conference. Unfortunately the paper is not available, however the most important conclusions can be summarized as follow:
- fibres show quite different behaviours according to manufacturer and composition: the radiation sensitivity decreases steadily ranging from blue to yellow fibres according to well known results.
- an oxygen rich atmosphere is of help during annealing, but can strongly enhance the damage during the irradiation.
- there are strong indications of an enhanced damage when fibres are irradiated at low dose rate, in comparisons with experiments performed using ^{60}Co sources or accelerators where dose rates are usually many order of magnitudes higher than the ones predicted in the LHC and SSC environment. Informations obtained after the Conference have confirmed that dose rate effects are well known (and also partially understood from the chemical point of view) for many organic materials. As an example enhanced damage has been observed for cable insulation in a reactor environment, resulting in failures at only 1/10 of the life time predicted according to the certification tests performed at high dose rate. This effect can have quite important consequences when extrapolating experimental results to the real LHC/SSC environment.

"Radiation Damage in Tile-Fiber Calorimeter Modules", presented by J. Earnwein: this talk is an interesting example of irradiation tests carried out on a full scale prototype (an EM module for SDC). The most relevant points I would like to stress are:

- at the SSC design luminosity there could be problems for the most exposed End Cap regions where the expected life time could be as short as 1 year (of course this limit could be improved using radiation hard scintillator tiles instead of the commercial ones tested here).
- there are interesting results about the effect of the atmosphere on the annealing properties of the irradiated tiles.
- the quoted results, obtained with a few short irradiations at high dose rate, could be seriously affected by the low dose rate effects discussed above.

"Effects of Radiation Damage on Scintillating Fibre Calorimetry", presented by A. Henriquez: in this talk the results concerning irradiations of fibres carried out in such a way to mimic the effect of an EM cascade have been discussed and limits on the possible applications of a "spaghetti" Calorimeter are derived using a simple model. These results have been already published so I would like only to stress that these limits make difficult the choice of such technology for and End Cap EM Calorimeter in the LHC environment, also considering that low dose rate effects can strongly lower these limits. Of course these considerations could be completely revised if much harder fibres will be available in the future.

"Measurement of Radiation Damage during Irradiation", presented by H. Grassmann: a facility has been built at the INFN National Laboratory at Legnaro for testing fibres and other equipments under exposure to ^{60}Co γ rays. One of the most interesting features is the possibility to perform on line measurements during the irradiations both on the irradiated samples and on a reference one. The first results obtained with this facility have been also presented.

3. Radiation Level Predictions

The last two talks were quite different: they dealt with the extension (D.E. Groom) and the limitations (G.R. Stevenson et al.) of the approach outlined in SSC-SR-1033 for estimating the radiation environment for the future colliders.

"Low-Energy Neutron Fluences in Idealized Collider Geometries", presented by A. Ferrari: a critical discussion, carried out using detailed Monte Carlo simulations, of previously reported models for estimating neutron fluxes has been presented in this talk. The conclusion is that, at least for Lead-dominated calorimeters, simple models like the one discussed in SSC-SR-1033 can lead to significant underestimation of the neutron fluxes at some positions in the detector.

"Extension of SSC-SR-1033 approach to Radioactivation in LHC and SSC Detectors", presented by D.E. Groom: in his talk Don Groom recalled that there are also other important radiation related effects besides the damage: using an extension of the approach outlined in that "famous" SSC report, he derived estimates for the radioactivaction levels and following doses for the inner cavity of a typical LHC or SSC detector. The predicted levels clearly show that radioactivation will be surely a serious problem at future colliders.

EXTENSION OF SSC-SR-1033 APPROACH TO RADIOACTIVATION IN LHC AND SSC DETECTORS*

D. E. Groom
Lawrence Berkeley Laboratory 50-308, Berkeley CA 94720, USA

ABSTRACT

In the SSC study of radiation levels in detectors (SSC-SR-1033), a very successful general approach was developed: The incident flux from the interaction point was described by the product of a flat pseudorapidity distribution and a transverse momentum distribution. A given effect (albedo neutrons, EM ionizing dose, ...) was studied as a function of incident particle momentum. The two functions were then combined to obtain the desired result. In the present study, the approach is extended to the radioactivation of detector components. Radioactivity in a given material is proportional to the average "star density" in hadronic cascades, as simulated in this case with FLUKA. Star density at each depth is found to scale with surprising accuracy as a power law in the incident momentum. At this stage only a "rule of thumb" is used to relate star density to activity. Results are presented for iron and lead.

1. The SSC-SR-1033 approach

The Task Force on Radiation Levels in the SSC Interaction Regions was able to approximate detector radiation levels by means of analytic functions.[1] The idea was to use a two-step approach:

1. Describe the flux incident on a detector element as a function of momentum and the position of the element.
2. Describe the effect of unit flux (one particle per unit area) on a detector element as a function of incident momentum; call this response $\phi(p)$.

The two distributions are then combined to obtain the desired response. The approach only breaks down in certain "crosstalk" situations. For example, the flux incident on a calorimeter at pseudorapidity $\eta \approx 0$ might contain significant albedo contributions from large-η regions; these are ignored. Similarly, there is an implicit approximation that the particle flux does not change significantly over the physical size of a cascade, which might not be true for a hadron calorimeter at very large η.

The production model is particularly simple, since it is known that for average events the number of particles per unit pseudorapidity interval (H) is constant in the central region to within about 10%. At the SSC this central region is expected to extend out to $|\eta| \approx 6$, or $\theta \approx 0.3°$. The distribution of the transverse component of momentum, $f(p_\perp)$, is independent of η, and so $d^2N_{ch}/d\eta\, dp_\perp \approx H\, f(p_\perp)$. Then the number of particles incident on a normal area element da at an angle θ and a distance r from the interaction point (IP) is given by

$$\frac{d^2N_{ch}}{da\, dp_\perp} = \frac{d\Omega}{da}\frac{d\eta}{d\Omega}\frac{d^2N_{ch}}{d\eta\, dp_\perp} \approx \frac{H}{2\pi r^2 \sin^2\theta} f(p_\perp) \ . \tag{1}$$

For the SSC, H is expected to be about 7.5, and the same extrapolation yields $H \approx 6.3$

* This work was supported by the U. S. Department of Energy under Contract No. DE-AC03-76SF00098.

for the LHC. The mean response of a detector element in a LHC or SSC interaction region for one (minimum bias) event then given by

$$R = \frac{H}{2\pi r^2 \sin^2\theta} \int f(p_\perp)\phi(p_\perp/\sin\theta)dp_\perp \,. \tag{2}$$

Some luck makes this integral particularly simple. In all cases studied so far, $\phi(p) \approx Ap^\alpha$. An acceptable form for the $p_\perp$ distribution is $f(p_\perp) \propto p_\perp \exp(-2p_\perp/\langle p_\perp\rangle)$. In the absence of a low-momentum cutoff (magnetic field), the integral in Eq. (2) is proportional to $\Gamma(2+\alpha)/2^\alpha$. Over the useful range $0 \leq \alpha \leq 1$, the same answer, within 6% in the worst case at $\alpha = 0.5$, is obtained by replacing $f(p_\perp)$ by $\delta(p_\perp - \langle p_\perp\rangle)$. It is thus sufficient to write

$$R = \frac{H\,A\,\langle p_\perp\rangle^\alpha}{2\pi r^2 \sin^{2+\alpha}\theta} = \frac{H\,A\,\langle p_\perp\rangle^\alpha}{2\pi r^2}\cosh^{2+\alpha}\eta \,. \tag{3}$$

We extrapolate existing data to estimate $\langle p_\perp\rangle = 0.60$ GeV/c at the SSC. With the same extrapolation $\langle p_\perp\rangle = 0.55$ GeV/c at the LHC.

2. Plan to include radioactivation

Cascade programs such as FLUKA[2] tally the density of nuclear interactions, or "stars," produced by hadrons above some cutoff energy, typically 50 MeV. In the "universal spectrum" approximation discussed elsewhere,[3] the rate of production of radionuclides of a given species is proportional to the star production rate in the same volume element. Some departure from this proportionality is to be expected early in the cascade, where many of the stars are produced in very high-energy collisions, but even here the departure is not thought be significant. The proportionality constant will depend on the program cutoff. The idea is that the yield per star can be established once and for all by detailed simulation, after which it can be used to relate star density to activation under arbitrary conditions in the same material.

One pitfall is that we tabulate the density of stars produced by fairly high-energy particles (mostly neutrons), whereas most of the stars are actually produced in lower-energy collisions. The lower energy neutrons have diffused somewhat further, so the activation profile is fuzzed out somewhat compared with the star density profile. As will be evident, only longitudinal smearing is relevant, and since the scale is about half of a nuclear interaction it is unlikely to be important.

Our program, then, is to describe star production as a function of depth z in typical materials and as a function of incident hadron momentum. Star density in detectors can then be calculated via Eq. (2). If we are fortunate enough to find $\phi(p,z) \propto p^\alpha$, then Eq. (3) can be used. In any case, the resulting response can be related to activation.

In this report (which has a 6-page limit), we carry this program through in detail for star production in iron and lead. The corresponding activation is only found at the level of a "rule of thumb" approximation, although the literature contains enough information for the completion of the program.[4,5]

3. Extension of the model to "star density"

FLUKA simulations of π^--induced cascades in iron and lead absorbers have been made at 9 momenta between 1 GeV/c to 500 GeV/c. In all cases the thickness was 100 cm, and the longitudinal binning was 2 cm. Transverse integrals were tablulated. (It can easily be seen that the transverse integral of such a quantity for one incident particle is the same as the density of the quantity for a uniform incident flux of one particle per unit area.) Results at two depth intervals in lead are shown in Fig. 1. The Fig. 1(b) results are typical: after the first few cm, the 10–500 GeV/c data at a given depth are very well described by a power law. At smaller depths the same thing seems to be true, but here the results are statistic-limited because of the smaller number of stars early in the cascades. In all cases, the 1 GeV/c points fall below the power-law fits. This may be real, or may be an artifact of the program. In any case it is relevant only if $\langle p \rangle$ is significantly lower than 10 GeV/c. This happens only for $|\eta| \lesssim 3$, where activation is less of a problem.

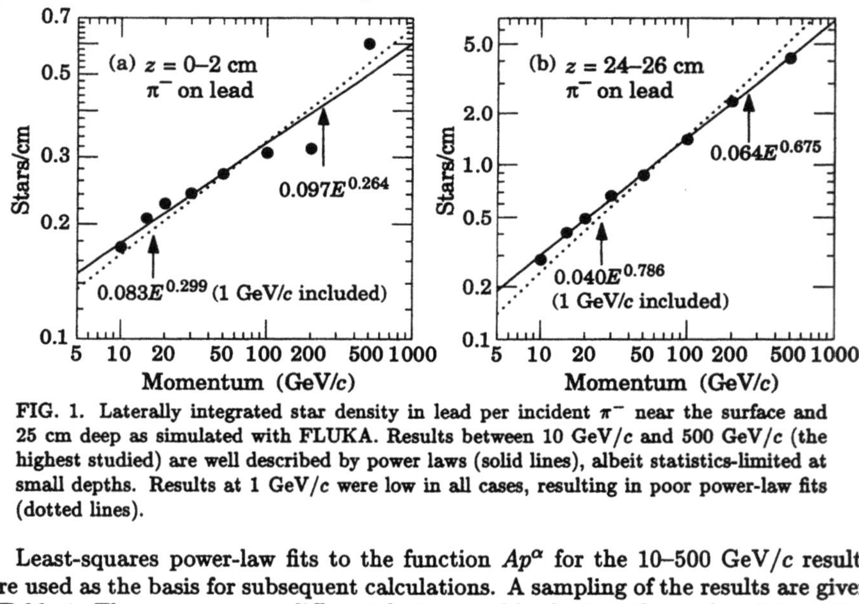

FIG. 1. Laterally integrated star density in lead per incident π^- near the surface and 25 cm deep as simulated with FLUKA. Results between 10 GeV/c and 500 GeV/c (the highest studied) are well described by power laws (solid lines), albeit statistics-limited at small depths. Results at 1 GeV/c were low in all cases, resulting in poor power-law fits (dotted lines).

Least-squares power-law fits to the function Ap^α for the 10–500 GeV/c results were used as the basis for subsequent calculations. A sampling of the results are given in Table 1. They are not very different for iron and lead, since the nuclear interaction lengths in cm are about the same for the two materials.

A worked example may be useful. We find from the table that the star density in lead is $0.097p^{0.264}$ near the surface. For a spherical lead shell "calorimeter" with an inside radius of 200 cm, Eq. (3) becomes

$$\text{Star Density} = \frac{6.3 \times 0.097 \times (0.55)^{0.264}}{2\pi \times (201\ \text{cm})^2} \cosh^{2.264}\eta = 2.06 \times 10^{-6} \cosh^{2.264}\eta\ \text{cm}^{-3}\ . \tag{4}$$

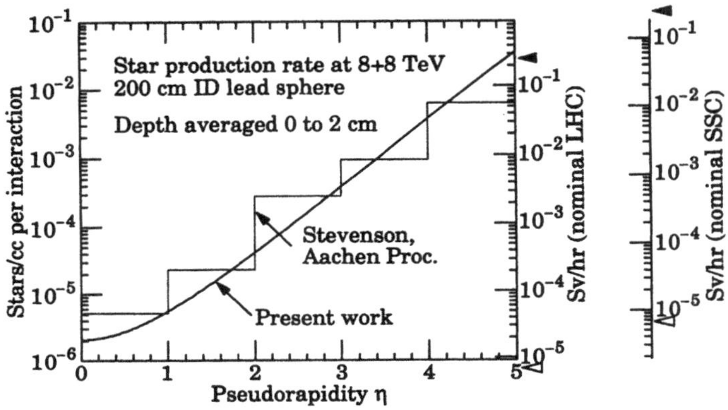

FIG. 2. Dose near the inside surface of a 2 m ID lead shell as a function of η, under nominal LHC conditions. The histogram is from a global simulation by Stevenson.[6] The right axes show conversion to dose in a scenario where a 30-day irradiation is followed by a 1-day cooldown. The open triangles indicate the CERN recommended limit to exposure (15 mSv yr^{-1} with a 2000 hr work year), and the solid triangles indicate the dose lethal to 50% of a population within 30 days of a 10 hr exposure.

Graham Stevenson reported global calculations for such a calorimeter at the the Aachen LHC workshop.[6] In Fig. 2 we compare the above result with his. As is the case at other depths, we find satisfactory agreement.

4. Relationship of star density to activation and dose

Let $f_{Z',A'}$ be the probability of producing nuclide (Z', A') per tallied star (not necessarily in that particular collision). Then, if the star *rate* per unit volume is S and the mean lifetime (halflife divided by ln 2) is $\tau_{Z',A'}$, then the time derivative of the number density of this particular nuclide is

$$dN_{Z',A'}/dt = f_{Z',A'}S - N_{Z',A'}/\tau_{Z',A'} \ . \tag{5}$$

The activity $\mathcal{A}_{Z',A'}$ due to this nuclide is just $N_{Z',A'}/\tau_{Z',A'}$, and it can be seen that in equilibrium $N_{Z',A'}/\tau_{Z',A'} = f_{Z',A'}S$. In principle one would integrate Eq. (5) over the detector time history to obtain the rate for each nuclide, then sum over all the nuclides to find the total activity. Rule-of-thumb estimates can be made by (a) calculating the equilibrium activities and correcting them for decay to estimate activity after a certain cooldown period, or (b) calculating activities on the basis of some particular exposure/cooldown scenario.

The relationship between activity and dose follows from energy conservation. Consider a volume element dV inside a uniformly irradiated material with density ρ, and let E be the energy of the photons and charged particles in an average decay. Then $\mathcal{A}E\rho dV$ is the power produced in dV and deposited nearby, where $\mathcal{A}$ is the activity in Bq g^{-1}. To conserve energy, dV must also *absorb* energy at a rate $\mathcal{A}E\rho dV$, so $\mathcal{A}E$ is the dose rate ($\dot{D}$). ("Uniformly irradiated" means over the absorption distance, or a few cm.) Now suppose that all of the material to the left of dV is removed.

Since the volume element is now irradiated only from the right, it receives only half the dose. It remains only to convert this dose rate, in Bq MeV, to more conventional units such as Sv hr^{-1}. The reader can verify that the surface dose rate in Sv hr^{-1} is $2.9 \times 10^{-7} \mathcal{A}E$, with $\mathcal{A}$ in Bq and E in MeV.

5. Results

For a 50 MeV FLUKA cutoff and a star rate of 1 $\text{cm}^{-3}\text{s}^{-1}$, the equilibrium activity in iron is 0.12 Bq g^{-1}.[6] In this case $\langle E \rangle = 0.8$ MeV. Using the above conversion factor and an assumed LHC rate of 6×10^8 interactions s^{-1}, and the iron analog of Eq. (4) (where the coefficient is 1.51×10^{-6} and $\alpha = 0.257$), one obtains

$$\begin{aligned} \dot{D} &= (2.78 \times 10^{-8})(6 \times 10^8)(1.51 \times 10^{-6}) \cosh^{2.257} \eta \text{ Sv hr}^{-1} \\ &= 2.52 \times 10^{-5} \cosh^{2.257} \eta \text{ Sv hr}^{-1} . \end{aligned} \tag{6}$$

For historical reasons (the fixed-target era), Stevenson[6] and Höfert[7] chose 30 days at constant luminosity followed by a 1-day cooldown as the standard history. Höfert has tabulated the coefficients corresponding to the first number in Eq. (6) for a variety of materials. He obtains 1.0×10^{-8} for iron and 1.5×10^{-8} for lead. The iron coefficient is lower by a factor of 2.8 because long-lived activities such as ^{54}Mn have not had time to build up. Nonetheless, we use these coefficients in calculating the coefficients B for the expression $\dot{D} = B \cosh^{2+\alpha} \eta / r^2$, for the SSC and LHC, in iron and lead and as a function of depth, which are given in Table 1. Details are given in the caption.

As can be seen from Eq. (4), the ratio of LHC and SSC doses under nominal conditions is given by

$$\frac{\dot{D}_{LHC}}{\dot{D}_{SSC}} = \left(\frac{6 \times 10^8 \text{ s}^{-1}}{1 \times 10^8 \text{ s}^{-1}} \right) \left(\frac{6.3}{7.5} \right) \left(\frac{0.55}{0.60} \right)^{\alpha} = 5.04 \times (0.92)^{\alpha} . \tag{7}$$

The LHC and SSC entries in Table 1 satisfy this ratio.

Vertical scales on the right of Fig. 2 show the corresponding dose rates for the nominal SSC and LHC in Höfert's scenario. For the reasons given above, rates after a long exposure will be higher, even after a long cooldown.

For the future experiments at the LHC and SSC, this scenario is unrealistic. Exposures will be longer, and it will be impossible to access the inner detector in as short a time as a day. For purposes of reporting calculations, Stevenson proposes standardizing to a 100-day exposure followed by several fixed cooldown intervals .

How does this compare with the dose from other sources? According to Ref. 1 and its erratum, the maximum electromagnetic dose (mainly due to photons from π^0 decay) is $124 \cosh^{2.93} \eta / r^2$ Gy yr^{-1}, for r in meters. The "yr" was defined as 10^7 s, so the coefficient is equivalent to 0.045 Gy hr^{-1}(or Sv hr^{-1}). EM shower maximum occurs close enough to the surface that we can compare directly: for lead the ratio of EM dose to radioactivity dose is $600 \cosh^{0.67} \eta$. For iron it is twice as large (less radioactivity), and for an interior dose the coefficient should be divided by two. At $\eta = 5$, $\cosh^{0.67} \eta = 18$. Thus, even allowing for longer exposures and realistic integrals over cooldown times, the dose from radioactivity is smaller than the direct dose by two to four orders of magnitude. It is a personnel problem, not a radiation damage problem.

Table 1

Detector radioactivation at the LHC and SSC under nominal operating conditions (6×10^8 interactions s^{-1} at the LHC, 10^8 s^{-1} at the SSC). Other machine assumptions may be found in RPP.[8] Following Stevenson[6] and Höfert,[7] we assume a 30-day irradiation time followed by a 1-day cooldown; these doses are smaller by factors of two or three than would be obtained with longer irradiations. For solid lead or iron at a depth z given in the table, the dose, in Sv hr^{-1}, is $B \cosh^{2+\alpha} \eta / r^2$, where r is the total distance to the interaction point in meters. It is a surface dose; that is, half the dose experienced by a piece of scintillator buried in the absorber. For example: in one version of the SDC detector, the endcap starts 4.2 m from the IP. If a solid lead calorimeter section at $\eta = 3$ were cut open at a depth of 25 cm after the radiation history given above, the surface dose at would be

$$8.08 \times 10^{-6} \cosh^{2.675} 3.0/4.45^2 = 200\ \mu\text{Sv hr}^{-1}$$

Similarly, at $r = 18$ m and $\eta = 5$, the surface dose from a SSC iron calorimeter is 400 μSv hr^{-1}.

	Iron				Lead			
z (cm)	α	A	B_{LHC}	B_{SSC}	α	A	B_{LHC}	B_{SSC}
0–2	0.257	0.0708	3.65×10^{-5}	7.41×10^{-6}	0.264	0.0970	7.47×10^{-5}	1.52×10^{-5}
2–4	0.299	0.0899	4.52×10^{-5}	9.21×10^{-6}	0.352	0.1000	7.31×10^{-5}	1.50×10^{-5}
4–6	0.343	0.0985	4.83×10^{-5}	9.87×10^{-6}	0.409	0.0987	6.98×10^{-5}	1.43×10^{-5}
14–16	0.518	0.0913	4.03×10^{-5}	8.37×10^{-6}	0.557	0.0904	5.85×10^{-5}	1.22×10^{-5}
24–26	0.653	0.0663	2.70×10^{-5}	5.67×10^{-6}	0.675	0.0637	3.84×10^{-5}	8.08×10^{-6}
34–36	0.747	0.0459	1.77×10^{-5}	3.74×10^{-6}	0.783	0.0382	2.16×10^{-5}	4.58×10^{-6}
44–46	0.831	0.0305	1.12×10^{-5}	2.38×10^{-6}	0.865	0.0244	1.31×10^{-5}	2.80×10^{-6}
54–56	0.885	0.0209	7.42×10^{-6}	1.59×10^{-6}	0.932	0.0152	7.87×10^{-6}	1.69×10^{-6}
64–66	0.946	0.0135	4.62×10^{-6}	9.95×10^{-7}	1.009	0.0089	4.40×10^{-6}	9.53×10^{-7}
74–76	1.009	0.0084	2.75×10^{-6}	5.96×10^{-7}	1.076	0.0052	2.45×10^{-6}	5.33×10^{-7}
84–86	1.072	0.0051	1.61×10^{-6}	3.50×10^{-7}	1.132	0.0032	1.46×10^{-6}	3.20×10^{-7}
94–96	1.135	0.0031	9.38×10^{-7}	2.05×10^{-7}	1.217	0.0016	6.77×10^{-7}	1.49×10^{-7}

6. References

1. "Report of the Task Force on Radiation Levels in the SSC Interaction Regions," ed. by D. E. Groom, SSC Central Design Group Report SSC-SR-1033 (June 1988).
2. A. Ferrari, A. King, and G. R. Stevenson, oral tradition (1991).
3. D. E. Groom, in *Proc. of the Workshop on Calorimetry for the Superconducting Super Collider*, Tuscaloosa, Alabama, 13–17 March 1989, ed. R. Donaldson and M. G. D. Gilchriese, World Scientific (June 1990), pp. 59–75.
4. T. A. Gabriel and R. T. Santoro, Particle Accelerators **4**, 169-186 (1973).
5. T. A. Gabriel, Nucl. Instrum. Meth. **91**, 67-72 (1971).
6. G. R. Stevenson, CERN Divisional Report TIS-RP/90-18/CF; also published in *Proc. of the ECFA Large Hadron Collider Workshop*, Aachen (4–9 October 1990), CERN 90-10 (1990).
7. M. Höfert and A. Bonifas, *Measurement of Radiation Parameters for the Prediction of Dose Rates from Induced Radioactivity*, CERN Divisional Report TIS/RP/90-10/CD (1990).
8. Particle Data Group, J. Hernández *et al.*, Phys. Lett. **B239**, 1 (1990), p. III.25; see also the erratum.

LOW-ENERGY NEUTRON FLUENCES IN IDEALIZED COLLIDER GEOMETRIES

Graham R. Stevenson, Alberto Fassò
CERN, Geneva, Switzerland
and
Alfredo Ferrari, Paola R. Sala
INFN Milan, Italy

ABSTRACT

The latest version of the FLUKA Program, which simulates the development of cascades induced by the interactions of high-energy particles in matter using Monte-Carlo techniques, is now also capable of simulating the production and transport of neutrons of energies between thermal and 50 MeV. This version was used to determine neutron fluences in a spherical shell geometry with lead as the absorber material and secondaries from high-energy p-p collisions as the source of incident hadrons. The results are compared with previous predictions made for neutron fluences in calorimeter detectors for the LHC based on data presented by an SSC Task Force. It is shown that in certain regions of the calorimeter the old predictions may have underestimated neutron fluences by up to an order of magnitude.

1 Introduction

At the 1990 ECFA Workshop for the LHC held in Aachen, a presentation was given of the derivation of the maximum neutron fluence in a calorimeter and the albedo (*i.e.* back-scattered) neutron fluence in the central cavity region from Monte-Carlo cascade simulations[1]. The calculations were made using an idealized detector geometry: that of a hollow lead sphere of 2 metres internal radius and 3 metres thickness from which the forward and backward cones in the high-rapidity (small polar angle) regions were removed.

In that paper the fluence of neutrons having energies above 100 keV, those which can efficiently damage silicon detectors and other semi-conductor devices, was derived from the assumption that the low-energy neutron component is proportional to the number of hadrons above 50 MeV in the cascade generated by a high-energy particle. The purpose of the present paper is to check the validity of this assumption by reporting on recent calculations made using a new version of the FLUKA code which generates and transports the low-energy neutrons produced in the intranuclear and evaporation phases of the inelastic nuclear collisions in addition to simulating the high-energy particle cascade. This version, FLUKAN, described in another paper at this conference[2], was used to determine neutron fluences in the same geometry as used in the previous calculations, that of a spherical lead shell of inner radius 2 metres and thickness 3 metres from which the cones which cover the regions of pseudorapidity $|\eta| > 3$, polar angle $\theta < 5.8°$, have been removed*. The spherical shell was chosen as a simplified but still realistic model of a collider detector from the point of view of neutron scattering in the central cavity region. Secondaries from the DTUJET model of high-energy p-p collisions were used as the source of incident hadrons[3].

*The reader is reminded of the relationship between pseudorapidity η and polar angle θ: $\eta = -\ln \tan\theta/2$.

2 The Calculations

If one were to place significant amounts of polyethylene inside the central cavity of a calorimeter close to its inner surface, the fluence of albedo neutrons with energies greater than 100 keV at a point in between the inner surface and the polyethylene will be affected only by neutrons from the region of the calorimeter close by since the point will be shielded from neutrons coming from other parts of the calorimeter. The albedo neutron fluence between the polyethylene and the calorimeter can then be determined from the "no-crosstalk" model presented in [1] for the operating conditions of the LHC (inelastic p–p cross-section = 60 mbarns and annual luminosity = 10^{41} cm^{-2}) but using constants taken from the SSC Task Group Report[4]. This is the curve shown in Figure 1. Also shown in Figure 1 are FLUKAN calculations of the average fluence of reflected neutrons having energies above 100 keV in different pseudorapidity regions. No reflections of neutrons were allowed in the simulation once the neutrons had left the calorimeter. The excellent agreement between the simple no-crosstalk model and the FLUKAN calculations is evident.

The variation of the neutron fluence above 100 keV with depth in the spherical shell, averaged over the three pseudorapidity regions $0 < \eta < 1$, $1 < \eta < 2$ and $2 < \eta < 3$ is given in Figure 2. The histogram is taken from [1], where the neutron fluence was derived from the density of the inelastic interactions of hadrons with energies greater than 50 MeV (which are the main source of low-energy neutrons), normalized to the maximum neutron fluence averaged over the same pseudorapidity interval calculated using the procedures of the SSC Task Group report, but using LHC parameters. It will be seen from this figure that the assumption of direct proportionality between the high-energy component and low-energy component of the cascade is not valid for incident hadrons within the 0–1 pseudorapidity region where the mean incident energy is about 400 MeV. The maximum of the fluence of hadrons with energies greater than 50 MeV occurs at the front face of the absorber in this angular region, but the low energy neutrons show a build-up before the normal exponential attenuation. This is because the neutron attenuation length is close to the hadron absorption length in lead. The net result is that at 20-30 cm depth there is an order of magnitude more neutron fluence than would be predicted by the simple model used in [1]. However the value of the maximum neutron fluence determined using the procedures of the SSC Task Group is identical to that found in the FLUKAN calculation. The neutron fluences for the other two pseudorapidity regions are well approximated by the assumption of proportionality between the low-energy and high-energy components of the cascade as proposed in [1], and again the maximum values of neutron fluence agree well with those determined using the Task Group procedures.

The results from similar calculations to those of Figures 1 and 2 but allowing multiple reflections inside the cavity are given in Figures 3 and 4. Figure 3 indicates that albedo low-energy neutron fluences in the low-rapidity region of a pure lead calorimeter will be very much higher than those derived from a) the no-crosstalk and b) the extrapolations from the Task Group report[4]. This is due to an increased number of reflections inside the cavity over and above the number estimates for the U/scintillator mixture assumed in this report. This increased number of albedo neutrons also increases the number of neutrons penetrating the calorimeter in the 0–1 pseudorapidity region as can be seen in Figure 4. The maximum fluence of low-energy neutrons is higher than that previously predicted and the decrease of fluence with depth in the calorimeter is less steep at low pseudorapidities. The fluence at depths of more than 10 cm is more than an order of magnitude higher than that suggested by previous estimations in this angular region. As expected, however, there is very little difference in the penetration curves at higher pseudorapidities in Figures 3 and 4 when multiple reflections are allowed inside the inner cavity.

References

[1] G. R. Stevenson, CERN Report CERN 90–10 (ECFA 90–133), Vol. 3, p. 566 (1990).

[2] A. Ferrari, P. R. Sala, A. Fassò and G. R. Stevenson, this conference.

[3] J. Ranft and K. Hahn, CERN Divisional Report TIS-RP/218 (1988).

[4] D. E. Groom (Ed.), SSC Central Design Group Report SSC-SR-1033 (1988).

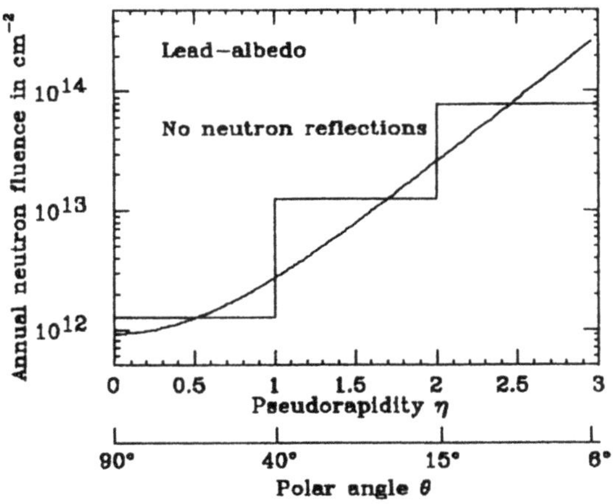

Figure 1: The albedo fluence of neutrons above 100 keV as a function of pseudorapidity using a no-crosstalk assumption. The solid curve is taken from Reference [1]. The histogram is a FLUKAN simulation using secondaries from DTUJET.

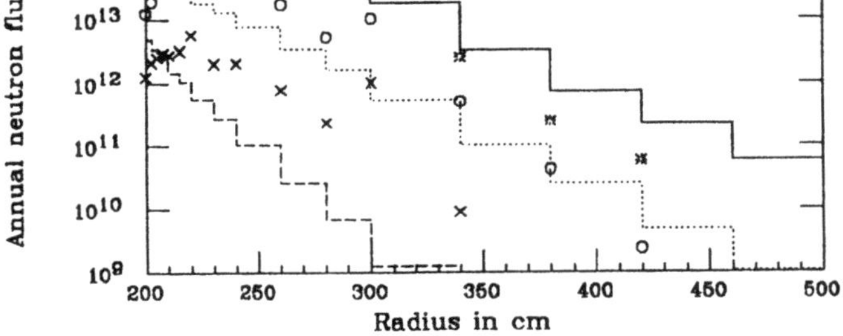

Figure 2: Variation of the fluence of neutrons having energies greater than 100 keV with depth in the sphere using a no-crosstalk assumption for the three angular regions in the sphere given by $\eta < 1$ ($\times$ and dashed histogram), $1 < \eta < 2$ ($\circ$ and dotted histogram) and $2 < \eta < 3$ ($*$ and solid histogram). The plotted points are from a FLUKAN simulation using secondaries from DTUJET, the histograms are from [1].

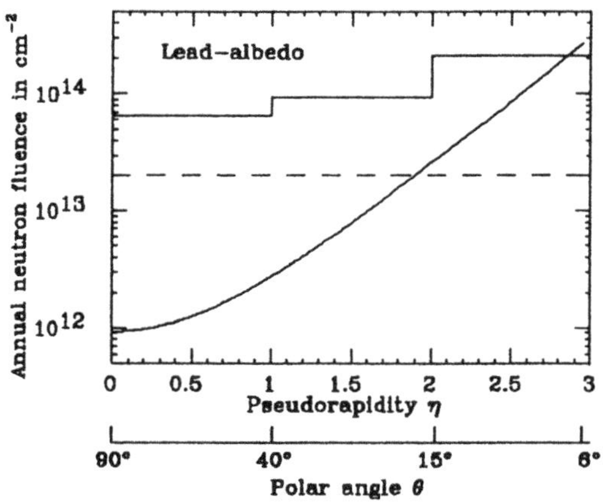

Figure 3: The albedo fluence of neutrons above 100 keV, close to the inner surface of the lead shell, as a function of pseudorapidity, allowing multiple reflections in the cavity. The solid curve is the no-crosstalk model taken from Reference [1]. The dashed line is the data of the SSC Task Group scaled to LHC conditions – average illumination model of [1]. The histogram is a FLUKAN simulation using secondaries from DTUJET.

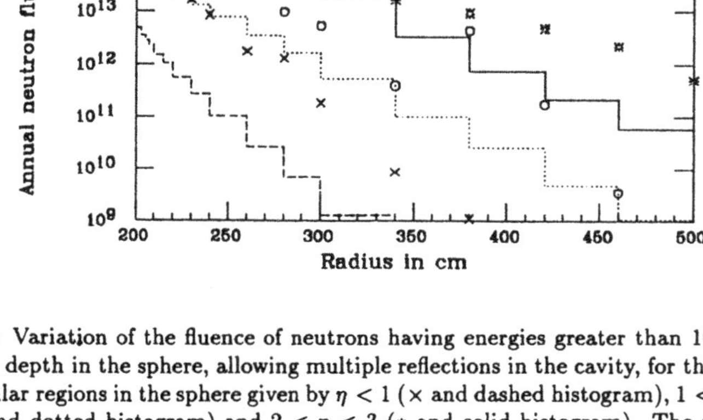

Figure 4: Variation of the fluence of neutrons having energies greater than 100 keV with depth in the sphere, allowing multiple reflections in the cavity, for the three angular regions in the sphere given by $\eta < 1$ (× and dashed histogram), $1 < \eta < 2$ (o and dotted histogram) and $2 < \eta < 3$ (* and solid histogram). The plotted points are from a FLUKAN simulation using secondaries from DTUJET, the histograms are from [1].

Effects of Radiation Damage on Scintillating Fibre Calorimetry

Presented by Ana Henriques
Cern, Geneva, Switzerland and LIP, Lisbon, Portugal

Representing the SPACAL Collaboration

Abstract

We report on measurements of radiation hardness of plastic scintillating fibres to be used for fibre calorimetry. Fibres were irradiated with a ^{60}Co γ-source and the effects on the emission and on the attenuation of scintillation light in a number of different fibre types were studied. A Monte Carlo study that simulates radiation damage in a compensating lead-fibre calorimeter and predicts the impact on the performance is described. We infer from these results that with the best fibres that are presently commercially available the effects on calorimeter performance appear to be within acceptable limits up to 7 Mrad doses.

1. Introduction

Detectors operating in future LHC or SSC experiments will encounter a considerable amount of radiation[1]. Most of damage is due to the production of electromagnetic showers in the calorimeter itself, through the 2γ decay of π^0's[1]. How such a radiation can influence the fibre calorimeter performance is the subject of this paper.It consists of two main parts. The first part (section 2) reports on a series of measurements of the effects of radiation on the fibre performance. The second part (section 3) gives the results of a Monte Carlo analysis of the effects of radiation on the calorimeter performance. Conclusions are given on section 4. This work was carried out by the SPACAL collaboration in the framework of the CERN-LAA project. For more details the reader is referred to the published paper[2].

2. Comparative Studies of Radiation Hardness

A radioactive ^{60}Co γ-source was used for the irradiation of fibres of 40cm length. An appropriate shielding was set up to simulate the narrow dose profile of electromagnetic showers in the calorimeter (see fig 1a). The maximum dose rate was 1.05 Mrad/h at 27 cm from the PM. Peak doses of 1.1, 3.2, 9.5 and 23 Mrad were applied. A yellow filter was used in these measurements. The typical reproducibility of the light output measurements was of the order of 5%. To quantify the local loss in light emission and light transmission for the different types of fibre as a function of the dose, we applied a simple mathematical model which reasonably fits the data[2]. The scintillation light produced in the fibre decreases with dose D in the following way:

$$I(0,D) = I_0(D=0)e^{-(D(z)/\gamma)} \qquad (1)$$

The local loss in light transmission is given by

$$k_{irr}(z) = k(D=0) + \alpha\, D(z) \qquad (2)$$

		SCSN38		SCSN81		S101A		3HF		3HF+PTP	
	$Dose_{peak}$ (Mrad)	α ($Mrad^{-1} cm^{-1}$)	γ (Mrad)	α ($Mrad^{-1} cm^{-1}$)	γ (Mrad)	α ($Mrad^{-1} cm^{-1}$)	γ (Mrad)	α ($Mrad^{-1} cm^{-1}$)	γ (Mrad)	α ($Mrad^{-1} cm^{-1}$)	γ (Mrad)
a)	3.2	6.5×10^{-3}	20.0	4.5×10^{-3}	25.1	—	—	2.9×10^{-3}	13.7	1.2×10^{-3}	18.2
	9.5	5.1×10^{-3}	28.9	4.6×10^{-3}	39.4	6.5×10^{-3}	21.26	1.8×10^{-3}	14.5	1.4×10^{-3}	27.9
	23.0	—	—	—	—	—	—	2.3×10^{-3}	25.8	1.4×10^{-3}	51.9
b)	3.2	6.1×10^{-3}	37.4	4.1×10^{-3}	54.7	—	—	2.6×10^{-3}	16.3	—	—
	9.5	5.0×10^{-3}	38.5	4.5×10^{-3}	71.4	6.3×10^{-3}	26.5	1.7×10^{-3}	15.7	1.0×10^{-3}	52.4
	23.0	—	—	—	—	—	—	2.3×10^{-3}	28.0	1.3×10^{-3}	71.8

Table 1 - Results of the fits to the measurements as a function of the peak dose.(a)The data were normalised at each position to the response of the non-irradiated fibre, with an overall normalisation that fixes an average ratio of 1 outside the region of the source. (b). The parameters 1/α and γ describe the resistance against a reduction of the light transmission and light emission, respectively, as a consequence of ionising radiation. γ is given in Mrad, and α in $Mrad^{-1}cm^{-1}$. The typical errors on α and γ are in the 5-10% range.

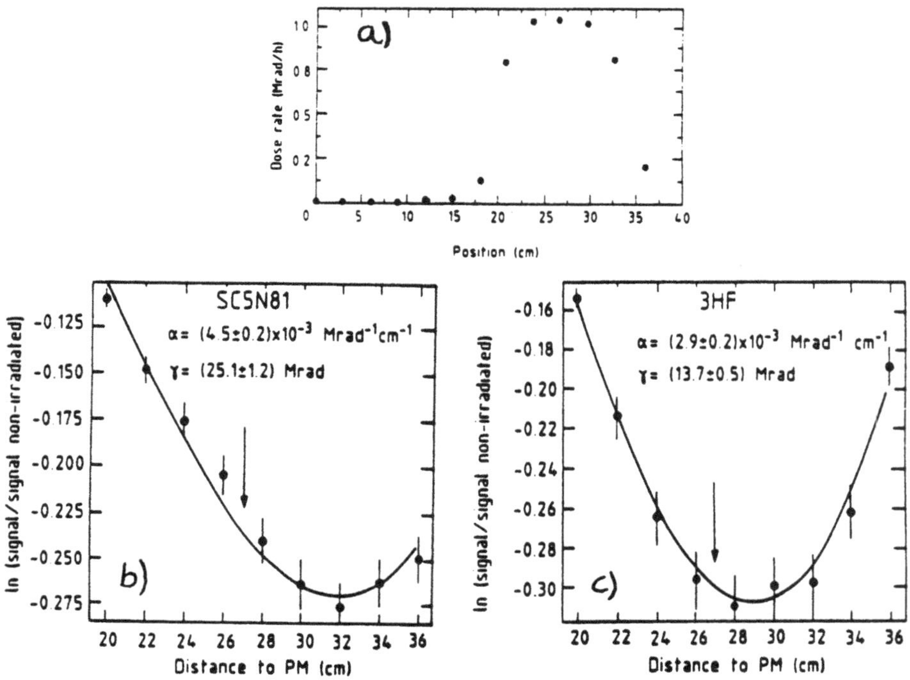

Fig. 1 - (a) Dose rate profile of the γ-Co source used for the irradiation of the fibres. (b) and (c) Example of two fibres irradiated with a dose of 3.15 Mrad at the peak of the 'very high intensity' profile, 3HF losing essentially emission and and SCSN81 losing essentially transmission, respectively. The curves represent the best fit to a model describing the decrease in light emission (γ) and in light transmission (α) as a function of the dose.

where $k(z)=1/\lambda_{att}(z)$, γ and $1/\alpha$ are a measure of the radiation resistance against a reduction of the light emission and light transmission, respectively.

A high radiation resistance corresponds to a small value of α and a large value of γ. The fitted values of α and γ for all fibres are presented in Table 1. The fits carried out included only data points in the irradiated region ($z > 18$ cm, see fig. 1a). In a first approach, the curves were normalised to 1 at the point closest to the photomultiplier (table 1a). As this overall normalisation influences the results in some cases (the fibre response was also affected outside the region of the source), the baseline was redefined for a second set of fits as the average ratio of the response for the irradiated and non-irradiated fibres outside the region of the source ($z<18$cm)(see table 1b).

From Tables 1a,b) one observes a systematic increase of γ with increasing dose, most likely as a result of the simplified model that does not include wavelength dependence. One sees that the uncertainty of γ due to the unknown baseline shift is much bigger. The fitted value of α changes very little, but the values of γ are up to a factor 1.8 larger in Table 1b.

For a single fibre-type a comparision was done with a measurement using a long (2.2m). It shows that α and γ are 15% more optimistic (less damage) for these long fibres due to the absence of cladding-light. In general we observed that the radiation effects in the green fibres (3HF and 3HF+PTP) mainly concerned the emission, whereas the blue fibres (SCSN38, SCSN81 and S101A) lost essentially in transmission. As an example, the response for two types of fibre, 3HF and SCSN81, both irradiated with 3.2 Mrad dose at the peak of the profile is shown in fig.1.

3. Monte Carlo Simulations

We summarize the Monte Carlo calculation, which simulates radiation damage in a fibre calorimeter. The damaging electromagnetic showers were parametrized by a longitudinal shower profile[3] of fixed energy E_{rad}.

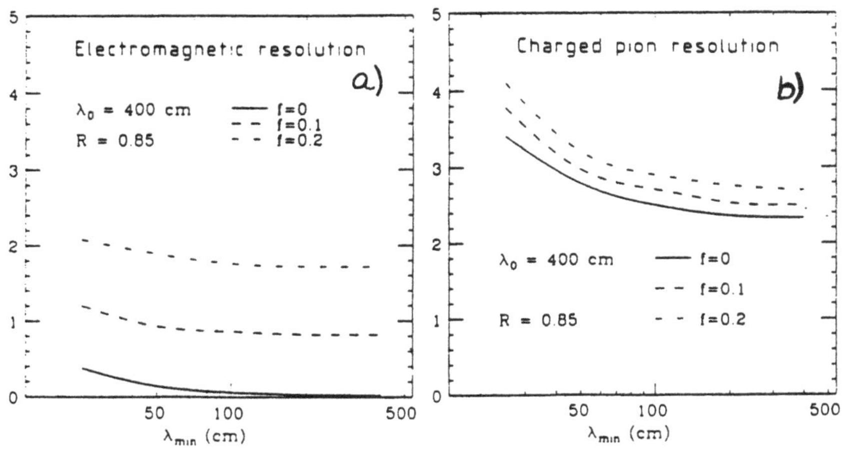

Fig. 2 Result of the Monte Carlo calculation. The impact of the fibre radiation damage on the (a) electromagnetic energy resolution (b) hadronic energy resolution for E = 10 GeV as a function of λ_{min}.

The charged hadronic contibution to the radiation damage profile was not included in this model. The incoming electrons or pions of energy E wich shower in the calorimeter were parametrized by their fixed longitudinal shower profile combined with a fluctuating starting point position. Concerning the fibre propreties, the reflectivity coeficient R=85% placed at the fibre end opposite to the read-out was included. λ_{min} is the local attenuation length after irradiation at the peak of the damage profile and f is the fractional emission loss at the peak of the damage. For the results described below, λ_0 was fixed at 4 m and R was fixed at 0.85, representative values for the present calorimeter prototypes.

Fig. 2a (2b) shows the contribution of the radiation damage to the electromagnetic (hadronic) energy resolution for E = 10 GeV and E_{rad} = 1 GeV as a function of λ_{min} and f. The dependence of λ_{min} on σ/E_{e-} tis very weak for $\lambda_{min} > 0.5$ m. A 10% drop in the light yield at the peak of the damage gives a 0.8% (0.2%+2.3%) contribution to the electromagnetic (hadronic) energy resolution at E = 10 GeV. By comparing Figs. 2a and 2b one sees that the hadronic energy resolution is more sensitive to transmission losses, whereas the electromagnetic energy resolution is mainly sensitive to emission losses.

Fig. 3 shows the ratio of the average calorimeter response for electrons between 150 GeV and 10 GeV as a function of f and λ_{min}. For f = 10% and $\lambda_{min} >$ 0.5 m the resulting nonlinearity amounts to 2.5%. For hadronic showers there is almost no nonlinearity effect due to the radiation damage.

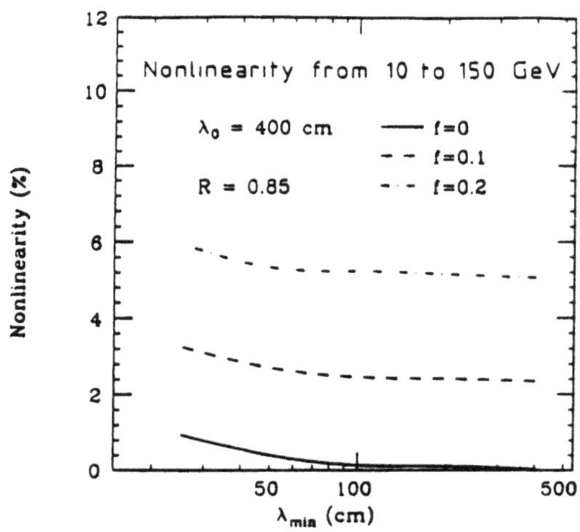

Fig. 3 - Result of the Monte Carlo calculation. The ratio of the average calorimeter response for electrons between 150 GeV and 10 GeV as a function of λ_{min} for 3 different values of f and E_{rad}=1 GeV.

By combining the results of the Monte Carlo study and the radiation damage tests one can determine which radiation level can be tolerated, while maintaining an acceptable calorimeter performance (f = 10%). Taking this into account the SCSN81 and 3HF+PTP fibres can presently be considered as the most radiation hard fibres for use in a calorimeter with a maximum dose around 7 Mrad and 6Mrad respectively.

It is interesting to see with which polar angle a given limit of 7 Mrad corresponds in a possible LHC or SSC experiment. Assuming a luminosity of 10^{34} $cm^{-2}s^{-1}$ at the LHC running 10^7s per year and a distance R to the interaction vertex of 4 m, a 7 Mrad limit corresponds to 4 years of running at pseudorapidity η = 2.7 (polar angle 8°). To further judge the suitability of the lead-fibre calorimeter technique we calculate a maximum total dose of 2 Mrad in a barrel calorimeter covering $\eta = \pm 1.5$ (polar angle $\pm 25°$) at a radius of 1.3 m over 4 years of LHC running. In terms of the parameters f and λ_{min} of the Monte Carlo simulation for the SCSN81 fibres this means λ_{min} = 90 cm and f = 5% (average of Table 1a and 1b and 2, 3.2 Mrad values). These values correspond to a degradation of about 0.4% in the electromagnetic energy resolution. This method assumes that the longitudinal damage profile of the fibres is measured in a real experiment and that the adequate correction to the calibration is applied.

4. Conclusions

For the use of SCSN81 fibres in a fibre calorimeter we can set the highest limit on the maximum allowable dose at the peak of the damage in a LHC experiment. Defining a maximum allowable degradation of 0.8% of the electromagnetic energy resolution, which corresponds to a degradation of about 0.2% of the hadronic energy resolution, a 2.5% nonlinearity effect on electrons between 10 GeV and 150 GeV (determined by a Monte Carlo simulation), the maximum allowable dose is 7 Mrad. This limit is probably underestimated by 15% due to the use of short fibres for the radiation tests and has an estimated $\pm$30% error due to the uncertainties in the baseline shift of the radiation measurements. We estimate from these results that lead/fibre calorimetry is feasible at the LHC (luminosity 10^{34} $cm^{-2}s^{-1}$, 4 years of running) up to pseudorapidity η = 2.7 at a distance of 4 m from the interaction vertex. For a barrel calorimeter ($\eta = \pm 1.5$) at a radius of 1.3 m, the corresponding deterioration of the electromagnetic energy resolution would be 0.4%.

Acknowledgments

I would like to thank the colleagues that have participated on the paper refered in 2 in which this presentation is based.I also thank the overall SPACAL Collaboration for the interest and assistance in the various tests described in this paper. We are also very much indebted to the CERN-TIS group for the dosimetry measurements. Special thanks are due to the Director and Staff of LNETI Sacavém for making the irradiations possible. We are indebted to the LAA project leader, Prof. A. Zichichi, for his interest and support for this project. The work described in this paper was supported in part by CFNUL/JNICT Portugal and by the U.S. Department of Energy and National Science Foundation.

References

1. *G.R. Stevenson*, CERN 90-10, ECFA 90-133, VOL. III, 566.
2. *D. Acosta et al.*, CERN-PPE/91-45, submitted to NIM and references included
3. *E. Longo and I. Sestili*, Nucl. Instr. and Meth. 128 (1975) 283.

Measurement of Radiation Damage during Irradiation

D.Bisello[1],A.Castro[1],M.Cobal[2],H.Grassmann[1],S.Leone[2],M.Loreti[1], N.Piacentino[3]
1) INFN Padova, 2) INFN Pisa , 3) Univ. Cassino

Presented by H.Grassmann

Abstract

We are constructing a computer controlled setup for measuring material properties while the material is irradiated. We aim at very small systematic errors and a high measurement precision. The device has to deliver measurements of several different parameters simultaneosly. The setup is compact and easy to handle and transport.
For a first measurement we investigate the transmission of optical fibers.

1. Experimental Setup

1.1. Basic design

The fundamental layout of our device is intended to be as follows : A signal is provided by one or more LED's of different colours. (Lasers or UV lamps could be used as well.) The light pulses emitted by the LED's are transmitted through the optical fibers to be investigated. Part of the fiber is exposed to the radiation field. The signal is than read by a photodiode. The current produced in the photodiode is amplified such that it can be read into a Camac ADC. We used photodiodes because their compactness, stability and robustness had been proven advantageous in similar applications (1). Of course one could use a photomultiplier instead of a photodiode, if wanted.

1.2 Monitoring system

There will be applications where the photodiode, the amplifier and the LED's cannot be shielded against radiation sufficiently. Therefore we also need to precisely monitor any possible change in those components. That can be done using one or several reference LED's which face the photodiode directly, without an optical fiber in between. Any change in the emission of the LED, the photodiode or its window or in the amplification of the electronics can be monitored in that way. To be able to distinguish between changes in the amplifier and the optical system we also can inject test pulses into the amplifier input through a capacitance. Photodiodes also act as solid state counters. If necessary one can correct also for that effect by reading signals without giving LED pulses.

1.3 Configuration for our first test run

Our setup was not yet completed for our first engineering run. We used a simplified monitoring system. The configuration was as follows :

Two long fibers view signals from a red and a green LED each. Part of these fibers is exposed to the radiation of a Co^{60} source. The fibers are read by a photodiode. LED's, photodiode and amplifier are protected agains radiation by several centimeters of lead. Also protected by this lead shielding are two shorter fibers which serve for monitoring. As the long fibers, also the short fibers view two LED's (green and red) and they are read by the same photodiode. A shematic view of the setup is shown in figure 1. Only one of the two long and one of the two short fibers are shown for simplicity.

Subsequently each of the LED's emits a light pulse. The transmitted signal is read by the photodiode for each of the different LED's separately. This measure - ment cycle is repeated each 10 seconds. The device is driven by an IBM PC controlling a CAMAC crate. We use the photodiode S1790 from Hamamatsu. Our amplifier is a very simple two-stage amplifier based on operational amplifiers LF357, it is not optimised for lowest possible noise.

In our plots we always combine 10 subsequent measurements to one data point.

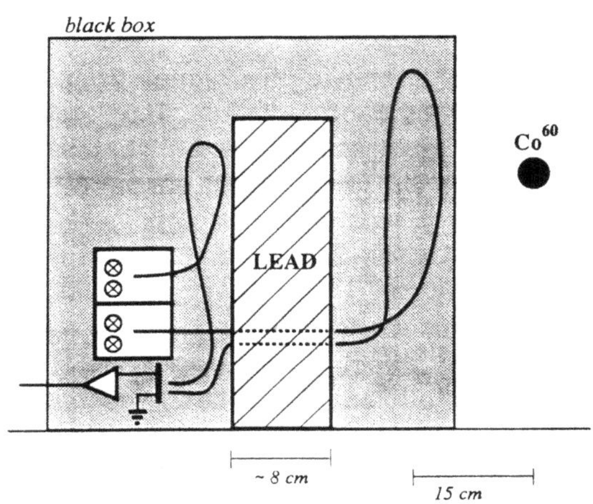

Figure 1 : schematic view of the test setup.

2. Measurement

2.1. The Co^{60} source

The data were taken at the INFN Legnaro, Italy. The Co^{60} source there has an intensity of 5780 Curie. The source is inside a lead container and can be brought out of that container by remote control. The source is installed in a room of about 3m*5m in size.

From the known source strength and the geometrical configuration we calculated that the exposed part of the fibers absorbed 140 krad/hour. The radiation had a non uniformity of ±25% over the length of the fiber. The total time of exposure was 18 hours, which corresponds to about 2.5 Megarad.

2.2. Measurement procedure

The fibers were green wavelength shifting plastic fibers from Bicron with 1mm diameter, as they are used in the CDF upgrade (2). The long fibers were 120 cm in length, 90 cm where exposed to radiation. The short fibers were about 40 cm long.

- We exposed the fibers to radiation for a time of t=45 minutes.
- The Co source was removed for t=17 minutes.
- The Co was activated again for 18 hours.
- We took data for another 30 minutes without Co.

2.3 Results

Figure 2 shows the signal from one of the long fibers over the first 240 krad of radiation. The vertical axis are the ADC channels, pedestal subtracted. The horizontal axis is in seconds.The source was switched on at t=2400 seconds. Between t=5050 and t= 6050 seconds it was switched off.

From figure 2 we see that the signal from the green LED decreases rapidly, while the signal from the red LED remains almost stable. Within our measurement precision we do not observe any recovery or anealing effect after the first 100 krad, when the source was switched off for 45 minutes. The other long fiber behaved the same.

We made the same plots for the short, shielded fibers (not shown): the transmission of green light gets only slightly worse, the signal from the red LED remains essentially stable.

That comparison between the transmission of red and green light in the shielded and unshielded fibers indicates that the system LED+photodiode+amplifier is stable within about 1%. For future runs we will have the full monitor system installed.

Figure 3 shows the change in transmission for red and green light over the full data taking period of 18 hours corresponding to 2.5 Mrad. Shown in figure 3 are the signals from the unshielded fiber number 2, the unshielded fiber nr. 1 has the same behaviour. We observe that the unshielded fiber has a clear maximum in Δtransmision/Δradiaton at 400 krad for both red and green light. We do not observe this effect in

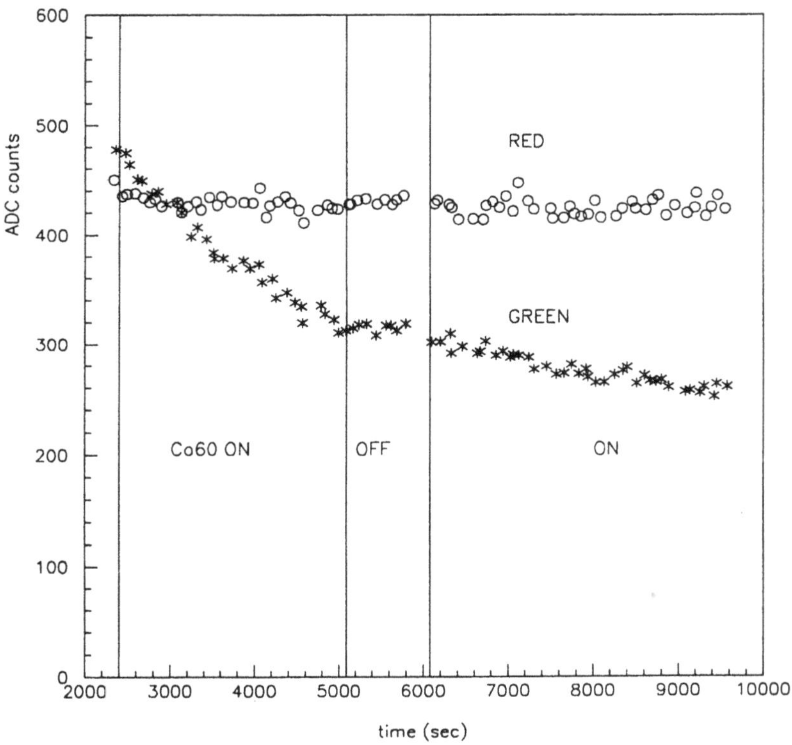

Figure 2 : Loss in transmission during the first 240 krad.

the shielded fiber. Therefore it must be caused by a response of the fiber itself to radiation, rather then by a problem in our electronics system. At the time being we don't know any explanation for this 'kink' at 400 krad. We will need confirmation.

After 2.5 Mrad the transmission left in the unshielded fiber 1 (2) for green light is 8%(11%) and for red light 57%(58%).

We also see a recovery effect after 2.5 Mrad of 10% in 30 min.

Therefore we expect that the transmission as a function of absorbed radiation will stronlgy depend on the intensity of the radiation.

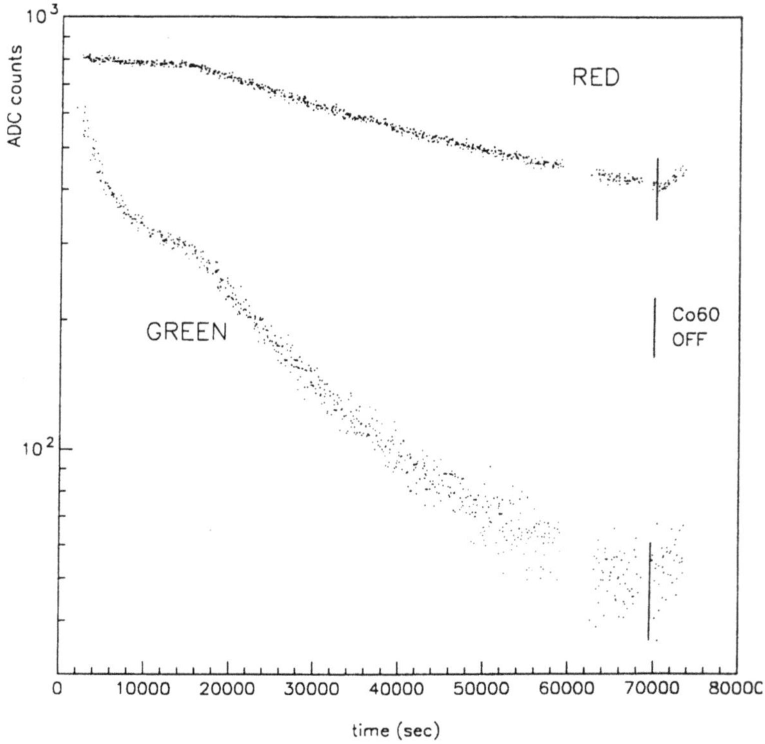

Figure 3 : Loss of Transmission over the total exposure of 2.5 Mrad

3. Conclusion

The first run with our online radiation monitor setup showed that the device is working. We are able to do detailed measurements with small sytematic errors. As first results we find evidence for a 'kink' in the transmission of red and green light at 400 krad. We also observe anealing effects. More measurements with a completed device will be done in the near future.

Literature

(1) H.G.Moser et al.,'a compact and inexpensive radiation monitoring device',DESY red report 84-063.

(2) J.Freeman, this conference, CDF plug upgrade.

RADIATION DAMAGE IN TILE-FIBER CALORIMETER MODULES

J. ERNWEIN
DAPNIA, CE Saclay
91191 Gif-sur Yvette, France

Abstract

A program of measurements has been carried out to determine the radiation hardness of the SDC central calorimeter at the SSC. First commercially available plastic scintillator plates and fibers were irradiated with a ^{60}Co source to identify the materials best suited for operation in a high radiation field. Then electromagnetic calorimeter modules were built with these materials and exposed to a 1 GeV electron beam. The behavior of the modules was studied for doses up to 2 Mrad in two different gas environments.

1 Introduction

The main area of concern for the tile-fiber option of the SDC central calorimetry is the radiation hardness of plastic scintillator placed in the region of electromagnetic shower maximum. We have conducted a program aimed at investigating how EM calorimeter modules will survive in the radiation field of the SSC.

In the first part of the program, commercially available scintillator plates and wave length shifting (WLS) fibers were measured after irradiation by a ^{60}Co source[1]. These measurements were used to select the material for the construction at Fermilab and Purdue Univ. of electromagnetic test modules to be tested in electron beams in France, Japan and China. We report here results obtained by the Saclay group[2] with two such modules exposed to a 1 GeV electron beam at Orsay.

2 Doses and Dose Rates

The total doses expected at Shower Maximum (SM) in the SDC calorimeter are given in Ref.[3]. For 10 years at 10 times the nominal luminosity ($L_{nom} = 10^{33}\text{cm}^{-2}\ \text{s}^{-1}$) the end of the barrel ($\eta = 1.3$) will receive $\sim$ 0.6 Mrad whereas the end cap will receive $\sim$ 60 Mrad at $\eta = 3$. These total doses can easily be reached in exposures of detectors to cobalt sources or to intense electron beams.

The exposures must necessarily be done in a relatively short time to yield useful information for the construction of the detectors. The dose rates at these test sites are therefore much larger (by a factor $\sim$ 100 to 1000) than the actual SSC dose rates[1].

3 Component tests with ^{60}Co

Samples of scintillator plates, scintillating fibers, WLS and clear fibers were exposed to a ^{60}Co source. Irradiation and annealing were done in air at room temperature. The absorbed doses were calculated from the known activity of the source and measured with analine dosimeters and radiochromic film. The dosimetries agree within 10 %.

[1] At shower maximum of the SDC EM calorimeter the dose rate is $\simeq$ 7 rad/h at the end of the barrel at nominal luminosity.

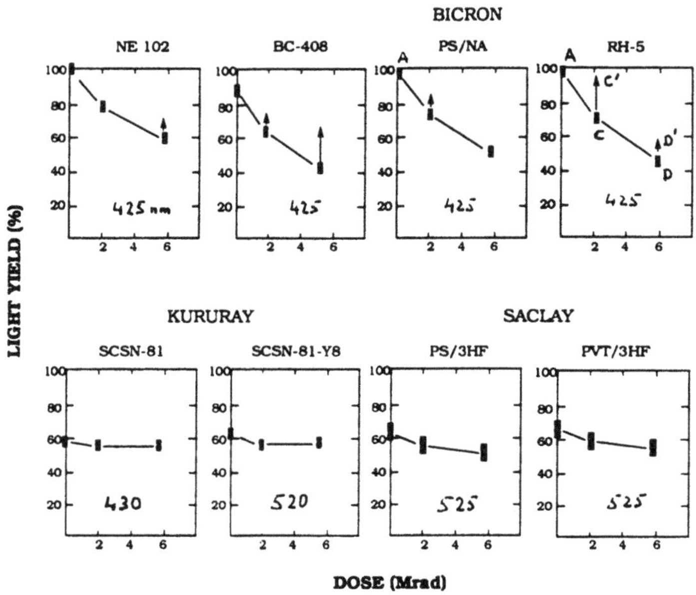

Figure 1: Relative light yield as a function of absorbed radiation dose for 8 types of scintillator plates. Light yields are normalized to NE102=100%. A) before irradiation; C) 2 Mrad at 9 krad/hr; C') 2 Mrad + 3 days of annealing in air: D) 5.7 Mrad at 9 krad/hr; D') 5.7 Mrad + 29 days of annealing in air. The wave length at peak emission is shown for each sample.

3.1 Scintillator plates

The light yield of scintillating plates was measured with small samples ~ 3 mm thick in a way which eliminates the effect of attenuation in the plate. Pulse height analysis of signals produced by electrons from a ^{106}Ru source was performed on samples of 8 different scintillators, before and after irradiation. Figure 1 shows that if one considers blue scintillators, two products stand out:

- RH-5 from BICRON, which shows a large initial light yield, a 30 % drop after 20 Mrad and an almost complete recovery after 3 days of annealing in air. A dose of 6 Mrad results in a 50 % light loss and no recovery after one month.
- SCSN-81 from KURURAY, which yields only 60 % of the light compared to RH-5 but shows no loss up to a dose of 6 Mrad.

The attenuation length in scintillator was obtained independently from the light yield measurement by the optical measurement of light transmission through scintillator bars using a spectrophotometer. The results are shown in figure 2 for two blue scintillators. SCSN-81 shows a significant decrease of the attenuation length for two Mrad, even after one month of annealing in air. RH-5 has an even larger drop of the attenuation length.

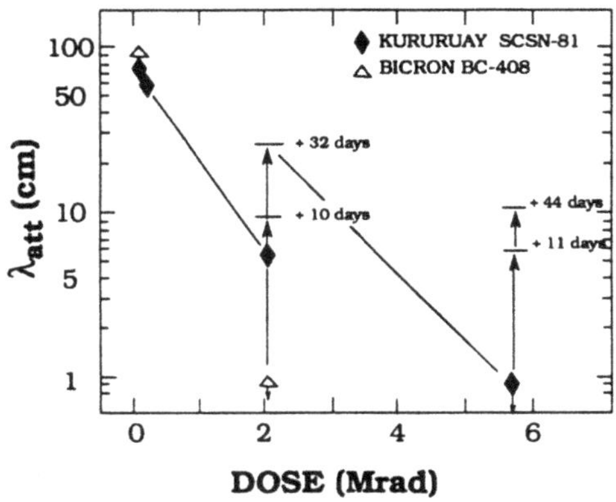

Figure 2: Attenuation length (λ_{att}) of two types of scintillator calculated from the measurement of light transmission along the axis of 25 cm long scintillator bars. The variation with absorbed dose and after annealing in air is shown.

3.2 Fibers

The attenuation lengths of different scintillating fibers were measured before and after irradiation. Figure 3 shows the signal measured by a PMT at one end as a β source is moved along the 2 m long, 1 mm diameter fibers.

$L(d)$ represents the light yield when the source is at a distance d from the PMT. The straight lines are lines of equal attenuation length. After irradiation (1.6 and 2.7 Mrad) the light yield and attenuation length are reduced significantly. These measurements verify that fibers emiting in the green (b) suffer less radiation damage than fibers emitting in the blue (a).

For the construction of the EMC modules, blue scintillator plates (SCSN-81 from KURURAY) and green WLS fibers (BCF-91A from BICRON) were chosen. The WLS fibers have the same plastic base and secundary emitters as the green scintillating fibers discussed above. Based on our measurements. we conclude that the loss of light in a tile-fiber assembly irradiated to $\sim$ 1 Mrad will come mainly from the reduced attenuation length in the plate (not from intrinsic light yield) and from the damage to the WLS fiber.

4 System Test in an Electron Beam

The modules (figure 4) are made of 20 scintillator tiles stacked between $\sim$ 6.4 mm thick lead plates ($\simeq 1.1X_0$). The total depth is $\sim$ 20 cm and represents $\sim$ 22 radiation lengths. The tiles are made of 3.2 mm thick SCSN-81 scintillator plates of area 12 cm x 12 cm. A machined groove holds a 1 mm diameter WLS fiber (BCF-91A). The fibers run on top of the module to the rear where they are connected to the PMT. Each module is enclosed in an aluminum container and can be flushed with gas.

Two modules were exposed to an electron beam at the LURE[2] in Orsay. The experiment

[2]Laboratoire pour l'Utilisation du Rayonnement Electromagnétique

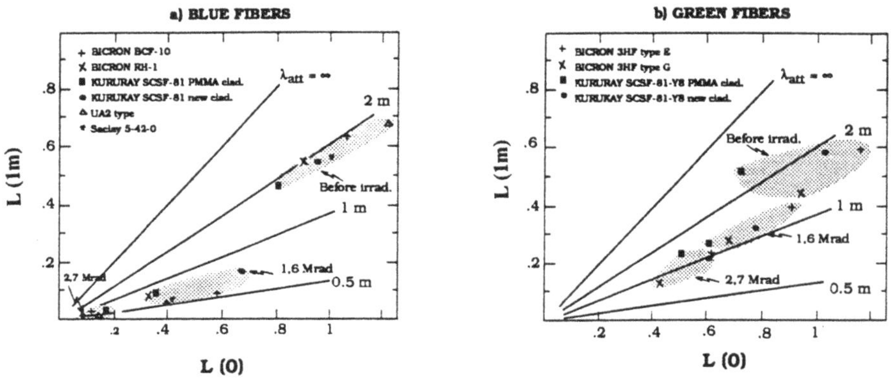

Figure 3: Relative light yield ($L(d)$) and attenuation length (λ_{att}) of fibers after various doses of irradiation. a) Blue bibers. L=1 corresponds to an absolute value of 7.7 photoelectrons. b) Green fibers. The absolute scale for L is not determined.

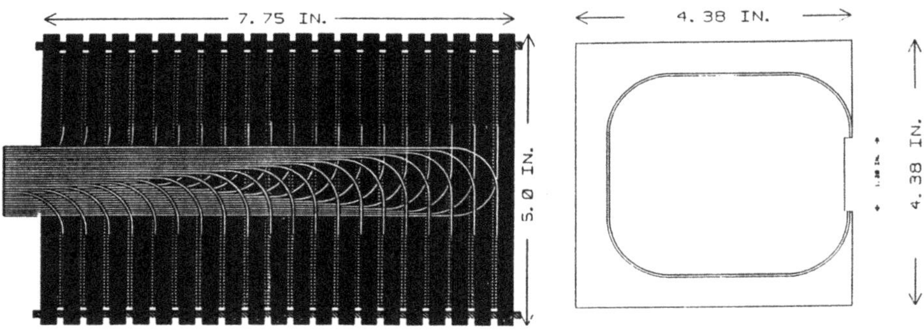

Figure 4: Lay out of the EMC module. The right hand side shows a tile with the groove holding the WLS fiber. The assembled module is shown on the left with the fiber path on top of the module.

took place in July 1991. The 1 GeV electron beam had a 30 ns spill with a 12.5, 25, or 50 Hz repetition rate. The intensity could be varied from 10^9 to 10^{10} electrons per spill. Since the beam size was small (~1 cm^2) compared with the module dimensions, we used a remote controlled table to move the modules across the beam during the 10 minute long exposures. The first 12 exposures were each ~ 20 krad and the last four ~ 400 krad. The time between exposures was ~ 30 hours so that the "average dose rate" was 0.7 and 13 krad/hour respectively.

We used three different dosimeter techniques during these beam tests:

- Alanine plastics, which develop free radicals under irradiation. The concentration of free radicals is proportional to the absorbed dose. Alanine dosimeters cover a large dose range: 10 krad to 10 Mrad.
- Radiochromic dye films, 50 μm thick from FWT[4]. These films change their optical density as a result of irradiation. They are sensitive in the range 0.1 to 50 Mrad.
- Radiochromic dye films 125 μm thick from FUJI[5]. They cover the range 0.5 to 30 Mrad.

We used arrays of all three kinds of dosimeters in front of the modules and at shower maximum. The dosimeters were left in place for several irradiations and the beam current monitor was used to determine the dose for each run. More than 160 dosimeters were measured during the experiment. The results obtained with the three methods agree and show a relative spread of 5%. We estimate the absolute uncertainty on the dose to be $\simeq$ 10%. The ratio of the dose measured at shower maximum and that measured in front of the modules is about 9, in agreement with the expectation from 1 GeV electron shower development. The irradiation was uniform across the modules with at most a $\simeq$ 10% drop at the edges.

The response of the EMC modules was measured before every irradiation. In this way there was always a ~ 30 hour annealing period which is taken into account for determining the average dose rate. The modules are equipped with transverse and longitudinal tubes. A radioactive source of ^{137}Cs located at the end of a wire can be moved through these tubes by using an automated source driver[6].

All tiles are connected to the same PMT, so the movement of the source in a longitudinal tube results in a peak-valley pattern of the PMT current. Figure 5 shows the signals before (a) and after (b) irradiation.

5 Results of Beam Test

The signals of all 20 tiles have been normalized to the first measurement made before irradiation. Figure 6 shows this relative signal for tiles # 2,5,10, and 20. The total dose accumulated at shower maximum at the time of each measurement is also shown. It is apparent that the damage is worst in tiles # 2 and 5. which are near the shower maximum.

Tile # 20 shows a 10% drop of signal after 2 Mrad. According to the shower profile (figure 7a), this tile should see very little radiation from the showers. We interpret this drop as the result of direct irradiation by the electron beam of the WLS fibers which run on top of the module. Figure 7b shows the damage profile measured after various doses absorbed at shower maximum. In this figure. the signal have been normalized to tile # 20 to take out the effect of direct irradiation of the fibers. The damage profile reproduces the shower development profile.

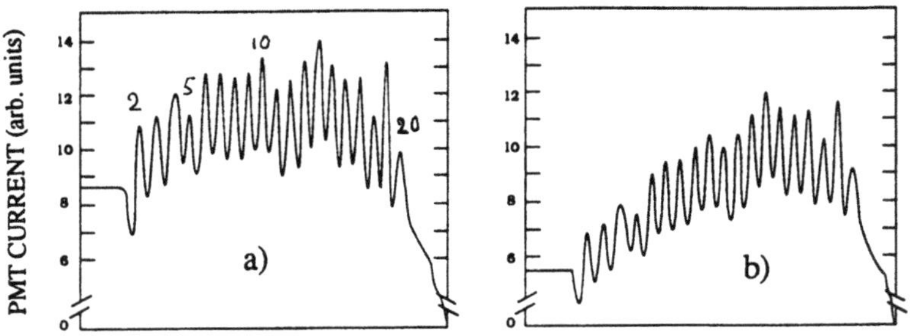

Figure 5: Measured PMT current induced by a ^{137}Cs source moving along a lontitudinal tube in the module before (a) and after (b) irradiation. The first tile seen by the beam is on the left. The vertical scale is arbitrary.

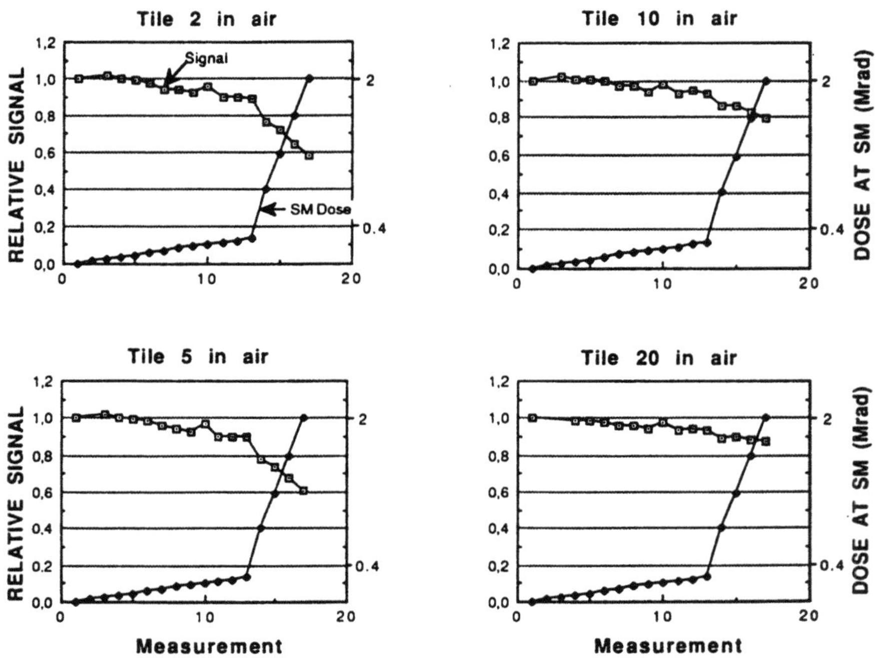

Figure 6: Relative signals as a function of the dose received at shower maximum for tiles # 2, 5, 10, and 20. Values are normalized to the measurements before irradiation.

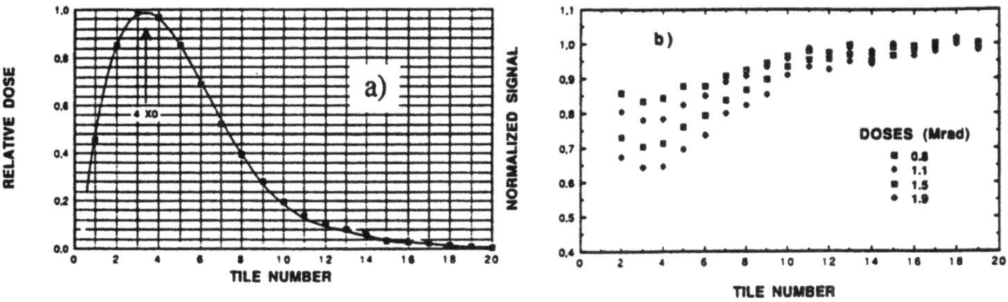

Figure 7: a) Simulated shower profile for 1 GeV electrons as a function of the tile number of the module. b) Damage profile as a function of the tile number for various doses.

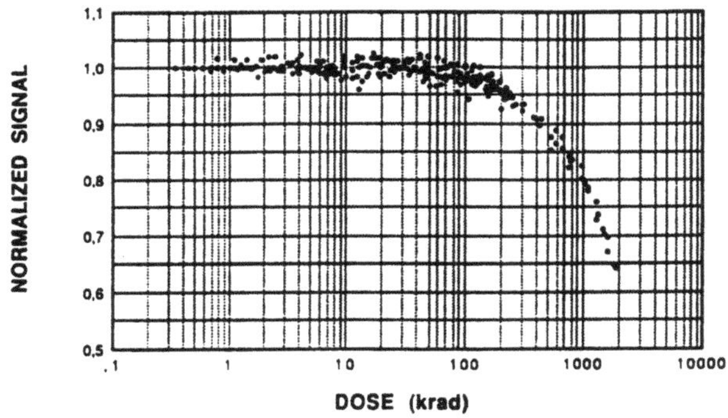

Figure 8: Relative signal. normalized to tile # 20 (see text), as a function of the dose received.

Let S_i be the measured relative signal of tile i made after irradiation j. The dose for each data point is calculated according to $D_{ij} = D_{SM(j)}.P_i$ where $D_{SM(j)}$ is the measured integrated dose at shower maximum and P_i the ordinate of the normalized shower profile for tile i. Figure 8 shows the relative decrease of the signal as a function of dose. One concludes that the light output of the tile-fiber assembly decreases by 20% for a total dose larger than 0.6 Mrad. This corresponds to the dose absorbed by the end of the barrel at shower maximum during 100 years at a luminosity of 10^{33} $cm^{-2}s^{-1}$.

All the above results concern an EMC module filled with air. An identical module was flushed with nitrogen during the entire duration of the experiment. Results are compared in figure 9 for tiles # 5 near shower maximum. The module in nitrogen deteriorates much more rapidly than the module in air. There is no annealing in the module filled with air, even after 60 days. The module in nitrogen recovers some signal when the flow of nitrogen is stopped at the end of irradiation.

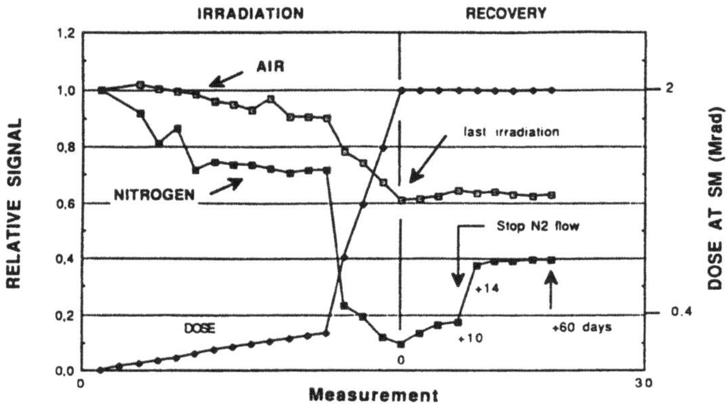

Figure 9: Relative signal of tile # 5. near shower maximum. for a module in air and a module in nitrogen. The recovery after the last irradiation is shown on the right. The dose at shower maximum is indicated on the right hand side scale.

6 Conclusion

We have performed radiation damage tests with a cobalt source on samples of plastic scintillators and fibers. Among the commercially available material, we have selected blue SCSN-81 from KURURAY for the plates and green BCF-91 from BICRON for the WLS fibers. We have found that there was no loss in the light yield of SCSN-81 after several Mrad. However, the attenuation length in the plate decreases by a factor $\sim$ 3 after 2 Mrad and 1 month of annealing in air.

We have tested the radiation resistance of two electromagnetic calorimeter modules built with these materials. The modules were exposed to a 1 GeV electron beam at the Orsay LINAC and their response to a radioactive source was measured before and after irradiations. We found that a tile-fiber unit irradiated in air loses $\simeq$ 20% of its light output after having received a dose of 600 krad. This dose corresponds to the dose accumulated at the end of the SDC barrel EM calorimeter (η=1.3) during 10 years at 10 times the nominal luminosity. According to a study of the effect of the damage profile on the detector performance[7], the resolution of the calorimeter is not significantly degraded for losses < 30%. The module in nitrogen degraded much more rapidly than the module in air.

References

[1] SDC-91-00011 and DPhPE Saclay Preprint 91-03

[2] The Saclay group in the SDC includes: P. Bonamy, G. Dubost, J. Ernwein, J.R. Hubbard, P. Le Dû, F. Rondeaux. G, Villet.

[3] Letter of Intent, SDC 90-00151, figure 4.

[4] FWT-60-00 films were obtained from Far West Technology, 330 D S. Kellog, Goletta, Ca 93117, USA. K. Johnson (FSU) suggested their use to us.

[5] FUJI films were provided by K. Takikawa and K. Yasuoka.

[6] See V. Barnes (these proceedings) for more details.

[7] D. Green and A. Para, Fermilab-FN-565, May, 1991.

RADIATION-TOLERANT PLASTIC SCINTILLATOR
Work in Progress

Kurtis F. Johnson, Florida State University

There are two groups working on radiation damage to plastic scintillator at FSU. One (DOE funded) concentrates on doing tests for SDC and on providing irradiations and dosimetry for others: E.Bartosz, M.Bertholdi, V.Hagopian, E.Hernandez, K.Hu, C.Immer, KFJ, C.Shaffer, J.Thomaston, H.Whitaker.

The other group (TNRLC funded) focusses on the chemistry of radiation damage: T.Herod, KFJ, M.Kasha, P.Trimmer, A.Sytnik.

Abstract:

The "better red than dead" strategy first proposed by S. Majewski several years ago is still the only successful method of fabricating radiation tolerant plastic scintillator. We describe our search for rad-hard, large Stokes' shift fluors. No greatly superior scintillator base material has been found to replace polystyrene or its derivatives; it therefore becomes very important to understand the kinetics of color-centers in PS. Measurements are described and some preliminary results presented.

Suggested readings:

1) J.B.Birks, The Theory and Practice of Scintillating Counting (Pergamon, London, 1964).

2) I.B.Berlman, Handbook of Fluorescence Spectra of Aromatic Molecules (Academic Press, New York, 1971).

3) A. Pla-Dalmau, "Design of Fluorescent Compounds for Scintillation Detection", doctoral dissertation, Northern Illinois University (1990).

4) Proceedings of the Workshop on Radiation Hardness of Plastic Scintillator, March 19-21, 1990, Tallahassee Florida, ed. K.F.Johnson

5) J.M.Flournoy, "An Introduction to Organic Scintillators", Proceedings of the Workshop on Scintillating Fiber Development for the SSC, Nov.14-16,1988, Fermilab.

6) S.Majewski, "Radiation Damage in Plastic Scintillators", Proceedings of the Workshop on Calorimetry for the Supercollider, ed. M.G.D.Gilchriese (World Scientific, Singapore, 1990) pp.451-485.

7) C.Zorn, "Studies in the Radiation Resistance of Plastic Scintillators, Reviews and Prospects", IEEE trans. Nuc. Sci. 37, No.2, April, 1990, p504.

SESSION SUMMARY: ELECTRONICS, TRIGGERING AND DATA ACQUISITION*

SERGIO RESCIA

Brookhaven National Laboratory, Instrumentation Division, 535-B
Upton, NY 11973, U.S.A.

ABSTRACT

The session focused on the requirements for calorimetry at the SSC/LHC. Results on new readout techniques, calibration, radiation hard electronics and semiconductor devices, analog and digital front end electronics, and trigger strategies are presented.

1. Session Outline

The electronics session was focused on the needs and specifications of calorimetry for experiments at hadronic colliders (SSC: Superconducting Super Collider and LHC: Large Hadron Collider).

The topic of "Electronics, Trigger and Data Acquisitions" encompasses a broad variety of issues from front end electronics which must cope with the high radiation environment of the calorimeter region to trigger strategies which must be able to select the exceedingly few events of interest while coping with a high rate of minimum bias events.

In organizing the contributions a "down to top" approach was followed moving from the detailed description of rad-hard front end electronics to the description of trigger strategies and systems.

2. Readout and Calibration

C. de La Taille (LAL, Orsay) described a new solution for the readout of liquid ionization chamber calorimeters. The current approach for the readout of calorimeters is to locate the front-end preamplifiers on the electrodes themselves to avoid the increase in both charge transfer time and in noise due to a transmission line connecting the electrodes to the front end electronics.[1] For liquid argon calorimetry such an approach requires the use of preamplifiers able to work at 90 K. A transmission line between the

---

*This research was supported by the U. S. Department of Energy: Contract No. DE-AC02-76CH00016.

detector and the front end electronics would exhibit many desirable advantages when compared to this solution:

- no need for cryogenic electronics, which allows the use of bipolar devices.
- no power dissipation inside the cryostat.
- easier maintenance of the front end electronics.
- preamplifiers could be located in a region with much lower radiation levels and neutron fluxes.

At fast shaping times (peaking times less than 30-40 ns) the transmission line cannot be modeled as a lumped parameter inductance and capacitance but its frequency behavior must be taken into account.

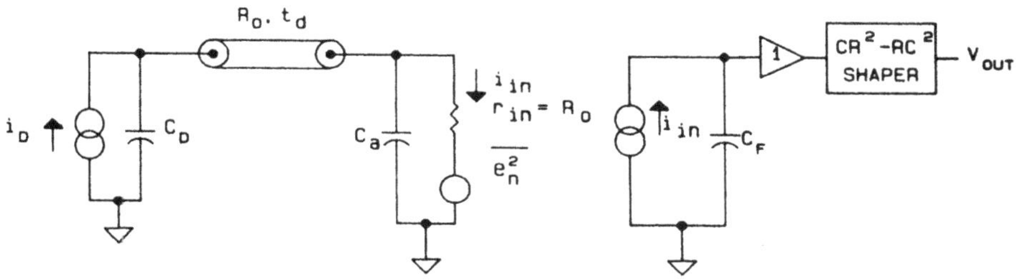

Fig. 1: Equivalent circuit for the analysis of the connection of a calorimeter tower to the preamplifier with a transmission line.

With reference to Fig. 1 a calorimeter tower, modeled as the capacitor C_D in parallel with the signal current generator, is connected with a transmission line of characteristic impedance R_0 to a charge sensitive preamplifier, modelled by its dynamic input impedance,[2] the "cold" resistance r_{in} in parallel to the preamplifier input capacitance. The controlled current generator models the integration in the preamplifier, which is followed by a CR^2 - RC^2 bipolar shaper (the use of a bipolar shaper is dictated by the need of avoiding the baseline shift at high rate). The preamplifier input impedance r_{in} matches the characteristic impedance R_0. The noise analysis is carried out by calculating the noise current flowing into r_{in}

$$\overline{i_n^2} = \frac{\overline{e_n^2}}{|R_0 + Z|^2} \tag{1}$$

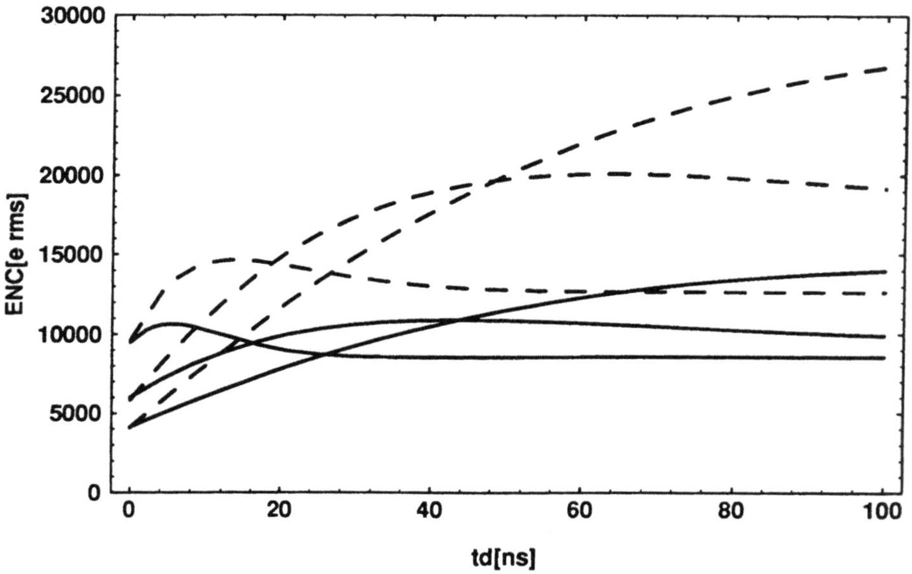

Fig. 2: Equivalent Noise Charge (ENC) versus line delay t_d for two different line impedances, 50 Ω (continuous line) and 25 Ω (dotted line) and three peaking times (20, 50 and 100 ns). The detector capacitance is $C_D = 400$ pF and the equivalent input noise is $e_n = 0.5\ \mathrm{nV}/\sqrt{\mathrm{Hz}}$.

where Z is the impedance seen at the receiving end of the transmission line. Figure 2 shows a plot of the equivalent noise charge (ENC) versus line delay t_d for a detector capacitance an equivalent voltage $e_n = 0.5\ \mathrm{nV}/\sqrt{\mathrm{Hz}}$, $C_D = 400$ pF, two values of line impedance, 25 and 50 Ω and for peaking times of 20, 50 and 100 ns. The point at $t_d =$ 0 gives a comparison point with the case with no line. It is apparent that the higher the impedance, the lower ENC is. For fast shaping times (25 ns) and for $R_0 = 50\ \Omega$ there is a slight reduction in noise with respect to the case with no line. A large toll in signal to noise ratio is to be paid at longer shaping times (100 ns). In this idealized case the noise quickly becomes independent of the line delay t_d, since the impedance seen at the end of the line converges to the characteristic impedance. There are some practical constraint on the cable which are to be taken into account. Given the large number of channels of the calorimeters proposed for the SSC/LHC the size of the cable must be minimized. This may cause additional attenuation and increased noise due to the skin effect of the conductors (about 0.7 Ω/m at 1 MHz for a 350 μm wide double-shielded flat cable). B. Chase (LAL, Orsay) proposed a calibration scheme for liquid argon calorimetry which avoids the problems associated with distribution of fast pulses, which is described in these proceedings.

3. Front End Preamplifier

The very first stage of front end electronics must be able to survive radiation levels of about 3×10^5 kGy (30 Mrad) and neutron fluences up to 10^{14} N/cm^2 in some part of the calorimeter (see article by A. Ferrari in these proceedings). Especially for cryogenic preamplifiers which are likely to be sealed inside the cryostat for the entire life of the detector, this poses a severe constraint on their reliability.

D. Camin (INFN, Milan) described a gallium arsenide preamplifier. The idea is to take advantage of the interesting properties of GaAs for fast, low noise applications at cryogenic temperatures. Table I gives some fundamental physical parameters of GaAs which are compared to Si, in both cases at a doping level of 10^{17} cm^{-3}, a typical value for a MESFET and JFET channel, respectively.

Table 1. Main Physical Parameters of GaAs

		GaAs	Si
Electron mobility (cm^2/Vsec)	300 K 77 K 4 K	5000 10000 3000	800 3000 nd
Electric field at peak velocity (V/μm)		0.7	3
Ionization energy of dopants (meV)		6	50
Energy bandgap (eV)		1.4	1.1

The high electron mobility and low electric field for carrier peak velocity make it possible to obtain high transconductance to input capacitance ratios at low power dissipation. This is a parameter of prime importance in charge-sensitive preamplifiers as it is related to the product of the charge sensitivity times the speed for a given detector capacitance.[3] In addition, the low ionization energy of dopant impurities prevents carrier freeze-out at cryogenic temperatures.

Regarding noise, the 1/f component is dominant at room temperature but it decreases by about two orders of magnitude when cooled to 77 K. Therefore its contribution to the total ENC is very much reduced at cryogenic temperatures particularly at short shaping times. A high value of transconductance can be reached at low bias voltage and current. For this reason a low white noise level can be obtained keeping the power dissipation low.

With this idea in mind, different versions of charge-sensitive preamplifiers based on GaAs MESFETs have been developed in Milan and tested in the prototype of the Accordion LAr calorimeter in the frame of the RD3 collaboration at CERN. The performance of the most recent preamplifier version used in the test of the calorimeter at 20 ns peaking time bipolar shaping are given in Table 2.

Table 2. GaAs Preamplifier at $t_p = 20$ ns

ENC @ $C_D = 400$ pF	13700 rms el. (54 MeV)
Matching capacitance	80 pF
Input resistance @ $C_F = 33$ pF	less than 9 Ω
Dynamic range	165 GeV
Power dissipation	54 mW

The circuit diagram is shown in Fig. 3.

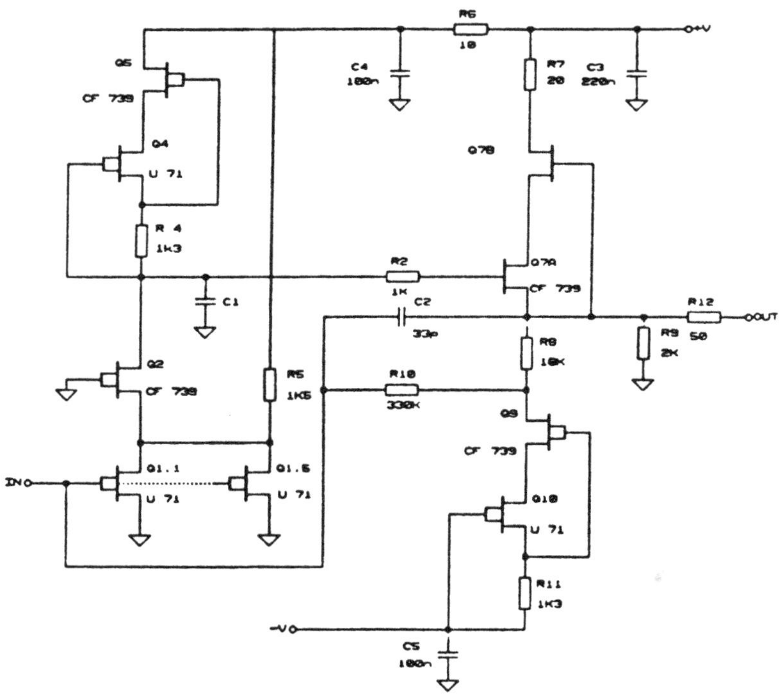

Fig. 3: Schematic diagram of the GaAs charge preamplifier used for the test of the accordion calorimeter at $t_p = 20$ ns.

The ENC as a function of peaking time for two different shapers is given in Fig. 4.

Preliminary measurements of radiation damage have been done on hybrids using commercial MESFETs. The results have shown a 5% increase in noise after 10^{14} N/cm^2, and a 7% decrease in transconductance after 10^7 rad of a ^{60}Co source.[4]

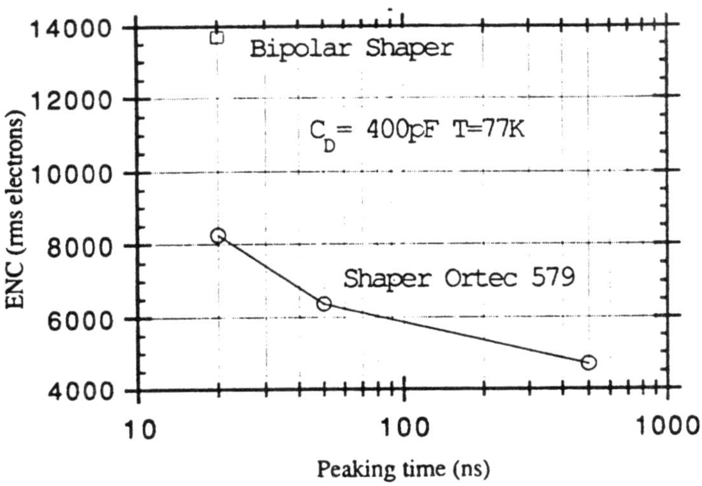

Fig. 4: ENC as a function of peaking time for two different shapers

As for the future, the Milan group intends to realize a monolithic version of a GaAs charge preamplifier in a rad-hard process. The semi-insulating characteristic of GaAs substrates make the problem of device isolation simpler. So far, a charge preamplifier based on a monolithic array of MESFETs have been successfully tested.

D. DiBitonto (University of Alabama) presented a monolithic preamplifier based on a transimpedance configuration. The circuit was built in a GaAs rad-hard technology with a radiation hardness of more than 100 GGy (1 Grad) and 10^{16} N/cm^2 for 5% variation of static parameters as measured by the manufacturer. Data on radiation hardness with respect to noise are being measured. The circuit is described in a separate article in these proceedings.

4. Front End Electronics

M. Levi (LBL) described the proposed front-end electronics system for the scintillating tile calorimeter of the SDC (Solenoidal Detector Collaboration) experiment at the SSC.

The key component of the system is a sixteen channel analog waveform sampler with 256 cells per channel able to operate at the 60 MHz sampling rate synchronized with the SSC clock. Analog information is stored on capacitors and a double set of switches allows simultaneous read and write of information. An ancillary circuit, the address list processor manages the various lists of available storage cells, pending level 1, level 2 and pending readout for a number of SCAs. The system is described in these proceedings. Given the complexity and sophistication of the trigger procedures necessary at SSC/LHC a second line of though advocates digital font end electronics and trigger system.[5]

N. Yamdagni (University of Stockholm) described a digital front end system (Front End Readout Microsystem: Fermi) and trigger level 1 and 2. Figure 5 shows a block diagram of the system. The key requirements are summarized in Table 3.

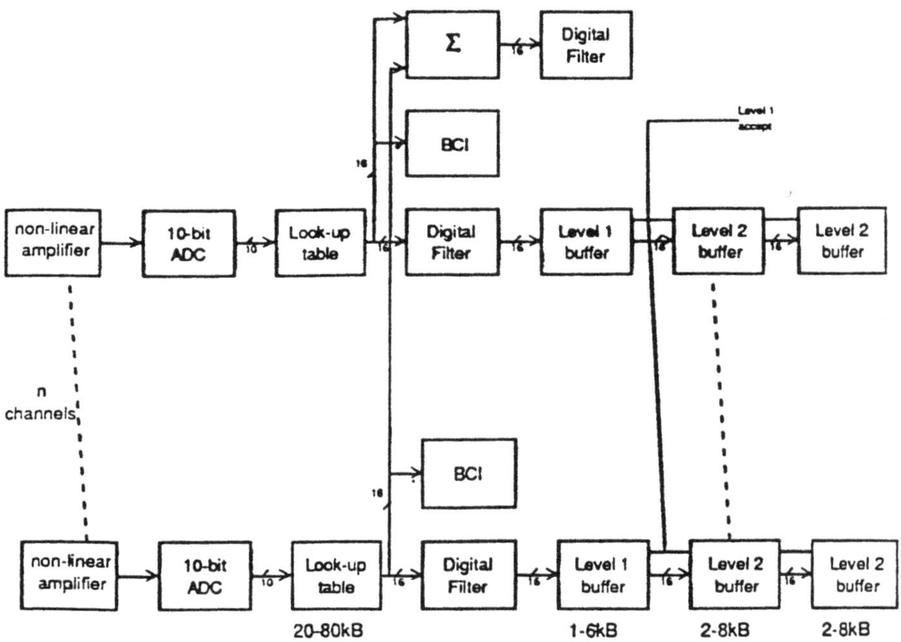

Fig. 5: Block diagram of the proposed Front End Readout Microsystem (FERMI).

Table 3: FERMI Data Acquisition Requirements (DA)

1.	Multiple (probably 9) parallel data channels
2.	15 ns/sample
3.	15 bit dynamic range, compressed to 10 bits by a nonlinear amplifier, then expanded to 15 bits by a look-up table after digitization
4.	Digital filtering of signals
5.	Sum output for first level trigger
6.	Event recognition
7.	On module level 1 and level 2 data storage
8.	No data loss for fixed level 1 processing time
9.	Low power consumption
10.	Fault tolerance

Particularly interesting in calorimetry is the idea of using a non-linear front-end stage which may allow handling of the wide dynamic range (from about 50 MeV noise level up to events in the few TeV range) with only one electronic channel instead of a split-range system as for example proposed in SDC.

5. Miscellaneous Devices and Circuits

W. Wallraff (RWTH, Aachen) reported on radiation damage tests and GaAs diodes. After 100 Mrad irradiation with 3 MeV electrons the reverse leakage current at -100 V bias increased from 100 nA to 450 nA and the pulse height from alpha particles decreased to about 50%. Neutron damage studies are under way.

J. M. Seixas (CERN) presented a novel technique for electron identification in a lead-scintillating fiber calorimeter based on the different time width of electron and pion signals. This technique is capable of providing a very fast trigger (response time about 50 ns) for electrons. The technique is described in detail in these proceedings.

L. Knudsen (CERN) described a small tungsten calorimeter built for beam monitoring purposes at CERN which makes use of advanced digital signal processing techniques.

6. Acknowledgements

The author wishes to thank all the participants in the session (too numerous to mention) for the friendly discussions which made the session most interesting and informative. A special thanks goes to the organizers of the conference and in particular to the chairman, A. Ereditato.

7. References

1. V. Radeka and S. Rescia, *Nucl. Instrum. Methods*, **A265**, (1988) 228.
2. V. Radeka, *IEEE Trans. Nucl. Sci.*, **NS-21 (1)**, (1974) 51.
3. D. V. Camin, G. Pessina and E. Previtali, "Front-End in GaAs," Proc. 5th Pisa Meeting on Advanced Detectors, to be published in , *Nucl. Instrum. Methods*.
4. D. V. Camin, G. Pessina and E. Previtali, *IEEE Trans. Nucl. Sci.*, **38 (2)**, (1991) 53.
5. A. Baumbag, et al., "Baseline Implementation of the Digital Phototube Readout System for the SDC Calorimeter," SDC Report (1991).

RADIATION TESTS ON GaAs PARTICLE DETECTORS

W. KARPINSKI, K. LÜBELSMEYER, M. TOPOROWSKY,
W. WALLRAFF
1. Physikalisches Institut, RWTH Aachen,
Sommerfeldstraße 28, D 51 Aachen, Germany

and

K. HEIME, R. WÜLLER
Institut für Halbleitertechnik, RWTH Aachen,
Sommerfeldstraße 24, D 51 Aachen, Germany

ABSTRACT

24 GaAs surface barrier diodes of 0.64 cm^2 active surface have been fabricated and tested as particle detectors. 4 of these diodes have been subjected to an irradiation test with 100 MRad of 3 MeV electrons.

1. Introduction

It is well known that GaAs devices can be far more more radiation resistant [1] (up to two orders of magnitude) than those made from Silicon. Operation near to the beamline of the LHC leads to energy deposits (neutron fluxes) exceeding 10 MRad / year (10^{14} / cm^2) [2]. Inner tracking detectors and especially those used for photon localization inside an electromagnetic calorimeter are only marginally safe to be operated over 10 years if they are made from Silicon. Therefore we have constructed 24 GaAs surface barrier diodes for test under high radiation load. Rather similar diodes have been used by the Glasgow group[3] for their tests with Co^{60} and neutrons.

2. Diode Parameters

12 diodes each (total / active area 10x10 / 8x8 mm^2) have been constructed on 1 commercial 350 μ wafer. The resistivity of the material used was ≈ 5 * 10^8 Ohm cm. The Schottky (ohmic) contact is made from Ti / Pt / Au [100 / 100 / 100 nm] (Ni / AuGe / Ni [20 / 200 / 100 nm]). In order to reduce the leakage current along the edges of the diodes we have surrounded the active area with a 1 mm wide passivating strip of SiO_2 on 75 % of the diodes.

3. Diode Performance

3.1 *Before Irradiation*

Fig. 1 shows the current - voltage characteristic (at room temperature) for a diode with (without) passivation. The current in reverse bias is clearly strongly influenced by edge effects. The passivated diode shows 100 nA less leakage current.

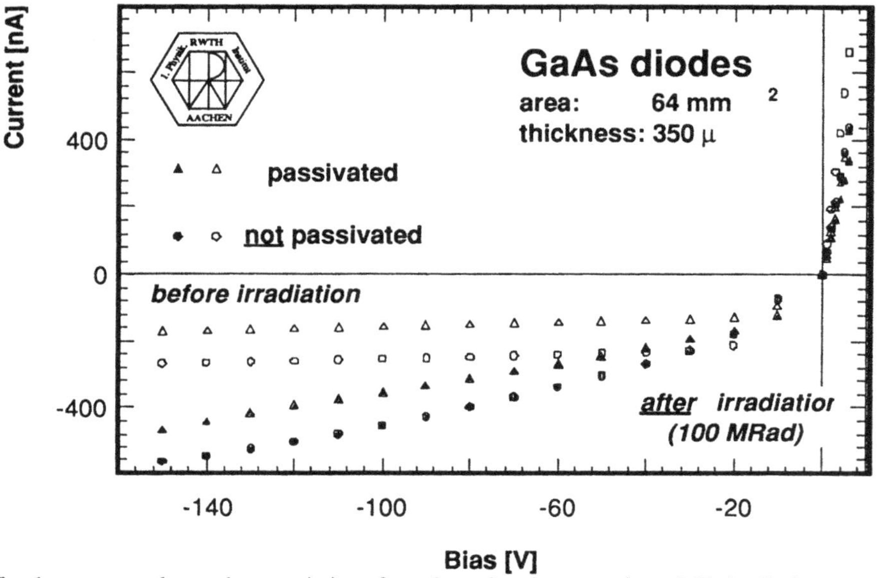

fig. 1. current voltage characteristics of passivated and not passivated GaAs diodes before and after irradiation

The reverse bias conductivity of the diodes is very low (170 pS) and independent of the passivation. As ususal for GaAs diodes these diodes are completely depleted at all reverse bias voltages. In addition to the electrical tests we have investigated the particle detection behaviour of these diodes , which will be reported elsewhere.

3.2 After Irradiation

The described diodes have been irradiated with 3 MeV electrons from a van de Graaf accelerator at the Forschungszentrum Jülich Gmbh. The electron current can deliver up to 1 MRad cm^{-2} sec^{-1} into a layer of 350 μ of GaAs.

The diodes were mounted on a 2 mm G10 plate and housed in a thermally controlled He atmosphere. The voltage - current characteristic after irradiation is displayed in Fig. 1 too. The fundamental change in the voltage - current characteristic is the greatly increased reverse bias conductivity of 2400 pS. Inspite of this rather large increase the noise figures from the diodes under test have changed by less than 10 %. The power dissipation in reverse bias (-100V) has increased from 300 mW/m^2 to 650 mW/m^2 only. Furthermore the mean pulseheight from alfa particles has decreased to ca. 50% of its initial value.

All these phenomena have been found to be independent of the mean doserate (0.2 -1 MRad cm^{-2} sec^{-1}).

4. Conclusion

GaAs surface barrier detectors have been tested at the 100 MRad level of irradiation. It is found that the electrical as well as the particle detection properties of GaAs surface barrier diodes are such that a safe longterm operation at the LHC can be envisaged.

In order to complete these investigations we will test diodes from this batch at high neutron flux densities in the near future. In addition we plan to investigate the behaviour of Si diodes in the same environment.

5. References

1. C.T. Wang, *Introduction to semiconductor technology, GaAs and related compunds* (John Wiley & Sons, New York 1989).
2. G. Stevenson, in *Large Hadron Collider Workshop, Aachen 4-9 October 1990* , ed. G. Jarlskog and D. Rein (CERN 90-10, vol. III) p. 566.
3. R. Bertin et al. NIM **A294** (1990) 211

Monolithic GaAs Electronics for Ionization Detectors

D. DiBitonto, R. Muñoz, T. Pennington, and K. Subhani

Department of Physics and Astronomy
The University of Alabama
Tuscaloosa, Alabama 35487, USA

ABSTRACT

We present a novel approach to primary ionization sensitive detection that utilizes a fast, low noise transimpedance amplifier fabricated in a monolithic industrial GaAs technology with a radiation hardness in excess of 1 Gigarad and 10^{16} n/cm^2. This technique exploits the leading edge of the induced current spike produced by ionizing radiation that contains information on the instantaneous total charge deposited, which can then be recorded without integrating this charge over the entire drift time. A principal advantage of this approach is that the speed and sensitivity of this amplifier are weakly affected by source capacitance, thereby allowing greater design flexibility for both hadronic and electromagnetic calorimetry, or even charged particle detection (tracking) integrated within the same device. The rise time measured in prototype circuits with this technology is < 5 nsec with an output referred noise of 125 μV_{RMS} (0.1 m.i.p.) up to .5 GHz.

1. Introduction

The instantaneuos total charge deposited by the passage of ionizing radiation through a dielectric medium is proportional to the initial current spike produced by this radiation, thereby allowing a measurement that does not require integration over the entire drift time. This current pulse consists of a fast (< 1 nsec) leading edge followed by a linear decrease to zero over a time determined by the free electron drift velocity and the gap length. The current is given by the product of the free electron/ion yield per eV deposited (G_{fi}), the energy deposited by this radiation (dE/dx), and the free electron drift velocity. Since this current depends mainly on the applied electric field and not on the actual length of the ionization gap, both the applied potential difference and ionization gap length need not be large for high applied electric fields. At roughly 50 kV/cm, the current pulse amplitude for a single (dual) gap in both TMS and LAr is about 12 (25) nA per minimum ionizing particle [1], although the total drift time in LAr is much longer. For a 1 mm gap in TMS, this corresponds to a maximum drift time of approximately 20 nsec. By contrast, the same ionization pulse in silicon is over an order of magnitude higher, both due to the increased equivalent free ion yield and faster drift velocity. For a 300 μm thick silicon gap across 70 V, approximately 24,000 free electrons drift in a pulse only 7.5 nsec long.

2. Design Simulation

The general readout electronics in Figure 1 consists of a transimpedance amplifier through which the input current is converted to a voltage at the output stage. The input decoupling capacitance, C_{in}, serves primarily to filter out low frequency noise and to provide an ac coupling to the amplifier. The open loop gain of the amplifier as determined from the semiconductor parameters of this technology is nearly 100 k. Across the feedback resistance network of 50 kΩ, the corresponding output voltage for a 25 nA current pulse is approximately 1 mV, which we choose as a design output voltage for a single charged track passing through the detector. The corresponding thermal noise in this network is less than 2 nA (room temperature at 1 GHz), as compared to 9 nA for a single feedback resistance of 50 kΩ. For a value of V_{CC} near 5 V the linear dynamic range of this device is at least 1:1000 for minimum ionizing charged particles. Electrical specifications are summarized in Table I. The general transfer characteristics of the circuit configuration in Figure 1 are plotted in Figure 2 using the HSPICE circuit simulation program for this opamp. The output response time is determined mainly by the open loop gain and not by source capacitance, C_D.

The circuit layout for this amplifier (without feedback) is shown in Figure 3. This circuit features a differential amplifier with Darlington input followed by a dual cascode and push-pull line driver. Each active device in the circuit consists of a complementary HIGFET (C-HFET) fabricated in a 1 μm GaAs technology and biased between ± 5 V. In particular, the technology used for the critical input stage differential pair transistors is a depletion mode C-HFET, which requires less than 2 nA into the gate for transistor action. This device is normally on (conducting) between source and drain, as compared to enhancement mode C-HFET's, which are used for all other active devices throughout the circuit. The very low input current required to operate this circuit is an important design specification for the signal current to follow the feedback path across the T-network. The high radiation resistance [2] and speed [3] of GaAs are well suited to the high radiation environments anticipated at the SSC and LHC.

3. Prototype Evaluation

Sample circuits fabricated in an industrial GaAs technology using doped channel MODFET devices (similar in performance to C-HFET's) have recently been tested using a 1 GHz digital oscilloscope, model HP54111D, configured with an HP9000 workstation for online analysis. Bandwidth characteristics of this technology were measured using a pulse generator with a 1 nsec fast square wave input without detector capacitance, C_D. In these circuits the feedback resistance network is replaced by a 1 pF feedback capacitance for an initial prototype evaluation of the bandwidth and sensitivity of this technology. The radiation hardness of these devices has been tested independently by the manufacturer, with negligible change in device performance up to 300 Mrad. The rise time of this circuit measured with this system is shown in Figure 4, which indicates a speed of faster than 5 nsec. To measure the corresponding noise levels in this circuit, we applied a fast Fourier transform (FFT) to the digitized output of this amplifier, as shown in Figure 5. Above 100 MHz the output spectral noise density is less than -165 db/$\sqrt{Hz}$, or less than 125 μV_{RMS} (0.1 mip) up to 0.5 GHz.

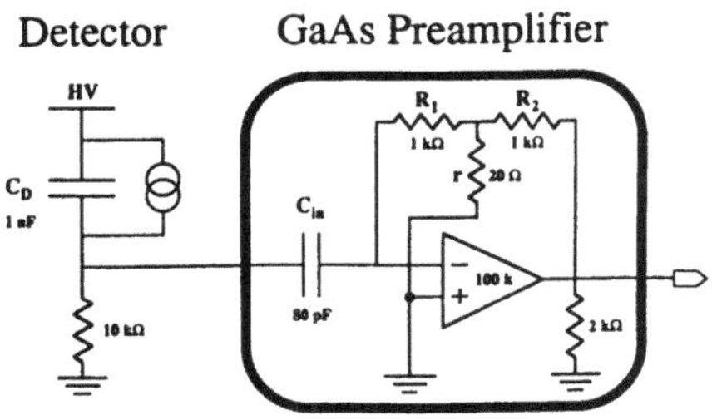

Fig. 1. Equivalent circuit showing ionization current source and source capacitance with preamplifier.

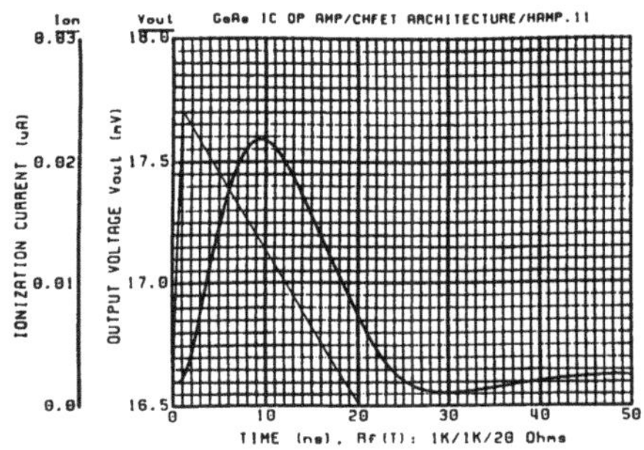

Fig. 2. 25 nA input current pulse (1 ns rise time) and preamp output.

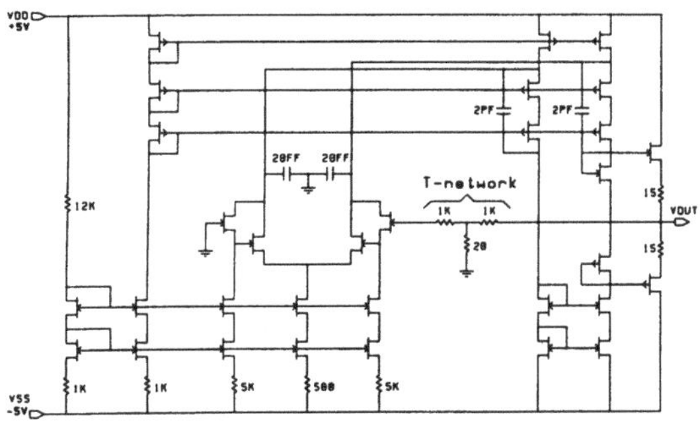

Fig. 3. Complete circuit schematic of transimpedance amplifier.

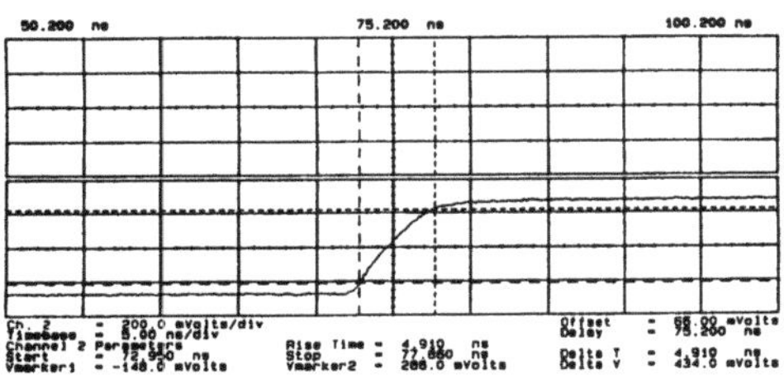

Fig. 4. Digital oscillograph of preamp output response to 1 ns input square wave.

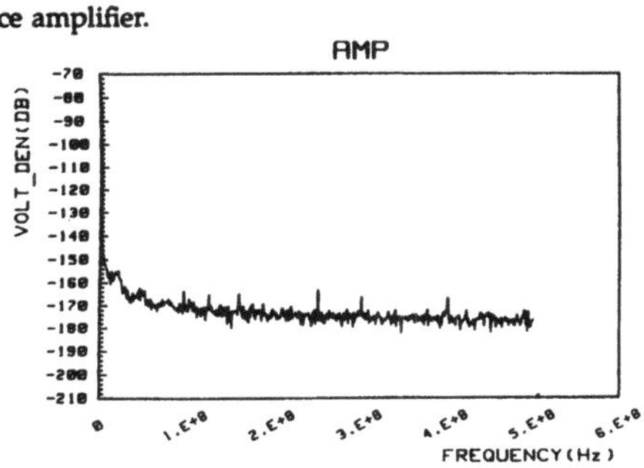

Fig. 5. Fast Fourier Transform of output waveform.

Table I

Circuit Specifications

Device Technology	GaAs Complementary HIGFET
Transimpedance	1 mV/20 nA
Configuration	Quad
Input Bias Current	< 2 nA
Open Loop Gain	95 db
Rise, Fall Time	5 nsec
Output Noise	< 0.1 m.i.p.
Linearity	< 0.1%
Dynamic Range	1:1000
Power Dissipation	25 mW per channel
Radiation Hardness	< 10% parameter change at 1 Grad and 10^{16} n/cm^2

4. Acknowledgements

We wish to thank our many collaborators in US industry for providing us with the circuits fabricated in a radiation-hardened industrial GaAs technology. This work was supported by the Texas National Research Laboratory Commission, the Superconducting Super Collider Laboratory, and the U.S. Department of Energy, grant number DE-FG05-84ER40141, Task B.

5. References

1. "Excess Electrons in Dielectric Media," Ed. Christiane Ferrandini and Jean-Paul Jay-Gerin, CRC Press (1991), Boca Raton, FL.

2. R. Zuleeg, J.K. Notthoff, and K. Lehovec, "Radiation Effects in Enhancement Mode GaAs Junction Field Effect Transistors," IEEE Trans. on Nucl. Sci. **24**, 2305 (1977).

3. M. Shur, "GaAs Devices and Circuits," Plenum Press, New York (1987).

A Switched Capacitor Array Based System for High-Speed Calorimetry†

M. Levi, C. Bebek, R. Ely, R. Jared, I. Kipnis,
F. Kirsten, S. Kleinfelder, T. Merrick, and O. Milgrome
Lawrence Berkeley Laboratory
1 Cyclotron Road, Berkeley, California 94720

ABSTRACT

A sixteen channel analog transient recorder with 256 cells per channel has been fabricated as an integrated circuit. The circuit uses switched capacitor array technology to achieve simultaneous read/write capability and twelve bit dynamic range. Combined with highly parallel analog-to-digital converter and readout control circuitry being developed this system should satisfy the demanding electronics requirements for calorimeter detectors at the SSC. The system design and test results are presented.

1. Introduction

We report here on the development of Integrated Circuit (IC) building blocks for an SSC calorimeter front end. We discuss the outline of the work on that part of the signal processing chain immediately following the amplification of the analog signals. It is contemplated that a single front end board will contain all the functions for event buffering, trigger sums, timing and control, calibration, and readout. This is made possible by the high level of integration achieved using custom integrated circuits that are currently in fabrication and are described in this report. This work is based on low power CMOS circuits and uses switched capacitor array technology to achieve a large dynamic range and high speed readout.

2. Calorimetry Readout and Signal Processing

Under the auspices of this project several specialized integrated circuits have been developed.[1] An integrated CMOS high dynamic range preamplifier/shaper for scintillation calorimetry has recently been fabricated. A low-power sixteen channel 12-bit ADC has been implemented on a single IC. A sophisticated control chip for managing the buffer contents during trigger decisions and readout is currently in fabrication. Most importantly, a switched capacitor array (SCA) which retains the data during trigger decision has had repeated successful fabrications. The SCA will provide for 256 memory cells per channel of electronics, allowing up to 4 μsecs of analog information to be stored at SSC crossing rates.

It is impractical (if not impossible) to move the data from every channel off the detector for every interaction. It is desirable to filter the data through multiple levels of triggering to reduce the bandwidth and digitization requirements. In the most likely scenario, data is stored during two successive levels of triggering prior to event readout. The general task to be accomplished is the temporary storage, followed by slower readout, of brief but high speed analog transients. Acquisition with sampling frequencies up to the beam crossing rate of 62.5 MHz are needed. Usually, a trigger decision of a 2-4 μsecs latency is made to judge the value of the stored signals. These signals are then either discarded or digitized, processed, and transferred to computer systems. Thus, these systems usually require time delay via temporary storage and time stretching of the stored signal to match the bandwidth of the data processing system.

The data are sampled at 16 nsec intervals and stored in analog form. No known techniques for analog storage and fast analog-to-digital conversion can cover the large dynamic range required (16 to 18 bits). Thus at least two data storage channels per signal channel with partly overlapping scales are needed, each covering 12 bits.

The analog pipelines are controlled by a specialized integrated circuit. This chip is responsible for keeping track of the stored data during the successive levels of trigger decision. The circuit is able to manipulate stored addresses (pointers) through a series of bookkeeping lists. This circuit, the address list processor, is described in section 5.

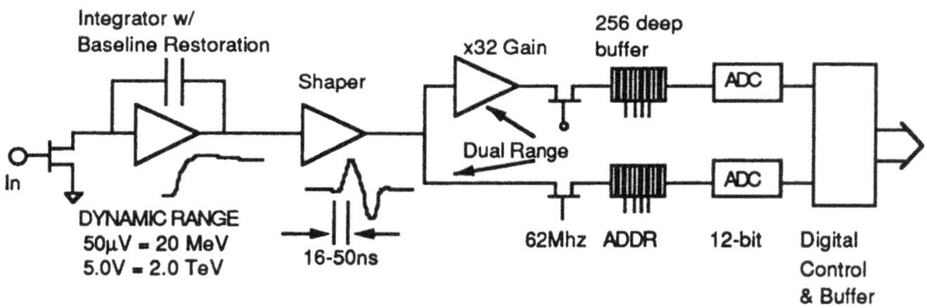

Fig. 1. Shown in the figure is a conceptual design for an SSC calorimeter front end electronics architecture. Signals are preamplified, shaped, split into high and low-range channels, and then stored during the level 1/2 trigger decision in an analog memory. After a Level 2 accept analog data are digitized and readout.

3. Switched Capacitor Array

The switched capacitor array (SCA) has 256 memory cells per channel to store up to 4 microseconds of analog information at the SSC. The number of cells be easily extended beyond 256. Two types of analog memory cells have been pursued: a single-ported, single-ended memory,[2] and a dual-ported, single-end memory with simultaneous read/write capability. Two variations of this cell, inverting and noninverting, have been pursued.[3]

The dual-ported, single-ended memory has four digital clock inputs, reset control input, and two sets of 8 address inputs. There are 16 channels on the IC. A channel consists of an input bus, 256 sample and hold cells, a readout bus, and a reconstruction amplifier, as shown in Fig. 1. Each sample and hold cell consists of a complementary CMOS transmission gate (complementary switch) and a 0.7 pF double polysilicon capacitor. The capacitors use a high quality silicon-oxide dielectric of 70 nm thickness. An externally supplied reference voltage is applied to the bottom plate of all capacitors. The voltage stored on each sample and hold capacitor then corresponds to the difference between the input signal and the applied reference at the time its sample and hold switches are opened. During the readout sequence each sample capacitor is placed in the feedback path of the reconstruction op-amp. A device with full simultaneous read/write capability has been fabricated at ORBIT semiconductor in a 1.2 μm double-metal double-poly technology. The device is 4 mm x 6 mm.

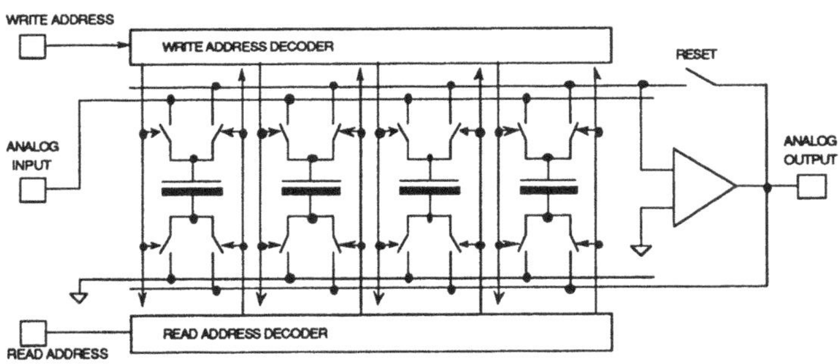

Fig. 2. A simplified block diagram of a single ended dual-ported analog memory cell structure as discussed in the text.

In order to achieve 12 bit dynamic range without correction the pedestal variation of each of the storage elements in a channel must be less than 1 mV of the 4 volt full scale voltage range. Variations in gain and offset between the various storage locations be sufficiently small that they may be ignored. Also, 8.3 time constants must fit in each 16 nsec write operation to charge the sample and hold capacitors. These are a very desirable goals for any system utilizing analog storage to avoid the need for calibration constants for each cell in the storage array. In a 1.2 μm CMOS version with 256 cells per channel with a 1.5 nsec charging time constant, the cell to cell variation at a 50 MHz acquisition rate was measured to be 0.6 mV rms. The cell to cell variation of the SCA is given by the quadrature sum of systematic noise sources (pedestal variation) and incoherent noise sources (op-amp noise, kT noise). Thus, the useable range is 0.6 mV to 6 V. This corresponds to a total dynamic range of 1:8000 or 13 bits.

The on-chip, reconstruction amplifiers developed were based on a published design in a different technology.[4] The single supply op-amp operates to within 20 mV of either voltage rail. The non-linearity of this op-amp is ± 1 mV out of 5 volts or $\pm 0.02\%$ of full scale. Currently the noise of the operational amplifier is one of the dominant incoherent noise sources in the SCA circuit. This noise can be easily rectified by a number of methods which are under study.

The non-linearity of the overall SCA circuit is dominated by the voltage dependent parasitic capacitance of the readout switches. This is measured to be ±4 mV or ±0.01% of full scale and is shown in Fig. 3. The largest inaccuracy is 0.5% of value. The measured parameters of the SCA are summarized in Table 1.

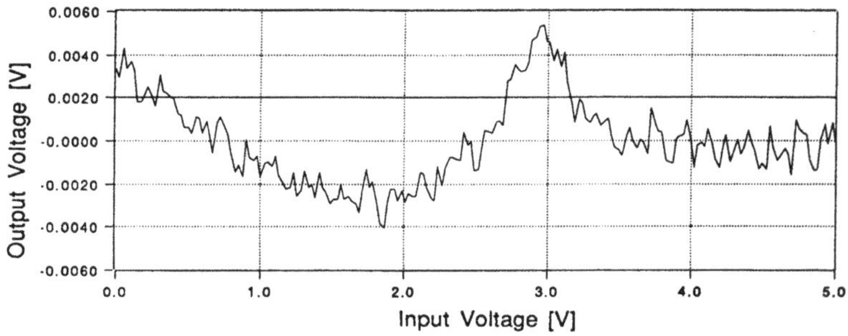

Fig. 3. Non-linearity of the 1.2 μm CMOS 256 cells per channel switched capacitor array. Shown is the deviation from a best fit line.

Analog Store Parameters	Achieved Performance
No. channels/chip	16
No. of elements/channel	256
Power consumption/channel	10 mW
Non-linearity	±0.1% (full scale) ±0.5% (value)
Cell to cell variation	0.6 mV rms
Charging time constant	1.5 ns
Channel dynamic range	8000
Maximum sample rate	90 MHz
Maximum readout rate	200 KHz
Input voltage range	0.0-6.0 Volts
Capacitor droop rate	0.1 mV/msec
Output settling time (0.02%)	1.2 μsec

Table 1. Measured performance parameters of a simultaneous read-write switched capacitor array integrated circuit. The array has been implemented in 1.2 μm CMOS, has 16 channels, and 256 storage cells per channel. Measurements were performed at a 50 MHz acquisition rate.

4. High Dynamic Range Analog to Digital Converter

A 12 bit linear analog to digital converter has been fabricated and tested in a 2 μm CMOS technology. The ADC uses a single slope conversion technique and therefore features small silicon area and low power consumption. The tested chip contains a reference voltage ramp, a comparator to detect when the input signal exceeds the reference ramp, a counter to provide a digital code versus time, and a latch to record the counter code when the comparator trips. The counter and ramp can be shared by all digitizer channels across the chip. A complete conversion on all channels requires 4096 clock cycles. Each incremental converter requires only an additional comparator and readout latch at 300 μm pitch (180 μm in 1.2 μm CMOS). Conversion frequency is above 40 KHz for 12 bits. Conversion rate is above 160 KHz for 10 bit conversions, and the input range is within 200 mV of either supply. The integrated circuit successfully combines high speed CMOS digital frequencies in excess of 70 MHz, with analog signals smaller than 1 mV. Crosstalk and substrate coupling effects have been successfully minimized in this mixed analog and digital chip. These interactions appear as noise in the ramp and comparators. Improvements in bias filtering in the analog section and synchronization between the counter and ramp are being implemented in the next layout.

5. The Address List Processor Circuit

The Address List Processor (ALP) manages a list of all possible SCA storage location addresses, sorting them into FIFO's that identify the contents of the address as empty, pending a Level 1 trigger decision, pending a Level 2 trigger decision, or pending readout to the data acquisition system. In essence, the ALP circuit substitutes movement of digital pointers for movement of analog data. The length of each pipeline is inherently unrestricted up to some fixed maximum so that the length of each pipeline adjusts dynamically, changing as trigger conditions change.

An ALP with 256 deep FIFO's and incorporating the design choices described has been submitted for fabrication. It is a 1.2 micron CMOS device, approximately 4 mm by 3 mm, and contains 60,000 transistors.

6. References

1. C. Bebek, *et.al.*, *Front End Electronics for SSC Calorimetery Progress Report*, Aug. 16, 1991. C. Blocker, *et.al.*, *Front End Electronics Development for SSC Detectors*, UPR-208E, October 1, 1991. A. Bhattacharya, *et.al.*, *Front End Electronics Development for SSC Detectors*, October 1, 1990. A. Bhattacharya, *et.al.*, *Front End Electronics Development for SSC Detectors*, October 2, 1989.

2. G. Rai, *et.al.*, IEEE Trans. Nucl. Sci **NS-37** (1990) 56.

3. S. Kleinfelder, IEEE Trans. Nucl. Sci NS-37 (1990) 1230. S. Kleinfelder, M. Levi, O. Milgrome, *Test Results of a 90 MHz Integrated Circuit Sixteen Channel Analog Pipeline For SSC Detector Calorimetry*, LBL-30222. S. Kleinfelder, M. Levi, O. Milgrome, *Toward a 62.5 MHz Analog Virtual Pipeline Integrated Data Acquisition System*, Nucl. Phys. B23A (1991) 382. O. Milgrome, S. Kleinfelder, M. Levi, *A 12 Bit Analog to Digital Converter for VLSI Applications in Nuclear Science*, submitted to IEEE.

4. J. Babanezhad, IEEE Journal of Solid State Circuits, 23 (1988) 1414.

†This work has been supported by the Director, Office of Energy Research, Office of High Energy and Nuclear Physics, Division of High Energy Physics of the U.S. Department of Energy under Contract No. DE-AC03-76SF00098 and by the Superconducting Super Collider Laboratory which is operated by URA, Inc. for the U.S. Department of Energy under Contract No. DE-AC02-89ER40486.

CALIBRATION OF LIQUID ARGON CALORIMETERS AT HIGH LUMINOSITY

R.L. CHASE, C. de la TAILLE, N. SEGUIN
LAL, 91405 - Orsay, France

and

L. ROGERS
Brookhaven Nat. Lab., Upton, N. Y., USA

ABSTRACT

With the short pulse shaping necessary for high luminosity operation of Liquid Argon Calorimeters, the criteria for precise calibration are different from those that apply to long shaping. A system is described that combines relaxed tolerances on circuit elements in the argon, combined with pulse distribution that is largely unaffected by imperfect, dispersive transmission lines.

Introduction

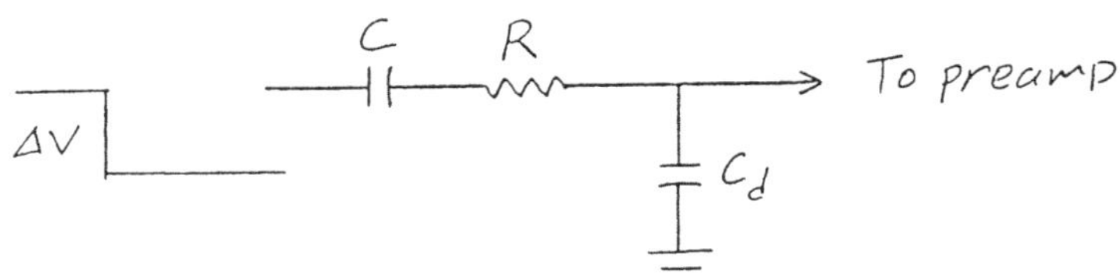

Fig. 1 Standard calibration configuration

Fig. 1 shows a standard calibration configuration that is often used when pulse shaping times are long compared with the charge collection time (~ 400 ns for 2 mm argon gaps). An exponential decay of the calibration current is a satisfactory approximation to the linearly decreasing current, typical of ionization chambers in which electron mobility is very much greater than positive ion mobility. Good accuracy is obtained if the exponential time constant is equal to 1/3 of the electron collection time. The total charge, determined by the voltage step and the calibration capacitor value, is critical to precise calibration, whereas the influence of the resistor on the time constant is of lesser importance.

In the case of shaping times that are short compared to the charge collection time, the initial value of the current, determined by the voltage step and the calibration resistance value, is critical while the influence of the capacitor on the time constant (whose optimum value is now equal to the total electron collection time) is much less important. For example, when the shaped signal peaking time is 45 ns and the electron collection time 400 ns, a 10% error in the capacitor value produces only a 1% error in the peak voltage amplitude.

This calibration configuration, however, has two major drawbacks. To simulate the maximum energy signals expected with the proposed new hadron colliders (SSC and LHC), substantial values of initial current are required (1 TeV corresponds to about 3 mA). If the maximum value of the voltage pulse to be distributed is to be reasonable (a few volts) the RC network will have to be low in impedance (small R, large C) and these elements will constitute a noise source when connected in parallel with the detector.

In addition, the voltage pulse will suffer attenuation and distortion due to skin effect resistance if reasonable size cables are used to distribute the pulses over the detector volume.

Proposed solution

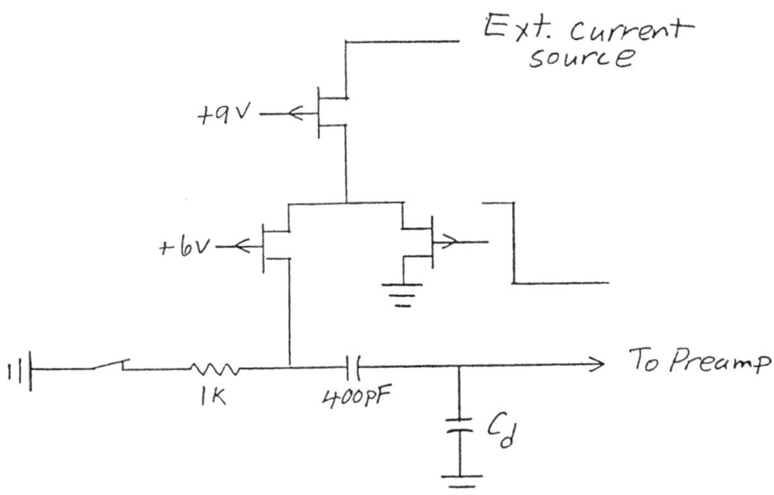

Fig. 2 Calibration configuration compatible with short shaping

In the configuration shown in Fig. 2, a precisely defined, externally generated current is supplied to a "long tailed pair" FET current switch. Initially, the current flows in the calibration resistor and is interrupted by the command signal. The current then transfers to the detector and preamplifier via the calibration capacitor, decaying exponentially with the time constant RC (chosen to be approximately equal to the electron collection time). Note that the initial value of the current is determined entirely by the external current source and is independent of both R and C, which only determine the decay time constant. This means that, when calibration is not being performed, R and C can be disconnected by an FET switch, whose "on" resistance is not critical, thus eliminating the associated noise.

The command voltage step serves to switch the calibration current from one side to the other of the long tailed pair. At least part of its leading edge should be fast, but its exact amplitude and shape are not critical, so that it can be distributed by miniature cables, each driving a number of closely spaced current switches.

Of course, the externally generated current source can be connected to only one switch at a time. The number of current sources and leads entering the cryostat can be kept to a reasonable value by distributing each current to a number of current switches via an addressable CMOS switch yard inside the cryostat.

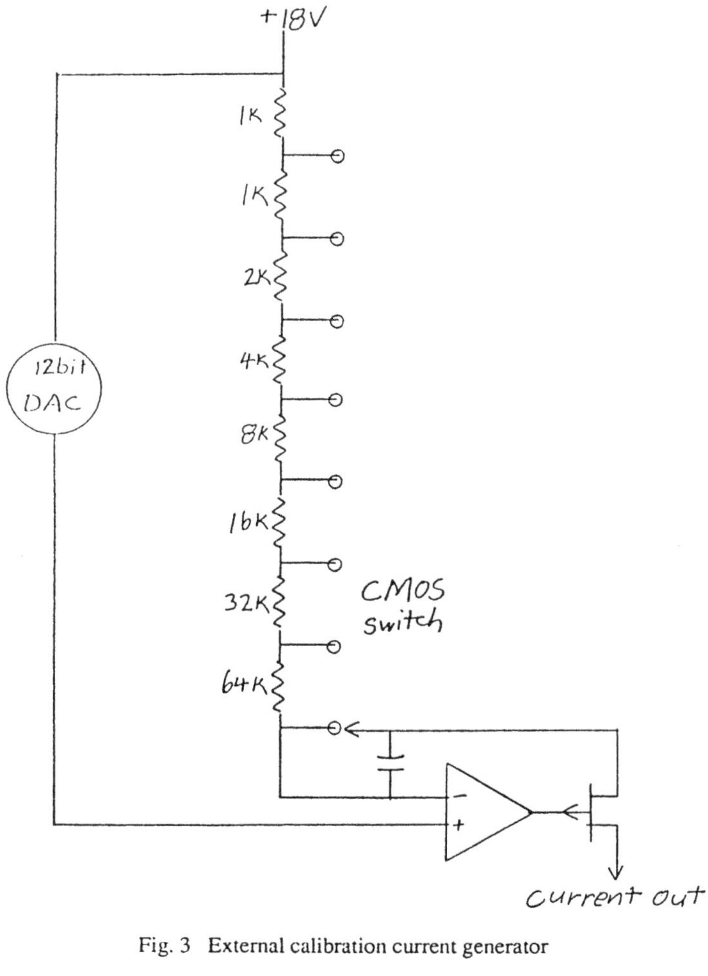

Fig. 3 External calibration current generator

Fig. 3 shows a possible current source configuration in which the value of the calibration current is determined precisely by a reference voltage from a digital-to-analog converter and a set of precision resistors.

The resistors, selected by the higher order bits of the calibration word, are chosen so that, over the very substantial calibration dynamic range, the voltage drop in the selected resistor is always very large compared with the operational amplifier offset voltage.

Fast Front-End Electronics for Electron/Pion Separation in a Lead/Scintillating Fiber Calorimeter

J.M. Seixas
CERN, PPE Division, CH 1211 Geneva, Switzerland
COPPE/EE/UFRJ, C.P. 68504, Rio de Janeiro 21945, Brazil

ABSTRACT

The methods for the electron/pion separation, based on the signal structure and intended for a first-level trigger operation on a LHC/SSC environment, are described . A constant fraction discrimination with a further refinement coming from analog filtering achieves pion rejection factors up to 1000, when an electron efficiency of 98% is kept.

1. Introduction

A lead/scintillating fiber calorimeter is being developed at CERN [1] having in mind its application in the new generation of multi-TeV high-luminosity colliders (LHC/SSC). Projective and non-projective prototypes were built using this technique.

For this kind of detector, a novel way for identifying electrons has been suggested [1, 2] . It is based on exploiting subtle differences of the calorimeter signal for the electromagnetic (e.m.) and hadronic shower developments. This technique can be considered interesting as a very fast trigger (response time $\leq$ 50 ns) for electrons can be provided.

In what follows we summarize this fast discrimination technique and show recent results concerning the application of analog filtering to refine the existing systems.

2. Calorimeter Signals

The calorimeter signals can be considered quite fast. Signals coming directly from the photomultiplier (without any further processing), have a Full Width Half Maximum $\sim$ 4 ns for electrons.

Figure 1 shows signals for electrons and negative pions. One can see that simply measuring the width of the calorimeter signal at a certain fraction (20% has been proved to be optimal) of its peak value is enough to reject a reasonable number of pions, as the signals belonging to this pattern exhibit a second peak structure due to light reflexion on the mirrored end of the fibers. Also, a considerable fraction of the hadronic signals is caused by soft neutrons which makes the signal broader even for many pions that don't exhibit the second peak structure. However some

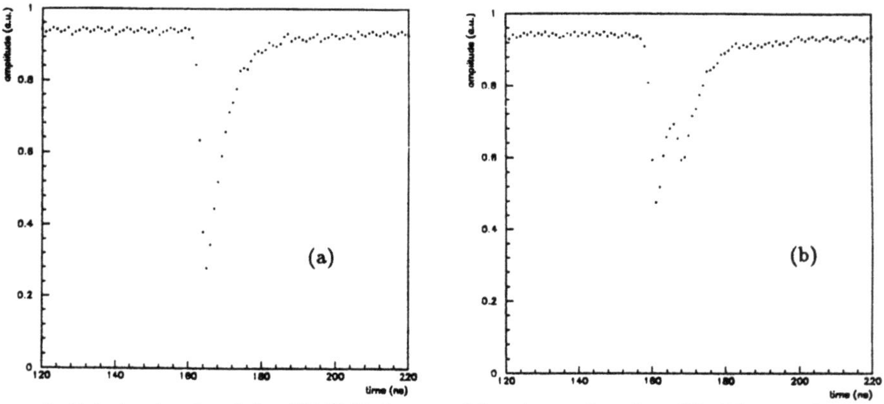

Figure 1: Calorimeter signals for 150 GeV electrons (a) and negative pions (b). The sampling rate is 1 ns.

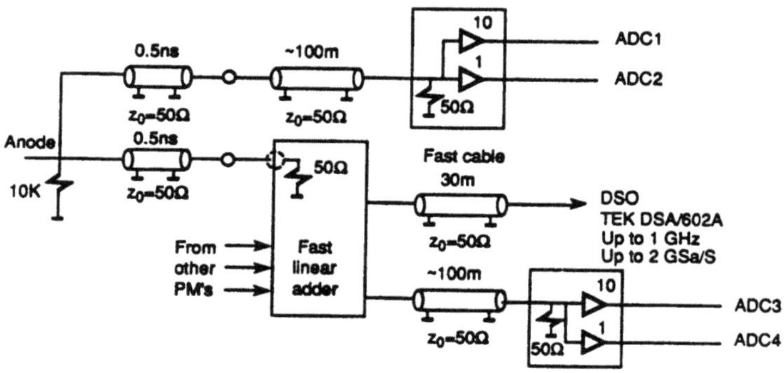

Figure 2: Acquisition chain for the calorimeter signals

pions with high π^0 contribution look like much closer to the e.m. pattern and make the discrimination based on the constant fraction method start failing. Analogue filtering, as described in Section 3, can then refine this constant fraction technique to get a further rejection on some of these electron-like pions.

The signal to be discriminated was selected to be the one coming from the sum of the seven innermost towers of the calorimeter, with respect to the interacting point. This signal contains very well the e.m. shower and around 80% of the hadronic one.

The acquisition of the signals from the calorimeter demanded a wide-bandwidth chain, due to the high-frequency contents involved. Figure 2 shows the chain used for the projective prototype. The line is matched through cables on both branches, keeping a 50 Ω matching all along the blocks of the chain. The signal from the photomultiplier (PM) base goes to a charge ADC (through an active splitter to circumvent dynamic range limitations on this branch) and a combination of such signals (through a fast active adder) is sent to a digital storage oscilloscope using a

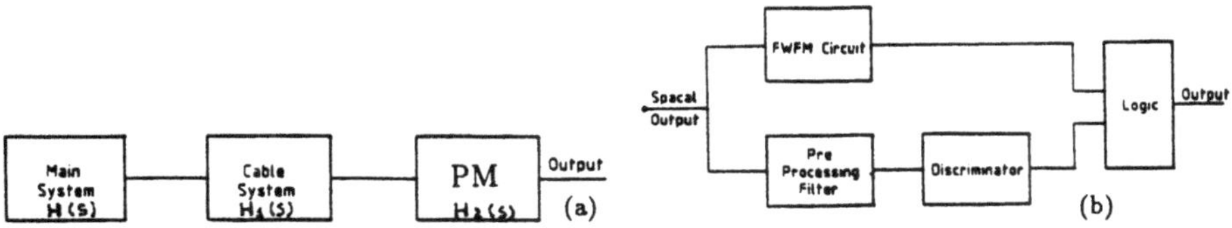

Figure 3: Block diagram of the calorimeter subsystems (a) and the discriminating scheme using the pre-processing filter (b).

fast cable to avoid long tail effects. The active adder was designed to support up to six input signals and to provide a split output (same gain), which allows monitoring the signals being acquired on ADCs 3 and 4. A pyramidal connection of these adders acquired the sum of seven signal.

3. Electron/Pion Separation Based on Signal Structure

The constant fraction (CF) discrimination, which is based on the measurement of the full-width at one fifth maximum (FWFM) of the signal, exhibited a rejection factor for pions of 800 (at 80 GeV) for a electron efficiency of 98% and different circuit realizations have been tested. This method is able to reject all pions with visible double peak structure, as the one shown in Figure 1b.

To achieve further pion rejection factors, analog filtering is being proposed. The idea is to identify subsystems that affect in the same way both e.m. and hadronic shower developments and filter their effects out of the calorimeter signals. Up to now, the photomultiplier tube and the signal cable subsystems have been identified (Figure 3a) and therefore the calorimeter transfer function can be written as:

$$\frac{V_O(s)}{V_i(s)} = H(s)H_1(s)H_2(s) = H(s)G(s) \tag{1}$$

The pre-processing filter - PPF - (with a transfer function $G^{-1}(s)$) can then be realized to compensate the effects of these subsystems and reveal the actual particle signatures, which allows further pion rejections. The function $G(s)$ has been achieved to be described as

$$G(s) = \frac{\prod_{i=1}^{n-2} s + z}{(\prod_{i=1}^{n} s + p_i)(\prod_{j=1}^{m} s + p_M)} \tag{2}$$

where n and m represent the number of sections for the cable and the PM subsystems respectively. For the cable, a characterization based on a recently introduced method [3] proved to be efficient with three sections, whereas for the PM we used a gaussian filter with two sections.

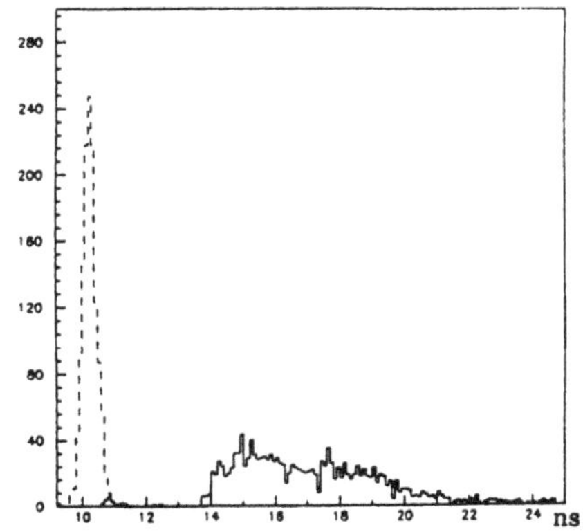

Figure 4: Performance of a CF-PPF combination for a projective prototype at 40 GeV.

The scheme of a discriminating system using this filtering technique can be seen in Figure 3b. In this scheme, one uses the CF to reject pions with relatively large width and makes the remaining pions to pass through the PPF and then another discriminator is used to analyse these pulses. This second discriminator can be another FWFM circuit, in which case an improvement of $\sim$ 30% on the performance of the standing alone CF system was measured for the non-projective prototype [1] . For the projective one, we analyzed the performance of such a discriminating scheme, using another FWFM system for the signals coming from the filter. The preliminary results (with the PPF including just the cable compensation block) can be seen in Figure 4, in which a rejection factor for pions of $\sim$ 800 was measured (for an electron efficiency of 98%) for isolated particles at 40 GeV. This represents the worst case measured [1] and an improvement of $\sim$ 40% on the performance of the CF system.

Another option is to build an optimum filter discriminator to be cascaded with the PPF. In this case, a filter is designed to match each pattern's main feature, which is extracted from the signal leaving the PPF. It's known [4] that being given two classes $(p_1(t), p_2(t))$ and a signal $p(t)$ to be classified, a criterion based on the distance between the signal and each pattern, $c = d_1^2 - d_2^2$, where d_i is given by

$$d_i^2 = \int_0^\infty (p(t) - p_i(t))^2 dt \tag{3}$$

can be made optimum in terms of the signal noise ratio. To achieve this goal one should use a filter described by an impulse response equal to $p_i(-t)$ for each pattern and perform the subtraction of the signals coming from each filter output, having the signal to be classified, $p(t)$, as a common input to the filters. In practice, such a scheme is not realizable, which leads to the approximation shown in Figure 5a, where each filter's impulse response now is described by $p_i(t)$.

In our case, the patterns for electrons and pions can be described in terms of exponential functions (at the PPF output), as one can see from Figure 5b. The dis-

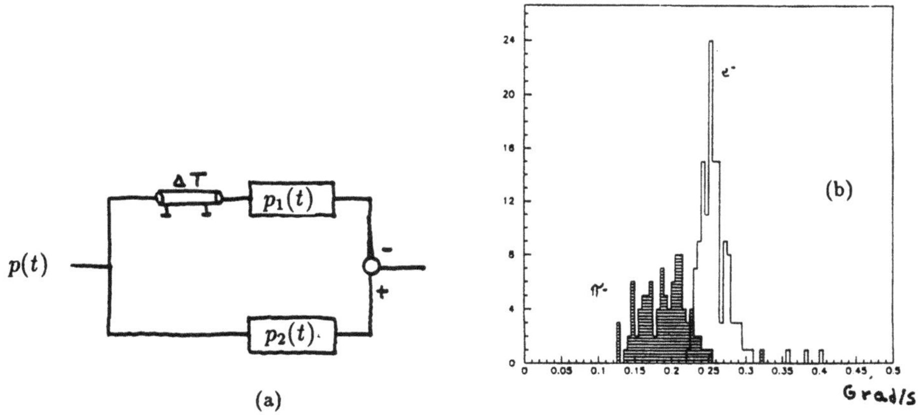

Figure 5: Realizable optimum filter scheme (a) and the exponential characterization of the e.m. and hadronic signals (b).

tribution of the exponential's frequency for each particle pattern shows that a further improvement can be expected from this discriminating system, as we observed that only 8% of these pions with smaller FWFM exhibited an exponential frequency overlapping with 98% of all electrons analyzed. An active filter realization was developed and is being currently tested.

4. Conclusions

The methods for achieving fast electron/pion separation were described. The techniques use the fast calorimeter signal development to measure differences observed on the particle' signal patterns. Rejection factors higher than 800 were measured using a constant fraction discrimination scheme and analog filtering.

Acknowledgements

I am thankful for the support that has been provided to this work from Conselho Nacional de Desenvolvimento Científico e Tecnológico (Brasília, Brazil), World Lab (Geneva, Switzerland) and CERN. I would like to thank the SPACAL group, with whom I started developing the work here described, and the RD1 Collaboration, where the work has been continued.

5. References

1. Acosta, D. et al. *NIM* **A302** (1991) 36.
2. Seixas, J.M. et al. **CERN-LAA-HC/90-013** (1990).
3. Seixas, J.M. et al. **CERN-PPE/91-84** (1991).
4. Turin, G.L. *IRE* **IT-6** (1960) 311.

REALIZATION OF LARGE SYSTEMS

Summary Talk

M. DERRICK
Argonne National Laboratory, Argonne, IL 60439

1. *Introduction*

We are facing the challenge of building four new detectors for the LHC/SSC of unprecedented size, cost and complexity. What can we learn from the experience of building the recently completed detectors that is relevant to this new challenge? To this end, a session was organized in which talks were given on the calorimeters of the four LEP detectors and on SLD, and on the three recently completed detectors that are not yet operating, D0, H1 and ZEUS. In addition, presentations were made about a system study for a liquid argon calorimeter for LHC and on the calorimeter readout system for ZEUS.

The individual contributions should be consulted for the details, but my own subjective conclusions from the presentations are as follows:

2. *Importance of Diversity*

The groups designing the four LEP detectors made quite different choices for their electromagnetic calorimeters: L3 emphasized outstanding energy resolution extending to quite low energies, OPAL chose to build a well-understood lead-glass array, building on the JADE experience, ALEPH emphasized excellent position resolution and DELPHI chose an novel, long-drifting technique that gives good position resolution but yet has a small number of readout channels.

All of the detectors have worked well and have done marvelous physics. As the LEP physics has turned out , with no top quark or new particles within range, a major motivation for the L3 BGO calorimeter has not materialized. The well-understood OPAL detector gave an early excellent limit on the mass of the Higgs. The ALEPH choice of granularity turns out to be important in τ decays, allowing, for example, a separation of decay of four photons from the $\tau^- \rightarrow \pi^- \pi^o \pi^o \upsilon_\tau$, decay. This is an example of an application that was not foreseen in the original design. The DELPHI calorimeter has difficulties with a rather complex aging problem and will need significant rework. Nevertheless, the fine-grained detectors promise to be very important in B physics, which is a future LEP program.

In the next few years, it will be interesting to see how the relative strengths of liquid argon and scintillator play out, both in 1.8 TeV hadron collisions (D0 and CDF) and in the HERA ep collisions (H1 and ZEUS).

The lesson for the LHC/SSC decision-making is clear - insist on diversity and complementarity to the extent that this is possible with two detectors.

3. *Simple Things*

This goes under the rubric "Systems Integration", but what it really means is paying attention to mundane things like cooling, cable routing, dead material, cracks both between towers and between the barrel and endcap. To date, these things have not been done well, perhaps because they often represent interfaces between different groups (effective communication is a major difficulty in these big collaborations), as well as different detector elements, but also because the tasks are not too intellectually challenging.

We now have the software tools necessary to study and understand these issues before the detectors are built, and we will presumably, in the future, avoid things like the ZEUS projective barrel and non-projective endcaps which makes triggering difficult, not to speak of the bizarre original CDF decision to use different technologies in different parts of the calorimeter. Even so, a group of physicists and engineers is needed to keep the overall detector in view and ensure that simple things are not forgotten. There is not much glory here, only the quiet satisfaction of avoiding screwups.

4. *Calibration*

This is an enormous task and the better the calorimeter, the more important it is to have a precision calibration system. The standard technique is to measure the response of the elements using a test beam. Experience shows that full-time availability of such a beam over many years is essential. Lack of availability of test beams at Fermilab is a major problem for the SSC detectors.

All of the crystals from OPAL and the L3 barrel were so calibrated as were all of the towers in the CDF barrel calorimeter. For ZEUS, the uranium radioactivity allows the response of each tower to be set at $\lesssim 1\%$ level, so it was not necessary to calibrate all modules in a test beam. Even so, test beam studies were essential, both in establishing the dimensions of the unit cell (e/h = 1) and in exploring the response across boundaries.

Time is needed to analyze the test beam data, to correct oversights, and to redo the calibration as necessary. For example, in the H1 beam tests, the argon was slightly contaminated, so the calibration cannot be carried over to the HERA environment at the required level, and the schedule did not allow a second round of testing.

Time pressures also did not allow the BGO endcaps of L3 to be beam calibrated, so the group must depend on the $Z^o \rightarrow e^+e^-$ events observed in situ. The same calibration is planned for the SDC detector at the SSC. It will be interesting to see how it works out in practice.

5. *Construction Issues*

The classical technique for constructing one of the large calorimeters is to purchase the components from industry and to assemble the detectors in one or more of the collaborating institutions. Laboratories such as ANL, RAL, and Saclay, with their existing assembly space, shop, and technician capabilities, can play a special role as assembly centers complementing the strengths of the universities.

Most of the time is taken up with preparations, starting with basic design concepts and continuing to physics prototypes, to engineering prototypes and to full-scale design. The actual construction time can be surprisingly short. For example, the OPAL and ZEUS barrel calorimeters were both built in two years or less, but three to four years was needed for the preparations. These preparations must be done thoroughly since once in production, changes are expensive, both in time and money. The schedule has to mesh both with the other components of the detector and with the accelerator; these are far from trivial requirements.

If the production is to go well, someone must take personal responsibility for seeing the project through. Meticulous attention to detail is needed. For example, the ALEPH calorimeter required $3 \cdot 5\ 10^6$ soldering connections, all of which had to be done correctly. These seem to be trivial comments, but the points are sometimes honored more in the breach than the observance.

6. *What Did We Learn*

This session was an experiment. It was only partially successful because of lack of time and because I chose a format of talks structured around given projects. For future conferences, perhaps a round table discussion of specific issues would be more fruitful. Nevertheless, I learned a lot and I wish to express my thanks to all of the speakers.

Work supported by the U.S. Department of Energy, Division of High Energy Physics, under contract W-31-109-ENG-38.

ALEPH CHOICES ON CALORIMETRY IN THE LIGHT OF LEP PHYSICS

HENRI VIDEAU
L.P.N.H.E. Ecole Polytechnique, Route de Saclay
Palaiseau, 91128 France.

ABSTRACT

The Aleph collaboration has made its choice of calorimeters emphasizing the spatial granularity against the energy resolution. What is the verdict of three years of physics. After a short reminder of the calorimeters structure and construction problems, the impact of our choice on different physics topics explored at LEP up to now is examined . It appears that we did not suffer from a lack of resolution but had a great profit from the spatial precision and resolution in most of the subjects and in particular in τ physics.

1. Introduction

The Aleph calorimeters have been designed in the first half of the eighties with some LEP physics in mind. Now that LEP has been operated for three years and that we have accumulated hundreds of thousand of Z_0's, can we draw some lesson from the construction phase and from the analysis phase. Very frankly phrased: "Would we do it again like that?"[1] Very specifically can we assess our gains or losses in physics with respect to our trade between granularity and spatial resolution on one side and energy resolution on the other.

2. Looking at the Aleph calorimeters

2.1. Description

A complete description of Aleph can be found in reference 1.. Aleph calorimetry is done by two different calorimeters separated by the magnetic coil. We have to notice that the presence of the coil in the middle of the calorimeters implies not only energy loss but showers disconnection due to a quite large space with field.

The two calorimeters are build with the same basic technique:

- The electromagnetic calorimeter (ECAL) is made out of a lead–wire chamber sandwich read by wire planes and by projective towers split into three storeys.
- The hadronic calorimeter (HCAL) is made out of iron–streamer tubes sandwich read by tube, digitally, and by projective towers.

1. It should be noted that the Ecal electronics upgrade we are currently discussing and designing is aimed at a better reliability and a better adequacy to high luminosity but does not introduce any conceptual change.

2.2. Structural problems

Both calorimeters have cracks in azimuth: 12 in ECAL and 24 in HCAL; we avoided aligning them between ECAL and HCAL to improve hermeticity. In ECAL the cracks amount to 6%: 4% in the barrel growing to 10% in the middle of the end caps, but their impact on analysis is nastier than such a small number suggests. Half of them are due to granularity because we needed space to get the numerous signals out. These cracks have no matter in it and we can recover some energy measurement in the HCAL. In HCAL they are needed to support the structure and filled with iron.

In the region where the barrel and end caps overlap there is a lack of radiation length in ECAL and the connection of information between corresponding parts is quite hard.

2.3. Construction problems

I will just mention here the outstanding problems we were confronted with:
HCAL:
– the homogeneity of the graphite coating ensuring the electrostatic field in the tubes,
– mistakes in cabling, the sensitive part of the calorimeter is made out of very numerous soft pieces which had to be put together only at installation time and cabled in situ,
– the ground connections made following the good principles had to be redone to ground all the tubes on the detector itself.
All these troubles were solved before the 1990 campaign.
ECAL
– making connections. The first headache was to connect the 3.5 million pads and check it. This has been done mostly by hand. We would definitely try to do it a different way.
– quality control. To ensure the quality and homogeneity of the detector at the percent level the control had to be extremely careful on connections, positions, gain variations in space and time due to the wires or to the gas, pedestal drift.
– front–end electronics reliability.This delicate electronics which multiplexes the 220000 channels by 32 is still plagued with reliability problems which impose frequent repairs.

3.Aleph calorimeters by numbers

In order to understand our requirements on the calorimeter resolutions we have to remember that the TPC momentum resolution ($\frac{\delta p}{p^2} = 6\ 10^{-4} GeV^{-1}$) gives 3% at full LEP1 energy.

3.1.ECAL

Granularity
Pads: 16 x 16 mrad^2 or 3 x 3 cm^2 for X_0 = 2 cm
74000 towers in 3 stacks: 220000 channels with 15 bits dynamic.
Wires: 1620 wire planes in 36 modules

Noise 6 MeV for stack 1, 8 for stack 2 and 12 for stack 3. We expect to gain a factor 2 with the new electronics.

Zero suppression values 20, 28 and 42 Mev for the different stacks. This gives a full efficiency on γ's at 200 MeV with the resolution folded in.

Defects
0.17 % disconnected pads, but a correction on the energy takes it into account.
0.17 % permanently disconnected storeys
between 0 and 0.2 % temporarily killed storeys, the ones repaired at each detector opening
0.12 % dead wires, note that we had 0.02 % last year
13 wire planes out, id est 0.8 % but a correction on the energies is made.

Energy resolution 3.5% observed on Bhabha's at half the Z_0 mass. It can be represented by $\frac{\delta E}{E} = \frac{.18}{\sqrt{E}} + .009$ or $\frac{\delta E}{E} = \sqrt{\frac{.187^2}{E} + .021^2}$. The calibration has been measured to change by .3% in one year.

Position precision $\frac{3mrad}{\sqrt{E}}$ which means that we need an alignment to better than 1 mm.

e / π rejection when isolated
about 10^{-3} for an efficiency around 90%. It is obtained by cutting on the ratio energy in the 4 central towers over momentum (10^{-2}) + longitudinal structure (1/4) + position (1/2). For specific physics see later.

3.2.HCAL

Granularity
Pads: angular size: 4x4 times the ECAL, 4600 towers
tubes 130000 digitally read.

Noise around 60 MeV or 1% of a muon.

Energy resolution HCAL alone on isolated pions $\frac{.8}{\sqrt{E}}$; with ECAL and coil in front on isolated pions it varies with energy from $\frac{.6}{\sqrt{E}}$ up to $\frac{.9}{\sqrt{E}}$; on jets, see the discussion later.

Position resolution .5 cm perpendicular to tubes, seen in matching muon tracks with tube pattern.

μ/h rejection above 5 GeV an efficiency of .99 is reached with a hadron misidentification probability of 1 %. On specific physics a much better rejection can be achieved for a smaller efficiency[3].

4.We gave up some energy resolution for a better granularity

4.1.What did we gain?

Isolation
This enables: – a good measurement of γ's and electrons in jets or collimated decays like

τ's, in particular it helps picking photons radiated by electrons to correct the electron energy,
– a good separation of particles such that an energy flow can be computed using the well measured momentum of the charged tracks without double counting. Note that for photons and electrons the energy measurement is done by keeping only the 4 central towers and correcting for the loss rather than trying to collect all the energy around.

Shape measurement
This is the basis of photon identification and a strong argument in identifying electrons. The muons are as well identified by their shape in the HCAL and K_{0L} can be seen much closer to charged particles.

Precision in position
This improves the e/π rejection by a factor 2 and the μ/π by a factor 3, gives an adequate position of the photons, in particular for the π^0 mass reconstruction and makes possible the reconstruction of the photon direction by measuring the position in different stacks.

4.2.What did we loose?

The measurement of photons below 250 MeV.

Hence, had been any fine spectroscopy to look at like toponium, we would have lost it.

Some e/π separation via p/E ratio but this amounts to a factor 2 which is just what we gain by the position measurement.

Jet resolution by pure calorimetry. The way we designed independent calorimeters for hadrons and electromagnetically interacting particles as well as the choice of non compensating calorimeters made us loose a lot on brute jet resolution. In fact a good deal is gained back by the granularity which gives the possibility to separate particles and use properly the TPC for the charged tracks and the ECAL for photons.

5.Effect of the quality of Aleph calorimeters on some known physics

Line shape and Higgs mass limit[4]. or the use of energy flow . In the measurement of an energy flow we have to consider not only an intrinsic resolution of the calorimeter but also the role of the cracks and the coil, the fluctuation in electromagnetic fraction, the sharing of hadron energy and the nature of the hadrons, the muons. The pure calorimetric resolution for Aleph would be larger than $\frac{1}{\sqrt{E}}$ but the granularity enables the use of the TPC and the separation of the electromagnetic component, remember that the typical fractions of energy are: .58 charged, .27 photons, .15 neutral hadrons. Our accuracy today for energy flow at the Z^0 is equivalent to $\frac{.6}{\sqrt{E}}$ and will go lower.

B physics. The ECAL granularity makes possible an electron identification in jets independent of the p_t relative to the jet giving good efficiency and quite flat acceptance [2.3]. The use of the position should increase the rejection by 2.

	$p_t < 1$ GeV		$p_t > 1$ GeV	
from ECAL only	$\varepsilon_e = .8$	$\varepsilon_\pi = .007$	$\varepsilon_e = .8$	$\varepsilon_\pi = .006$
dE/dx and pair subtraction	$\varepsilon_e = .6$	$\varepsilon_\pi = .001$	$\varepsilon_e = .7$	$\varepsilon_\pi = .0004$

photons and π_0's in jets. The overall γ reconstruction efficiency in jets amounts to more than 70 % above 1 GeV, the loss is due to cracks, threshold at 250 MeV and contamination by charged hadrons. The mean π_0 efficiency is 33 % with a strong dependence on energy at low energy but remains quite flat up to 20 GeV.

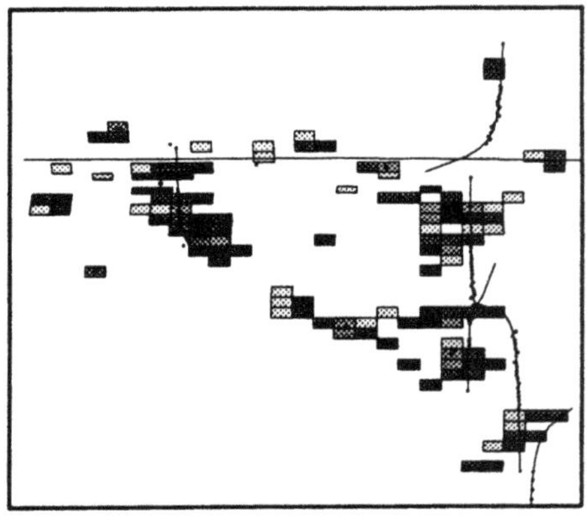

Jet containing 5 charged tracks and 15 γ's seen in θ, ϕ coordinates in a square of .4x.4 rad^2.

τ physics

This is the place where the possibilities of our calorimeters show best. Remark that a τ appears as one charged particle or a very collimated set of particles. Our published[5.] selection efficiency is 73 % due to geometrical acceptance and cut in momentum at 2 GeV, with a background of 2.5%. With such efficiencies the branching fractions can be determined with an accuracy comparable to what is done in experiments at the τ threshold.[6.] The efficiency for reconstructing τ decays in ρ ν is 50 % and for decays in pions has been published[7.] at 38 % but is now close to 60 %. This gives a very neat analysis of the polarisation of the τ and a precise value of $\sin^2\theta_w$. An evident use of the calorimetry is in the analysis of π^0 from ρ, multi π^0's and as well in the separation of π from μ.

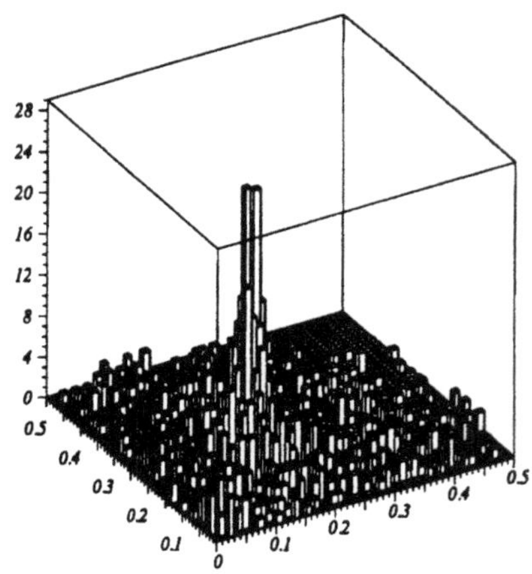

$\gamma\gamma$ mass versus $\gamma\gamma$ mass in τ events with one charged π and 4 γ's in one hemisphere .

$\underline{t\bar{t}}$ would have been missed, but it was not there.

$\underline{\gamma\nu\bar{\nu}}$. A good electromagnetic resolution would help at low energies but there the trigger is anyway swamped with true γ's from $e^+e^-\gamma$ events. The granularity from planes and towers gives us here, not only the shape and position but the angle of the photon equivalent to a precision of $\frac{6cm}{\sqrt{E}}$ on the extrapolated vertex.

6.Conclusion

The construction has been rather tedious and painful.

The quality control has been essential at each step of construction and still is during data taking, but we got the bad numbers at the 1% level or below.

The choice of high granularity versus resolution does not appear as a mistake and did not jeopardized any physics up to now. On the contrary, it gives us a remarkable handle, at reasonable price, on subjects like τ physics. We still did not get all out of it and I know that we could benefit from more granularity.

REFERENCES

1. D. Decamp et alii, *Nuclear Instruments and Methods* **A 294** (1990) 121–178.
2. D. Decamp et alii, *Physics Letters* **B 244** (1990) 551–565.
3. D. Decamp et alii, *Physics Letters* **B 263** (1991) 325–336.
4. D. Decamp et alii, **Searches for New Particles in Z decays Using the Aleph Detector**, *submitted to Physics Reports.*
5. D. Decamp et alii, **Improved measurements of electroweak parameters from Z decays to fermion pair**, *submitted to Zeitschrift für Physik.*
6. D. Decamp et alii, **Measurement of tau branching ratios**, *submitted to Zeitschrift für Physik.*
7. D. Decamp et alii, *Physics Letters* **B 265** (1991) 430–444.

THE DELPHI HPC CALORIMETRIC SYSTEM

PAOLO MORETTINI
Dipartimento di Fisica, Università di Genova, via Dodecaneso 33
16146 Genova, ITALY

ABSTRACT
Different aspects of the DELPHI electromagnetic calorimeter are reviewed after two years of operation at LEP.

1. Introduction

The High Density Projection Chamber[1] (HPC) is the barrel electromagnetic calorimeter of the DELPHI detector. It is a large (208 cm radius, 496 cm length) sampling calorimeter designed to obtain high spatial resolution. The converter used is lead, and the sampling device is TPC-like. Using this techique it is possible to have a good granularity (9 longitudinal samplings, 10 to 30 mrad in the R-ϕ plane, 3 mm along the beam direction) and a full three-dimensional shower reconstruction using a limited number (18432) of electronic channels.

In the following I will analyse the characteristics of the different detector components, the related problems and the solutions used.

2. The mechanics

The complexity of the HPC mechanics is mainly related to the choice of using the lead to produce the electric drift field. To do that the lead has been segmented along the drfit direction in small (1.5 mm) slices. Practically this is obtained glueing lead wires on a vetronite ribbon, and then folding this "lead ribbon" to build the converter–gap structure (we have 8 mm wide gaps alternated to 3.4 mm of lead). The precision needed for the positioning of the lead wires (200 μm, in order to avoid electric field distortions) and the large quantity of ribbon needed to build the detector made necessary the developement of specific tools as well as the use of carefull quality checking procedures before the installation.

Another problem arises from the fact that the HPC is made of 144 indipendent units; even if special care has been used in reducing dead spaces, the cracks cover 7 % of the inner surface of the detector.

3. The Drift System

Beside the electric field, mentioned in the previous section, also the quality of the gas plays an important role in determining the performances of the drift system. The gas mixture chosen for the HPC is Ar (80%) + CH_4 (20%). The drift velocity is sensitive to several gas parameters, like the percentage of water content, the temperature and the pressure of the gas. In the HPC running conditions:

$$\Delta V_{drift} = 1\% \quad \Longleftrightarrow \quad <\Delta z>= 4.5\ mm$$

For this reason an efficient gas monitoring is required, in order to obtain a precise space mesurement.

The gas system is made of 48 independent lines (every line feeds three modules). On each line the temperature, the pressure, the gas composition, the water and oxygen contamination and the flow rate are monitored. Moreover, each line is fed on a reference drift system, providing a drift velocity and a chamber gain measurement. The HPC volume is filled with 25.000 litres of gas; 10 % of the total volume is continuously recycled, while the remaining 90 % passes through a purifier especially designed to remove unwanted pollution. The measurement of V_{drift} we can perform are precise at the level of 1 %. This can be verified looking at the distance between the shower starting point measured by the HPC (z_{HPC}) fitting the energy depositions in the 9 layers and the impact point of the extrapolated TPC track (z_{TPC}). The uncertainty on this measurement comes from the TPC extrapolation erros, from the HPC intrinsic resolution and from the error on V_{drift} in the 144 different modules; we have $\sigma(|Z_{HPC} - Z_{TPC}|) = 6\ mm$

4. The Proportional Chambers

The aim of the HPC is to detect MIP as well as 45 GeV EM showers. The signal produced by the detector is rather clean, even for high energy showers. The noise level is low if compared to a muon signal; the tails can be suppressed, and an efficient noise suppression is feasible. Nevertheless three decades of dynamic range are required to be efficient in muons detection. The main factor limiting the dynamic range turns out to be the gas saturation inside the proportional chambers. The gas saturation can of course be eliminated reducing the chamber operating voltage, but this implies a reduction of the efficiency for MIPs to the level of $\simeq$ 60 %.

Another problem comes from the fact that the HPC chambers show a fast ageing (0.4 % in 24 hours of operations). There are two main sources for this effect:

- The large material surface exposed to the chamber (150 m^2 of material vs 0.3 m^2 of chambers).
- The natural radioactivity which turned out to be rather high in a large fraction of the lead used to build the detector; this source produces an heavy chambers activity (the average current drawn by an HPC chamber is of the order of 60 nA).

5. The Electronics Chain

The HPC electronics performs well: we have 0.1 % of dead channels (over a total of 18432 channels) after two years of operation; the gain and the pedestals remain stable in time (within 1 %) and the level of the noise is very low (0.6 ADC counts). Also, thanks to the tail and noise suppression performed online by the digitization system, the average size of the data logged on tape is small (8000 pixels set over $5 \cdot 10^6$ possible in a typical Z^0 event).

We had stability problems of the data acquisition system, mainly due to intrinsic fastbus limitations in treating large set of boards as single logical crates.

6. The Monitoring and Calibration System

Several thousand parameters must be monitored in order to exploit the HPC performances; for this reason we had to design a sophisticated slow control system, providing a continous monitoring of all the relevant detector parameters, and the storing of the fluctuation in a database system accessible from the offline programs. The same system provides also a fully automatic support for controlling and reporting the detector status and for detecting hardware faults.

A relative pad calibration is usefull both to improve the energy resolution and to monitor the ageing. The technique we use (as proposed by ALEPH[2]) consists in measuring the signal induced by the decay products of Kr_{83}^{m} (essentially two photons of 9 and 32 KeV each). The Krypton, produced by a source of Rb_{83}, can be mixed to the gas and then fed into the detector through the normal gas lines. Four hours of running are sufficient to collect enough statistic to fit the position of the 41 KeV peak in all the 18432 pads. The position of the peak can be measured with a precision of the order of 1 % (Fig. 1–a). An average over the pads gives a relative module calibration; the precision of this measurement is better than 1 %. Adjusting the chamber high voltages, the response of all the 144 modules can be equalized within 2 %.

7. The Software

The complexity of the detector affects the off-line programs in several ways. As far as the Montecarlo is concerned, it is extremely difficult to simulate all the details of the drift and of the chamber amplification. This may become important under particular circumstances (e.g. large gas saturation). Also, the simulation of EM showers is CPU consuming. This implies that it is pratically impossible to generate large MC samples with different detector conditions. The MC always reproduces an ideal and uniform calorimeter.

For the pattern recognition, the main problem is the slightly different behaviour of each HPC unit; the reconstruction must combine the information of all the modules, taking care of the various operating conditions. The presence of cracks makes sometimes difficult to reconstruct properly showers spanning between two modules. Moreover, several factors (the presence of cracks, the material in front, the differences between the modules) may spoil the HPC intrinsic resolution (for 45 GeV electrons coming from Z^0 decay we have $\sigma(E)/E = 7$ % considering the 144 modules, to be compared to a single module test beam resolution of 4.4 % at the same energy).

The PR program is essentially made up of two parts:

- A first pass, using the HPC information alone, is devoted to electrons and photons reconstruction.

- A second pass, combining HPC and TPC information, reconstructs MIPs and hadronic showers, and identifies charged and neutral showers.

The particular structure of the detector has required the development of "ad-hoc" algorithms to exploit the granularity. Tipical examples are the electron identification (Fig. 1–b) performed analysing the three-dimensional structure of the shower, and the π^0 reconstrucion, that we are able to perform not only when the two gammas are completely separaded, but also, thanks the spatial resolution, in case of partial overlapping.

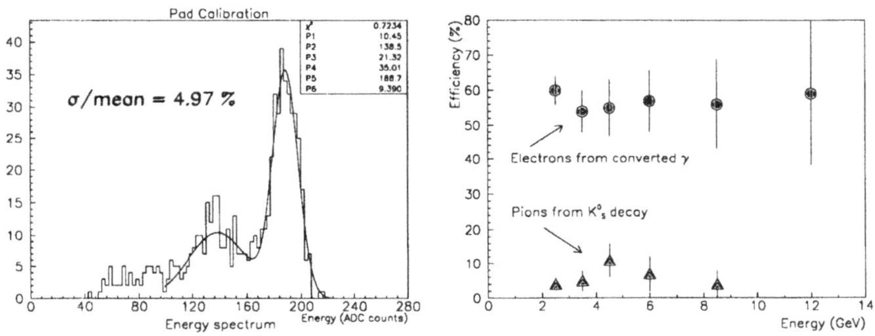

a) b)

Fig. 1 – a) Energy spectrum of Kr^m_{83} decay products as mesured in an HPC pad. b) HPC identification efficiency for electron coming from γ conversion inside a quark jet

8. Conclusions

The HPC is a complex and new device. To exploit its granularity we need to control several thousands of parameters. Mechanical constraints, inherent to the large size of the detector, affect its global behaviour.

The present knowledge of the detector has guaranteed satisfactory results in the physics analysis of the first running periods, but further studies will improve the HPC performances.

9. Acknowlegments

I would like to thank the whole HPC group for the support in collecting and organising the material needed to prepare this report.

10. References

1. DELPHI Technical Proposal, *CERN/LEPC* 83-3 (1983)
   DELPHI Progress Report, *CERN/LEPC* 84-16 (1984)
   DELPHI Collaboration, *Nucl. Instr. and Meth.* **A303** (1991) 233.
2. ALEPH Collaboration, *Nucl. Instr. and Meth.* **A294** (1990) 121.

DESIGN, REALIZATION AND PERFORMANCE OF THE L3 BGO ELECTROMAGNETIC CALORIMETER

FERNANDO FERRONI
Dipartimento di Fisica, Università di Roma "La Sapienza"
INFN, Sezione di Roma, Roma, Italy.

ABSTRACT

The experiment L3 at LEP is using a unique electromagnetic calorimeter consisting of bismuth germanate (BGO) crystals. The performances achieved in the daily running of the detector are discussed in this paper. We recall the design considerations and the realization steps which made possible the operation of this very performant calorimeter at LEP.

1. Introduction

L3[1] is one of the four LEP experiments at CERN. It is a detector concentrating its efforts on measuring with the best possible resolution electrons, photons and muons.

It consists of a 7800t solenoidal magnet providing a 0.5T field containing all the detectors. From the beam axis toward outside one finds a small central detector based on *time expansion principle* used to track charged particles, the BGO electromagnetic calorimeter followed by an hadron calorimeter made of depleted uranium plates interspersed with proportional chambers. Muons are measured by a set of large drift chambers placed in the empty volume between calorimeters and magnet.

In this paper the BGO performances are discussed. We first remind the design considerations and expectations (section 2) , then we review in section 3 the key issues in the realization process crucial to the calorimeter resolution and finally we discuss in section 4 the actual performances of the calorimeter both in single particle measurement and in its ability to discriminate electromagnetic objets from hadronic ones. A first look to the behaviour of the recently installed end caps is also given (section 5)

2. Design considerations

At the time of L3 proposal[2] the main requirements to the electromagnetic calorimeter were very simple:

a to measure at the best e/γ from 50 MeV to 50 GeV;
b to have good angular resolution for γ down to 50 MeV;
c to allow an efficient finding of e/γ showers in jets.

These requests were dictated by the physics at that time considered of **highest** priority, namely:

i $e^+e^- \rightarrow H^0Z^0(\rightarrow e^+e^-)$;

ii $e^+e^- \rightarrow' new'(\rightarrow e^+e^-)$;

iii Toponium, with its radiative decays.

The equivalent emphasis given to the muon resolution resulted in the request of a very compact calorimeter ($\leq$ 35 cm) having $\sigma(E)/E \leq 1\%$ and $\sigma(x/y) < 5$ mm.

Only a material looked attractive at that time for building such a calorimeter: BGO[3]. Table 1 shows the main properties of the material. Amongst them the most important is the short X_0 = 1.1 cm and the correspondingly small Moliére radius.

Table 1: BGO properties

Parameter		BGO
Density	(g/cm³)	7.13
Hardness	(Mho)	5
Melting point	(°C)	1050
Radiation length	(cm)	1.13
Absorbtion length	(cm)	22
Critical energy	(MeV)	8.8
Moliere radius	(cm)	2.7
(dE/dx)	(MeV/cm)	9.2
$< \lambda_{scint} >$	(nm)	480
Decay constant at 20°C	(ns)	300
Photons/MeV		10^4
Temperature gradient (%/°C)		-1.55
Index of refraction		2.19

The calorimeter was designed in two different parts, the central (barrel) and the forward-backward one (end cap). The barrel is composed by 7680 crystals arranged in 48 ϑ times 160 ϕ cells, in such a way that a crystal covers a solid angle of 2° * 2° at 50 cm distance from the beam axis. The crystals have been chosen 24 cm long (22 X_0) for a good containment of even the most energetic showers. They are tapered for keeping a pointing geometry. The solid angle covered by the barrel extends from 42° to 138° in ϑ.

The end cap system is composed by two symmetric halves each made of 1536 crystals. With this system the solid angle covered by the electromagnetic calorimeter extends down to 11° in ϑ. The geometry is somewhat complicated and the crystals are no longer very similar each other. The system is in place since the 1991 run.

3. Realization of the system

All the crystals were produced by the Shangai Insitute of Ceramics. The crystals were cut and polished in order to meet the stringent tolerance required ($\leq 300\mu m$ and soon after shipped to CERN. Once in CERN the crystals underwent an extensive series of acceptance tests. The main steps were: visual inspection, transparency measurement, dimension control, light output level check and, most important of them, uniformity in light collection measurement[4].

Due to the tapered shape, the light emitted at the end far from the photon detector is strongly favoured in the collection process respect to the one emitted close to it. Such a non uniformity (a factor 2 nearly) is not tolerable in a device of such a potential precision since the non linearity in the response would consequently affects the energy resolution . The effect was corrected by painting all the faces of the crystals (exception made for the one where the photodetector is attached) with a white diffusing paint.

At the end of this quality control chain all the curable cases were fixed, the others (very few indeed) rejected.

At this point all the crystals were precalibrated in a cosmic ray bench.

Another crucial issue for minimizing the resolution is the use of a mechanical structure supporting the crystals as *virtual* as possible. The gap between two adjacent crystals has to be minimized as weel as the amount of material in front of the calorimeter. A carbon fiber structure of a total weight of 250 kg (for 10 tons of BGO !) did the job. The wall between crystals is of 200μ m of carbon. The total solid angle lost would be of 1.5% but in effect it is much less since the calorimeter is tilted by 10mrad in ϕ to avoid particles entering and continuing their path along a gap.

Of the utmost importance is the Xe-lamp monitoring system. A calorimeter which is build once, calibrated for two years and which is supposed to work for at least a decade ought to have a system allowing a fast, continuous and reliable check of any single element. Xe system is used to determine on daily basis the actual number of faulty channels and to correct the crystal to crystal calibration fluctuations.

Finally the hope of measuring energies with a precision of 1% or better imposes a calibration of all the individual cells of the detector with a fairly better precision.

L3 BGO group spent few years on electron beams (200 MeV< E< 50 GeV) in order to master calibration methods[5] and to actually measure calibration constants, their stability and transportability. We remind the reader that any care put in any previous step could have been washed out at this stage. The results of the calibration in terms of energy resolution are summarized in fig. 1.

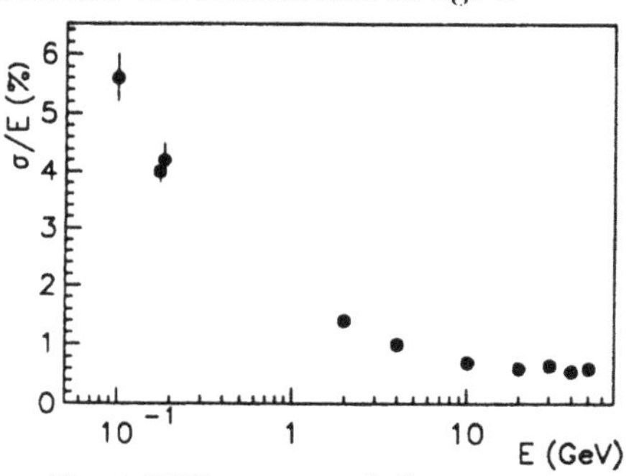

Fig. 1 BGO energy resolution.

4. Performances at LEP

The basic elements of the calorimeter response can be checked at LEP making use of simple events like Bhabha's. High energy resolution can certainly be studied this way.

The 'low energy' performances can be somehow inferred reconstructing the invariant masses of π^0 and η.

The capability of hadronic rejection is measured in the selection of $Z^0 \to b\bar{b}$ events, followed by a $b \to eX$ decay.

From large angle Bhabha events we measure an energy resolution (fig. 2) of 1.2% and a position resolution of 600μ m. The observed value is the convolution of the single crystal resolution measured in the test beam (0.5%) with the calibration constant dispersion (0.5%), the natural width of the line due to radiation (0.6%), the flat illumination of the entire crystal face (0.5%) and the Xe recalibration intrinsic precision (0.6%). Xe lamp system is vital at this step, since it tracks a very slow but continuous decay (-1%/year) of the BGO signal presumably due to ageing effects on paint or photodiode glue.

In figure 3 you see the π^0 invariant mass reconstructed[6] in the BGO calorimeter. Th energy range used is (700 MeV< E_{π^0} < 6 GeV). This corresponds to 65% of the potential acceptance. In this interval the reconstruction efficiency ranges from 6% to 2%. The invariant mass spectrum shows a 7 MeV wide peak, in fair agreement with MonteCarlo expectation.

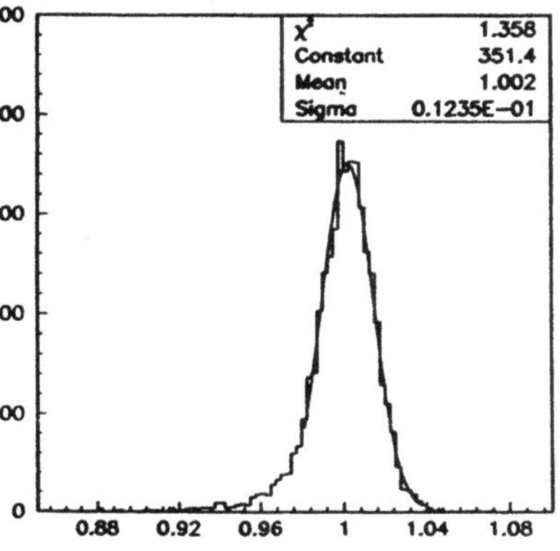

Fig. 2 Bhabha electron energy divided by beam energy.

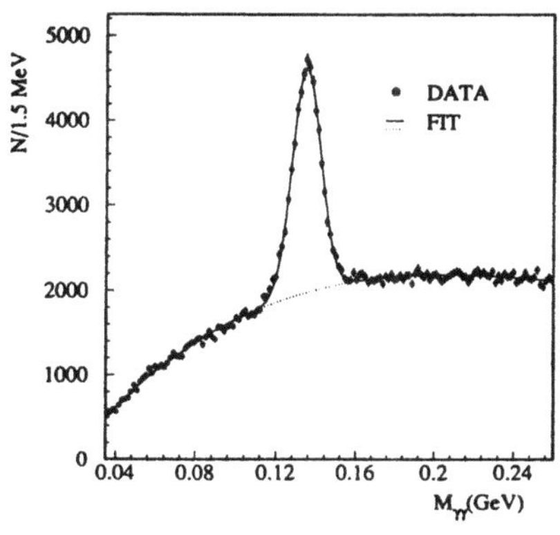

Fig. 3 π^0 invariant mass.

The search for $b\bar{b}$ events calls for a good electron identification in jets. L3 makes use of an algorithm almost entirely based on the electromagnetic calorimeter. The main request is that when a charged track points to an energy cluster in BGO we control that the energy deposited in a matrix of 3 times 3 crystals (Σ_9) centered around the most energetic crystal is consistent with that deposited in an enlarged 5 by 5 (Σ_{25}) one in the hypothesys of an electromagnetic shower. The algorithm is quite powerful in terms of rejection. The resolution on the ratio (Σ_9/Σ_{25}) is infact better than the energy resolution itself, representing only the fluctuations on the 4% energy contained in $(\Sigma_{25} - \Sigma_9)$. Purity higher than 90% are achieved for $p_e > 3$ GeV with efficiencies ranking around 20%.

5. End caps status

The end cap system is operative since the beginning of this year run. The crystals mounted in the end caps were not calibrated, mainly for lack of time, before installation. A combination of physics event (Bhabha, whose rate is abundant in the forward region) and Xe system is used to calibrate the crystals *in situ*. Bhabha events fix the scale and Xe system gives the intercalibration between channels illuminated by the same mixer. In figure 4 you see how things improved from the beginning (5% due to cosmic bench measurement) to three months later (1.5% already achieved).

This procedure somehow anticipates the LHC scenario, where electron pairs could be used to calibrate the calorimeters.

Another important use of the Xe system in the end cap detector is to measure and correct for the effect of radiation damage in this hot region. Differences up to 5% are observed in the crystal ring closer to the beam pipe in average, local effects up to 20% have also been spot and corrected.

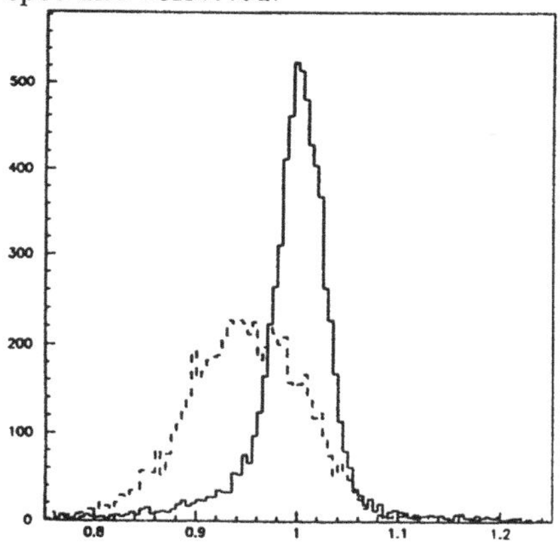

Fig. 4 End cap energy resolution in Bhabha events. Dashed line with cosmic ray precalibration, solid line after Bhabha plus Xe calibration.

6. Conclusion

The L3 BGO calorimeter is running since more than two years at LEP. Its performances are very close to the design ones. This achievement is due to a combination of three factors; the care put at the realization stage, the complete calibration of the system and the very efficient monitoring system.

7. Acknowledgements

I wish to thank my collegues M.N. Kienzle and J. Wenninger who provided me with a lot of detailed information on BGO actual performances.

8. References

1. B. Adeva et al., *NIM* **A289** (1990) 35.
2. L3 Collaboration, *Technical proposal* CERN/LEPC 83-5 (1983).
3. R. Nitsche, *J. Appl. Phys.* **36B** (1965) 2358.
4. M. Schneegans, *NIM* **A257** (1987) 528.
5. J.A. Bakken et al., *NIM* **A275** (1989) 25.
6. B. Adeva et al., *Phys. Lett.* **259B** (1991) 199.

REALIZATION OF OPAL ELECTROMAGNETIC BARREL CALORIMETER

Tomio Kobayashi
ICEPP, University of Tokyo
7-3-1 Hongo, Bunkyo-ku, Tokyo, 113, Japan

Abstract

The barrel part of OPAL electromagnetic calorimeter, which consists of 9440 lead glass counters, has worked well from the start of LEP till now, and has played a crucial role in many analyses of OPAL data. Our experience on the design, construction and performance of the detector is described.

1. Introduction

The general concept of the OPAL detector[1] (*Omni-Purpose Apparatus for LEP*) dates back to 1980. The main features of OPAL are a homogeneous coverage of large solid angle with a high resolution tracking detector in a solenoid and a high resolution electromagnetic shower calorimeter as well as a hadron calorimeter and a muon detector. A special emphasis was put on using established techniques for the detector, so that one could achieve anticipated performance and be ready for the first operation of LEP.

The barrel part of electromagnetic calorimeter, which was a responsibility of the University of Tokyo, is made of lead glass Čerenkov counters (9440 pieces in total). Our first beam test of lead glass counters for OPAL took place in 1981, while the conceptual design was still ongoing. After the approval of OPAL by the LEP Committee in 1982, the technical design and the beam tests continued until 1985. The actual construction took place from October 1985 to May 1987. We calibrated all the counters at least twice between 1986 and 1988. It took about 3 months for the installation of the OPAL electromagnetic barrel calorimeter in the LEP underground experimental hall. Since the observation of the first Z^0 event at LEP in August 1989 by our detector, all the 9440 lead glass counters have worked well, and play an important role for most OPAL analyses.

In the following, I describe briefly how we realized OPAL electromagnetic barrel calorimeter, including the design concept, construction and performance of the detector. Some more details about the OPAL detector are described elsewhere[2].

2. Design

We chose lead glass counters for the OPAL electromagnetic calorimeter because of their excellent shower energy resolution, uniform coverage, ease of operation, stability and reliability. We attached particular importance to the energy resolution, because it is crucial for the Higgs particle search in e^+e^- reactions such as the toponium decaying into a photon + Higgs or the Higgs production associated with an e^+e^- pair.

As we used photomultiplier tubes, the barrel part of the electromagnetic calorimeter had be placed outside the coil. In order to limit the degradation of energy resolution caused by the material of the coil, a presampling device was installed immediately in front of the lead

Table 1: Parameters of OPAL barrel lead glass counters

Lead glass	Material	SF57 (Schott)
	Density & rad. length	$\rho = 5.54$ g/cm^3, $X_0 = 1.50$ cm
	Block dimensions	~(10cm × 10cm) × 37cm(24.6X_0)
	Angular coverage	$\lvert\cos\theta\rvert < 0.82$
	Block geometry	semi-pointing ($\Delta\varphi = 0.6°$, $\Delta\theta = 1.1°$)
Photomultiplier tube	Type	R2238 (Hamamatsu), 3 inch diameter
	Photocathode	bialkali, Q.E. $\simeq$ 20% at 420nm
	Dynode	12-stage proximity mesh
Gain monitoring system	Xe flash lamp	L2360 (Hamamatsu)
	Optical fiber	quartz, step-index (Fujikura)
	PIN photodiode	S1722-02 (Hamamatsu)
	Light pulse output	20~30 GeV shower equivalent

glass counters. A pointing geometry was adopted for the configuration of lead glass blocks, of which advantage was considered to be a better electron identification and π^0/η detection.

We then optimized the design of lead glass counters by considering the following points backed up by beam tests. First, the material and length of the lead glass block were determined so as to minimize the shower leakage and achieve a good energy resolution and linearity up to 100 GeV within a limited space. Next, the lateral size of the block was fixed by considering the position resolution, the diameter of photomultiplier tubes and the total number of blocks. We tried also to have good uniformity and e/π discrimination. The characteristic parameters of the lead glass counter are listed in table 1.

The electromagnetic barrel calorimeter consists of a cylindrical array of 9440 lead glass counters, whose mechanical structure is illustrated in fig. 1 (only one half is shown). All the lead glass blocks are slightly tilted away from a perfectly pointing geometry to prevent photons from escaping through the gaps between the blocks.

Each half ring contains a gain monitoring system for the counters, which is composed of a xenon flash lamp and optical fibers. Some properties of the monitoring system are listed in table 1. The light intensity of the xenon flash lamp is monitored by a PIN photodiode.

3. Construction

In constructing a large detector system, it is necessary to pay attention to the quality control, labor efficiency and safety. Some key points on the construction of our calorimeter are given below.

We decided to glue photomultiplier tubes on to the lead glass blocks for the sake of mechanical stability. It was therefore necessary to make repeated checks on the photomultiplier tubes before gluing. The lead glass block was supported on the mechanical structure through a stainless steel flange, which was glued to the lead glass block, so that we had to take special care on the temperature change as well as the mechanical shock.

Since the assembly work of the lead glass counters was a mass production effort, we tried to save labor and simplify construction techniques. We asked the companies to glue light guides on photomultiplier tubes and to glue flanges on lead glass blocks as well as the wrapping of the lead glass blocks for the optical isolation.

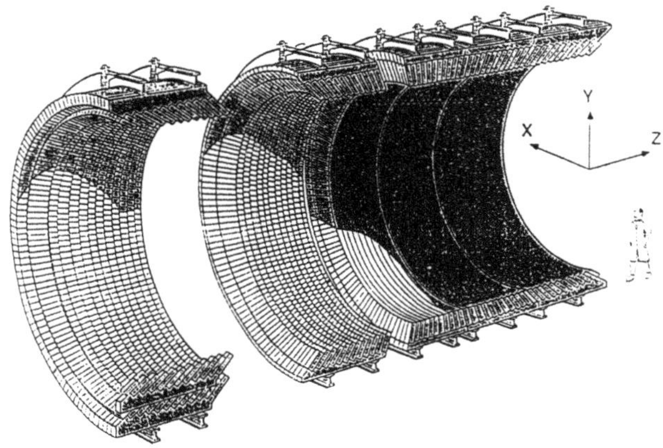

Figure 1: OPAL electromagnetic barrel calorimeter

Due to the pointing geometry, the gap between lead glass blocks was aimed to be as small as 1mm including the thickness of wrapping foil and adhesive tapes, which left only ~0.3mm for the actual play. Hence the mechanical tolerance of the support mechanics was required to be kept within 0.1mm typically.

Special care was taken on the optical fibers of the gain monitoring system. Because the light transmission of the fibers is very sensitive to their bending radius, the fibers should be well fixed to the support structure with the bending radius as large as possible.

We made a gain calibration for all the counters in a 50 GeV electron beam. For the sake of stability of the gain monitoring system as well as for saving time, the calibration was carried out for complete half ring units. This obliged us to build a special device, which held the half ring and aligned any given counter perpendicular to the beam line. The whole calibration procedure was fully automatic.

4. Performance

4.1. Beam Test Results

The energy resolution of the lead glass counters was measured to be $\sigma_E/E(\%) = 6.3/\sqrt{E} + 0.2$ (E in GeV) without material in front. See ref. 2 on the degradation of the energy resolution by a material and its recovery by the presampler. The linearity of the counters was found to be much better than 1% over the entire energy range of LEP.

The performance of the gain monitoring system was studied during the beam calibration. The counter gains could be followed by the gain monitoring system with an accuracy of 0.2% in a short term (~2 days), and better than 1% in a longer term (~1 year).

4.2. LEP Physics Analyses

Fig. 2 is a plot that has been used to select the charged lepton pair channels from the Z^0 decays for the Z^0 line shape analysis[3]. It is a scatter plot of energy sum in the electromagnetic

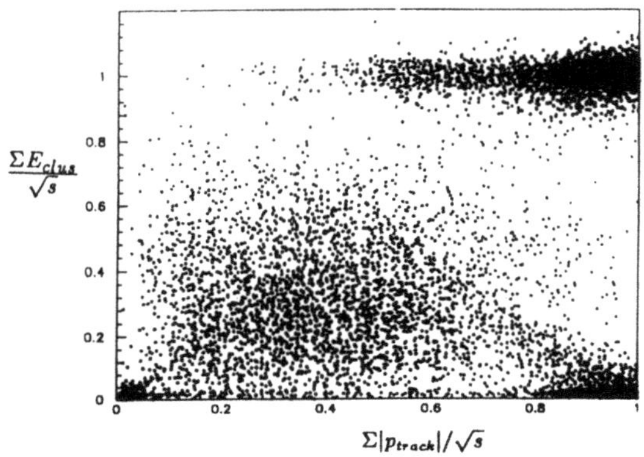

Figure 2: Plot used for $Z^0 \rightarrow l^+l^-$ event selection

calorimeter and momentum sum of the charged tracks in the angular region, $|\cos\theta| < 0.7$, after the rejection of some background and multihadronic events, where both quantities are normalized to the center of mass energy ($\sqrt{s}$). The three leptonic channels (e^+e^-, $\mu^+\mu^-$, $\tau^+\tau^-$) and the two-photon backgrounds are clearly separated each other. Especially, the e^+e^- channel is well separated from $\tau^+\tau^-$ channel owing to the good energy resolution of the lead glass counters.

In the tau polarization analysis[4], the electromagnetic calorimeter played an important role to select $\tau \rightarrow e\nu\bar{\nu}$ and $\tau \rightarrow \pi\nu$ channels using a quantity E/p, i.e., comparing the energy and momentum of a particle. This quantity is also quite useful to the electron identification for the analysis of $Z^0 \rightarrow b\bar{b}$. For the reactions $e^+e^- \rightarrow \gamma\gamma$ or $\gamma\gamma\gamma$, the electromagnetic calorimeter was indispensable[5]. The low noise of the electromagnetic barrel calorimeter, and hence the low trigger threshold, made it possible to detect single photons from the reaction $e^+e^- \rightarrow \gamma\nu\bar{\nu}$[6].

5. Conclusion

Since the start of LEP, the OPAL electromagnetic barrel calorimeter has worked well, with the expected high performance, and has played an important role in the physics analyses. For the construction of a large detector system like ours, or for even larger system, mass production techniques are required. It would be advisable to pay attention to quality control, labor efficiency and safety; in other words, build a "factory"-like setup.

References

1. OPAL Letter of Intent, CERN/LEPC/82-4; Technical Proposal, CERN/LEPC/83-4.
2. K. Ahmet et al., *Nucl. Instr. and Meth.* **A305** (1991) 275.
3. G. Alexander et al., CERN-PPE/91-67, submitted to *Z. Phys. C.*
4. G. Alexander et al., CERN-PPE/91-103, submitted to *Phys. Lett.*
5. M.Z. Akrawy et al., *Phys. Lett.* **B257** (1991) 531.
6. M.Z. Akrawy et al., *Z. Phys.* **C50** (1991) 373.

THE LIMITED STREAMER TUBES SYSTEM FOR THE SLD WARM IRON CALORIMETER

A.C. BENVENUTI

INFN, Sezione di Bologna, I-40126 Bologna,Italy

B. CAMANZI, L. PIEMONTESE, P. ZUCCHELLI

INFN, Sezione di Ferrara and Universitá di Ferrara, I-44100 Ferrara, Italy

A. CALCATERRA, R. DE SANGRO, P. DE SIMONE, S. DE SIMONE
M. GALLINARO, I. PERUZZI and M. PICCOLO

Lab. Nazionali di Frascati dell'INFN, I-00044 Frascati, Italy

P.N. BURROWS, W. BUSZA, S.L. CARTWRIGHT, S. FUESS, S. GONZALEZ,
T. HANSL-KOZANECKA, A .LATH, T. LYONS, L.S. OSBORNE, L.ROSENSON,
U. SCHNEEKLOTH, F.E.TAYLOR, R. VERDIER, D.C. WILLIAMS and J.M. YAMARTINO

Massachusetts Institute of Technology, Cambridge, MA 02139, USA

N. BACCHETTA, D. BISELLO, A. CASTRO, S. GALVAGNI, M. LORETI,
L. PESCARA and J. WYSS

INFN, Sezione di Padova and Universitá di Padova, I-35100 Padova, Italy

R. BATTISTON, M. BIASINI, G.M. BILEI, B. CHECCUCCI, G. MANCINELLI,
G. MANTOVANI, M. PAULUZZI, A. SANTOCCHIA and L. SERVOLI

INFN, Sezione di Perugia and Universitá di Perugia, I-06100 Perugia, Italy

M. CARPINELLI, R. CASTALDI, U. CAZZOLA , R. DELL'ORSO
E. PIERONI, C. VANNINI and P.G. VERDINI

INFN Sezione di Pisa and Universitá di Pisa, I-56010 S.Piero a Grado, Italy

B.L. BYERS, J. ESCALERA, D. KHARAKH, R.L. MESSNER and R.W. ZDARKO

Stanford Linear Accelerator Center, Stanford University, Stanford, Ca 94309, USA

J.R. JOHNSON

University of Wisconsin, Madison, WI 53706, USA

Presented by G.M. Bilei

ABSTRACT

The SLD detector at the Stanford Linear Accelerator Center is a general purpose device for studying $e^+ e^-$ interaction at Z^0. The SLD calorimeter system consists of two parts: a lead Liquid Argon Calorimeter (LAC) with both electromagnetic (22 radiation lengths) and hadronic sections (2.8 absorption lengths) housed inside the coil and the Warm Iron limited streamer tubes Calorimeter (WIC) outside the coil which uses as radiator the iron of the flux return for the magnetic field. The WIC completes the measurement of the hadronic shower energy ($\sim$ 85% on average is contained in the LAC) and it provides identification and tracking for muons on 99% of solid angle.

In this note we report on the construction, test and commissioning of such a large system.

1. Construction and test

The WIC surrounding the magnet, consists of eight iron barrel sections and two end-caps which complete the solid angle coverage [1]. The iron is segmented into 14 layers 5 cm thick, separated by 3.2 cm gaps instrumented with plastic limited streamer tubes chambers [2] (see Fig.1).

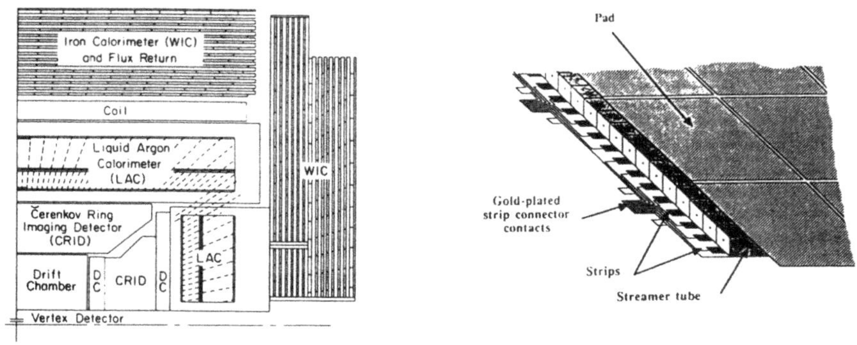

Fig.1 One quadrant of a cross-section of SLD detector and the streamer tube chamber

This type of detector has been chosen because it offers several advantages: it allows x&y strips and pad readout, the signals are comfortably large (~ 8 pC from a strip), the modules are robust and rather easy to produce and to assemble, it has fully efficiency for minimum ionizing particle and it is reasonably inexpensive.

A chamber consists of a single layer of streamer tubes sandwiched between two external readout electrode sheets, which pick up the signals induced by streamer discharge. The bottom sheet is composed of long narrow strips parallel to the individual tubes (1 cm pitch). Signals on these strips are recorded as digital hits by a low cost discriminator/shift register electronics. The opposing electrode sheet is subdivided by quadrilateral pads. Pads of different layers are daisy chained to form projective towers pointing to the interaction region, and the collected signals are recorded through analog amplifiers and digitized.

Quality control was the main issue for such a large system. Besides the high voltage testing described below, all critical parts of a chamber were checked carefully for quality at each step of manufacturing and assembly. The large number of components (~ 8700 modules required) together with the inaccessibility of modules after the installation into the SLD iron gaps, were critical points. The devices had to survive inside the iron for many years with as few failures as possible. It was crucial to have a device which could be produced on a large scale with high reliability. Thus it was decided the modules had to undergo through a series of stringent high voltage tests during the construction, after the chamber assembly and before the final installation. The whole testing procedure required several days [3].

Preliminary tests on the modules indicated that there is a strong correlation between eventual failure and earlier abnormal behaviour of the anode current (higher than normal values either continuously or in bursts). We established testing procedures which, on the basis of such short term behaviour, could weed out devices which had inadequate longevity. During these tests anode currents were monitored, and we applied a stringent set of acceptance criteria based on the results. The overall module rejection rate was $\sim$ 30 %. Some of these tests included:

- High Voltage Conditioning (ramping up–down to 4.9 KV for 3 days).
- Preacceptance (applying a constant 4.8 KV high voltage for 6 days and checking the total current of a group of modules every 12 hours).
- Acceptance (applying a constant 4.8 KV high voltage for 3 more days and checking each individual module current continuously).

Finally we set–up a longevity test which gave us confidence that the adopted acceptance criteria were good. The test gave the reassuring result that the modules we were building and selecting would fail at a measured rate of less that 1.8 % per year at 90% confidence level. In addition, every completed chamber was tested with cosmic rays under operating conditions of gas and high voltage to ensure the correct operation of the finished detector.

2. Electronics

The WIC electronics is designed to acquire digital data from $\sim$ 80,000 strip channels and $\sim$ 8,600 calorimetric towers.

A block diagram of the digital readout system [4] is shown in Fig. 2. Thirtytwo channels of preamplifiers, discriminators, one–shots and shift–registers are packaged in a single board (by SGS–Thompson), and all of the boards of each layer of the detector (typically 10 boards/layer) are daisy–chained together. Subsequently, groups of nine layers (half octant) are further multiplexed by an intermediate controller (Splitter board), and up to eight Splitter boards (half barrel) are read by one of the six WIC Digital Read–out Fastbus Modules. The overall read–out time is $\sim$ 0.8 ms.

Data from the detector is heavily multiplexed and transmitted by optical fibers leading to a very small cable plant. All 40,000 barrel strips signals are carried on 36 fibers (control and trigger included) with a cable cross section less than 10 cm^2. As most of all the other SLD subsystems, we have redundant control and data fibers and also redundant control logic in the controllers inside the detector. In this way, during a run a single controller or a cable failure can be isolated switching (software controlled procedures) to a different path, avoiding the loss of important tracking information in any solid angle.

The small space available for the installation of the SGS front-end electronics inside the iron together with cables and many connectors, made the commissioning a heavy task. Nevertheless today after about 1 year of operation the number of dead and noisy channels is only ~ 1 %.

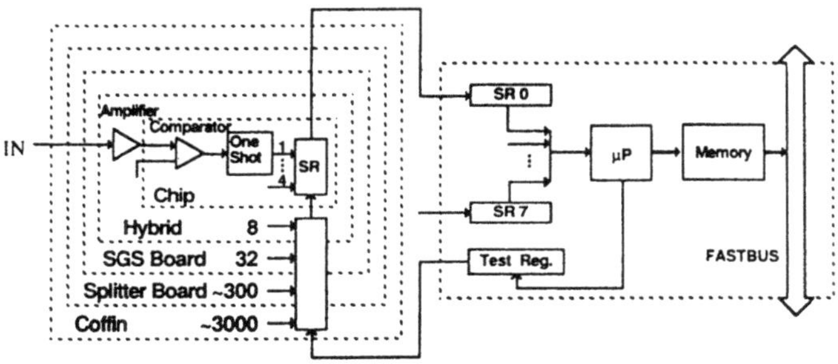

Fig.2 The strip digital electronics data chain

For the calorimetry read-out [5], the signals from the towers are sent via shielded cable to 64 VME cards, mounted on the periphery of detector in 16 very easily accessible VME crates. The signal induced on a tower is integrated with a 5 μs time costant by a hybrid preamplifier, read by a sample-and-hold circuit and digitized by an ADC (see Fig.3). The digitized signals are then serially sent through fiber-optic cables to two Fastbus Calorimetry Data Modules. The pad system is very stable, well integrated, with a high degree data multiplexing and supported by a powerful menu-driven, user friendly, diagnostic software. During the commissioning these tools allowed us to track-down and dramatically reduce the number of problems such as shorted pads, loose connectors and bad channel hybrids. Today the number of unrecoverable pad channels is just ~ 0.3 % and it's stable.

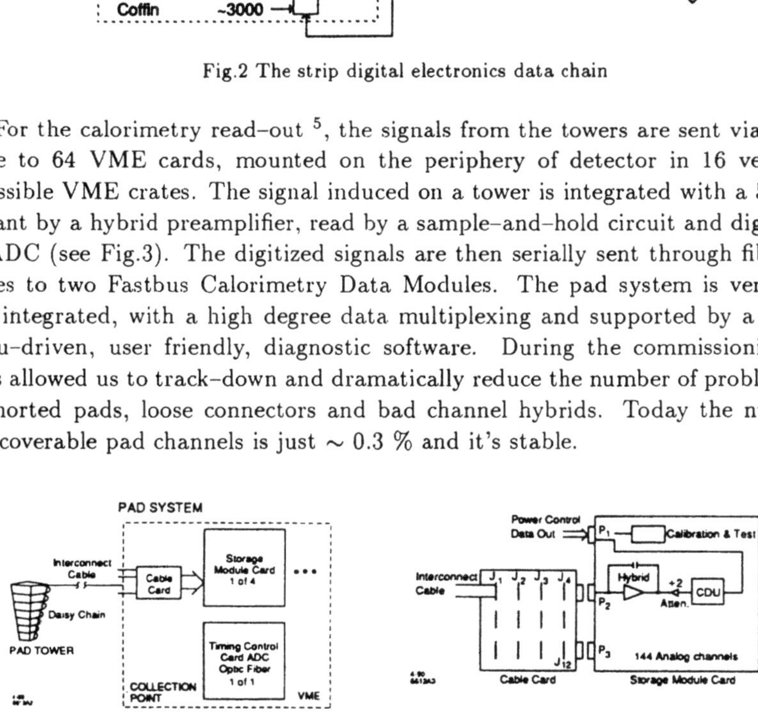

Fig 3. The pad electronics block diagram

3. WIC status

Results from the test beam [6] have confirmed that the energy resolution is about $\sigma_E/E = 80\%/\sqrt{E}$ and is in agreement with the expected performcance.

The detector is completely debugged and fully operational since ~ 1 year. Cosmic ray runs have been taken routinely to debug the whole apparatus and study its performance [7] using a new nonflammable gas mixture [8] consisting of 88 % CO_2, 9.5% isobutane and 2.5% argon. The fraction of modules that have been disconnected from the HV up to date is ~ 2.5 %. This number is fully compatible with the life test results.

In the SLD engineering run during the summer of 1991, the WIC demonstrated to be in excellent shape and it performed as expected in the proposal, both as a calorimeter and as a muon identification and tracking device. Nearly 400 Z^0 have been logged on tape and data analysis is in progress.

4. Conclusions

The Warm Iron Calorimeter built for SLD is fully operational since ~ 1 year and it successfully participated to this summer physics run. Overall it works well and it's easy to operate and monitor.

References

1. SLD Design Report, *SLAC Report SLAC-273* (1984).
2. A.C. Benvenuti et al., *Nucl. Instr. and Meth.* **A290** (1990) 353.
3. A.C. Benvenuti et al., *Nucl. Instr. and Meth.* **A276** (1989) 94.
4. A.C. Benvenuti et al., *Nucl. Instr. and Meth.* **A289** (1990) 577.
5. P.N. Burrows et al., *SLAC-PUB-5305*, 1990.
6. G.Callegari et al., *IEEE Trans. Nucl. Sci.* NS-**33** (1986) 201.
7. A.C. Benvenuti et al., *SLAC-PUB-5332* 1990.
8. A.C. Benvenuti et al., *Nucl. Instr. and Meth.* **A284** (1989) 339.

The Uranium Liquid Argon Calorimeter of the DØ Experiment: Experience in Realizing a Large System

Włodek Guryn

Brookhaven National Laboratory
Upton, New York 11973
USA
for the DØ Collaboration*

ABSTRACT

The major aspects in realizing the calorimeter system of the DØ experiment are discussed. They include: technologies developed for calorimeter production, schedule, and experience with module production.

This manuscript has been authored under contract number DE-AC02-76CHO-0016 with the U.S. Department of Energy. Accordingly, the U.S. Government retains a non-exclusive, royalty-free license to publish or reproduce the published form of this contribution, or allow others to do so, for U.S. Government purposes.

* The DØ collaboration includes: University of Arizona, Brookhaven National Laboratory, Brown University, University of California - Riverside, Columbia University, Fermi National Accelerator Laboratory, Florida State University, University of Florida, University of Hawaii, Indiana University, Lawrence Berkeley Laboratory, University of Maryland, University of Michigan, Michigan State University, New York University, Northern Illinois University, Northwestern University, University of Rochester, CEN Saclay France, Institute for High Energy Physics Protvino USSR, State University of New York at Stony Brook, Tata Institute of Fundamental Research Bombay India, Texas A&M University and Yale University.

Introduction

The DØ experiment is one of the two major detectors at Fermilab's Tevatron collider, and DØ is scheduled to take its first data in spring 1992. Proposed and approved in 1983, the DØ detector was designed to emphasize good lepton identification and jet energy measurement. As a result, calorimetry has been the focus of this experiment.

The DØ calorimeter covers the rapidity range $-4 < \eta < 4$ and the full azimuthal angle. Depending on rapidity, the thickness of the calorimeter ranges from 8-10 interaction lengths.

The calorimeter is divided into three: the central calorimeter (CC) and the two end calorimeters (EC). Each of these calorimeters consists of three types of modules: the electromagnetic modules (EM), the fine hadronic modules (FH), and the coarse hadronic modules (CH). The major differences among the module types are in absorber-plate materials and thicknesses: 3 mm and 4 mm uranium plates were used for the EM modules in the CC and EC, respectively; 6 mm plates of uranium-niobium alloy, with 1.5% Nb, were used in the FH. The CH sections have 46.4 mm copper plates in the CC, and stainless steel plates of the same thickness in the EC.

With the gap structure common to all modules, the unit sampling cell is composed of: an absorber plate, 2.3 mm liquid argon (LAr) gap, signal board, and another LAr gap.

From design through construction, the DØ calorimeter cost \$22.8 million. Installation added another \$2.4 million and electronics another \$2.8 million to the cost, making the total \$28.0 million. More details on the calorimeter may be found elsewhere [1–7].

Technologies

While LAr calorimeters have been built before, the DØ calorimeter is one of the largest built in terms of size and the number of channels. The total weight of the three calorimeters is 900 tons, and the number of electronics channels is about 50,000.

The calorimeter's requirements of 4π geometry and good missing transverse energy resolution challenged the design to minimize dead areas. As a result, physics and engineering considerations placed many new constraints on well known LAr technology.

For example, because of the lack of access to the calorimeters essentially for the duration of the DØ experiment, no major repairs can be made during data taking. Therefore, the system had to be designed and built robustly. In addition, though the modules operate at LAr temperatures, they were built and tested at room temperature; hence room-temperature testing had to be developed that would assure that the modules would work at LAr temperatures without repair for long periods.

To simplify construction of various types of calorimeter modules, common technologies were employed where possible. Besides the calorimeter components made by industry such as calorimeter plates, connectors, cables, and cryostats, the following technologies and methods were developed within DØ:

1. **Resistive coating.** To provide a high-voltage electrode, the surface of the signal boards was screen-printed with a resistive coating composed of carbon-loaded epoxy. This arrangement allowed the blocking capacitance to be built into the board structure, therefore eliminating the need for space-consuming high-voltage blocking capacitors. The resistivity (40 megohms/square at room temperature) was chosen to minimize the cross talk between adjacent channels and the voltage drop due to the uranium-leakage currents or due to the presence of the ionization from the signal.
2. **Connections.** To minimize the space needed for connections, insulation-displacement connections (IDC) were used in most of the calorimeter. To reduce electron noise due to the capacitance and inductance of the connections, what is called local ganging was employed. Local ganging is accomplished by placing the signal return path as close as possible to the signal path. For signal return, percussion welding of niobium wire to the uranium absorber plates was developed. A signal return was provided for every 2-4 signals.
3. **Signal and readout board fabrication.** Most signal collection boards were fabricated by DØ; the exceptions were the small angle modules for which commercially built multilayer boards with internal signal trace layers were used. Signal electrodes were made from two laminated 1-mm G10 sheets The board fabrication procedure included five major steps: screen printing of the resistive coating; routing the pad patterns; routing the high voltage patterns; laminating the two G10 sheets; and cutting the boards. Following their fabrication, the boards were tested under high voltage, and the continuity of the signal connections was checked with a capacitance meter.
4. **Cooldown procedures.** So as to assure that the calorimeter modules would survive the cooldown mechanically and electrically to LAr temperatures (87K), the modules were first tested in liquid nitrogen. Then, to minimize mechanical stress due to the temperature differences in the module array, the cooldown to LAr temperatures was carefully monitored and controlled. This was achieved by instrumenting the modules and support structures with more than 250 temperature sensors in each of the three cryostats. These sensors provided information about the temperature differences within the modules, the module interconnections, and the module-support structure. As a result, it was easy to maintain a maximum cooling rate consistent with structural limitations, which corresponded to maximum temperature difference of 50K between any two temperature sensors in each cryostat. For example, the cooldown of the CC took only 14 days.
5. **LAr purity monitoring.** The purity of LAr is one important factor that determines its response uniformity and stability. To prevent LAr poisoning, all materials used in the calorimeter were tested to see their effect on charge collection in LAr. Poisoning tests were done in a special test cell, in which the signals from alpha and beta sources were measured for one week or more. Materials were accepted if signal-loss rates corresponded to less than 2%/year in the actual calorimeters. For most materials, this was the limit of the test sensitivity. In addition, production modules were tested in LAr, and the signal-loss rate was less than few parts per thousand over 100 days. More information about the LAr purity monitoring can be found elsewhere [8–9].

Experience With Module Production

The DØ calorimeter modules were produced at four different labs: the CC modules were built at Brookhaven; the EM modules for the EC were constructed at LBL; the stainless steel frames for the coarse EC modules were provided by IHEP, Protvino, USSR; and the remaining EC modules were produced at Fermilab. This production arrangement was beneficial, because the different production teams collaborated on solutions to common production problems, but developed unique solutions for problems particular to their specific calorimeter.

Module-production factory startup was rather slow because initial procedures were based on experience with prototypes. As experience was gained with the production modules themselves, procedures were refined and, so, production increased. The production of calorimeter modules took about two years.

Quality control was an important aspect of module production from the beginning: all modules to be installed in the calorimeters, including those first built, had to satisfy the same criteria. Quality inspections were performed upon receiving materials for module production, as well as at the end of each production step. A quality-assurance document traveled with each module, and steps had to be checked off on that form as the module was built.

Testing, as a part of quality assurance, included the measurement of cell capacitances to verify that all connections were intact and a high-voltage (HV) test (at 3 kV) during which leakage cuurents were measured. High leakage currents indicated shorts, which had to be removed before progressing with the module. Capacitance and HV tests were also done in liquid nitrogen to make sure that the modules could operate in LAr. Before installation, modules were retested in orientations corresponding to their possible locations within the cryostat.

To keep track of physics parameters and module-testing results, data bases were established to follow the history of the modules from production through installation. The most important data-base items were tower capacitances, leakage currents, and mechanical measurements relating to sampling fraction determination. The latter included plate thicknesses, module height, signal and and readout board thickness, and weights.

During the course of module production, a number of surprises occurred, including high-voltage discharges, and uranium oxidization of the absorber plates:

High-voltage discharges. In 1985, in the test beam of the DØ prototype[6], it was shown that the uranium noise could be reduced by 25% by drifting electrons towards the absorber plates held at ground potential. In this test, the signal boards had no resistive coat. In the next beam test[7] in 1987, the CC EM production modules were studied using resistive coat boards. With negative polarity HV applied to the boards to minimize the uranium noise, discharges were found to be present. These pulses occurred at rates $\sim$ 10-100 Hz for signals corresponding to >5 GeV; the rise times were $\sim$5-50μs. In Fig. 1, the rate of discharges as a function of time is shown. Around day 1 and 28, the HV was turned off. The rate of discharges increased initially, reaching maximum after two to three days, and then decreased with a very long time constant of about 20 days. The spectrum of pulses fell off slowing with increasing amplitude. The rate of discharges varied significantly between the modules and sections of a module.

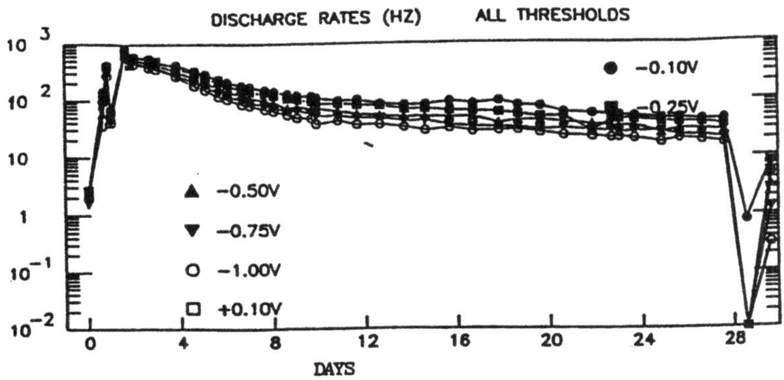

Figure 1: The rate of discharges (Hz) as a function of time for different threshold settings.

It was found experimentally that, if the positive HV was applied to the resistive coat so that ionization electrons drift away from the absorber plates, the discharges were eliminated. This HV polarity, however, resulted in anomalous time-dependent HV leakage currents larger than those expected from the uranium radioactivity by a factor of three to eight. This was a concern because of the resistance built into the HV planes. Subsequent measurements of the HV plateau and the better than 2% uniformity of response in the EM modules proved that those anomalous leakage currents were not a problem for the operation of the calorimeter.

Though the sources of the discharges or the leakage currents have not been conclusively determined, interesting explanations have been suggested [10] within DØ: Discharge and anomalous current phenomena are both believed to be related to the Malter effect [11] through which charge buildup on insulating layers on the negative polarity electrodes result in large local fields and consequent injection of electrons into the argon gap.

Uranium oxide. The oxidization of uranium is a well known phenomenon: it proceeds through the absorption of water $U + 2H_2O \rightarrow UO_2 + 2H_2$ and is accelerated in an oxygenless atmosphere [12]. This was rediscovered while testing modules in liquid nitrogen: during the warm-up phase of the test, the modules were kept in the gaseous nitrogen, which resulted in the formation of UO_2 powder on the uranium plates' surface. To minimize the oxidization during the warm-up stage as well as during module storage and shipping, a small amount of O_2, about 2%, was mixed with the gaseous nitrogen.

Summary

The realization of a large liquid argon calorimeter system poses a many challenges. The excellent experience and tests to date with the DØ calorimeters demonstrate that it can be done. The major lesson learned, according to this author, is that the robustness of large systems is most important, and therefore that prototyping and testing are crucial and should be an integral part of any schedule.

Ackowledgements

The author would like to thank his DØ colleagues for their help in preparing this paper and for the very exciting seven years of working together.

References

1. Design report, The DØ Experiment at the Fermilab Antiproton–Proton Collider, November 1984.
2. P.D. Grannis, Proceedings of Les Rencontres de Physique de la Vallee d'Aoste on Results and Perspectives in Particle Physics, La Thuile, March 1987, M. Greco (Editions Frontieres, 1987) p.253.
3. N.A.Amos, Proceedings of the First International Conference on High Energy Physics Calorimetry, FNAL, USA, eds. D. F. Anderson et al.,(World Scientific, Singapore, 1991), p.8, A.L. Spadafora, same proceedings p. 77, M. Demarteau, same proceedings p.91.
4. S. Feher (DØ collaboration), these proceedings.
5. N. Roe (DØ collaboration), these proceedings.
6. S. Aronson et al., Nucl. Instr.&Meth. A269 (1988) 492.
7. M. Abolins et al.(DØ collaboration), Nucl. Instr.&Meth. A280 (1989) 36.
8. G.C. Blazey, Proceedings of the First International Conference on High Energy Physics Calorimetry, FNAL, USA, eds. D. F. Anderson et al.,(World Scientific, Singapore, 1991), p.101.
9. G.C. Blazey et al., "Liquid Argon Purity Monitoring in DØ," to be submitted to NIM).
10. R. McCarthy, DØ Note 780 (unpublished), M.L. Stevenson, DØ Note 1154 (unpublished).
11. L. Malter, Phys. Rev. 50, (1936) 48.
12. M.McD. Baker et al., Trans Farad Soc 62, (1966) 2525.

Construction and Beam Test of the ZEUS Calorimeter

T. Tsurugai
II Inst. für Exp-physik der Universität, Hamburg, BRD
for
the ZEUS Calorimeter Group

Abstract

The calorimeter of the ZEUS experiment consists of 1 X_0 uranium plates interleaved with plastic scintillator tiles read out via wavelength shifters and photomultipliers. Ten calorimeter modules have been tested with beam particles. Our main result is the achieved calibration accuracy of about 1% obtained by using the signal from the uranium radioactivity due to a strict quality control of the calorimeter components. Recent results of the magnetic field effect on the calibration will be also discussed.

1 Introduction

ZEUS [1], one of experiments at HERA, has selected a sandwich type calorimeter consisting of uranium plates with scintillator tiles in order to fulfil the requirements about good energy resolution and good angular resolution for jets.

The ZEUS collaboration has performed several prototype tests in order to check the mechanical design and performance of the calorimeter [2,3]. The main results of the prototype test are:

1. Equal response for electrons and hadrons (compensation mechanism);
2. An energy resolution of $\sigma(E)/E = 18\%/\sqrt{E}$ for electrons and $\sigma(E)/E = 35\%/\sqrt{E}$ for single hadrons;
3. An intercalibration of individual calorimeter sections to about 3 % and a stability of calibration to better than 1 %.

Ten of ZEUS forward and rear calorimeter modules have been tested in a beam with electrons, muons ahd hadrons in the momentum range of 15 to 100 GeV/c. It has been shown that a section-to-section intercalibration accuracy is close to 1 % for final modules by means of the uranium radioactivity signal (UNO) alone due to a strict quality control on the module components and assembly [4].

Production of the calorimeter modules is complete, and all modules have been installed in the ZEUS experiment. Recently the first test run has been taken with cosmic ray triggers successfully. During this cosmic run influence of magnetic fields on the calorimeter calibration was measured. The preliminary results will be also discussed here.

2 ZEUS calorimeter

The ZEUS calorimeter surrounds the solenoid and tracking detectors hermetically. It is divided into three parts: FCAL, BCAL and RCAL covering forward, barrel and rear regions. A full size FCAL module is shown in fig.1. The depleted uranium (DU) plate (3.3 mm thick) is used as an absorber and the scintillator (SCI) tile (2.6 mm thick) is used as a detector. The module consists of 20×20 cm^2 towers and is segmented longitudinally into an electromagnetic part (EMC, 25 X_0, 1λ deep) and two hadronic sections (HAC1 and HAC2, each 3λ deep). The EMC part is further segmented into four 5×20 cm^2 sections. In RCAL the EMC part is segmented into two 10×2010 cm^2 and the HAC consists of a single hadronic section. Each section is read out on the left and right side by the WLS bars and PMs. More details about the mechanical structure of the calorimeter can be found in ref.4.

The quality of each components (the DU plates, the SCI tiles, the WLS bars, the PMs and electronics) was well controlled. Figs. 2 show distribution of thickness measurements for the DU plates of the various calorimeter sections as one of example. We note that the r.m.s. thickness is around 2%.

3 Beam calibration of FCAL/RCAL

Six FCAL (corresponding to 36 % of all EMC towers and 33 % of all HAC towers of FCAL) and four RCAL (22 % of EMC towers and 20 % of HAC towers of RCAL) modules have been calibrated at the CERN SPS with electron, muon and pion beams. Electrons of 15 GeV/c and 100 GeV/c, and muons of 100 GeV/c have been directed into the 74 EMC sections of each module and the ratio between electron (muon) response and the calibration signal from uranium radioactivity (e/UNO, μ/UNO) has been determined. Fig.3 shows the result of the e/UNO values for one FCAL module. Table 1 shows the section-to-section spreads of e/UNO and μ/UNO within a given module. They are typically 1% for FCAL modules at 15 and 100 GeV/c and somewhat larger for RCAL modules. If all sections of all modules are put together, the spread of e/UNO is 1.1% (FCAL) and 1.5% (RCAL). These values measure the calibration uncertainties for the case that the calibration is derived from the UNO currents alone.

Table 1. Spread of e/UNO and μ/UNO values in EMC and HAC0 sections

module	muons 100 GeV/c	electrons 15 GeV/c	electrons 100 GeV/c
FCAL1	1.3%	1.1%	1.0%
FCAL2	1.4%	0.7%	1.0%
FCAL3	0.9%	1.2%	1.3%
FCAL4	0.8%	0.8%	1.0%
FCAL5	1.1%	0.9%	0.9%
FCAL6	1.1%	1.0%	1.1%
all channels	1.2%	1.1%	1.1%
RCAL1	1.1%	1.4%	1.7%
RCAL2	0.8%	1.1%	0.9%
RCAL3	0.9%	1.1%	1.4%
RCAL4	0.9%	1.5%	1.2%
all channels	1.0%	1.5%	1.6%

4 Influence of Magnetic Fields on the calibration

In ZEUS experiment, the calorimeter will operate in the fringe field of a superconducting solenoid. Where the calorimeter is located, the field will range typically from 0 to 0.3 Tesla. Prototype studies have shown that magnetic fields increase the light output of plastic scintillator [5]. The effect was measured during the last cosmic run at ZEUS. Fig. 4 shows the UNO signal variation from 0-field and the magnet field at 3700 A solenoid current as a function of calorimeter y position in one of the RCAL modules. The scintillator light output increases up to 4 % depending on the position of the calorimeter.

5 References

1. ZEUS, a Detector for HERA, Letter of Intent, DESY, June 1985; The ZEUS Detector. Technical Proposal, DESY, March 1986; The ZEUS Detector, Status Report 1987, DESY, September 1987; The ZEUS Detector, Status Report 1989, DESY, March 1989.
2. U.Behrens et al, *Nucl. Instr. and Meth.* **A289** (1990) 115
3. A.Andresen et al, *Nucl. Instr. and Meth.* **A290** (1990) 95
4. A.Andresen et al, *Nucl. Instr. and Meth.* **A309** (1991) 101
5. J.Mainusch et al, *DESY 91-027*

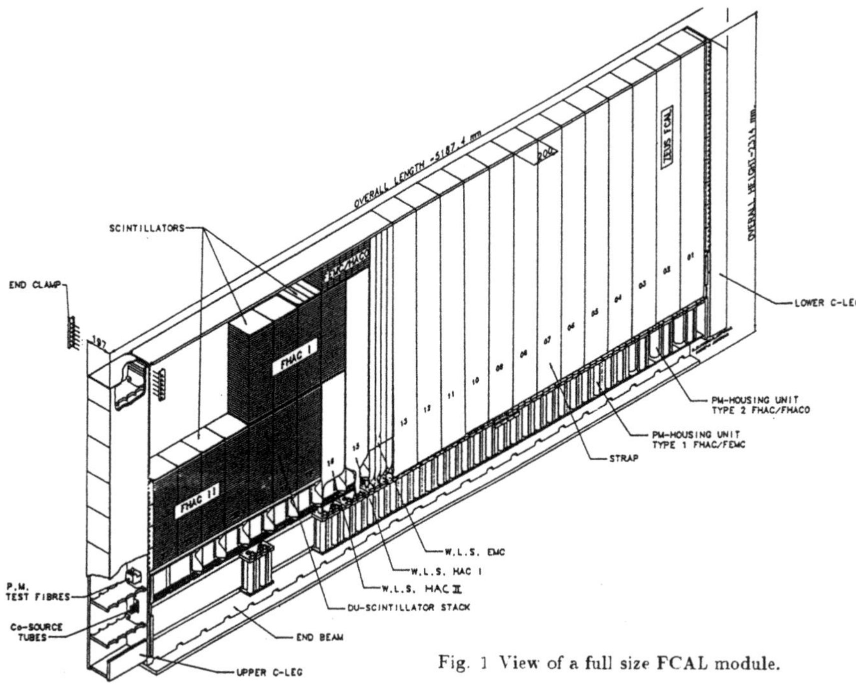

Fig. 1 View of a full size FCAL module.

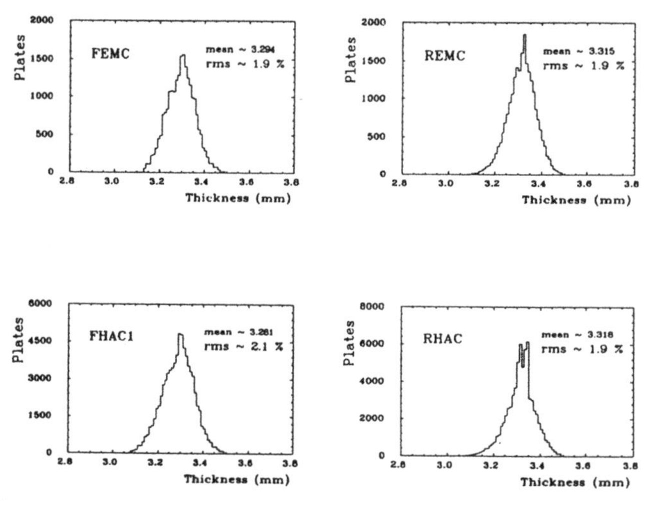

Fig. 2 Thickness distributions of DU plates for a) EMC sections from FCAL, b) HAC1 sections from FCAL, c) EMC sections from RCAL and d) HAC sections from RCAL.

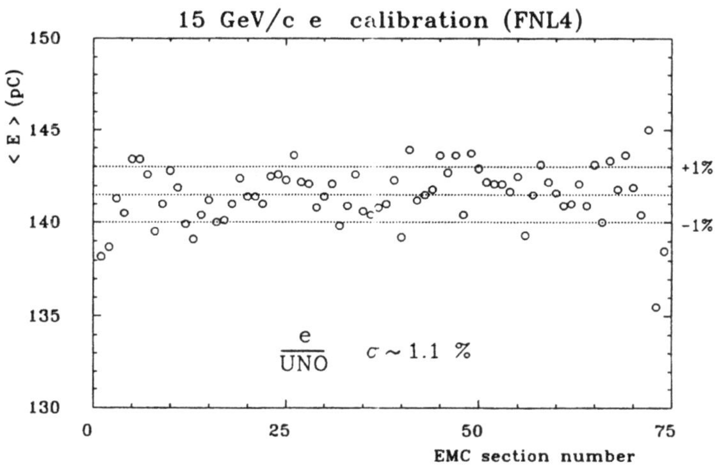

Fig. 3 The average response to 15 GeV/c electrons of the 74 EMC sections of FCAL module after UNO calibration.

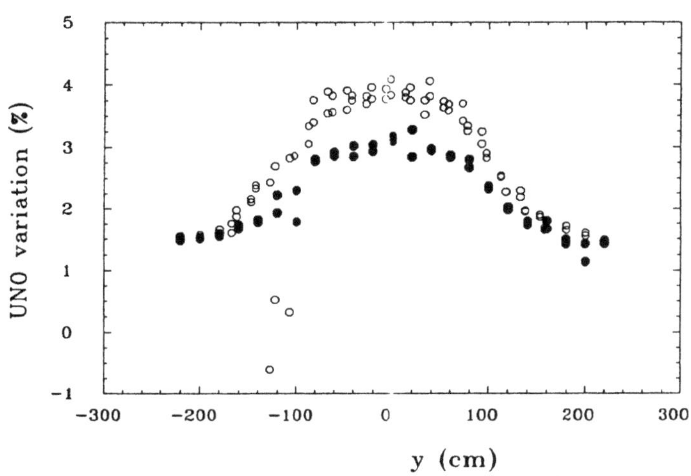

Fig. 4 The UNO signal variation from 0-field and the magnet filed at 3700 A solenoid current as a function of the calorimeter y position. Open (close) circles show the result of EMC (HAC) sections.

STUDY OF AN ALL-LIQUID ARGON CALORIMETER WITH INTEGRATED SOLENOID FOR LHC

the RD3 collaboration
presented by

Bruno MANSOULIE
DAPNIA/SPP, CE-SACLAY, FRANCE

Abstract
Recent developments in liquid -argon calorimetry ("accordion" structure) have made possible to get fast signals. In this study we propose solutions to the other main problem of the technique ,which is to limit the amount of dead material and dead space due to cryogenic requirements.

1) Introduction

We have studied the feasibility of an all-liquid-argon calorimeter based on three elements which, taken together, provide an elegant solution for an experiment at LHC (although several aspects of the study can be used separately)(Fig 1):

- Accordion structure à la RD3 [1].
- Minimization of the amount of dead material and dead space between barrel and end-caps by use of thin flat walls for the cryostats.
- Integration of a small-radius solenoid in the cryostat vacuum.

The goal of this study is to solve the most critical issues and set the basis of a reasonable design; it is not intended to provide final design numbers, and can be further optimized.

2) The RD3 hadron calorimeter

The barrel calorimeter (8m long) is divided in 64 modules : 2 times along the z-axis, then in 32 modules in phi. Each module (Fig. 2, 3) is made by 4 sectors, and contains 5 continuous accordion shaped steel sheets (ground plates) : 3 internal 10 mm thick, 2 side plates 5 mm thick. Each sector, between 2 ground plates, contains 6 or 8 (depending on the radius) read-out plates, which are made of steel plates segmented in radius and rapidity, sandwiched between two epoxy sheets. Additional layers on this epoxy provide a conducting plane (coppered kapton) for signal transmission, and a resisitive plane (carbon charged prepreg) for applying the high voltage necessary to the electric field in the argon gap.The connection of the gaps for the read-out is done in a series-parallel mode (electrostatic-transformer [2])

The continuous, accordion-shaped ground plates provide the module with an excellent rigidity.

3) The end-walls (Fig 1,4)

In most design of large liquid argon calorimeters, the necessity to hold the hydrostatic pressure of the liquid leads to having either flat thick walls, or thinner but bulged walls. In our design, we tie the cold wall to the ends of the modules, taking advantage of their good mechanical properties. Thus the cold wall can be flat and thin (less than 10 mm of steel for a spacing of the fixation points of 280 mm). The vacuum is kept by again a thin wall (<10 mm) spaced from the cold wall by insulating pods.The thickness of this warm wall can be further decreased by inserting an epoxy lattice between the wall and the pods.

When the calorimeter cools down, the ends of the inside barrel move by 12 mm along the z axis towards the center, whereas the warm tubes inside and outside the calorimeter keep fixed lengths. As the end-walls will follow the movement of the cold inside part, there must be a flexible connection between the warm end-wall and the warm tubes. This is provided at the inside and outside radius, by two rings of $\Delta R = 200$ mm, slightly bulged for holding the pressure, and thin (2mm steel) for flexibility. This delicate part has been checked by a detailed mechanical calculation.

4) The coil and its vacuum

In order to minimize the amount of material and radial space in front of the electro-magnetic calorimeter, the super-conducting coil is placed in between the warm and cold tubes of the calorimeter cryostat. The coil and its support is a 5 cm thick Al tube for a 2 T field.It is hanged at the ends by tie-rods.Horizontal rods hold it in place along the z-axis.

The coil connections (He, power, vacuum) run along the end-walls of the calorimeter inside the vacuum. It is proposed to have only a thermal screen between the coil and the warm tube, not between the coil and the argon vessel. There is a separate vacuum for the coil, in order to reach a lower pressure than in the calorimeter vacuum.This enclosure is shut at its ends by a flexible membrane similar to the one discussed above, but with the additional concern that its outside radius is cold and its inside radius is warm. Again a detailed calculation shows that this is indeed possible.

5) The central tubes

Special care must be devoted to minimizing the thickness (in radiation lengths) of the central tubes. The warm tube can be made of aluminum (10 mm)without special problems.However the use of aluminum for the cold tube is more difficult, because of two main reasons :

- the connection between the aluminum tube and the steel walls, at low temperature, is difficult : Al/steel welding on such a diameter is unknown, and a metal seal would bring bulky flanges and reliability problems.

- the differential contraction of aluminum and steel has to be taken care of.

Our proposal is to use a bi-metal tube with a bore of 27 mm of aluminum and a skin of 2mm of steel. This dissociates the two functions of the tube : the aluminum provides the structure for holding the pressure, and the steel can be welded at the ends to provide leak-tightness to the liquid.

6) Conclusion

A lot of work is still ahead of us to turn these ideas into a design for construction. However, we think that the main problems which were considered as tied to the technique of liquid argon calorimetry can be overcome : the accordion structure (and EST for the hadronic part) allows a fast read-out, and careful design of the cryostat brings the amount of dead material and dead space to the level of or below that of other calorimetric techniques. For example in this study the total amount of material (including the 2T coil) in front of the e-m calorimeter is 1 radiation length at rapidity 0; the total distance between the active part of the barrel and the active part of the end-caps (in the e-m region) is of the order of 30 cm, including 5 cm space for the cables from the central tracking.

References

1) Hadronic and electro-magnetic liquid argon LHC prototype calorimeter with pointing geometry : CERN/ DRDC/ 91-21, DRDC P5 Add. 1, 13th March 1991.

2) J. Colas, M. Pripstein and W.A. Wenzel : The electrostatic transformer NIM A294 (1990) 583

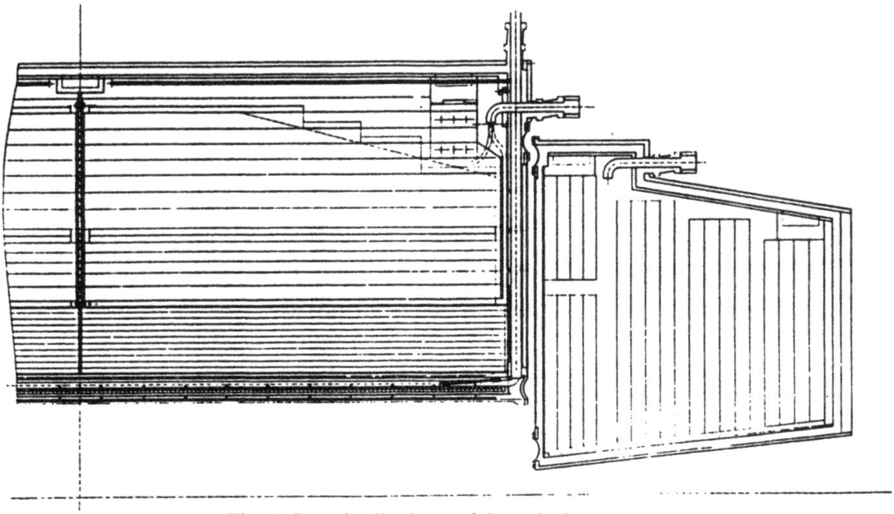

Fig. 1 Longitudinal cut of the calorimeter.

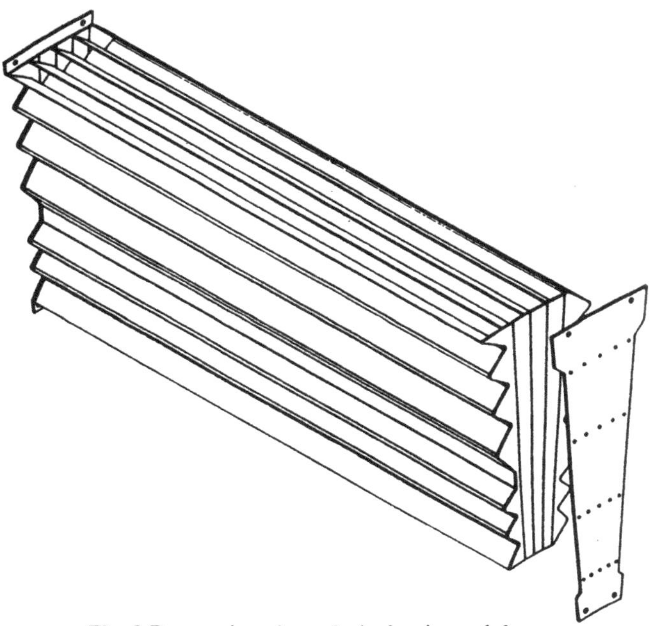

Fig. 2 Perspective view of a hadronic module.

Fig. 3 R-ϕ cut of a hadronic module.

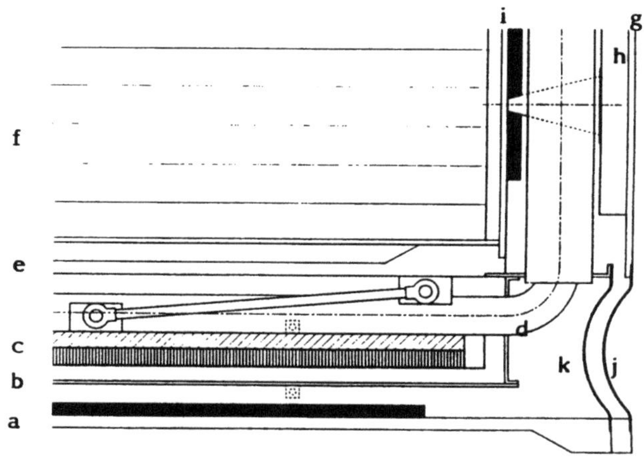

Fig. 4 Close-up view of the end-corner of the barrel (longitudinal cut)
a warm tube, b thermal screen, c coil, d He collector, e cold tube, f e-m calorimeter, g warm end-wall, h epoxy lattice, i cold end-wall, j flexible ring (warm wall) k flexible ring (coil vacuum separation)

SUMMARY AND CONCLUSIONS

Klaus Pretzl

Laboratory for High Energy Physics, University of Bern, Sidlerstrasse 5, 3012 Bern, Switzerland

Some historical remarks

In the 1950'ies and 1960'ies calorimeters were developed mainly for the use in cosmic ray physics. With a new generation of high energy accelerators, which became operational at Fermi-Lab and CERN in 1974 and 1976, calorimeters began to be important instruments in high energy physics. Since their energy resolution is following $1/\sqrt{E}$, it was realized that above ~ 50 GeV the particle energies are measured more accurately with calorimeters than with magnetic spectrometers. Calorimeters played a central role in the discovery of the QCD-jets. For this purpose tower structured granular calorimeters were constructed to measure the transverse energy flow $E_T = E\cdot\sin\theta$ and the transverse energy clustering in deep inelastic events. It is perhaps appropriate to give here credit to Richard Garwin [1], who invented the principle of the wave length shifter optical readout in 1960 providing the basis for the construction of tower structured scintillator calorimeters. This principle has been further developed by many people and is today the most frequently used readout technique for scintillator sampling calorimeters.

Calorimeters play a key role in astro- and particle physics experiments. Their importance is underlined by the fact, that an entire conference is devoted to the subject.

Calorimeter performances and limitations

Two questions were discussed in detail at this conference:

1.) How precise can we measure energy (see contribution of T. Virdee)?
2.) Do we measure the energy correctly (see contribution of R. Wigmans)?

The energy resolution of calorimeters can be parametrized in the following way:

$$\frac{\sigma}{E} = \sqrt{\frac{c_1 + c_2 + c_3}{E}} \oplus c_4 \oplus \frac{c_5}{E}$$

with the parameters c_1 for intrinsic, c_2 for sampling and c_3 for photoelectron fluctuations. c_4 represents the energy independent constant and c_5 the noise term.

Intrinsic fluctuations are usually very small in electromagnetic (e.m.) calorimeters. Provided the photoelectron statistic is high, the energy resolution depends then mainly on the sampling fluctuations. By increasing the fraction of the visible energy in a calorimeter (sampling fraction) the energy resolution can be considerably improved. This is illustrated in Fig.1, which shows the achievable energy resolutions for sampling and homogeneous calorimeters. For a given sampling fraction the energy resolution can be improved further by increasing the sampling frequency. This has been demonstrated with scintillating fiber calorimeters (Fig.2).

In contrast to electromagnetic calorimeters, the intrinsic fluctuations in hadron calorimeters are not small, since a fair fraction (~ 42 %) of the initial energy is dissipated in undetectable form (invisible energy). In addition the electromagnetic component of the hadronic shower (π^0's and η's) is changing with energy (~ lnE), which leads to "non-Gaussian" tails and nonlinearities in the energy response of the calorimeters. In order to measure the hadron energy correctly one needs a compensating calorimeter with equal electron and hadron response (e/h = 1). The ratio e/h is approximately given by the response ratios e/mip and n/mip (n = neutrons):

$$\frac{e}{h} = \frac{\frac{e}{mip}}{0.41 + 0.12\,\frac{n}{mip}}$$

Compensation can be achieved by decreasing e/mip and (or) increasing n/mip. Both ratios are material dependent. It has been demonstrated (Fig.3) that a linear energy response for sampling calorimeters can be obtained with hydrogeneous active materials and low sampling fractions (the ratio R = thickness of absorber layer/thickness of active layer has to be 1 for U/Scint. and 4 for Pb/Scint. calorimeters). Due to the correlation between the invisible energy in the shower and the kinetic energy carried away by neutrons, compensating calorimeters with efficient neutron detection have a considerably better intrinsic energy resolution than noncompensating calorimeters. However, because of the required low sampling fraction, compensation and good energy resolution for electromagnetic showers seem to be mutually exclusive.

The question whether or not compensating calorimeters are required for LHC and SSC physics was intensively discussed at the conference. For making the right choice the physics priorities as well as factors other than the calorimeter resolution have to be considered. In addition high radiation levels may restrict the choice of hydrogeneous readout materials. At HERA, for example, priority was given to accurate QCD-jet measurement in charged and neutral current interactions. Both experiments, ZEUS and H1, have therefore aimed for e/h = 1 (see contribution of H. Oberlack). It is remarkable that for the apriori noncompensating Pb/LAr calorimeter of the H1 collaboration an e/h ~ 1 response was obtained by an appopriate shower weighting technique. This has been made possible by a sufficiently fine granularity and depth segmentation.

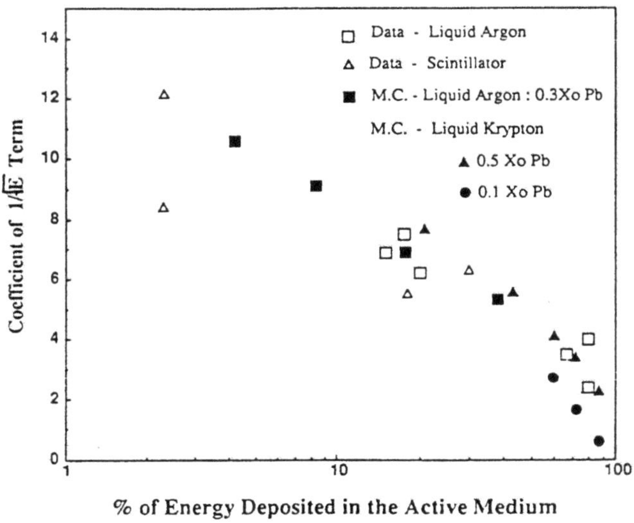

Fig.1 Achievable energy resolutions for e.m. calorimeters as a function of the energy fraction deposited in the active material.

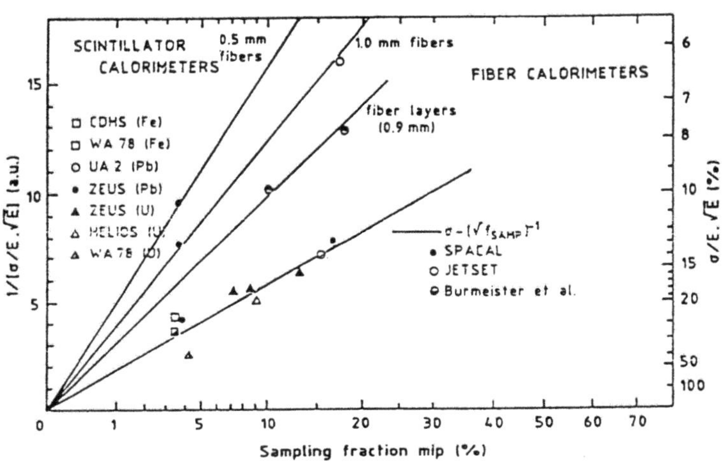

Fig.2 Energy resolution for e.m. showers in scintillator sampling calorimeters as a function of the sampling fraction and the sampling frequency.

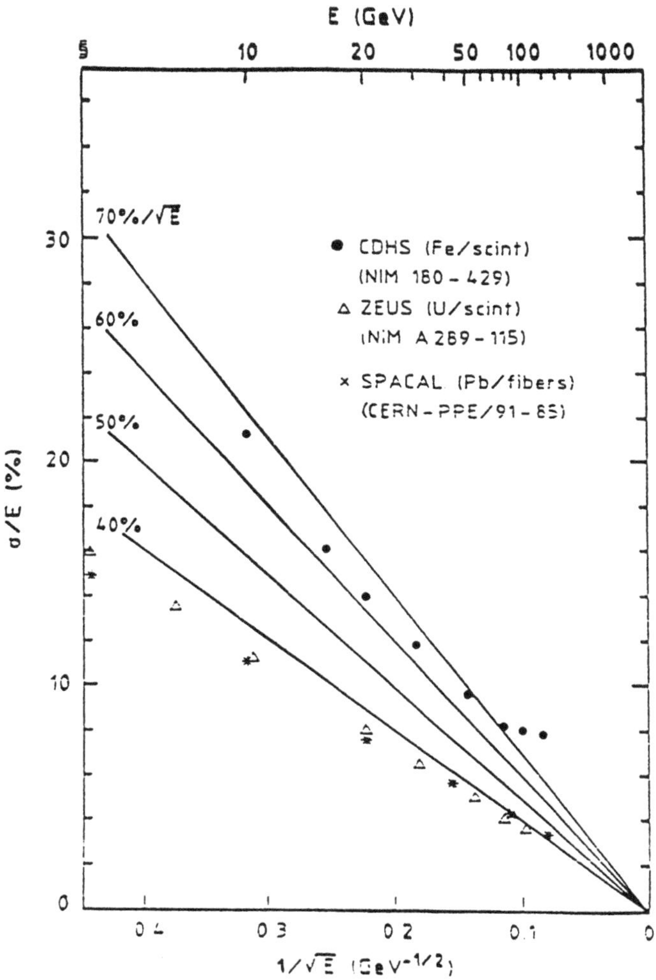

Fig.3 Hadronic energy resolutions for compensating (ZEUS, SPACAL) and noncompensating (CDHS) calorimeters.

The physics priorities at LHC and SSC are clearly focused on the detection of the Higgs boson, supersymmetric particles and higher mass vector bosons (Z'). Simulation studies have shown that the jet-jet mass resolutions are mainly influenced at low p_T by the pileup of overlying events and at high p_T by the fluctuations in the jet fragmentation. Whereas the calorimeter resolution does not seem to make a major inpact on the jet resolution (see contribution of D. Green and M. Della Negra). This is illustrated in Fig.4, where the effects on the mass resolution of compensating (e/h = 1) and noncompensating (e/h = 1.3) calorimeters are compared in the presence of jet fragmentation.

Considering the detection of the Higgs through the leptonic channel $H \rightarrow \ell\ell\nu\nu$ and the detection of supersymmetric particles, the necessary missing p_T measurements are influenced in the following order of importance by: 1.) the rapidity coverage of the calorimeter ($|\eta| \sim 4$ is required for LHC and $|\eta| \sim 5$ for SSC), 2.) the hermeticity, 3.) the pileup of overlying events. 4.) the calorimeter resolution (see contribution of L. Poggioli).

From this one concludes that factors other than the calorimeter resolution dominate the measurement accuracies. Therefore moderate hadron calorimeter resolutions of the order of $\sigma/E \sim 50\ \%/\sqrt{E} + 2\ \%$ seem to be adequate for LHC and SSC physics. One should also keep in mind that with sufficient depth segmentation and granularity a e/h = 1 response can be achieved even with an apriori noncompensating calorimeter as was demonstrated by the H1 collaboration.

A particular problem, however, is the extrapolation of the calorimeter performance in the multi TeV energy region, which is not accessable by test beams. In addition the jet fragmentation models cannot be reliably extrapolated into this region. These difficulties will set limits on the determination of the Λ-contact term, which can be extracted from inclusive jet measurements at high transverse energies.

The detection of an intermediate mass Higgs through the decay channel $H \rightarrow \gamma\gamma$ requires a high quality performance of the electromagnetic calorimeter (see contribution of Ch. Seez). As demonstrated in Fig.5 the detection significance depends very crucially on the calorimeter resolution as well as on the capability to locate the longitudinal vertex position. For a vertex localisation with $\sigma = 1$ cm the shower direction has to be measured with an accuracy of $\sigma \sim 7$ mrad. In the case of associated Higgs production ($pp \rightarrow t\bar{t}\ H$; $pp \rightarrow WH$), however, the vertex can be determined sufficiently accurate by associated t, $\bar{t}$ or W particles. For an effective background rejection of π^o's and η's the calorimeter needs high granularity and the capability to determine the shower position with high accuracy (shower position detector at $5X_o$). In order to reach a high level of signal significance (for a significance larger than five the signal starts to be distinguishable from the background) the electromagnetic calorimeter should have an energy resolution $\sigma/E = < 3\ \%/\sqrt{E}$ with a constant term $< 0.5\ \%$. There was a general consensus at the conference that at least one of the planned detectors for LHC or SSC should have these features.

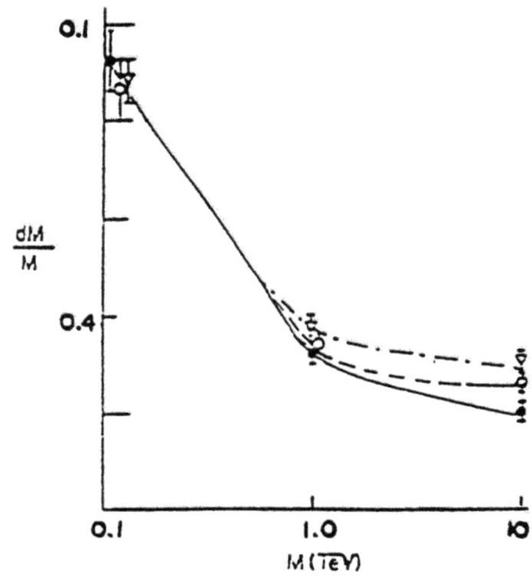

Fig.4 Dijet mass resolutions for • compensating (e/h = 1) and noncompensating (e/h = 1.3) hadron calorimeters in the presence of jet fragmentation fluctuations. e.m. and hadron calorimeter calibrated to equal response at all energies and at asymptotic energies.

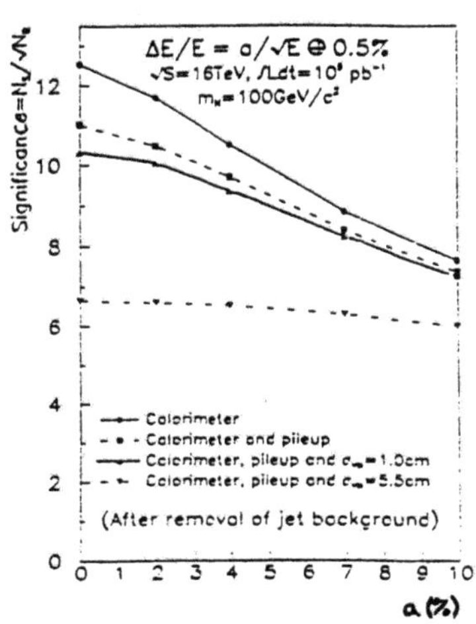

Fig.5 Signal significance for H → γγ as a function of the e.m. calorimeter resolution.

The calorimeter performance is strongly influenced by the constant term which depends on the uniformity of the signal generation, the signal loss in nonactive materials (support structures, magnet coils ...) and on the calibration uncertainties (cell to cell calibration, monitoring ...). For example a 2 % thickness variation in the scintillator material already leads to a 0.5 % constant term. We heard encouraging results from the HERA calorimeters, which managed to control their calibration accuracies to 1 % (see contribution of H. Oberlack). At LHC and SSC the calibration of the electromagnetic calorimeters can perhaps be monitored in situ by $Z_o \to e^+ e^-$ decays, which should deliver approximately 100 events per cell and per weak of running time.

Design tools for large detector systems

Electromagnetic and hadronic shower simulation codes (GEANT, EGS4, GHEISHA, FLUKA, CALOR, HETC, MORSE ...) are essential tools for prototype and detector designs. In the future their validity ranges (in terms of energy) will be extended by combining the positive features of the various codes. Also parametrisation techniques are vigorously persued since there is a shortage of a factor 1000 in CPU-time for shower simulation studies (see contribution of R. Brun).

Another area, where Monte Carlo simulations turned out to be extremely usefull, is the prediction of radiation levels in calorimeters and other detector components. Neutron fluxes and radiation doses were evaluated for specific detector designs using the FLUKAN code. From these simulations one learned the following: 1.) The magnetic field has practically no effect on the neutron flux. 2.) The neutron fluence depends only weakly on the materials in the hadron calorimeter. 3.) Hydrogeneous readout material in the e.m. calorimeter helps to reduce the neutron flux considerably. 4.) A neutron moderator (10 cm polyester layer) in front of the e.m. calorimeters as well as a transition radiation detector can provide effective neutron shieldings in the central barrel regions of the detector. Along with these studies impressive work has been reported at the conference on the evaluation of the radiation hardness of detector components and electronics (see contribution of A. Ferrari).

System engineering using modern CAD design tools and outside companies (like for example Martin Marietta), has been lively discussed at the conference (see contributions of J. Pohlen and M. Derrick). Although large scale engineering of the LEP and HERA experiments was successfully done without outside professionals it was felt that for the even larger LHC and SSC detectors the experience of outside engineering companies could profitable be used. CAD systems have also proved to be very powerfull in debugging complicated system designs.

Some very new results and trends

Many very fine R & D results as well as new ideas have been presented. The author appologizes for not beeing able to mention them all.

One of the important requirements for LHC and SSC calorimeters is to reach a signal response time which corresponds to the time intervall of the bunch crossings (15 nsec). Considering this aspect interesting testbeam results have been presented by the RD3-Collaboration (see contribution of G. Costa) on the "accordeon" LAr calorimeter. A fast response time was reached by keeping the calorimeter cell capacity (accordeon cell structure) and the amplifier input resistance (amplifiers were mounted directly at the calorimeter cells) very small. With suitable pulseshaping a signal integration time of 20 nsec was obtained. With this arrangement an energy resolution for electromagnetic showers of $\sigma/E = 10\ \%/\sqrt{E} + 0.2\ \% + 0.3/E$, a spatial resolution of $\sigma_x = 3.6/\sqrt{E} + 0.1$ [mm] and an angular resolution of $\sigma_\theta \simeq 80/\sqrt{E}$ [mrad] were achieved. The linearity of the energy response was 1 %. Preliminary results on a pointing geometry were also reported. These results clearly demonstrated, that liquid Argon qualifies as a suitable device in a high luminosity environment.

Progress has also been made in the development of spaghetti type calorimeters. With 500 μm diameter scintillating fibers, embedded in a lead structure, an energy resolution of $\sigma/E = 9.3\ \%/\sqrt{E} + 0.3\ \%$ for electromagnetic showers was achieved. In order to reach better uniformity and linearity in the signal response of such a calorimeter the uniformity and the attenuation length of the fibers still have to be improved (see contribution of L. Pogiolli). First encouraging results on shower depth sampling with a wedge shaped spaghetti calorimeter (see contribution of G. Polesello) and some other interesting schemes for depth sampling (see contribution of C. Bromberg) were also presented at the conference.

Another promissing calorimeter readout, which was originally developed by the UA1 Collaboration, seem to be provided by wave length shifter (w.s.) fibers embedded in scintillating tiles. A prototype of such a calorimeter was successfully tested by the CDF Collaboration (see contribution of J. Freeman). The main advantages of this technique are, that it provides the possibility for shower depth sampling in a compact calorimeter design and that it allows to guide the light in long undoped fibers outside magnetic field regions. Scintillator tiles will also be developed for the SDC calorimeter at the SSC. An energy resolution of $\sigma/E = 9.5\ \%/\sqrt{E} + 0.4\ \%$ for electromagnetic showers was obtained with a prototype scintillator tile calorimeter (see contribution of L. Camillieri). Such an electromagnetic calorimeter is planned to be used for a neutrino oscillation experiment at CERN. An interesting new way to integrate scintillators into an iron structured hadron calorimeter combined with a w.s. fiber readout was also described at the conference (see contribution of Otto Gildemeister).

Homogeneous, high resolution electromagnetic calorimeters are strongly demanded for the search of $H \to \gamma\gamma$. Two main techniques are presently under development: crystals and liquid Xenon. Due to their small Molière radius (~ 2 to 3 cm) crystals provide the optimal basis for highly granular systems (see contributions of P. Lecoq, M. Schneeganz and E. Divitsin). For the future the following lines for crystal developments are considered: 1.) The properties of CeF_2 crystals (their temperature independence, radiation hardness, fabrication etc. ...) are already better known today than they were of BGO at the time the L3 experiment was approved. Therefore CeF_2 crystals are good candidates for being used in a high resolution electromagnetic calorimeter. 2.) PbF_2, being a nonscintillating material, can perhaps be converted in a scintillator by replacing some of the Pb ions in the lattice by scintillating materials (Ce ...). 3.) Fluoroluminate and fluorozirkonate glasses are scintillating materials of great future interest, because they have rather high densities (~ 6 g/cm^3) and good production properties. Crystals have in general a rather fast response time (20 - 30 nsec) and a large light yield in a region where photomultipliers and silicon diodes provide a high quantum efficiency.

The properties of liquid Xenon for calorimetric measurements in nuclear physics have been vigorously investigated in the 1950's. However, their development has been slowed down considerably with the introduction of Germanium detectors. Recently liquid Xenon has gained new interest in the development of homogeneous high resolution calorimeters. Although both, the ionisation and the scintillation features of liquid Xenon, can be used for high resolution calorimetry, the scintillation properties are mainly considered for LHC and SSC calorimeters, because of the fast response time (~ 30 nsec). The main advantages of liquid Xenon are its short radiation length (2.77 cm) and its radiation hardness. Its disadvantages are the cost (~ SFr. 3200.-- per liter liquid) and its availability (5 m^3 liquid per year is the world production). Its light emission at 175 nm requires the development of readout systems with a high quantum efficiency in this region. Although test beam measurements with large scale prototype calorimeters are still missing some of the properties of liquid Xenon have been experimentally evaluated and were reported at this conference: 1.) The light attenuation length in Xenon was measured to be ~ 40 cm. 2.) The light yield came out to be $(1.5 \pm 0.5)\ 10^4$ photons per MeV. 3.) No saturation effects (according to Birks law) have been observed (see contributions of T. Virdee and W. Wallraff).

A variety of new calorimeter developments have been reported at this conference including: liquid Argon, scintillator tiles and w.s. fibers, scintillating fibers, crystals, warm liquids, gas, silicon and liquid Xenon. It seems that rather conventional calorimeters, with certain improvements to make them compatible with the requirements for radiation hardness and for short bunch crossings, are favoured for SSC and LHC detectors. A certain trend towards the first four of the above mentioned techniques was noticeable. Although very attractive, silicon and liquid Xenon calorimeters are still questioned by their price and availability.

The quest for solar neutrinos, the neutrino mass and the nature of dark matter has motivated the development of new unconventional detectors which would be able to measure energies in the range between 1 eV and 1000 eV. Cryogenic detectors are of specific interest since at low energies (< 1 keV) most of the energy lost by a particle traversing a detector is transformed into heat, causing a temperature change $\Delta T = \Delta E/C$ in the detector, with ΔE the energy loss of the particle and C the heat capacity of the detector. At low temperatures (where C is very small) a small energy loss can be transformed into a measurable signal with superconducting and bolometric devices. The interest in superconducting devices for particle detection is based on the very small quantum energies involved (it takes $\lesssim$ 1 meV to break a Cooper pair) as compared to conventional ionization (~ 20 eV) and semiconductor detectors (~ 2 eV). Consequently, one hopes to reach very much lower energy thresholds and better energy resolutions with such devices (see contribution of M. Spiro).

If one is not enough honored to be associated with an enterprize of the LHC or SSC magnitude (see cartoon), interesting new developments in astroparticle physics offer plenty of opportunities for calorimeter fans.

Thanks to the organizers of the conference

We would like to thank very warmly Antonio Ereditato for the superb organization of this conference. We are also very gratefull to the secretaries for their kind support during the conference and to all those people, who have organized the transportation so well.

References:

[1] R.C. Garwin, **Rev.Sci. Instr.** *31* (1960) 1010

"Oh, stop complaining! It's an honor
to be associated with an enterprise of this magnitude."

List of Participants

JEAN BADIER
E. POLITECHNIQUE
FRANCE

VIRGIL BARNES
UNIV. PURDUE
USA

ANTONIO BARONCELLI
INFN SANITA'
ITALY

EDMOND BELLAMY
INFN PISA
ITALY

GIORGIO BELLETTINI
UNIV. PISA
ITALY

GIANMARCO BILEI
INFN PERUGIA
ITALY

BRUNO BORGIA
UNIV. ROMA I
ITALY

CARLO BOSIO
INFN SANITA'
ITALY

CARL BROMBERG
MICHIGAN STATE UNIV.
USA

RENE BRUN
CERN
SWITZERLAND

ALLEN CALDWELL
UNIV. COLUMBIA
USA

LESLIE CAMILLERI
CERN
SWITZERLAND

DANIEL CAMIN
INFN MILANO
ITALY

ANTONIO CAPONE
UNIV. POTENZA
ITALY

CLAUDE CHARLOT
E. POLITECHNIQUE
FRANCE

ROBERT CHASE
LAL ORSAY
FRANCE

IGOR CHIRIKOV
JINR DUBNA
USSR

MARIA PAOLA CLARA
UNIV. TORINO
ITALY

MARINA COBAL
INFN PISA
ITALY

JAQUES COLAS
LAPP ANNECY
FRANCE

GIUSEPPE COSTA
INFN MILANO
ITALY

YURY DAVYDOV
JINR DUBNA
USSR

CHRISTOPHE DE LA TAILLE
LAL ORSAY
FRANCE

NORA DE MARCO
INFN TORINO
ITALY

MICHEL DELLA NEGRA
CERN
SWITZERLAND

MALCOM DERRICK
ANL
USA

EUGENI DEVITSIN
LEBEDEV
USSR

DARYL DI BITONTO
UNIV. ALABAMA
USA

UBALDO DORE
UNIV. ROMA I
ITALY

JOACHIN ENGLER
KARLSRUHE
GERMANY

ARNOLD ENGLER
CARNEGIE-MELLON UNIV.
USA

ANTONIO EREDITATO
INFN NAPOLI
ITALY

JEAN ERNWEIN
SACLAY
FRANCE

FRANCO FABBRI
INFN FRASCATI
ITALY

LYDIA FAYARD
LAL ORSAY
FRANCE

SANDOR FEHER
SUNY STONY BROOK
USA

ALFREDO FERRARI
INFN MILANO
ITALY

FERNANDO FERRONI
UNIV. ANCONA
ITALY

DANIEL FOURNIER
CERN
SWITZERLAND

JAMES FREEMAN
FERMILAB
USA

DANIEL FROIDEVAUX
CERN
SWITZERLAND

JEAN MARC GAILLARD
CERN
SWITZERLAND

YURY GALAKTIONOV
ITEP MOSCOW
USSR

IOANNIS GIALAS
UNIV. COLUMBIA
USA

OTTO GILDEMEISTER
CERN
SWITZERLAND

ALAIN GIVERNAUD
CERN
SWITZERLAND

HANS GRASSMANN
INFN PISA
ITALY

DANIEL GREEN
FERMILAB
USA

DANIEL GROOM
LBL
USA

ATUL GURTU
CERN
SWITZERLAND

WLODEK GURYN
BNL
USA

JOHN HAUPTMAN
UNIV. IOWA
USA

CHRISTOFER HEARTY
LBL
USA

ANA MARIA HENRIQUES
CERN
SWITZERLAND

CHARLES HURLBUT
BICRON
USA

PETER JENNI
CERN
SWITZERLAND

KURTIS JOHNSON
UNIV. FLORIDA
USA

VALERII KHOVANSKY
ITEP MOSCOW
USSR

FRANS KNIEST
BICRON
THE NETHERLANDS

LARS KNUDSEN
CERN
SWITZERLAND

TOMIO KOBAYASHI
UNIV. TOKIO
JAPAN

LAI KWAN-WU
INFN PADOVA
ITALY

KATI LASSILA
CERN
SWITZERLAND

JEAN MARIE LE GOFF
CERN
SWITZERLAND

PAUL LECOQ
CERN
SWITZERLAND

MICHAEL LEVI
LBL
USA

LUCIE LINSSEN
CERN
SWITZERLAND

MICHELE LIVAN
UNIV. CAGLIARI
ITALY

AMELIA MAIO
LIP LISBON
PORTUGAL

BRUNO MANSOULIE
CEN SACLAY
FRANCE

ALDO MENZIONE
INFN PISA
ITALY

THOMAS MEYER
AMES LAB.
USA

MASANORI MISHINA
FERMILAB
USA

PAOLO MORETTINI
UNIV. GENOVA
ITALY

MARZIO NESSI
CERN
SWITZERLAND

FRIEDER NIEBERGALL
UNIV. HAMBURG
GERMANY

HORST OBERLACK
MPI MUNICH
GERMANY

VITTORIO PALLADINO
UNIV. NAPOLI
ITALY

DEMETRIS PANDOULAS
AACHEN
GERMANY

JOHN PARSONS
UNIV. COLUMBIA
USA

VLADIMIR PESKOV
CERN
SWITZERLAND

GIOVANNI MARIA PIACENTINO
UNIV. CASSINO
ITALY

FRANCESCO PIETROPAOLO
INFN PADOVA
ITALY

HARTMUTE PLOTHOW-BESCH
CERN
SWITZERLAND

LUC POGGIOLI
CERN
SWITZERLAND

JOE POHLEN
MARTIN MARIETTA
USA

GIACOMO POLESELLO
INFN PAVIA
ITALY

KLAUS PRETZL
UNIV. BERN
SWITZERLAND

ERNST RADERMACHER
CERN
SWITZERLAND

JEAN PAUL REPELLIN
LAL ORSAY
FRANCE

SERGIO RESCIA
BNL
USA

NATALIE ROE
LBL
USA

ALESSANDRA ROMERO
UNIV. TORINO
ITALY

ANTONIO ROSSI
UNIV. CALABRIA
ITALY

CHRISTINE SCHEEL
NIKEF
THE NETHERLANDS

MARC SCHNEEGANS
LAPP ANNECY
FRANCE

CRISOSTOMO SCIACCA
UNIV. NAPOLI
ITALY

CHRIS SEEZ
IMPERIAL COLLEGE
UK

JOSE MANUEL SEIXAS
CERN
SWITZERLAND

ARMAND SIMON
UNIV. HEIDELBERG
GERMANY

MICHEL SPIRO
CEN SACLAY
FRANCE

PAOLO STROLIN
UNIV. NAPOLI
ITALY

SHOJIRO SUGIMOTO
UNIV. TOKYO
JAPAN

LUDOVICO TORTORA
INFN SANITA'
ITALY

ILSE TWEER
FIZ ENERGIE
GERMANY

VALERIO VERCESI
INFN PAVIA
ITALY

TEJINDER VIRDEE
CERN
SWITZERLAND

VACLAV VRBA
INFN SANITA'
ITALY

RICHARD WIGMANS
CERN
SWITZERLAND

NARENDRA YAMDAGNI
UNIV. STOCKHOLM
SWEDEN

BERNARDO TOME'
LIP LISBON
PORTUGAL

TATSUMASA TSURUGAI
DESY
GERMANY

TERESA TYMIENIECKA
UNIV. WARSAW
POLAND

HENRI VIDEAU
E. POLYTECHNIQUE
FRANCE

LUCIA VOTANO
INFN FRASCATI
ITALY

WOLFGANG WALLRAFF
AACHEN
GERMANY

JOERG WOTSCHACK
CERN
SWITZERLAND

STEFANO ZUCCHELLI
INFN BOLOGNA
ITALY

www.ingramcontent.com/pod-product-compliance
Lightning Source LLC
LaVergne TN
LVHW051924120926
847902LV00009B/70
* 9 7 8 9 8 1 0 2 0 9 1 6 2 *